FORCE IN NEWTON'S PHYSICS
The Science of Dynamics
in the Seventeenth Century

History of Science Library
Editor: MICHAEL A. HOSKIN
Lecturer in the History of Science, Cambridge University

THE ENGLISH PARACELSIANS
Allen G. Debus
Professor of the History of Science, University of Chicago

WILLIAM HERSCHEL AND THE CONSTRUCTION OF THE HEAVENS
M. A. Hoskin
Lecturer in the History of Science, Cambridge University

A HISTORY OF THE THEORIES OF RAIN
W. E. Knowles Middleton

THE ORIGINS OF CHEMISTRY
R. P. Multhauf
Director of the Museum of History and Technology at the Smithsonian Institution

THEORIES OF LIGHT FROM DESCARTES TO NEWTON
A. I. Sabra
Reader in the History of the Classical Tradition, The Warburg Institute

MEDICINE IN MEDIEVAL ENGLAND
C. H. Talbot
Research Fellow at the Wellcome Institute of the History of Medicine, London

THE EARTH IN DECAY
A HISTORY OF BRITISH GEOMORPHOLOGY, 1578–1878
G. L. Davies
Lecturer in the Department of Geography, Trinity College, University of Dublin

THE CONFLICT BETWEEN ATOMISM AND CONSERVATION THEORY 1644–1860
W. L. Scott
Professorial Lecturer, School of Business Administration and Department of Mathematics and Science, American University

THE ROAD TO MEDICAL ENLIGHTENMENT 1650–1695
L. S. King
Professorial Lecturer in the History of Medicine, University of Chicago

THE SCIENTIFIC ORIGINS OF NATIONAL SOCIALISM:
SOCIAL DARWINISM IN ERNST HAECKEL AND THE GERMAN MONIST LEAGUE
Daniel Gasman
Assistant Professor of History, John Jay College, City University of New York

WILLIAM HARVEY AND THE CIRCULATION OF THE BLOOD
Gweneth Whitteridge
Lecturer in the History of Medicine, Edinburgh University

History of Science Library: Primary Sources

VEGETABLE STATICKS
Stephen Hales
Foreword by M. A. Hoskin
Lecturer in the History of Science, Cambridge University

SCIENCE AND EDUCATION IN THE SEVENTEENTH CENTURY
THE WEBSTER-WARD DEBATE
Allen G. Debus
Professor of the History of Science, University of Chicago

FUNDAMENTA MEDICINAE
Friedrich Hoffmann
Translated and introduced by Lester S. King
Professorial Lecturer in the History of Medicine, University of Chicago

THE PNEUMATICS OF HERO OF ALEXANDRIA
Introduced by Marie Boas Hall
Reader in History of Science and Technology, Imperial College, London

FORCE IN NEWTON'S PHYSICS

THE SCIENCE OF DYNAMICS
IN THE SEVENTEENTH CENTURY

RICHARD S. WESTFALL
*Professor of the History of Science,
Indiana University*

MACDONALD · LONDON
AMERICAN ELSEVIER · NEW YORK

© Richard S. Westfall 1971
First published 1971

Sole distributors for the United States and Dependencies
American Elsevier Publishing Company, Inc.
52 Vanderbilt Avenue
New York, N.Y. 10017

Sole distributors for the British Isles and Commonwealth
Macdonald & Co. (Publishers) Ltd
49–50 Poland Street
London W1A 2LG

All remaining areas
Elsevier Publishing Company
P.O. Box 211
Jan van Galenstraat 335
Amsterdam
The Netherlands

Library of Congress Catalog Card Number 76-139572
Standard Book Numbers
British SBN 356 02261 7
American SBN 444 19611 0

All Rights Reserved. No part of this publication may be reproduced, stored in a retrieval system, or transmitted in any form or by any means, electronic, mechanical, photocopying, recording or otherwise, without prior permission of the publishers.

Printed in Great Britain by Hazell Watson & Viney Ltd
Aylesbury, Bucks

*To my mother, Dorothy M. Westfall, and
the memory of my father, Alfred R. Westfall.*

Societas nostra id cum primis nunc agit, ut Natura et leges motus penitius, quam hucusque factum, vestigentur et innotescant.... Cum ignorato motu ignoretur Natura, eo diligentius scrutinio ejus Philosophis est incumbendum ...

Our Society is now particularly busy in investigating and understanding Nature and the laws of motion more thoroughly than has been done heretofore.... Since Nature will remain unknown so long as motion remains unknown, diligent examination of it is the more incumbent upon philosophers ...

> Henry Oldenburg to Hieronymo Lobo,
> 27 May 1669

Contents

	Preface	ix
I.	Galileo and the New Science of Mechanics	1
II.	Descartes and the Mechanical Philosophy	56
III.	Mechanics in the Mid-Century	99
IV.	Christiaan Huygens' Kinematics	146
V.	The Science of Mechanics in the Late Seventeenth Century	194
VI.	Leibnizian Dynamics	283
VII.	Newton and the Concept of Force	323
VIII.	Newtonian Dynamics	424
	Appendices	526
	Bibliography	551
	Index	567

Preface

MORE years ago than I care now to recall, this book began as a study of Newton's concept of force, and the title, *Force in Newton's Physics*, the suggestion of Michael Hoskin, Editor of the History of Science Library, was tentatively bestowed on the project at that time. Historian that I am, I understood from the beginning that any examination of Newton's concept of force worthy of the name would need to see it in the setting of seventeenth-century mechanics as a whole, but in my innocence I scarcely realised what such a project would involve. I was still primarily at work on Newton himself when the invitation to deliver a paper on Galileo at UCLA's quadricentennial celebration impelled me to examine Galileo's mechanics anew from the point of view of the concept of force. The revelation of how much the seventeenth century before Newton could teach me in illuminating his dynamics set the pattern for the rest of my investigation. To the degree that Newton still occupies more than a quarter of the space, the work maintains its original intent. To a considerable extent, however, it has also become a history of dynamics in the seventeenth century, and I have acknowledged as much by adding the subtitle that the book now bears.

If the study is a history of dynamics in the seventeenth century, I like to think that it is a history with a difference. Although I explore all of the major advances in dynamics in some detail, I have not concerned myself primarily to catalogue them. Rather I have tried to understand the obstacles that cluttered the path leading toward modern dynamics, or to employ a different figure, the conceptual knots that had to be loosed. It would be correct to say that I have tried to see seventeenth-century dynamics through the eyes of the men engaged in creating the science – I have attempted to define the problems on which they expended themselves in their terms, and to see their proposed solutions in relation to the intellectual equipment at their disposal. Equally, however, I have sought to exploit the advantages that three centuries of perspective provide. I do not believe that any scientist of the seventeenth century could have stated the set of questions around which this study revolves with the clarity possible for the historian of the twentieth century. If it is necessary to see dynamics through their eyes in order not merely to recount those achievements that still seem correct to us, so also it is necessary to examine their confusions from our distance in order fully to understand what their problems were. I began the study of seventeenth-century dynamics predisposed to be impressed, and I have not been disappointed. Basic dynamics has now been rationalised and systematised to the extent that an intelligent schoolboy can master all that the seventeenth century produced in a few weeks. A careful study of how they produced it is calculated to remove any suspicion that the task was, after all, rather simple. Seen from the dual perspective I have proposed, the creation of modern dynamics in the seventeenth century appears as one of the supreme conquests of the human spirit, a triumph wholly worthy of the 'century of genius'.

Should any doubt remain, let me say straight out that this is a history of ideas. I have paid almost no attention to the social and economic setting in which dynamics emerged. When the book closes, the bourgeoisie have not risen through my efforts one inch above the level they occupied on page one. Undoubtedly the fact that many men of ability devoted serious study to natural science in general and mechanics in particular during the seventeenth century depended to a high degree on the social and economic state of Europe. If my experience is any guide, however, it is impossible to conclude from seventeenth-century literature on mechanics that practical considerations, technological problems set by the economic system, guided and determined the conceptual development of the science. The most important applications of dynamics at the time were to problems in pure science – the celestial dynamics of Newton's *Principia* epitomised the uses to which it was put – and as far as the seventeenth century was concerned, it was more by accident than design that the engineers of a later age would exploit its conclusions to such effect. The story I have to tell, then, concerns itself, not with social contexts or with practical considerations, but with conceptual developments initiated by a new idea of motion proposed near the beginning of the century. Whereas I devote no attention to social factors, I devote very little more to technical mathematical questions. I do not mean to deny in any way the importance of mathematics in seventeenth-century dynamics. With the calculus, for example, a whole new range of problems hitherto beyond the grasp of quantitative mechanics became amenable to exact treatment. My central concern has focused on conceptual issues, however; and during the development of dynamics up to Newton, such matters appear to me to have been central to the science of dynamics. My attempt has been to follow the conceptual development within the limits imposed by the body of inherited ideas on the one hand, and on the other by the prevailing philosophy of nature.

Somewhat to my surprise, I realised as I was completing the manuscript that I was perhaps commenting implicitly on a question now exercising the discipline of the history of science. Is scientific change a matter of gradual development or of sudden paradigm shift? Let me only state, as my text will amply document, that the study of seventeenth-century dynamics has impressed me with the incredible tenacity of received ideas in a conceptual context that appears antithetical to them from a distance of three centuries. Looking back, we want to find Newtonian dynamics in Galileo. In fact, nearly a century's labour by a host of able men was required to extract from the new conception of motion consequences that appear obvious to us. In the meantime, a set of ideas descended from medieval and ancient mechanics, ideas that continued to appear familiar and therefore valid, despite the new conception of motion, dominated attempts to construct a quantitative dynamics. However one may interpret the change from Aristotelian dynamics, it is difficult to find evidence of a dramatic paradigm shift in the seventeenth-century literature on mechanics.

Because the change was gradual, a considerable ambiguity of technical language existed. Without exception, I believe, the technical terms of Newtonian dynamics descended from those in common use in earlier mechanics. Words that acquired a precise meaning with Newton were part of the coinage in which the commerce in mechanics had long been carried on, and for want of alternate coinage, they continued in use during the seventeenth century. Those who employed them undoubtedly thought they were clear enough; in relation to the new dynamics that eventually appeared, however, they were charged with ambiguities. I must warn the reader not to impose the precise modern meaning on such words. When *'velocitas'* appeared in a Latin text, or *'gravité'* in a French one, I could do no other than to translate the words as 'velocity' and 'gravity', and I could scarcely alter an English text to replace the word 'force'. A fair part of the history I recount has to do with the development of precise terminology, and the indication of prevailing ambiguities is essential to my task. Let the reader then exercise care in the meanings he imposes. As for me, the growing realisation of the problem of terminology in seventeenth-century dynamics led to extensive essays on usage, especially of the word 'force'. These essays compose the appendices referred to in the footnotes.

In connection with usage, I need also to remark on translations. Rather than translate anew every passage I include, I have utilised standard accepted translations where they are available. Whenever specific terms have been at issue, I have checked with the original texts to be sure of the exact word or phrase.

Like every author who completes an extended work, I am fully aware of the help I have received in many forms from many sources. Without the National Science Foundation, I could never seriously have embarked on the project, and I wish to acknowledge its generous support. A grant-in-aid from the Office of Research and Advanced Studies of Indiana University facilitated the preparation of the final manuscript. Most of the research was conducted in three libraries – University Library Cambridge, the Widener Library at Harvard, and the library of Indiana University. To the expert assistance of the staffs in all three I extend my thanks. The Syndics of Cambridge University Library have graciously permitted me to quote extensively from the Newton papers in their collection. Members of the secretarial staff of the Department of History and Philosophy of Science at Indiana University applied skill and willingness beyond any reasonable demands of their positions in the preparation of the manuscript, and I wish to thank especially Joyce Chubatow, Jean Coppin, and Ina Mitchell. And finally I owe a debt beyond calculation to my wife Gloria and my children who long endured a husband and father frequently engrossed in matters three centuries removed. It is not at all clear that I shall return to the twentieth century now that the book is done, but it is equally unclear that I should be better company if I did.

Chapter One

Galileo and the New Science of Mechanics

IN the introductory passages of the *Mathematical Principles of Natural Philosophy*, published near the end of the seventeenth century, in 1687, Isaac Newton set down the three laws of motion, which became the foundation of the science of mechanics and of the entire structure of modern physical science.

> Law I. Every body continues in its state of rest, or of uniform motion in a right line, unless it is compelled to change that state by forces impressed upon it.
>
> Law II. The change of motion is proportional to the motive force impressed; and is made in the direction of the right line in which that force is impressed.
>
> Law III. To every action there is always opposed an equal reaction: or, the mutual actions of two bodies upon each other are always equal, and directed to contrary parts.[1]

As far as the third law was concerned, Newton believed that its formulation was original with him – although, of course, he also believed that the phenomena cry out for it. The first two laws he attributed to Galileo as a matter of common knowledge.

> By the first two Laws and the first two Corollaries, Galileo discovered that the descent of bodies varied as the square of the time and that the motion of projectiles was in the curve of a parabola . . .[2]

In this book, I am primarily concerned with the second law, the concept of force and its role in rational mechanics. Inevitably I find Newton's attribution suggestive.[3] Early in the seventeenth century, Galileo transformed the science of mechanics and introduced the problem of force in its present form. His mechanics and his philosophy of nature seem to demand the second law of motion – so much so that more than one historian of science has followed Newton by implicitly ascribing it to him. A handful of crabbed precisionists excepted, everyone agrees that Galileo contributed decisively to the first law, even though he did not formulate it exactly. The first law in turn demands the second, to the point that we can hardly imagine them apart. Ernst Mach wondered why Newton had bothered to state the first since it is implicit in the second. Fortunately for Newton's role in the history of physics, he gave Galileo more than his due. However suggestive we may find Galileo's mechanics, he did not formulate the second law of motion. Its formulation was Newton's work and not the least of his claims to immortality.

It was not solely Newton's work, however, and Galileo cannot be neglected in the history of the concept of force. Indeed, Galileo looking forward early in the century was better as a prophet than Newton was as an historian looking back near its end. There 'have been opened up to this vast and most excellent science, of which my work is merely the beginning, ways and means by which other minds more acute than mine will explore its remote corners.'[4] Except for the suggestion of other minds more acute (which we may understand as a rhetorical flourish, and not as a serious expectation on Galileo's part), his words forecast the course of mechanics in the seventeenth century. Newton's *Principia* was the culmination of more than half a century's labour, which took its start from Galileo. After the new conception of motion, to which Galileo's contribution was basic, the second law was the major achievement of seventeenth-century mechanics. Galileo yoked the new conception of motion to a new ideal of mechanics which set as its goal the mathematical description of the phenomena of motion. Until the formulation of the second law, the new ideal remained more hope than fact. Realised at last in the *Principia*, it became at once the basis and the example of mathematised science.

Galileo's initial work in mechanics predated the seventeenth century by about a decade. During the period 1589–92, when as a young man he occupied the chair of mathematics at the University of Pisa, he composed an essay, *De motu*, which remained unpublished until the nineteenth century; and in 1593, shortly after he had moved to the University of Padua, a treatise on the simple machines displayed another facet of his interest. Together, the two works provide the basic evidence of Galileo's early understanding of mechanics. Beginning in the first decade of the seventeenth century while he was still at Padua, Galileo fundamentally revised his earlier views, partly under the stimulus of his commitment to Copernican astronomy, the justification of which entailed a new conception of motion. As a result of the fame that his telescope brought, he was able to return to Tuscany and Florence in 1610 with the grandiloquent title of Chief Mathematician and Philosopher to the Grand Duke, and in Florence he completed the two books that accomplished a revolution in the science of mechanics – *Dialogo sopra i due massimi sistemi del mondo* (the *Dialogue*, 1632) and *Discorsi e dimostrazioni matematiche intorno a due nuove scienze* (the *Discourses*, 1638). Galileo died in 1642, blind and confined to his home as a prisoner following his trial before the Inquisition in 1633 for his defence of Copernican astronomy in the *Dialogue*. His *Discourses* had to be spirited abroad in manuscript and published in the Netherlands.

Since Galileo's mechanics transposed the problem of force by placing it in a new context, the history of the second law of motion effectively begins with him. The Aristotelian and medieval science of mechanics had not recognised any effective distinction between statics and dynamics. The simple machines, such as the lever, obviously served, not to hold bodies in equilibrium, but to move them; and in the Aristotelian tradition, the simple machines were analysed in dynamic terms. The most sophisticated conception of force in medieval mechanics is found in the treatises on statics ascribed to Jordanus. What we call dynamics, the consideration of motions not constrained by simple machines, was built on principles directly transposed from statics and justified by the basic proposition of peripatetic mechanics, that *omne quod movetur ab alio movetur* (everything that is moved is moved by something else). Aristotle's analysis of motion in the *Physics* reveals its full meaning when we imagine the two bodies to be placed on the ends of a lever.

> If, then, A is the moving agent, B the mobile, C the distance traversed and D the time taken, then A will move $\frac{1}{2}B$ over the distance $2C$ in time D, and A will move $\frac{1}{2}B$ over the distance C in time $\frac{1}{2}D$; for so the proportion will be observed. Again, if A will move B over distance C in time D and A will move B over distance $\frac{1}{2}C$ in time $\frac{1}{2}D$, then $E\;(=\frac{1}{2}A)$ will move $F\;(=\frac{1}{2}B)$ over distance C in time D; for

the relation of the force $E\left(\frac{1}{2}A\right)$ to the load $F\left(\frac{1}{2}B\right)$ in the last proposition is the same as the relation of the force A to the load B in the first, and accordingly the same distance (C) will be covered in the same time (D).[5]

With Galileo's denial that every body in motion requires the continued action of a mover, the concept of force began a new career which culminated in Newton's second law.

Two major achievements of Galileo's mechanics, worked out gradually during the first three or four decades of the seventeenth century and made accessible to the scientific world in his two great books, the *Dialogue* and the *Discourses*, immediately strike the modern reader. The first is the new conception whereby motion (at least horizontal motion) is held to be, not a process requiring a cause, but a state to which a body is indifferent, a state in which it remains, as it remains in rest, until some external agent causes it to leave. Once motion on a frictionless horizontal plane is acquired, Galileo declared, 'it will continue perpetually with uniform velocity.'[6] Only recently have we been taught to ask what Galileo meant by a horizontal plane. Since a horizontal plane was one equally removed at every point from a gravitational centre, his perpetual motion with uniform velocity took place on a spherical surface like that of the earth. We cannot then properly say that he stated the principle of inertia, which holds for rectilinear motion.

Once we have seized that limitation, we must be careful not to ignore the extent to which Galileo approached the concept of inertia. He identified uniform horizontal motion and rest, considering rest merely as a special case, what he called an infinite degree of slowness.

> Motion, in so far as it is and acts as motion, to that extent exists relatively to things that lack it; and among things which all share equally in any motion, it does not act, and is as if it did not exist.

To the relations of things that move together, motion is 'idle and inconsequential'; it is 'operative' only in relation to other bodies that lack it.[7] Since a body is indifferent to motion, it can participate freely in more than one motion at the same time. Thus by considering two perpendicular but not mutually obstructing motions, Galileo demonstrated that the trajectory of a projectile is a parabola. Small wonder that Newton mistook such a conception for his first law of motion.

Galileo's second major achievement in mechanics was the identification of free fall as a uniformly accelerated motion and the detailed exposition of its role in nature. 'A motion is said to be uniformly accelerated, when starting from rest, it acquires, during equal time-intervals, equal increments of speed.'[8] Galileo insisted that we understand the definition, not as an arbitrary one, such as a mathematician might construct in order to examine its logical consequences, but as a definition answering to natural phenomena. If we neglect the small resistance of the atmosphere, every body falls with uniformly accelerated motion, and Galileo used the phrase 'naturally accelerated motion' as fully synonymous.

> When I consider that a stone, which falls from some height starting from rest, constantly acquires new increments of velocity, why should I not believe that these additions are made in the simplest and easiest manner of all? The falling body remains the same, and so also the principle of motion. Why should the other factors not remain equally constant? You will say: the velocity then is uniform. Not at all! The facts establish that the velocity is not constant, and that the motion is not uniform. It is necessary then to place the identity, or if you prefer the uniformity and simplicity, not in the velocity but in the increments of velocity, that is, in the acceleration.[9]

The passage appears to be intimately related to the new conception of motion. Having concluded that uniform motion is a state which requires no cause, Galileo was free to identify a new dynamic product of weight. When a heavy body falls, its weight generates, not a uniform velocity, but uniform increments of velocity, that is, a uniform acceleration.

The concept of uniform acceleration was only half of Galileo's revision of the understanding of free fall. Aristotelian mechanics had set velocity proportional to weight. What would have been more natural in revising Aristotle than to set acceleration proportional to weight? Since Galileo put such an opinion into Sagredo's mouth in the *Dialogue*, we may be sure that he considered it.[10] As everyone knows, he concluded ultimately that all bodies fall toward the earth with the same acceleration, the variation produced by the resistance of the atmosphere being ignored. It is necessary, Salviati explained to Sagredo,

> to distinguish between heavy bodies in motion and the same bodies at rest. A large stone placed in a balance not only acquires additional weight by having another stone placed upon it, but even by the addition of a handful of hemp its weight is augmented six to ten ounces according

> to the quantity of hemp. But if you tie the hemp to the stone and allow them to fall freely from some height, do you believe that the hemp will press down upon the stone and thus accelerate its motion or do you think the motion will be retarded by a partial upward pressure? One always feels the pressure upon his shoulders when he prevents the motion of a load resting upon him; but if one descends just as rapidly as the load would fall how can it gravitate or press upon him? Do you not see that this would be the same as trying to strike a man with a lance when he is running away from you with a speed which is equal to, or even greater, than that with which you are following him?[11]

The dynamic effect of weight is not the generation of a uniform increase of velocity after all. It is rather the generation of a uniform increase of motion, where 'motion' is understood as the product of velocity and the size of the moving body (which is understood implicitly to be proportional to its weight). Again, it is hardly surprising that Newton saw the second law of motion in such passages.

Galileo's mechanics appears even more familiar when we see it against the backdrop of his conception of nature. He denied the anthropocentricity of medieval cosmology, according to which the whole universe exists for the benefit of man, and he denied as well the uniqueness of the earth, or of any other body, as a centre in the cosmos. He affirmed the homogeneity of matter, holding that all bodies are composed of the same material, packed more or less densely together. Matter is inert, and bodies composed of it obey the compulsion of external agents and move according to fixed laws. On a horizontal plane, every body 'finds itself in a condition of indifference as to motion or rest; has no inherent tendency to move in any direction, and offers no resistance to being set in motion.'[12] Because all matter is identical, the distinction of heavy and light in vertical motion vanishes, and all bodies, being heavy, move downward when not restrained. Galileo suggested that generation and corruption, the most fundamental changes in Aristotelian science, are nothing more than 'a simple transposition of parts . . .'[13] The well-known passage in the *Assayer* that denies the reality of many qualities appears to assert that physical nature is composed solely of particles of matter. 'I do not believe,' he stated, 'that for exciting in us tastes, odours, and sounds there are required in external bodies anything but sizes, shapes, numbers, and slow or fast movements; and I think that if ears, tongues, and noses were taken away, shapes and numbers and motions would remain but not odours or tastes or sounds.'[14] These are familiar notions to the student of seventeenth-century science. Powerfully they evoke the mechanical philosophy of nature and render Galileo's analysis of free fall still more suggestive of the second law of motion.

Nevertheless, neither the second law of motion nor a satisfactory conception of force is to be found in Galileo's mechanics. The problem is not one of kinematics versus dynamics. Although there are extensive kinematical passages in Galileo, his mechanics rested on a consciously dynamic analysis of free fall. At the beginning of that analysis, he insisted that his definition of uniformly accelerated motion was not an arbitrary one, such as a mathematician might propose, but one that corresponded to a motion nature continually employs. He also reminded his readers frequently that motion on a horizontal plane is uniform because there is neither a tendency to increase motion nor a repugnance that works to diminish it. A crucial proposition of his mechanics, that the total change of velocity is determined by a body's vertical displacement independently of the path it follows, rested frankly on a dynamic analysis of motion on inclined planes. Far from restricting itself to kinematics, Galileo's mechanics returned to the dynamics of free fall at every critical point, and attempted to illuminate the whole of mechanics with its light. Moreover, in his analysis of free fall, weight functions exactly as force in Newton's second law, so that his analysis established the paradigm for the treatment of force in classical mechanics, the simplest case, in which a constant force produces a uniform acceleration. For all of that, neither the second law of motion nor a satisfactory conception of force can be found in Galileo's mechanics.

It is not a quibble on words in which I am interested. A problem of terminology certainly existed in seventeenth-century mechanics as a set of new concepts struggled to achieve precision. Viviani was so overborne by the verbal anarchy when he expanded a passage in the *Discourses* that he referred to '*l'impeto, il talento, l'energia*, or we might say *il momento*,' of a moving body.[15] Indeed we might say! We might also say, as Galileo said more than once, '*la virtù*,' and '*la propensione al moto.*' Among the other terms used more or less synonymously was the one that Newton chose and defined, *la forza* (in Latin, *vis*).[16] The chaos of terminology must have helped to obstruct the emergence of a concept of force, but terminology was not the decisive aspect of the problem. Galileo never suggested that the acceleration of free fall is produced by a force acting on a body. He never considered free fall as a special case of a general phenomenon, such that its analysis could be applied to the understanding of all changes of motion.

A problem of major dimensions for the history of science in the seventeenth century is involved in the second law of motion. The second law made possible the perfection of the mathematical science of mechanics, the supreme achievement of the scientific revolution. In the *Principia*, rational mechanics produced the masterpiece of seventeenth-century science. What is more important than the *Principia* itself and the law of universal gravitation, rational mechanics established the pattern on which physical science has modelled itself ever since. Galileo's mechanics proposed the paradigm of the second law in the analysis of free fall, a paradigm apparently so clear that Newton himself attributed the second law to Galileo. Obviously it was not that clear. Half a century elapsed between the *Discourses* and the *Principia*, and the labour of the age's ablest men did not suffice to produce a workable conception of force before Newton. In following the story of the concept of force, we follow the steps of seventeenth-century science as a whole and face the basic problems with which its creators grappled.

*

Undoubtedly, the very nature of Galileo's mechanics was a major cause of the difficulty the seventeenth century experienced in seeing that it implied the equivalent of the second law of motion. With Newton to guide us, we do not find it hard to discover insightful passages; the others we know how to interpret. His mechanics must have presented a different appearance to those who did not yet have Newton to guide them. Galileo himself belonged to the latter group, of course. He did not think of himself as struggling to clarify a concept of force. When we attempt to examine his mechanics from a pre-Newtonian point of view, instead of pregnant suggestions of the second law of motion, it appears to be dominated by quite a different set of ideas. If they tend toward any conception of force, it is one quite different from Newton's.

Most prominent is the idea of natural motions. Perhaps nothing in Galileo's mechanics separated him more emphatically from Newton, from whose universe natural motions had been banished. Galileo's universe, in contrast, contained two of them, and they made an appearance on nearly every page that he wrote. The uniform acceleration of free fall was to Galileo the 'natural acceleration downwards common to all bodies.'[17] Again, he declared that acceleration occurs when a body 'is approaching the point toward which it has a tendency, and retardation occurs because of its reluctance to leave and go away from that point...'[18] However much we want to read his analysis of free fall as the description of a uniform acceleration produced by a constant force, Galileo never treated it in such terms. Weight was not a force acting on a body to accelerate it. On the contrary, weight was more a static force which a body exercises on another that restrains it from its natural motion. The interpretation of what Galileo meant when he spoke of the natural tendency of bodies to move downward presents problems, as I shall wish to discuss later. Suffice it to say at this point that the idea of a natural tendency cannot be explained away from Galileo's mechanics since it expressed his conviction that the universe is an ordered one in which bodies have natural places. Another consequence of that conviction was his use of the word 'force' to express that which opposes the natural order.[19] Motion upward was 'forced' motion. Following a long tradition, he frequently used the words 'violence' and 'violent' in the same context. Forced motion did violence to the natural order. As a natural motion, the uniform acceleration of free fall resisted identification with other changes of motion.

If free fall was a natural motion, so also was the uniform motion of terrestrial objects around the centre of the earth. Uniformly accelerated motion toward a centre was the natural motion of bodies seeking their place; uniform motion around a centre was the natural motion of bodies that had found their place. Galileo employed the concept of (circular) inertia to solve the problems raised by Copernican astronomy. How is it possible for phenomena of motion to appear as they do if the earth is turning daily on its axis? When a ball is dropped from a tower, the motion of the tower from west to east should leave it far behind, and it should appear to fall well to the west. In fact, it appears to fall perpendicularly along the side of the tower. How is this possible on a moving earth? 'Keeping up with the earth,' Galileo explained, 'is the primordial and eternal motion ineradicably and inseparably participated in by this ball as a terrestrial object, which it has by its nature and will possess forever.'[20] Elsewhere he referred to the diurnal motion, which all terrestrial bodies share whether they are attached to the earth or not, as 'natural and eternal,' as 'a thing indelibly impressed upon them by nature,' as 'a natural propensity,' as 'an inherent and natural inclination.'[21]

If the diurnal rotation of terrestrial objects around the axis of the earth is natural, it would appear that this motion is distinct from other horizontal motions, just as natural acceleration is unique. Indeed, Galileo did make the distinction more than once, but the predominant tendency of his mechanics was to identify all uniform horizontal motions with the eternal motion of terrestrial bodies. In the *Dialogue*, Simplicio brought up the example of a ship in motion. Dropping a stone from the top of its mast should be equivalent to dropping it from a tower on a moving earth, and since he was convinced that the stone falls to the rear of the moving ship, Simplicio argued that its fall parallel to the tower proves that the earth must be at rest. In replying to Simplicio, Salviati insisted at first that the artificial motion of the ship cannot be compared to the natural motion of the earth. He quickly dropped the distinction, however, asserted that the stone falls at the foot of the mast, and explained the phenomenon in terms almost identical to those employed for the natural motion. The stone carried by the ship acquires 'an ineradicable motion' as fast as that of the ship; the ship's motion remains 'indelibly impressed' on the stone when it is dropped.[22] Another thought experiment in the *Dialogue* places two men in the cabin of a ship. They jump; they play catch; butterflies flutter about; fish swim in a bowl; water drips from a bottle; smoke rises – and everything that happens when the ship is at rest proceeds exactly the same when it is in motion because 'the ship's motion is common to all the things contained in it . . .'[23] Yet again, a piece of wax so little heavier than water that it descends only a yard in a minute is placed in a vase on a ship that moves one hundred yards a minute. To the men on board it appears to descend vertically.

> Now these things take place in motion which is not natural, and in materials with which we can experiment also in a state of rest or moving in the opposite direction, yet we can discover no difference in the appearances, and it seems that our senses are deceived. Then what can we be expected to detect as to the earth, which, whether it is in motion or at rest, has always been in the same state? And when is it that we are supposed to test by experiment whether there is any difference to be discovered among these events of local motion in their different states of motion and of rest, if the earth remains forever in one or the other of these two states?[24]

Whenever he was concerned with problems analogous to the rotation of the earth – problems which he raised in order to explain how the rotation of the earth can be reconciled to the observed phenomena of motion – Galileo emphasised the ultimate identity of all uniform horizontal motions, natural and artificial. For the sake of simplicity, we can call such motion inertial motion, though in using the phrase we must remember that inertial motion to Galileo was a circular motion.

Another class of problems in Galileo's mechanics introduced motions which Galileo treated in different terms although we consider them indistinguishable from inertial motion. The words '*momento*' and '*impeto*', used interchangeably, frequently designated a body's motion. In Galileo's system, the *momento* or *impeto* of a body cannot be lost in horizontal motion from which all obstacles have been removed. It can also not be generated. Motion downward generates velocity; motion upward destroys velocity; motion on a horizontal plane conserves velocity. Hence in Galileo's own terms, the *momento* or *impeto* of a body at any time should only express its inertial motion, the state which can be perceived only by a body's relation to others that do not share it, the state to which it is indifferent. Unfortunately, the verbs that Galileo used with *momento* and *impeto* suggest anything except indifference. The *impeto* which the bob of a pendulum acquires in descending with a natural motion is able to drive it upward by a forced motion (*sospignere di moto violento*) through an equal ascent. In general, when a heavy body falls from any height, it acquires just as much *impeto* as was necessary to carry it (*tirarlo*) to that height. A point on the circumference of a moving wheel has the *impeto* with which to hurl (*scagliare*) a stone. When water in a barge is set in oscillation by an uneven motion, it rises at one end, falls again because of its weight, and, pushed (*promossa*) by its own *impeto*, goes beyond equilibrium. He also imagined cases in which the *impeto* acquired by one body in falling was transferred to another and functioned to drive (*cacciare*) the second upward.[25] Since the entire tradition of mechanics before Galileo had treated force as that which is opposed to the natural order, that which compels a body to move contrary to its natural inclination, a student of mechanics in the seventeenth century was more likely to see a concept of force in these passages than in Galileo's analysis of naturally accelerated motion. Such a concept of force would not only have differed from Newton's; it would have been quantitatively incompatible with it as well. I shall refer to it as the paradigm or model of impact. The fact that a notion harking back both to the medieval theory of impetus and to the medieval science of weights survived in Galileo's mechanics can help us both to comprehend the basic thrust of his mechanics and to understand the problems which he left to be solved by the seventeenth century after him. We must remember constantly that he presented, not one, but two models of force, and the one we are least prepared to recognise was the one most adapted to the comprehension of his age.

*

Inevitably the ideal of *momento* or *impeto* recalls the treatise *De motu*, Galileo's first independent effort in mechanics, which was composed sometime around 1590, while he was teaching at Pisa. To understand his mechanics and his contribution to the concept of force, we must first analyse the structure of *De motu*.

At much the time when he was composing it, Galileo jotted down a note which illuminates the purpose of the treatise and presents a theme which dominated his mechanics through all its changes. 'A fragment of Euclid asserts that heavy and light are to be handled mathematically.'[26] Galileo's treatment of motion underwent a fundamental revision after he completed *De motu*, but his mechanics never veered from the direction he took at the beginning. Although he adopted the theory of impressed force in *De motu*, he gave it a twist not to be found among its medieval exponents. By defining impressed force as accidental lightness, he endeavoured to render it measurable by the weight to which it is equal. Which is to say, the mathematical treatment of heavy and light was equivalent to the development of a mathematical science of mechanics.

Fundamental to the structure of *De motu* was the assertion, which also remained unaltered in his mature mechanics, that all bodies are heavy. What conditions are necessary, Galileo asked, for a motion to be called natural? First, it cannot be infinite and indeterminate. Those things which are moved by nature 'are carried toward some goal where they can be at rest naturally.' Second, the body moved must be moved by an intrinsic and not by an extrinsic cause. Motion downward has a goal which is definite, the centre of the earth. Motion upward, in contrast, must always have an external cause. Where is the end of up? When would a body arrive there?

> The matter of all bodies is the same, and it is heavy in all of them. But the same heaviness cannot have contrary natural inclinations. Therefore, if there is one natural inclination, the contrary inclination must be against nature. The natural inclination of heaviness, however, is toward the center. Therefore, that which is away from the center must be against nature.[27]

Since every body strives to move toward the centre of the earth, which is its natural place, it can remain at rest outside its natural place only if a force equal to its weight restrains it.

In this concept Galileo saw the means to quantify impressed force. What is the impressed force that makes a heavy body move upward? It is a taking away of heaviness, Galileo replied. As fire deprives a piece of iron of its coldness, impressing heat on it, so hurling a stone upward deprives it of heaviness and impresses lightness on it. Why call impressed force lightness? Because we say that bodies that move upward are light, and there is no assignable difference between the stone as it rises and a light body. The stone is not naturally light, of course; it is accidentally or preternaturally light because its motion upward is against nature and forced. Galileo compared it to a piece of wood, which is naturally heavy, but which rises with a preternatural motion when it has been submerged in water. As the heat in a piece of iron gradually dissipates, so the impressed force fades. When it equals the weight of the stone, the top of the trajectory has been reached – an equation which only repeats the proposition that a body at rest outside its natural place must be restrained by a force equal to its weight. As the impressed force continues to fade, the weight of the stone preponderates, and it begins to fall.[28]

The comparison of the stone to a piece of wood in water was important. In *De motu*, Galileo employed Archimedean fluid statics to revise Aristotle's treatment of the role of the medium in motion. On the one hand, Aristotle had assigned the continuation of projectile motion to the medium. According to the basic proposition of his mechanics, a body moves only if something moves it, and in the case of a projectile separated from the original projector, the role of mover was assigned to the medium. Impetus mechanics transferred the motive force from the medium to the projectile – a body placed in motion acquires an impressed force or impetus, an internal motive power which continues the body's motion after it has separated from the projector – and Galileo's *De motu* repeated what had virtually become the accepted orthodoxy. Aristotle's mechanics also assigned a second role to the medium; it functioned as a resistance as well, and the velocity of motion was held to depend on the proportion of motive force to resistance. In the special case of a heavy body falling, Aristotle said that its velocity is directly proportional to its weight and inversely proportional to the resistance (or density) of the medium. If we commit the anachronism of putting Aristotle's analysis in functional form, we can represent it by the formula

$$v \propto \frac{F}{R}$$

where F represents the weight acting as motive force and R the resistance of the medium. Within the tradition of impetus mechanics, the contradictory nature of the two roles assigned to the medium was a well-established criticism of Aristotle's theory, and Galileo's repetition of it held nothing original.

What was original in *De motu* was his quantitative critique drawn from fluid statics. By its means, Galileo showed that Aristotle's treatment of fall leads to inconsistencies. Take bodies of two different materials, a piece of lead, say, and a piece of wood. According to the formula we used to express Aristotle's position, the proportion of their velocities (whatever the proportion may be) should be the same in all media. If their velocities in air are in the ratio of two to one, their velocities when they fall in water should be in the same ratio, although both fall more slowly. In fact, of course, wood does not fall in water at all.

To resolve the discrepancy, Galileo employed a concept of effective weight which altered the mathematical role of the medium's resistance. By the Aristotelian formula, which places the resistance in the denominator, the density of water would have to be infinite to account for the fact that wood falls through it with a speed of zero. In the case of lead, on the other hand, the density of water would be finite. To reconcile the two cases, Galileo proposed that the weight of an equal volume of the medium be subtracted from the weight of the body and not divided into it. All bodies are heavy, but some are heavier specifically than others. Why does wood rise when it is submerged in water? Because an equal volume of water has a greater weight than the wood. The wood moves up with a 'force' equal to the amount by which the weight of an equal volume of water exceeds its weight. Correspondingly, the piece of lead moves down with a 'force' equal to the amount by which its weight exceeds that of an equal volume of water.[29]

De motu proposed a revision in the basic formula of dynamics. The resistance of the medium affects the speed of a falling body by subtracting from the maximum speed, in effect, instead of dividing into it. The suggestion was not original with Galileo; it had a long history stretching back at least to John Philoponus in the sixth century. Even the merger of Archimedian fluid statics with that tradition had been proposed before Galileo by Giambattista Benedetti. The originality of *De motu* lay in weaving the concept of impressed force into the same fabric. The buoyant effect of the medium is mathematically identical to the artificial lightness of an impressed force. The motion of wood in water is identical to the motion of a stone thrown upward, except that the artificial lightness of the wood remains constant whereas the artificial lightness of the stone decays.

Among other things, Galileo's concept of effective weight allowed him (as it had allowed others before him) to accept motion in a void. By the Aristotelian formula, velocity in a void would be infinite because resistance would be nil. The absurdity of an infinite velocity had been one of Aristotle's arguments against the possibility of a void. Obviously that absurdity evaporated in the reformulation of the role of resistance.

> For in a plenum the speed of motion of a body depends on the difference between its weight and the weight of the medium through which it moves. And likewise in a void [the speed of] its motion will depend on the difference between its own weight and that of the medium. But since the latter is zero, the difference between the weight of the body and the weight of the void will be the whole weight of the body. And therefore the speed of its motion [in the void] will depend on its own total weight. But in no plenum will it be able to move so quickly, since the excess of the weight of the body over the weight of the medium is less than the whole weight of the body.[30]

More interesting than the treatment of the void was the treatment of velocity. Like Aristotle, like the Scholastics, the Galileo of *De motu* believed that the maintenance of a constant velocity implies the action of a constant motive power. It is true that the other terms of the equation had been modified. When Galileo said that speed in a void depends on the total weight of a body, he meant its specific weight. As all pieces of pine, large and small, rise in water with the same force and velocity, so they all fall in the void with the same force and velocity. If lead and wood move with different velocities, and if the difference increases in a denser medium which subtracts a greater proportion of the specific weight of wood than of lead, all pieces of lead, whatever their size, fall with the same velocity through the same medium, and all pieces of wood do the same, though their velocity differs from that of lead. Behind the point he disputed with Aristotle, however, looms the larger fact that Galileo still shared the conviction that a constant velocity of motion implies a constant motor. *Omne quod movetur ab alio movetur.* With its effort to measure impressed force by weight and its concentration on vertical motion, *De motu* displayed in its starkest form a fundamental aspect of this dynamics. The dynamics of *De motu* was a direct transposition of statics. Velocity was set directly proportional to (specific) weight. Speed, Galileo asserted, cannot be separated from motion.

> For whoever asserts motion necessarily asserts speed; and slowness is nothing but lesser speed. Speed therefore proceeds from the same [cause] from which motion proceeds. And since motion proceeds from heaviness and lightness, speed or slowness must necessarily proceed from the same source. ... For if the motion is downward, the heavier substance will move more swiftly than the lighter; and if the motion is upward, that which is lighter will move more swiftly.[31]

The primary thrust of *De motu*, the attempt to quantify the concept of impressed force, served further to confound statics and dynamics. By defining impressed force as lightness, he set velocity proportional to specific weight at every point in the trajectory of a rising or falling body. 'Since, then, a heavy falling body moves more slowly at the beginning, it follows that the body is less heavy at the beginning of its motion than in the middle or at the end.'[32] The state of a body at the top of its trajectory is identical to that of a suspended one. In each case, the impressed lightness exactly balances the body's weight, with the result that the speed is nil. As the impressed lightness decays, the weight increases and with it the speed. Thus the mathematical treatment of heavy and light implied a complete mechanics, not only statics, but dynamics as well.[33]

Within the conceptual framework of *De motu*, a quantitative dynamics was possible because the long tradition of statics, stretching back to the ancient world, to Archimedes and Aristotle, had prepared the way for Galileo's conception of static force. The very word 'force', deriving from the Latin *'fortis'* (strong, powerful), came to Galileo from mechanics, that is, from the science of machines. 'Force' generally designated what was applied to one end of a lever to raise a 'weight' or 'resistance' at the other.[34] Among other things, the usage expressed the distinction of natural and forced; the lever and the other simple machines were devices to lessen the strain of lifting weights against their natural tendency. In his early treatise on simple machines, Galileo displayed a general conception of static force. Whatever the force applied might be, the strength of a man, say, or that of a beast, he found no problem in replacing it mentally with a weight hung over a pulley.[35] Weight arises from the tendency of all bodies to move toward the centre of the earth. As such, weight was not conceived to be a force. Since force was held to oppose and balance weight, however, weight could serve as the measure of force. In *De motu*, Galileo sought further to generalise by making weight the measure of dynamic action as well.

Behind the whole of *De Motu* stands the image of the balance. If fluid statics provided the rationale for the concept of artificial lightness, the understanding of fluid statics depended on the balance, which itself supplied another image to explain unnatural lightness. 'Upward motion is caused by the extruding action of a heavy medium,' Galileo stated. 'Just as, in the case of a balance, the lighter weight is forcibly moved upward by the heavier, so the moving body is forcibly pushed upward by the heavier medium.'[36] The balance provided the ultimate foundation for the measurement of force by weight. If two weights are in equilibrium on a balance, and an additional weight is added to one side, that side moves down, not in consequence of its whole weight, but only in consequence of the added amount. 'That is the same as if we were to say that the weight on this side moves down with a force measured by the amount by which the weight on the other side is less than it. And, for the same reason, the weight on the other side will move up with a force measured by the amount by which the weight on the first side is greater than it.'[37] In the case of natural motions, that is, the motions of heavy bodies through media that also are heavy, the weight of the body corresponds to one side of the balance and the weight of an equal volume of the medium to the other.

> And since the comparison of bodies in natural motion and weights on a balance is a very appropriate one, we shall demonstrate this parallelism throughout the whole ensuing discussion of natural motion. Surely this will contribute not a little to the understanding of the matter.[38]

To the scientist seeking to develop a mathematical mechanics at the beginning of the seventeenth century, the law of the lever represented the one secure quantitative relationship. Surely, as Galileo said, it would contribute not a little to understanding motion. When *De motu* attempted to transpose statics into dynamics, the statics it used derived from the simple machines. *De motu* took the law of the lever and attempted to expand it into a whole mechanics. If his mature system of mechanics was more subtle, the role played by the balance and lever was scarcely less central, and so it continued to be in mechanics through the rest of the century.

*

Sometime in the decade and a half following its composition, Galileo rejected the mechanics of *De motu*. No doubt his growing commitment to Copernican astronomy helped to dictate the rejection. Ironically, one thing that the quantified theory of impressed force could not explain was the phenomenon that the qualitative theory had been developed to explain, the phenomenon that had to be explained if the Copernican system were to be vindicated. Taken in its own terms, *De motu* had nothing to say about the horizontal motion of projectiles. It did discuss such projectiles briefly, but it did not face the issue of how an impressed force that is artificial lightness can cause any motion except one straight up. The quantification of impressed force was tied to its being lightness; if it were lightness, however, it was confined to explaining a class of motions so restricted as to be practically non-existent. *De motu* also mentioned that horizontal motion, like the rotation of a sphere at the centre of the universe, is neither natural nor forced, so that the smallest force can make a body move on a horizontal plane. In the suggestion, we can see Galileo's first step toward the concept of inertia. In *De motu*, however, it stood irreconcilably in conflict with the underlying premises on which the treatment of vertical motion rested. In Galileo's mature system of mechanics, expressed in his two master works, the *Dialogue Concerning the Two Chief World Systems* (1632) and the *Discourses and Demonstrations Concerning Two New Sciences* (1638), inertial motion, or uniform motion on a horizontal plane, did not replace *De motu*'s concept of impressed force. Quite the contrary, it addressed itself to a range of phenomena that *De motu* had been powerless to handle adequately.

Instead of impressed force, uniform motion replaced the concept of rest as it was expounded in *De motu*. Perhaps I should rather say that Galileo expanded the brief suggestion that horizontal motion is like rest in its equilibrium of forces. He defined rest as infinite slowness, a special case of uniform motion. Within the context of Copernican astronomy the idea that the rotation of a sphere is neither natural nor forced took on added significance. Rotation on an axis involves no change of place; revolution in a closed orbit entails no rearrangement of positions. Rectilinear motion, in contrast, translates a body from one place to another, with the consequence either that the body was outside its proper place before its motion began or that it now moves away from its proper place. Circular motion, and circular motion alone, Galileo asserted, is compatible with an ordered universe.

> This being the motion that makes the moving body continually leave and continually arrive at the end, it alone can be essentially uniform. For acceleration occurs in a moving body when it is approaching the point toward which it has a tendency, and retardation occurs because of its reluctance to leave and go away from that point; and since in circular motion the moving body is continually going away from and approaching its natural terminus, the repulsion and the inclination are always of equal strengths in it. This equality gives rise to a speed which is neither retarded nor accelerated; that is, a uniformity of motion.[39]

In circular motion, the tendency toward the centre is exactly balanced by the repulsion from it. *De motu* had described rest as a similar state of equilibrium. Within Galileo's ordered, circular universe, all points equally removed from their relevant centres were equivalent. A body moving on the spherical surface through such points was effectively at rest; it could continue forever without introducing disorder into the cosmos. So overpoweringly obvious to the seventeenth-century mind was this conception of circular motion as dynamically equivalent to rest that it frustrated successful analysis of the mechanics of circular motion long after the nicely ordered cosmos that Galileo pictured had dissolved away.[40] Circular motion, which appeared so natural in the context of the Aristotelian world view as to be the symbol of perfection, became an enigma in the mechanical universe. Until its riddle was solved, a workable dynamics was impossible.

The ideal of a quantified, mathematical science of mechanics, the animating notion behind *De motu*, did not become less prominent in the reformulated mechanics of the *Dialogue* and the *Discourses*. The concept of uniform horizontal motion played a major role in its realisation. Without it, Galileo could not have demonstrated that the trajectory of a projectile is parabolic. The reformulation of his views on vertical motion played an even more important role, however, and one more directly concerned with the problem of force.

The conviction that uniform motion requires no cause bears some relation to the concept of uniformly accelerated motion, of course, although the relation is closer in our minds, with Newton's laws before us, than it was in Galileo's. The relation cannot be eliminated in Galileo's case, however. 'The stronger the cause, the stronger will be the effect,' he had asserted in the original version of *De motu*. 'Thus, a greater, that is, a swifter, motion will result from a greater weight, and a slower motion from a smaller weight.'[41] Compare the assertion with another, quoted above, made some ten years later in an essay on accelerated motion. Though the principle of motion in free fall remains constant, he declared, the velocity does not. 'It is necessary then to place the identity, or if you prefer the uniformity and simplicity, not in the velocity but in the increments of velocity, that is, in the acceleration.'[42] To us, the statement almost cries out with the assertion that force operates to change uniform motion so that a constant weight produces a uniform acceleration; although it did not cry out in the same way to Galileo, he probably could not have written the statement without the prior conviction that uniform motion requires no cause. Certainly it contradicts the earlier statement that a swifter motion requires a stronger cause. Equally his realization that a falling body cannot press on another falling with it, as it would if they were at rest, so that the weight of a falling body exerts itself entirely, as it were, in generating a uniformly accelerated motion, seems successfully to distinguish dynamics from statics and to state the relation of the two in acceptable form.[43] The appearance of both passages alters somewhat, however, when they are considered, against the background of *De motu*, as aspects of Galileo's correction of the internal contradictions in his early effort to frame a quantitative mechanics.

Even aside from the question of horizontal motion, the problems internal to *De motu* were acute. In constructing a mechanics that could describe free fall in quantitative terms, Galileo had at once accepted and amended Aristotle. When he set velocity directly proportional to weight, he reasserted the basic proposition of Aristotelian dynamics. If the proposition were granted, if he meant the assertion stemming from it that a stronger cause produces a stronger effect, there appears to be no way of avoiding the conclusion that a heavier body falls more swiftly than a lighter one. Galileo was convinced, however, that all bodies of the same material, regardless of their size, fall with the same velocity. Two bodies of the same material, identical in size, obviously fall side by side with the same speed. According to Aristotle's mechanics, Galileo said, their speed will be doubled if, instead of falling side by side, they are joined together into one body twice as big. This is absurd; clearly the larger body will fall with the same velocity as its two parts. To meet the problem, *De motu* defined 'weight' in a peculiar way. Weight referred, explicitly, to specific gravity. Pieces of wood, he concluded, fall with one characteristic speed, pieces of lead with another, greater speed. At the same time, he held that all bodies are heavy because matter as such has a natural tendency toward the centre of the earth. Moreover, there is 'a single kind of matter in all bodies'; denser bodies merely enclose more particles of the same matter in equal spaces.[44] If that is the case, the argument which established that all pieces of lead fall with the same speed should have demonstrated as well that all bodies whatever fall with the same speed. The admission of that conclusion would have demolished the entire structure of *De motu*. On the one hand, it would have contradicted the principle that a stronger cause produces a stronger effect. On the other hand, it would have negated the explanation of acceleration in free fall, one of the central purposes of *De motu*. Impressed force was an artificial lightness, like the buoyant effect of water which alters the effective specific weight of a body; and the body's speed at every moment was set proportional to its effective specific weight. To conclude that all bodies fall with the same speed was to abolish acceleration in free fall, an unlikely foundation on which to base a mathematical dynamics.

In a word, the mechanics of *De motu* was shot full of irresolvable contradictions, and it is possible to see the mature analysis of uniformly accelerated motion, not primarily as a consequence of the concept of inertia, but rather as a reformulation of the imperfect *De motu*, a reformulation directed to the same ends the early treatise had pursued. When we think of uniformly accelerated motion in the context of inertia, we think immediately of a constant force constantly producing new increments of motion which are added to those produced before and conserved as inertial motion. Inevitably, such a view suggests Newton's second law, and when scientists came to view accelerated motion in these terms, Galileo's analysis of free fall was seen to furnish the prototype of the action of force. Galileo himself did not express his conception of accelerated motion in such a way. He spoke instead of the natural tendency of heavy bodies to move toward the centre of the earth with a uniformly accelerated motion.

> Every body constituted in a state of rest but naturally capable of motion will move when set at liberty only if it has a natural tendency toward some particular place; for if it were indifferent to all places it would remain at rest, having no more cause to move one way than another. Having such a tendency, it naturally follows that in its motion it will be continually accelerating. Beginning with the slowest motion, it will never acquire any degree of speed without first having passed through all the gradations of lesser speed . . . Now this acceleration of motion occurs only when the body in motion keeps going, and is attained only by its approaching its goal.[45]

In briefer form, 'a heavy body has an inherent tendency to move with a constantly and uniformly accelerated motion toward the common centre of gravity . . .'[46] The natural tendency toward the centre was identical to the natural tendency of *De motu*, although the motion it generated and the centre toward which it inclined had changed. In the Copernican world, the centre of the earth could not be the centre of the universe as *De motu* had implied, and Galileo duly insisted that there are multiple centres. The motion deriving from the tendency toward a centre was now uniformly accelerated. Acceleration was seen less as a dynamic effect than as a logical consequence of some implicit principle of continuity whereby a body at rest can acquire a given velocity only by passing through all the lesser degrees of velocity. Whereas there was one natural motion in *De motu*, there were now two, and Galileo was consistent in referring to them as two different motions. Thus his view of inertial motion contained a basic ambiguity. On the one hand, it was the motion that is neither natural nor forced, the motion involving neither a propensity nor a repugnance. On the other hand, it was something more: as a 'natural motion' it was a unique phenomenon with a cosmic purpose. The natural motions were necessary to an ordered universe built on Copernican lines. Circular motion could persevere forever without change, but could never generate itself; rectilinear motion toward (or away from) a centre was necessary to generate new motion (or to destroy it).

> [A] falling body starting from rest passes through all the infinite gradations of slowness; and ... consequently in order to acquire a determinate degree of velocity it must first move in a straight line, descending by a short or long distance according as the velocity to be acquired is to be lesser or greater, and according as the plane upon which it descends is slightly or greatly inclined.... In the horizontal plane no velocity whatever would ever be naturally acquired, since the body in this position will never move. But motion in a horizontal line which is tilted neither up nor down is circular motion about the center; therefore circular motion is never acquired naturally without straight motion to precede it; but, being once acquired, it will continue perpetually with uniform velocity.[47]

God himself apparently chose the natural means of imparting orbital velocities to the planets. Galileo claimed to be following Plato in stating that there is a single point from which all the planets, in falling toward the sun, would attain precisely the right speeds at the levels of their orbits. If they were deflected into circular (horizontal) motion around the sun at that point, they would continue in it naturally forever.

Galileo insisted that a naturally accelerated body passes through every degree of velocity. Although the conception of accelerated motion implied no limit, his belief in an ordered universe did, and he spoke, for example, of the preternatural speed of a cannon ball which the resistance of the air destroys. In *De motu*, the existence of an upper limit to natural motion was implicit in the very structure of the treatise. What was more important, every degree of speed below that limit corresponded to an effective specific gravity. Since the gradations of specific gravity were infinite, so were the gradations of speed. Galileo's insistence that an accelerating body passes through every degree of velocity replaced the scale of weights in the earlier treatise.

In nothing is the continuity of the mature physics with *De motu* more evident than in Galileo's conviction that the natural tendency of heavy bodies to fall can provide the standard of measurement for a quantitative mechanics. Clearly the system of *De motu*, in which weight was the measure both of impressed force and of velocity, had to be revised. The major step forward in the *Discourses* was to replace the earlier statical mechanics with a true dynamics. Heavy bodies at rest, he had concluded, must be distinguished from heavy bodies in motion. The naturally accelerated motion of heavy bodies, however, could also be adapted to measuring other quantities in mechanics. Whereas velocity had been measured by weight, he now suggested that it be measured by sublimity, the vertical distance through which a body must fall in order naturally to attain that velocity. By this means a standard of velocity, which is constant everywhere, could be provided,

> since this velocity increases according to the same law in all parts of the world; thus for instance the speed acquired by a leaden ball of a pound weight starting from rest and falling vertically through the height of, say, a spear's length is the same in all places; it is therefore excellently adapted for representing the momentum acquired in the case of natural fall.[48]

In uniform motion, such a speed will carry a body through twice the distance of fall in the same period of time.

He sought as well to use the fact of naturally accelerated motion to establish a standard for measuring the force of percussion. The problem of percussion bothered Galileo, as it was to bother many others until Huygens restated it in more tractable terms and solved it. As Galileo first attacked the problem, he imagined a stake being driven into the ground. One blow of a hammer drives it a certain distance, and a dead weight considerably heavier than the hammer drives it an equal distance. Is the dead weight then a measure of that force of percussion? Galileo decided it is not, since a second blow of the hammer, identical to the first, will drive the stake the same distance again whereas more weight will have to be added to drive it further. Moreover, he found ambiguity in the role of the stake as a resistance. Suppose that the ground becomes harder with greater depth so that apparently equal blows drive the stake different distances. Can we say that the blows are in fact equal? Galileo decided they are not. The motion of the stake takes off some of a blow's force, and since the stake moves different distances, the blows cannot be equal. In searching for a standard by which to measure both resistance to a blow and the force of a blow, Galileo returned to the natural motion of heavy bodies. What force is constant? The force of a body which has fallen from a given height. What resistance is constant? The resistance of a heavy body to being raised. Galileo's final arrangement to examine the force of percussion consisted of a rope over a pulley with a weight on either end. The larger weight rests on a table. The other is dropped, and the force of its blow when it snaps the rope is transmitted over the pulley to the other weight. Because velocity can compensate weight, Galileo concluded, and because the one body is at rest in the beginning, the ratio of weights is always smaller than the ratio of velocities. However small the falling body may be, it cannot fail to move the larger body at rest – a conclusion drawn from an analysis containing at least as much ambiguity as that which he sought to avoid.[49] Be that as it may, the natural tendency to descend had furnished the means to measure the force of percussion.

The fact that Galileo tried first to compare the force of percussion with a dead weight suggests that the system of *De motu* may have been less fully rejected than we have believed. By and large, however, the quantity which he referred to as *impeto* or *momento*, a quantity also connected with the natural tendency of heavy bodies, as the analysis of percussion shows, replaced weight as the measure of force. The quantity is similar to what is called 'momentum' today, but it differs in two respects. For one thing, Galileo did not possess a concept of mass distinct from that of weight. The second difference is far more important. As I indicated earlier, *impeto* (or *momento*) was usually employed with an active verb. Its appearance as the 'force of percussion' illustrates that it was central to the problem of force in Galileo's mature mechanics. Let me repeat that the word 'force' (or *'forza'* or *'vis'*) is not the crucial consideration, but rather the conception employed whatever the word. If his own age saw a conception of force in Galileo's mechanics, it was one expressed by *impeto* and *momento*. Certainly it recalls the medieval doctrine of impetus and reminds us how profoundly that doctrine embodied our perception of the force of a moving body. Nevertheless, Galileo's use of *impeto* and *momento* was different from the medieval use. Like the impressed force of *De motu*, *impeto* was directed, not toward horizontal motion, but toward vertical motion by which it was measured. In this it revealed that its true forebear was less the concept of impetus than the concept of force applied to a simple machine to move a heavy body against its nature. Perhaps nothing is more revealing of seventeenth-century mechanics than the fact that the basic dynamic concept of the man who formulated the new conception of motion and defined uniformly accelerated motion referred, not to an external action changing a body's state of motion or rest, but to the capacity of a body in motion, its force, to raise itself again to the height from which it fell in acquiring that motion.

One of the most important propositions in Galileo's final system was the assertion that 'a heavy body falling from a height will, on reaching the ground, have acquired just as much *impeto* as was necessary to carry it to that height; as may be clearly seen in the case of a rather heavy pendulum which, when pulled aside fifty or sixty degrees from the vertical, will acquire precisely that speed and force which are sufficient to carry it to an equal elevation save only that small portion which it loses through friction on the air.'[50] As one consequence of the proposition, the path of descent was seen to be irrelevant to the *impeto* acquired (in the ideal case, of course, in which friction is excluded). Vertical displacement alone governs the speed acquired, so that bodies descending by different inclined planes between the same pair of horizontal planes acquire the same speed. Initially, Galileo introduced the proposition as an assumption; the second edition of the *Discourses* included a passage, written by Viviani under Galileo's instruction, which attempted to demonstrate it from dynamic considerations.[51] Why should it have occurred to him in the first place? Because, I suggest, it embodied a dynamic equivalent of the commensurability of impressed force and weight in *De motu*. A body is at rest, the early treatise had asserted, when the impressed force striving to raise it equals the weight urging it down. The *impeto* of a moving body, the *Discourses* held, is just enough to overcome the natural inclination to descend while it raises the body to its original position. Because circular motion is equivalent to rest, any place on the horizontal plane around the gravitational centre is equivalent to its original position. Thus *De motu's* measurement of impressed force by weight was replaced in the mechanics of the *Discourses* by a measurement of *impeto* which was also based on the natural inclination of heavy bodies to descend.

The measurement of force as *impeto* is inconsistent with the concept of force we see implicit in the analysis of uniformly accelerated motion. To express it in our terms, force as *impeto* would be equal, not to the rate of change of momentum, but to the momentum itself or to the total change of momentum, Δmv. Both usages, despite their inconsistency, appear side by side in Galileo. The first is implicit in his concept of the percussive force of a body, and the second appears with the assumption that this force must be equal to the force which can generate it. When a ball is rolled up an inclined plane, 'a given movable body thrown with a given force [*forza*] moves farther according as the slope is less.' In discussing the range of projectiles fired at different elevations, Galileo demonstrated that 'less momentum' [the Latin is "*impetus*"] is required 'to send a projectile from the terminal point *d* along the parabola *bd* [with an elevation of 45°] than along any other parabola having an elevation greater or less . . .' He spoke also of the difference of '*impeti*' and '*forze*' when projectiles are fired over the same range at different elevations.[52] Force in this context is measurable only as Δmv.

The notion of *impeto* or *momento*, so prominent in Galileo, illustrates the continued domination of mechanics by the law of the lever, even in the case of the man who first effectively distinguished statics and dynamics. In his *Discourse on Bodies in Water*, he justified the definition of *momento* by reference to the balance.

> As for example, two weights equall in absolute Gravity, being put into a Ballance of equall Arms, they stand in Equilibrium, neither one going down, nor the other up: because the equality of the Distances of both, from the Centre on which the Ballance is supported, and about which it moves, causeth that those weights, the said Ballance moving, shall in the same Time move equall Spaces, that is, shall move with equall Velocity, so that there is no reason for which this Weight should descend more than that, or that more than this; and therefore they make an Equilibrium, and their Moments continue of semblable and equall Vertue.[53]

If a lesser weight is to raise a greater, it must be placed further from the fulcrum so that it moves more swiftly.

> Generally then we say that the moment of the lighter body equals the moment of the heavier when the velocity of the smaller stands in the same proportion to that of the larger as the weight of the larger to that of the smaller. . . .[54]

The passages above indicate how our concepts of momentum and moment both derived from the single term *momento* as Galileo and many others used it.

In Galileo's exposition, the concept of *momento* continually strained to free itself from bondage to the balance and lever. In his *Discourse on Bodies in Water*, he employed the principle to demonstrate that a beam of wood can float in a volume of water one hundredth as great. The beam, of course, is in a narrow tank surrounded by a curtain of water, so that a small vertical motion by the beam causes a large fluctuation in the narrow sheets of water along its sides. The demonstration might seem paradoxical, he concluded. 'But he that shall but comprehend of what Importance Velocity of Motion is, and how it exactly compensates the defect and want of Gravity, will cease to wonder . . .'[55] When he treated the force of percussion, he employed the same principle, now wholly separated from any constraining mechanism. In his early treatise *On Mechanics*, he suggested that the 'resistance to being moved' of the hammer bears the same proportion to the 'resistance to being moved' of the object it strikes as the distance the force would drive the hammer if it did not strike to the distance the object is driven.[56] By the time of his final deliberations on percussion, Galileo no longer spoke of a general resistance to being moved, but the tentative conclusion at which he arrived repeated the law of the lever. In the arrangement of two bodies over a pulley, he compared the ratio of the bodies to the ratio of their speeds, and decided that a body no matter how small will move another since the velocity of the other is initially zero.

The balance and the lever contained a basic ambiguity which further confused the conception of force. The problem was present already in his early *Mechanics*, where he treated the lever on the principle that

> whatever is gained in force is lost in speed. For the force C elevating the lever and transferring it to AJ, the weight is moved through the interval BH, which is as much less than the space CJ traversed by the force as the distance AB is less than the distance AC; that is, as much as the force is less than the weight.[57]

What is the significant factor, the speed or the displacement? In the passage above, and in many others, Galileo used the two interchangeably, as of course they can be used interchangeably in the case of simple machines. The problem lay in the possibility of forgetting that the interchangeability is valid only when the two motions being compared are virtual motions constrained to a simple machine. On the whole, Galileo kept the condition in mind, but the possibility of forgetting was built into the very structure of his crucial proposition, that a body in falling acquires *impeto* sufficient to carry it back to its original height. Here he was dealing with real, accelerated motion, not with virtual motion, but he yoked displacement and velocity together in a way all too likely to suggest their interchangeability. On at least one occasion, Galileo demonstrated how readily the two might be confused. In the so-called sixth day of the *Discourses*, he applied the law of the inclined plane (which he derived directly from the lever) to the question of percussion. A body of ten pounds descending vertically balances one of a hundred pounds on an inclined plane the length of which is ten times its elevation. Therefore, he continued, drop a body of ten pounds through any vertical distance; the *impeto* it acquires, when applied to a body of one hundred pounds, will drive it an equal distance up the inclined plane, which involves a vertical rise equal to one tenth the length of the inclined plane. It was concluded above that a force able to drive a body up an inclined plane is sufficient to drive it vertically a distance equal to the elevation of the inclined plane, in this case a tenth part of the distance traversed on the incline, which distance on the incline is equal to the fall of the first body of ten pounds. Thus it is manifest that the vertical fall of a body of ten pounds is sufficient to raise a body of one hundred pounds vertically, but only through a space equal to one tenth of its descent.[58]

Galileo applied the conditions of equilibrium on an inclined plane to the accelerated motions of two independent bodies and tried to connect the two by the concept of *impeto*. What is the measure of force in the case above? Is it the *impeto* (in our terms, mv) or is it the product of weight times vertical displacement, what we call work, and what Leibniz used to measure *vis viva* (mv^2). The two are not equivalent to each other. From the ambiguity of the lever emerged the controversy over *vis viva*. To us, with over two centuries to digest both Newton's achievement and Leibniz's, the ambiguity seems obvious enough. To the age that created modern mechanics, however, the problem of force was so complicated that the ambiguity only began to be manifest, much less to dissolve, in the second half of the century, a generation after Galileo's death.

Of course, a third use of force, incompatible both with mv and with mv^2 was also present in the analogy of the lever and in the examples above. This was the idea of static force, identical to that employed in *De motu*. Consider a passage from the 'Sixth Day' of the *Discourses*. Since a body in falling acquires enough *impeto* (mv) to raise it to its original height, a certain amount of *forza* (mv^2) is necessary to raise the body a given vertical height whatever the plane by which it rises. To pull it up different planes, however, a smaller *forza* (F) is needed as the inclination is less.[59] In discussing percussion, he asserted that the *forza* which can raise a weight is equal to the *forza* with which it presses down, and with that quantity he tried to measure the force of percussion.[60] Perhaps nothing expresses the ambiguity so well as the passage by Viviani, interpolated at Galileo's request in the second edition of the *Discourses* to justify the assertion that a given change of *impeto* corresponds to a given vertical displacement, whatever the path followed may be. Here Viviani referred to 'the impetus, ability, energy, or, one might say, the momentum' [*l'impeto, il talento, l'energia, o vogliamo dire, il momento*] of a body on an inclined plane.⁶¹ In the context, he can only be referring to the component of weight along the plane. The use of the very words employed elsewhere to express a different conception of force, or a conception of force which appears different to us, suggests again that Galileo had not freed himself entirely from *De motu*'s pattern of thought. And already in Galileo, in the continuing pattern of *De motu*, we can perceive the two ambiguities in which so much of seventeenth-century dynamics floundered – on the one hand the uncertain message of the lever with its confusion as to the measure of force, on the other hand the uncertain distinction of statics and dynamics.

*

The pattern of *De motu* also endured in the continuing central role of vertical motion in Galileo's mechanics. The centrality of vertical motion was significantly modified, of course. The concept of inertia not only enabled him to justify Copernican astronomy, but it ultimately became the basis of a wholly new idea of motion. With Galileo, however, inertial motion was always horizontal, perpendicular to accelerated vertical motion, and as far as vertical motion was concerned, equivalent to rest.⁶² In this form, he was able to compound it with vertical motion and to demonstrate the parabolic shape of a trajectory. What he never attempted was an extension of the analysis of naturally accelerated motion to accelerated motions (violently accelerated motions, he would have said) on horizontal planes. For Galileo, the naturally accelerated motion of free fall was not a paradigm of the action of force. As in *De motu*, he was prepared to use it as a standard to measure other mechanical phenomena, such as the force of percussion. Nevertheless, naturally accelerated motion remained a unique phenomenon in Galileo's mechanics, albeit one that was very common in nature. The bulk of his mathematical science of motion was devoted to its analysis, and the analysis was, as by definition, applicable solely to the unique motion and its modifications.

In two special cases, motion down an inclined plane and fall through a resisting medium, Galileo came closest to generalising the analysis of free fall. Since he had concluded that all bodies fall with the same acceleration, free fall itself presented an invariant phenomenon. In the

GALILEO AND THE NEW SCIENCE OF MECHANICS

cases both of inclined planes and of fall through resisting media, however, the effective moving weight decreases while the body it moves remains the same. When Galileo set acceleration proportional to effective weight in both cases, he appeared to employ the equivalent of Newton's second law. Special attention must be given to these two cases if we are to understand Galileo's dynamics.

Already in *De motu*, Galileo had considered the inclined plane. By employing the device of a bent balance, and imagining an inclined plane perpendicular to the bent arm and tangent to the circle its virtual motion would trace, he derived the law of the inclined plane from the law of the balance or lever. On any plane, a body tends to move down with as much force as is necessary to prevent its motion. On a vertical plane, the force with which it tends downward is equal to its weight. On an inclined plane, it tends down with less force, the ratio of forces being equal to the ratio of the length of the plane to its vertical rise.[63]

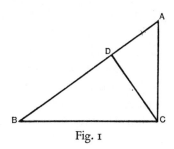

Fig. 1

That is to say, in *De motu* Galileo saw the inclined plane in terms identical to those with which he analysed fall through a dense medium. It was not an accident that his treatment of both cases depended on the law of the lever. Both the inclined plane and the medium were seen to subtract part of the downward tendency, thereby decreasing the velocity. In the *Discourses*, he treated the inclined plane in exactly the same terms, except, of course, that acceleration replaced velocity. Whereas the lower velocity along an inclined plane was, in *De motu*, equivalent to the lower velocity of a less dense body, the common matter of all bodies was seen now to dictate a common acceleration in free fall. That common acceleration, the maximum acceleration a body can naturally undergo, is diminished on an inclined plane in the same proportion as he had found velocity to decrease earlier. Let *AB* be an inclined plane and *AC* its vertical rise (see Fig. 1); then the force (*impeto*) along AC is to that along AB as AB/AC. On AB lay off AD such that $AD/AC = AC/AB$. Then the forces are as AC/AD. 'And therefore the body will traverse the space AD, along the incline AB, in the same time which it would occupy in falling the vertical distance AC (since the forces [*momenti*] are in the same ratio as these distances); also the speed at C is to the speed at D as the distance AC is to the distance AD.'[64] The acceleration decreases with the inclination until, along the horizontal, a body experiences no tendency whatever to motion. Galileo saw motion on an inclined plane in terms of the analysis of the role of resistance which he had incorporated in *De motu*. Like the resistance of a medium, the inclined plane opposes the natural tendency of a body and thereby diminishes it. Unlike Newton's second law, the analysis did not see the change of acceleration as a result of the changing ratio of force and mass – or, since Galileo did not have a concept of mass, the changing ratio of the effective component of weight to the size of the body moved.

From what has been said of the inclined plane, the *Discourses*' treatment of fall through a medium will be obvious. In fact, Galileo managed to obscure it by slipping back into the terminology of *De motu*. The buoyant effect of the medium, he said, 'diminishes the inherent and natural speed' of fall.[65] He cast the whole discussion in such language despite the fact that he introduced it with a stinging rebuke to Simplicio for not realising that bodies fall with a constant increase of velocity. Nevertheless, we can treat this as an oversight, for Galileo's examples and other discussions adequately demonstrate that he was thinking in terms of acceleration.[66] With bodies starting from rest, velocities at the end of unit time are proportional to acceleration. Like the inclined plane, a medium operates to reduce the effective tendency of heavy bodies and hence to reduce the acceleration.

Fall through a medium has one other characteristic. Whereas motion down an ideal inclined plane accelerates at a constant rate, fall through a medium approaches a terminal velocity (although Galileo did not use that phrase). The medium checks the acceleration of bodies however dense they be; it also reduces the velocity of bodies that have been accelerated by whatever means beyond their natural limit in that medium. Water quickly destroys most of the momentum of a cannon ball that falls into it. Air can do the same; a cannon ball fired straight down from the top of a tower makes less impression on the ground than one fired down from a height of only four or six cubits.[67] (Imagine performing

GALILEO AND THE NEW SCIENCE OF MECHANICS

that experiment with a muzzle loader! The Society to Save the Tower of Pisa would have sprung into action had he tried.) The resistance of the medium is proportional to the rapidity with which it must give way; hence the resistance increases as the velocity of the body increases until it equals the natural tendency of the body downward, at which point the body accelerates no further and continues to fall with a uniform motion.[68] As far as the falling body was concerned, Galileo saw the resistance as similar to the buoyant effect of the medium – it diminished the downward tendency of the body. The interest of the passage lies in the explicit recognition that the acceleration is the result of an effective tendency at every point, another direct heritage of the *De motu*. It is the one case in Galileo's mechanics in which he accepted non-uniform acceleration.[69] In terminal velocity, moreover, he saw an upper limit to the velocities that occur in nature, as the *De motu* had set an upper limit.

From what does the resistance of the medium arise? The medium is also matter, composed of matter identical to that of the falling body, and if resistance arises from thrusting it aside, that is, if force must be exerted to set it in motion, have we not found another case analogous to the natural acceleration of bodies? Galileo's treatment of the resistance of media illuminates the limits of his system of mechanics. In the paragraph immediately following his exposition of the principle of terminal velocity, he attempted to explain the source of the resistance.

> Assuming this principle, that all falling bodies acquire equal speeds in a medium which, on account of a vacuum or something else, offers no resistance to the speed of motion, we shall be able accordingly to determine the ratios of the speeds of both similar and dissimilar bodies moving either through one and the same medium or through different space-filling, and therefore resistant, media. This result we may obtain by observing how much the weight of the medium detracts from the weight of the moving body, which weight is the means employed by the falling body to open a path for itself and to push aside the parts of the medium, something which does not happen in a vacuum where, therefore, no difference [of speed] is to be expected from a difference of specific gravity. And since it is known that the effect of the medium is to diminish the weight of the body by the weight of the medium displaced, we may accomplish our purpose by diminishing in just this proportion the speeds of the falling bodies, which in a non-resisting medium we have assumed to be equal.[70]

The fact of terminal velocity is, of course, incompatible with this analysis. So also is resistance to upward motion. Galileo never succeeded in separating adequately two different effects, the buoyancy due to the weight of the medium, and the resistance it offers to motion through it. 'The fact that any fluid medium diminishes the weight of a mass immersed in it,' he asserted, 'is due . . . to the resistance which this medium offers to its being opened up, driven aside, and finally lifted-up.'[71] He agreed that smaller bodies of the same material are retarded more than larger ones, but he referred this to the roughness of the surface, which is proportionately greater in smaller bodies. Apparently he meant that the rough projections strike a greater quantity of the medium and cause it to be lifted.

What Galileo said about resistance in the *Discourses* relied heavily on his earlier analysis of the same problem in the *Discourse on Bodies in Water*. There he took Aristotle to task for asserting that bodies that ascend in water will ascend more swiftly in air. The source of Aristotle's mistake, he concluded, lay in his reference of velocity to diversity in figure and to 'the more or less Resistance of the greater or lesser Crassitude, or Rarity of the Medium; not regarding the comparison of the Excesses of the Gravities of the Moveables, and of the Mediums: the which notwithstanding, is the principal point in this affair . . .'[72] Galileo insisted that fluid media have no resistance to being divided. Hence any body heavier than water, no matter how small its excess may be, sinks ultimately to the bottom, and a great beam of wood can be drawn through water by a woman's [*sic* – not only a man's, but even a woman's] hair. Nevertheless, if grains of dust settle to the bottom, they settle slowly, and great ships are driven, not by a woman's hair, but by oars and sails. Although water does not resist division, Galileo explained, it resists sudden division.

> Now, because all Motions are made in Time, and the longer in great time: and it being moreover true, that those Bodies that in a certain time are moved by a certain power such a certain space, shall not be moved the same space, and in a shorter time, unless by a greater Power: therefore, the broader Ships move slower than the narrower, being put on by an equal Force: and the same Vessel requires so much greater force of Wind, or Oars, the faster it is to move.[73]

Why is this so? Not because water resists division, but because the parts of the water must be moved out of the way. Weight was the only cause of resisting movement that Galileo ever recognised. When he implicitly referred the effect to weight, Galileo again interjected the issue of relative specific gravity, in effect, the image of the balance, into a problem to which it was irrelevant. His entire approach to motion, even after he decided that natural motion downward is accelerated, continued to be coloured by *De motu*'s correction of Aristotle on the role of the medium. The problem is to find the difference of specific gravities, which is the effective motive force. Hence his analysis of accelerated motions was confined to motions with a vertical component, in which the resistance arises from the reluctance of heavy bodies to be moved away from the centre.

The lack of a concept of mass, or perhaps the use of weight in place of mass with the consequent unconscious intrusion of the vertical into every problem, continually thwarted the generalisation of Galileo's analysis of free fall into something like the second law.

> At present the only internal resistance to being moved which I see in a movable body is the natural inclination and tendency it has to an opposite motion. Thus in heavy bodies, which have a tendency toward downward motion, the resistance is to upward motion.[74]

A body 'has neither a resistance nor a propensity' to motion on a horizontal plane.[75] As Galileo's famous phrase expressed the idea, a body on a horizontal plane 'finds itself in a condition of indifference as to motion or rest . . .'[76] On an inclined plane, a body descends with increasing speed. If it is given an impetus upward, its speed diminishes until it stops. On a horizontal plane, however, a body

> will do what pleases us; that is, if we place it at rest, it will remain at rest; and if we give it an impetus in some direction, it will move in that direction, always conserving the same velocity which it received from our hand, lacking the power either to increase it or to decrease it . . .[77]

Indifference to motion, the foundation stone of the concept of inertia, was equally the stumbling block in the path to a generalised concept of force applicable to all changes of motion. Galileo's physics contained no concept whatever by which to determine the acceleration that a given motive force imparts to a given body on a horizontal plane.

The point, of course, is not to question Galileo's awareness that more effort is required to generate a given velocity in a large body than in a small one. Even if we had no explicit texts, it would be impossible to imagine that a student of motion so acute could have failed to recognise a phenomenon so obvious. The point rather is to examine the conceptual tools in his system of mechanics and to ask whether the instruments at his command enabled him to handle all accelerations in quantitative terms similar to those he applied to vertical accelerations.

> Now fix it well in mind as a true and well-known principle that the resistance coming from the speed of motion compensates that which depends upon the weight of another moving body, and consequently, that a body weighing one pound and moving with a speed of one hundred units resists restraint as much as another of one hundred pounds whose speed is but a single unit. And two equal movable bodies will equally resist being moved if they are to be made to move with equal speed. But if one is to move faster than the other, it will make the greater resistance according as the greater speed is to be conferred upon it.[78]

The passage concluded an exposition of the law of the lever, in which the motions envisaged were both vertical and virtual, and the resistance to motion was the resistance of a heavy body to being lifted.

*

The passage on resistance to motion cited immediately above suggests the problem of impact. In fact, the principle which Galileo applied to impact was almost identical, and his discussion of impact illustrates again the limits inherent in his mechanics. Near the beginning of the 'Sixth Day' of the *Discourses*, Aproino stated that the great power of impact differs from the multiplication of force by simple machines. Nevertheless, in impact as in the simple machines there is motion, and indeed two different motions, one in the body that strikes, and one in the body that is struck and moved. Hence he would try to apportion the force of percussion between the weight of the percussant and its velocity.[79] From the beginning, then, Galileo tried to measure the force of percussion by a static weight, the first of a long series of such attempts during the century. Ultimately, Huygens found that the problem could be handled more readily in kinematic terms which avoid the question of force entirely, and today we prefer to follow his approach. When we employ a dynamic analysis, however, we use the concept of impulse, which is set equal to the change of momentum. Impulse is defined as the integral of force over the time it acts:

$$\int F dt = \Delta mv.$$

The integral itself is highly complex since the time is very short in most impulses and the variation of force during the impulse difficult to ascertain. The integral is readily evaluated by the change of momentum, however, and in form it is identical to the action of gravity on a freely falling body during some interval of time. That is, Galileo's analysis of free fall is adaptable to the analysis of impact. How far he was from a workable dynamics is revealed by his effort to measure the force of impact by static weight.

The first device suggested to measure the force of impact was a large balance. On one arm hung a bucket full of water with an empty bucket slung under it, and initially the counterweight on the other arm established equilibrium. Then a hole was opened in the bottom of the upper bucket so that water flowed from it, falling into the lower one. The weight required to restore equilibrium was expected to measure the water's force of percussion. When that device proved unsatisfactory, he returned to the example, used in earlier writings, of a stake being driven into the ground by a hammer on the one hand and by a dead weight on the other. Finally, as a means of obtaining a standard measure both of the blow and of the resistance, he devised the arrangement of two bodies connected over a pulley. The ultimate result led to the conclusion that no resistance that is less than infinite can resist a blow without giving way, and further that 'it is not possible validly to assign a determinate measure to such a percussion.'[80] In one sense, the conclusion that the force of percussion is potentially infinite was similar to the law of the lever, in that any resistance less than infinite can be moved by a given force when the lever arms are adjusted. Galileo apparently considered the two cases to be different, however. With a lever of given ratio and a given force, a resistance too big to be moved can always be found. The same is not true of a given body striking with a given velocity. It will move any resistance that is not infinite, Galileo decided. In concluding that the force of percussion is potentially infinite, that is, potentially bigger than any static force, Galileo approached the conclusion that the two are incommensurable. When he added that it is impossible to give a determinate measure to the force of percussion, however, he confessed that he did not know how to bring it within the boundaries of rational mechanics.

In some fragments associated with the 'Sixth Day', Galileo took a step toward a resolution of the dilemma. In every mobile, he declared, there are two types of resistance. The first is that by which a thousand pounds resists being lifted more than one hundred pounds does. The second resistance is that which requires the use of more force to hurl a stone a hundred feet than to hurl it fifty feet. Two different motors correspond to the two resistances; the one moves by pressing and the other by striking. The first moves a small resistance, but moves it through an unlimited distance, continuing to press it with the same force. The second moves any resistance, but only through a limited interval. To pressure, the resistance is proportionable, but not the interval; to percussion, the interval is proportionable, but not the resistance.[81] Although the suggestion remained obscure, we may perhaps see in his first resistance the downward tendency of weight, and in his unlimited motion, the continuing product of a continuing disequilibrium. The second resistance gropes still more obscurely toward the concept of mass as the measure of resistance to any accelerating force.

In other fragments, Galileo suggested that the *momento* of a body in motion is composed of an infinite number of *momenti*, each of which is equal to a single *momento*, whether natural and internal (the weight of the body at rest), or extrinsic and violent (a moving force). Thus a falling body accumulates the *momenti* from instant to instant as its velocity increases. Some time is required for the act of percussion, so that the *momenti* are dissipated through time as they are accumulated through time, and the briefer the percussion the greater the effect.[82] Here was an insight of immense potential, although Galileo lost it from view almost at once when he compared the dissipation of the *momenti* to the repeated blows of a hammer on the end of a log. Nevertheless, not only had he suggested that the accumulation of *momenti* is infinite and therefore incommensurable with a single *momento* (we might say the accumulation is the integral of *Fdt*), but also he had virtually treated uniform acceleration as the effect of a uniform force, in this case weight. In this context, the conception of weight as the measure of other forces was relevant to accelerated motion, and he extended the analysis to other moving forces. Those who close the heavy bronze doors of San Giovanni cannot do it with one push, but with 'continued impulse' they impress so much force that the whole church trembles when the doors strike the sill. The same thing happens with bells. Each repeated pull on the rope adds to the force already accumulated. The heavier the bell, the more force it can hold; a smaller bell loses its *impeto* more quickly since it cannot be saturated (*imbevuta*) with as much force as a larger one. So also a ship is not put into motion all at once, but by repeated pulls of the oars or by the continued impression of the force of the wind on its sails. A longer catapult shoots farther than a shorter one because it has a longer time to impress force on the missile.[83] In these fragments, Galileo most nearly approached a generalised conception of force. Torricelli was to take up their idea and exploit it more fully. Meanwhile, with Galileo, the fragments remained separated from the main body of his mechanics – brief insights with which he could have extended his system, but which he came upon too late to explore.

Galileo's correction of the Aristotelian treatment of resistance played an ambiguous role both in his own mechanics and in that of the age following him. On the one hand, it led to a correct understanding of the resultant effective force acting on a body. By examples combining two different bodies and two different media, he was able to show that the Aristotelian ratio involved inconsistencies that could not be avoided in a quantitative mechanics. When the resistance is subtracted from the moving force, however, instead of divided into it, the difficulty is solved. Wood falls through air because its buoyancy in air subtracts only a small part of its weight; it does not fall through water because its buoyancy more than offsets its weight, and when it is fully submerged, the resultant force is upward rather than down. Because of the peculiar way in which Galileo viewed the acceleration of gravity, this approach yielded correct results when applied to simple examples of motion downward – acceleration down an inclined plane, fall through a medium the buoyant action of which is considered but the friction of which is ignored. For the sake of comparison, we can say that he worked with the formula $a \propto F-R$. Since he believed that the tendency of a body to fall with accelerated motion is the sum of the tendencies of all its particles, he concluded that all bodies fall with the same acceleration in a vacuum. From our point of view, the mass to be accelerated increases in proportion to the accelerating force; the constant acceleration of gravity expresses the constant ratio of the two:

$$a = \frac{F}{m}.$$

This ratio has exactly the form of the one he had rejected, and Galileo's insistence on the indifference of matter to motion was part of that rejection. One of the resistances adduced by medieval philosophy had been the resistance of matter to motion as such. What Galileo rejected on one level, he could not re-admit on another. Thus he did not succeed in applying his analysis of free fall to a generalised problem of acceleration. Until the idea of the indifference of matter to motion was modified to admit its resistance to changes of motion, until the (correct) formula for calculating effective force was supplemented by a new ratio of effective force to resistance, a major conceptual obstacle stood in the way of seeing the analysis of free fall as a paradigm of the action of force.

*

Taking at once a retrospective view of Galileo's mechanics, and a look forward at the problems mechanics would need to solve before the second law could be formulated, we can ask ourselves what factors stood in the way of the generalised conception of force which his mechanics seems so obviously to demand. One factor certainly, beyond those mentioned already, was the limitation his mathematics imposed. When we think of force today, we think immediately of the formulae which describe it:

$$F = ma \quad \text{or} \quad F = \frac{d}{dt}(mv).$$

Even to discuss the problems of Galileo's mechanics, I have found myself compelled to introduce the formulae. They are, of course, deliberate anachronisms of which Galileo knew nothing. His mathematics was not algebra, much less calculus, but Euclidean geometry, and it is difficult for us to imagine the limitations it imposed on mechanics. To compute the distance that an inflated bladder falls from the top of a tower while an ebony ball falls to the ground, for example, he had to choose a unit of time which would allow time to cancel out of the ratio. For this problem, the unit was the time required for the ebony ball to reach the ground, whatever that time might be. For that period, and for that period alone, the distance covered by the ebony ball is to the distance covered by the inflated bladder as the effective weight of the ebony ball in air is to the effective weight of the bladder. (Obviously, in this problem, he ignored the resistance of the air.) The example also illustrates that Galileo continued to adhere to the precept that only similar quantities can be compared by geometric ratios, another restriction that his mathematics imposed on his mechanics. For the problems Galileo handled, his method of geometric ratios was adequate. Which may be equivalent to saying that he was limited to such problems.

With one exception, Galileo's diagrams were built on the trajectory of the moving body, usually a straight line either vertical or inclined. Since he held that time flows uniformly, he could lay out distances proportional to it along the line. In uniformly accelerated motion, velocity increases in proportion to time, and the same units along the line could represent velocity. The one relation involving a squared quantity was the increase of distance with time. Had geometry not included the device of the mean proportional, it is hard to imagine how Galileo could have realised one of his proudest achievements. Nowhere on his diagrams, however, was there a place to show acceleration or force; his mathematics confined him to problems in which acceleration is constant. The one diagram which differed from the common pattern was the familiar triangular representation of time, velocity, and distance; again it had no place to show acceleration directly. Galileo had good reasons, therefore, to avoid problems involving variable acceleration; whenever he did consider such, he fell into error. The pendulum is an example; by extrapolating from motion along the chords of a circle, on which acceleration is uniform, Galileo fooled himself into thinking he had demonstrated that all vibrations in a circle are isochronous. The fall of a body through a resisting medium which reduces the motion to uniformity at the terminal velocity was one he avoided examining in detail. 'Of these accidents of weight, of velocity, and also of shape, infinite in number,' he admitted, 'it is not possible to give any exact description; hence, in order to handle this matter in a scientific way, it is necessary to cut loose from these difficulties; and having discovered and demonstrated the theorems, in the case of no resistance, to use them and apply them with such limitations as experience will teach.'[84] In order to deal with complicated motions in the quantitative manner that Galileo himself introduced, mechanics had to develop the conception of force. A mathematics more powerful than traditional geometry was also required. Lacking the latter, Galileo was effectively restricted to problems involving a constant acceleration.

It would be a serious mistake, however, to conclude that the development of dynamics was an inevitable by-product of the development of mathematics. Quite the contrary, dynamics had to cope with major conceptual difficulties which were already present, as I have indicated, in the work of Galileo. As far as Galileo himself was concerned, nearly every conceptual difficulty involved the word 'natural'. The more I examine Galileo's conception of nature, the more I am forced to conclude that it was an impossible amalgam of incompatible elements, born of the mutually contradictory world views between which he stood poised. To adapt the famous undergraduate judgment of Dante, he stood with one foot in the neatly ordered cosmos of medieval philosophy and with the other greeted the dawn of the mechanical universe. The natural motions which appeared so frequently belonged to the ordered cosmos.

> I therefore conclude that only circular motion can naturally suit bodies which are integral parts of the universe as constituted in the best arrangement, and that the most which can be said for straight motion is that it is assigned by nature to its bodies (and their parts) whenever these are to be found outside their proper places, arranged badly, and therefore in need of being restored to their natural state by the shortest path. From which it seems to me that one may reasonably conclude that for the maintenance of perfect order among the parts of the universe, it is necessary to say that movable bodies are moved only circularly; if there are any that do not move circularly, these are necessarily immovable, nothing but rest and circular motion being suitable to the preservation of order.[85]

The composition of natural motions with other motions was a problem which Galileo could not quite ignore although he refused to face it directly. The central problem in the composition of motions was the assumption that a body can participate freely in two motions because motions do not interfere with each other. Galileo avoided discussing the assumption directly,[86] but in each of his major works he introduced composition first in the context of two natural motions. When the question came up in the *Dialogue*, the apparent vertical fall of a body on the rotating earth was under discussion; hence the two motions being compounded were both natural.[87] When he derived the parabolic shape of the trajectory in the *Discourses*, he chose to start with a body moving along a horizontal plane 'with a motion which is uniform and perpetual'; after the body passes over the edge of the plane, the second natural motion compounds with the first.[88] Having demonstrated the parabola in this special case, he applied it to all trajectories, such as those of cannon balls. A serious problem arises, however, when one of the motions is not natural. If bodies do have natural motions, can they possibly be indifferent to them? Is it possible that a natural motion not obstruct a violent one and in turn not be obstructed by it? We no longer believe in natural motions, and we accept the composition of motions without question. Galileo did believe in natural motions, and when he refused to face the issue of their relation to other motions, we are forced to conclude that he solved the problem of composition by nothing more convincing than bald assertion.

As far as natural motion on a horizontal plane is concerned, he surreptitiously equated it with all uniform horizontal motions. He used the same language in describing them all, and although such sleight of hand is not philosophically convincing, we can accept it as his opinion that all uniform horizontal motions are identical in being inertial. In fact, we are so pleased with the result that we are grateful for his sparing us the longwinded discussion.

In the case of vertical motion, the natural tendency of heavy bodies to fall with a uniformly accelerated motion remained a metaphysically privileged motion, precluded as such from serious comparison with other motions. No single element in his physics stood more directly in the path of a generalised conception of force. Consider friction, for example. Its action in retarding motion is similar to that of a plane sloping gently upwards. That is, friction is a force which changes motion; the negative acceleration it causes can be compared with the acceleration caused by gravity. Such a comparison did not occur to Galileo. Far from presenting itself in such terms, friction was to him only an unpleasant nuisance which disturbed the ideal motions of an ordered cosmos.

As in *De motu*, Galileo continued to think of the natural tendency downward as the universal constant which made a quantified mechanics possible. It could be used to measure velocity and the force of percussion. But this was different from *De motu*, in which weight measured force directly when the two were in equilibrium. In this sense, it was difficult to bring another accelerated motion into direct comparison with free fall. Because he considered free fall as a natural motion, Galileo never saw his analysis of its quantitative elements as the paradigm of the action of force.

Descartes' judgment on Galileo's mechanics deserves more sympathy than it has usually received. Without having considered the first causes of nature, he said, Galileo 'has merely investigated the causes of some particular effects, and thus he has built without foundation.'[89] Certainly it appears to me that Galileo stretched mechanics to the very ilmit of its capacity within his context. He was, in fact, bogged down in a philosophic morass. In order to advance further, mechanics would require the solid footing of a consistent philosophy of nature and a re-examination of motion. It would require, in a word, the work of a Descartes, in whose philosophy natural motions would cease to exist and all motion as motion would be placed on the same metaphysical plane. In that context, there would be at least the possibility that the uniform acceleration of free fall be seen as one example of the broader class of accelerations in general.

Even then, an interrelated set of conceptual problems remained to be solved. It is hardly surprising that the seventeenth century did not discover a systematic dynamics in Galileo, for none was there. Rather he presented unwittingly the tangle of conceptual problems, not of his own making, which emerged in full view only in his work, and which absorbed the energy of rational mechanics for the rest of the century. First there was the question of dynamic models, free fall or percussion. If we see the model of force in his analysis of free fall, Galileo himself saw it more in percussion, and most of the century agreed. Second was a pair of technical questions, the role of mass in dynamic action and the dynamics of circular motion. Without a comprehension of the first, Galileo lacked the means to generalise his analysis of free fall to horizontal accelerations. A generalised dynamics had also to embrace circular motion. If Galileo's concept of inertia did not raise the problem as such, it revealed the assumptions about circular motion that made the problem so intractable. Third, a set of ambiguities had to be resolved – the measure of force and the distinction of statics and dynamics. Associated with the simple machines, the ambiguities derived from the almost universal attempt to extend the law of the lever into dynamics. The mere birth of the mechanical philosophy of nature did not remove the conceptual difficulties internal to rational mechanics.

As for Galileo, the peculiar problem of his picture of nature lies in the fact that it contained, not only natural motions, but also many characteristics of the mechanical philosophy. The very extent to which he approached the mechanical philosophy, moreover, raised another obstacle to the concept of force. Like the mechanical philosophers, Galileo denied the anthropocentricity of the medieval universe. His world was multi-centred; the utility of man was not the purpose of its existence. It was filled with the homogeneous matter of the mechanical world, subject all to the same laws of motion, not entirely inert in Galileo's philosophy, but animated (in the case of terrestrial matter) only by its natural inclination toward the earth. The idea of natural motions appears so wholly unnatural in this context that one cannot avoid asking what Galileo meant by it. One passage in the *Dialogue* approached the problem by denying the reality of one of the natural motions. Galileo examined the trajectory of fall of a body on a rotating earth and demonstrated to his own satisfaction, albeit erroneously, that it is a segment of an eccentric circle on which the body moves with uniform motion. The appearance of fall is entirely relative to an observer on the surface of the earth; the true motion of the 'falling' body is not accelerated at all.[90] Whether it is accelerated or not, such a body does approach the earth, and in a number of passages Galileo explained the approach in terms of the mutual cooperation of the parts to form a whole, 'from which it follows that they have equal tendencies to come together in order to unite in the best possible way and adapt themselves by taking a spherical shape . . .' When separated from its whole, a part returns 'spontaneously and by natural tendency.'[91] As he said elsewhere, separated parts 'move to their whole, their universal mother.'[92] On the one hand, this notion repeated the ancient belief that like joins with like. Within the atomic tradition, which influenced Galileo's conception of matter so profoundly, the idea expressed itself in terms of mutually adaptable shapes.[93] On the other hand, the union of like with like meshed perfectly with his conviction of an ordered cosmos. Tendencies like those of terrestrial bodies could be assigned to the parts of other planets. The well-ordered cosmos is also the well-sorted universe, in which all particles of matter have been placed in their proper bins. Even the sphericity of the universe and its components appeared to follow from the natural tendency of parts to unite with their kind. However incompatible with the mechanistic elements of his thought the ordered cosmos may appear, it was obviously a basic datum of his philosophy of nature. Natural motions were closely connected to it, and it is impossible to explain them away.

At the same time, other passages do suggest that the very presence of two competing philosophies of nature in his mind led Galileo to regard the downward motion of heavy bodies from a second point of view. In the *Discourses*, he had Sagredo expound the cause of accelerated fall in terms somewhat like those of *De motu*. Salviati refused to pursue the subject.

> The present does not seem to be the proper time to investigate the cause of the acceleration of natural motion concerning which various opinions have been expressed by various philosophers, some explaining it by attraction to the center, others by the decreasing amount of medium to

45

be penetrated, while still others attribute it to a certain stress in the surrounding medium which closes in behind the falling body and drives it from one of its positions to another. Now, all these fantasies, and others too, ought to be examined; but it is not really worth while. At present it is the purpose of our Author [a reference to Galileo himself] merely to investigate and to demonstrate some of the properties of accelerated motion (whatever the cause of this acceleration may be) . . .[94]

Positivists have not failed to praise Galileo for proceeding with its description. It appears to me, however, that the significant aspect of the discussion is the word he applied to the explanations that had been offered. They were 'fantasies'. The same attitude, if not the same word, appeared frequently in his writings. The notion that the sun and the moon cause the tides, he said, 'is completely repugnant to my mind; for seeing how this movement of the oceans is a local and sensible one, made in an immense bulk of water, I cannot bring myself to give credence to such causes as lights, warm temperatures, predominances of occult qualities, and similar idle imaginings.' He called such theories the 'wildest absurdities' and he was astonished to note that Kepler assented to the 'moon's dominion over the waters, to occult properties, and to such puerilities.'[95]

Here spoke the mechanical philosopher and his revulsion from the animate universe of Renaissance naturalism. Given the attitude, how was he to view the accelerated motion downward of heavy bodies? Later mechanical philosophers were to attribute it to the multiple impacts of tiny particles. Aside from the fact that Galileo never mentioned such an opinion, it appears incapable of reconciliation with the uniform acceleration of free fall which is constant for all bodies. Perhaps one could explain gravity by magnetism. Not only did this opinion, which some adopted, embody insoluble problems, but magnetism was the very epitome of the occult attractions which mechanical philosophers were concerned to reject. Galileo's positivism in regard to gravity may have covered his embarrassment as much as it expressed a deliberate stance.

By calling the uniform acceleration of heavy bodies a natural tendency, he was able at once to avoid the pitfalls of the occult and to maintain the basic quantitative relations which were the heart of his mechanics. But in adopting this solution, he made it impossible for himself to see free fall as the paradigm of the action of force. As the product of a unique internal tendency, it was distinct from every other motion in nature. Galileo was tossed here on the horns of the century's dilemma. On the one hand, the completion of a mathematical mechanics required a generalized conception of force. On the other hand, metaphysical considerations labelled the concept of force absurd in its classic application, the case of free fall. The apparent incompatibility between the demands of mathematical mechanics and those of the mechanical philosophy of nature stood athwart the path of the second law of motion. Until the dilemma was resolved, its formulation was impossible.

NOTES

1. *Mathematical Principles of Natural Philosophy*, trans. Motte-Cajori, (Berkeley and Los Angeles, 1934), p. 13, reprinted by permission of the Regents of the University of California.
2. Ibid., p. 21.
3. Let me note in passing that Newton probably never read Galileo's principal work in mechanics, *Discourses on Two New Sciences*. (Cf. I. B. Cohen, 'Newton's Attribution of the First Two Laws of Motion to Galileo,' *Atti del simposio su «Galileo Galilei nella storia e nella filosofia della scienza»*, (Firenze-Pisa, 14–16 Settembre 1964), pp. xxiii–xlii. There were numerous sources from which he could have learned of Galileo's mechanics second hand. He did read the *Dialogue Concerning the Two Chief World Systems* when he was an undergraduate at Cambridge.
4. *Dialogues Concerning Two New Sciences*, trans. Henry Crew and Alfonso de Salvio, (New York, 1914), pp. 153–4, reprinted through permission of Dover Publications, Inc., New York. The mistranslation of the title is only the first of the many infelicities in this the only English version. I shall refer to it, as it should be referred to, as the *Discourses*.
5. *Physics*, trans. P. H. Wicksteed and Francis M. Cornford, 2 vols. (London, 1934), VII, v, reprinted by permission of The Loeb Classical Library and Harvard University Press.
6. *Dialogue Concerning the Two Chief World Systems – Ptolemaic & Copernican*, trans. Stillman Drake, (Berkeley and Los Angeles, 1953), p. 28, reprinted by permission of the Regents of the University of California.
7. Ibid., p. 116. Cf. p. 171: 'With respect to the earth, the tower, and ourselves, all of which all keep moving with the diurnal motion along with the stone [which is dropped from the top of the tower], the diurnal movement is as if it did not exist; it remains insensible, imperceptible, and without any effect whatever. All that remains observable is the motion which we lack, and that is the grazing drop to the base of the tower.'
8. *Discourses*, p. 162.
9. 'Dum igitur lapidem, ex sublimi a quiete descendentem, nova deinceps velocitatis acquirere incrementa animadverto, cur talia additamenta, simplicissima atque omnium magis obvia ratione, fieri non credam? Idem est mobile, idem principium

movens: cur non eadem quoque reliqua? Dices: eadem quoque velocitas. Minime: iam enim re ipsa constat, velocitatem eandem non esse, nec motum esse aequabilem: oportet igitur, identitatem, seu dicas uniformitatem, ac simplicitatem, non in velocitate, sed in velocitatis additamentis, hoc est in acceleratione, reperire atque reponere.' *Le opere di Galileo Galilei*, direttore Antonio Favaro, ed. naz., 20 vols. in 21, (Firenze, 1890–1909), **2**, 262. The passage is similar to one in the *Discourses*, but it differs significantly in some details.
10. *Dialogue*, p. 202. The Latin treatise, *De motu locali*, which provided the basis of Books Three and Four of the *Discourses*, and which was apparently completed long before the *Discourses* was composed, discussed uniformly accelerated motion without mentioning the constant acceleration of gravity which pertains to all bodies; presumably Galileo had not yet arrived at the conclusion of constant acceleration when he wrote it.
11. *Discourses*, pp. 63–4.
12. Ibid., p. 181.
13. *Dialogue*, p. 40.
14. *The Assayer*, in *The Controversy on the Comets of 1618*, trans. Stillman Drake and C. D. O'Malley, (Philadelphia, 1960), p. 311, reprinted by permission of the University of Pennsylvania Press.
15. *Discourses*, p. 181.
16. See Appendix A.
17. *Discourses*, p. 215.
18. *Dialogue*, p. 31.
19. Ibid., p. 264. Cf. Appendix A.
20. Ibid., pp. 177–8.
21. Ibid., pp. 154, 142.
22. Ibid., pp. 148, 154.
23. Ibid., p. 187.
24. Ibid., p. 250. Cf. *History and Demonstrations Concerning Sunspots and Their Phenomena*, in *Discoveries and Opinions of Galileo*, trans. & ed. Stillmann Drake, (Garden City, N.Y., Doubleday and Company, Inc. (1957), pp. 113–14: 'For I seem to have observed that physical bodies have physical inclination to some motion (as heavy bodies downward), which motion is exercised by them through an intrinsic property and without need of a particular external mover, whenever they are not impeded by some obstacle. And to some other motion they have a repugnance (as the same heavy bodies to motion upward), and therefore they never move in that manner unless thrown violently by an external mover. Finally, to some movements they are indifferent, as are these same heavy bodies to horizontal motion, to which they have neither inclination (since it is not toward the center of the earth) nor repugnance (since it does not carry them away from that center). And therefore, all external impediments removed, a heavy body on a spherical surface concentric with the earth will be indifferent to rest and to movements toward any part of the horizon. And it will maintain itself in that state in which it has once been placed; . . . if placed in a state of rest, it will conserve that; and if placed in movement toward the west (for example), it will maintain itself in that movement. Thus a ship, for instance, having once re-

ceived some impetus through the tranquil sea, would move continually around our globe without ever stopping; and placed at rest it would perpetually remain at rest, if in the first case all extrinsic impediments could be removed, and in the second case no external cause of motion were added.'

25. *Dialogue*, p. 227; *Discourses*, p. 94; *Dialogue*, p. 213; ibid., p. 428; *Opere*, **8**, 338. See also *Dialogue*, p. 156: when a ball is thrown, the '*moto*' (a few lines later the '*impeto*') is conserved in it and continues 'to urge it' (*condurlo*) on. On occasion, Galileo forgot himself and applied similar language to inertial motion. In discussing the hypothetical case of a ball rotating around the earth once a day, he referred to the '*virtù*' which makes it go around (ibid., p. 233). In the famous passage about the experiment on a ship, he said at one point that the horizontal motion is caused by an 'impressed force' (ibid., p. 149).
26. 'De gravi et levi tractationem mathematicam esse, testatur fragmentum Euclidis.' *Opere*, **1**, 414. Undoubtedly Galileo referred to the *Book on the Balance* attributed to Euclid.
27. 'Omnium corporum una est materia, eaque in omnibus gravis: sed eiusdem gravitatis non possunt esse contrariae inclinationes naturales: ergo, si una est naturalis inclinatio, ut contraria sit praeter naturam opus est: naturalis autem gravitatis inclinatio est ad centrum: ergo necesse est, quae a centro praeter naturam esse.' *Opere*, **1**, 353, 362.
28. *De motu*, Chapter 17; *On Motion and On Mechanics*, trans. I. E. Drabkin and Stillman Drake, (Madison, The University of Wisconsin Press; © 1960 by the Regents of the University of Wisconsin), pp. 78–81.
29. Ibid., pp. 38–9.
30. Ibid., p. 45.
31. Ibid., p. 25.
32. Ibid., p. 88.
33. Galileo himself appears ultimately to have arrived at a similar assessment of *De motu*. At the end of the Third Day of the *Discourses*, Salviati referred in markedly different terms to the same fragments from Euclid mentioned above (n. 26). 'There is a fragment of Euclid which treats of motion, but in it there is no indication that he ever began to investigate the property of acceleration and the manner in which it varies with slope. So that we may say the door is now opened, for the first time, to a new method fraught with numerous and wonderful results which in future years will command the attention of other minds.' (pp. 242–3.)
34. See Appendix A for a discussion of Galileo's use of '*forza*'.
35. *Mechanics*; *Motion and Mechanics*, pp. 160–1.
36. *De motu*; ibid., p. 22 fn.
37. Ibid., p. 39.
38. Ibid., p. 23.
39. *Dialogue*, pp. 31–2.
40. It appears to me that centrifugal force did not present any problem within this context. For Galileo, circular motion was not force-free motion; rather it was equivalent to the rest of a heavy body which demands an equilibrium of forces. Centrifugal force could contribute to the repulsion from the centre just as the support of a perfectly smooth plane could. Later in the century, when the same

view was applied to orbital motion, centrifugal force replaced the smooth plane entirely. In the *Dialogue*, Galileo constructed an argument to show that in circular motion a centrifugal tendency can never overcome a natural tendency toward the centre where such exists; that is, the rotation of the earth at whatever speed cannot project a heavy body off the surface of the earth (pp. 188–203).

41. *De motu*; ibid., p. 31 fn.
42. *Opere*, **2**, 262. The Latin text is quoted in note 9.
43. *Discourses*, pp. 63–4. Cf. Sixth Day of the *Discourses* in which he attempted to measure the force of percussion of water falling from one bucket hanging on a large balance into another bucket slung beneath it. The water that is in the act of falling does not affect the balance 'perchè, andandosi continuamente accelerando il moto della cadente acqua, non possono le parti più alte gravitare o premere sopra le più basse . . .' (*Opere*, **8**, 325).
44. *De motu*; ibid., p. 15.
45. *Dialogue*, pp. 20–1.
46. *Discourses*, p. 74.
47. *Dialogue*, p. 28.
48. *Discourses*, p. 264.
49. *Opere*, **8**, 332–3.
50. *Discourses*, p. 94. Cf. Sixth Day of the *Discourses*: We must accept 'l'impeto acquistato in *A* dal cadente dal punto *C* esser tanto, quanto appunto si ricercherebbe per cacciare in alto il medesimo cadente, o altro a lui eguale, sino alla medesima altezza; onde possiamo intendere che tanta forza bisogna per sollevar dall'orizonte sino all'altezza *C* l'istesso grave, venga egli cacciato da qualsivoglia de' punti *A*, *D*, *E*, *B*.' In the diagram, *A*, *D*, *E*, and *B* are all on the same horizontal line. *CB* is perpendicular to it, and *CE*, *CD*, and *CA* are three inclined planes with the same vertical drop (*Opere*, **8**, 338).
51. *Discourses*, pp. 180–5.
52. *Dialogue*, p. 147; *Discourses*, pp. 275, 286. Cf. a discussion of pendulums in the *Discourses*: A large pendulum can be put into motion by blowing on it; one puff gives it a small motion to which successive puffs add. 'Continuing thus with many impulses [*impulsi*] we impart to the pendulum such momentum [*impeto*] that a greater impulse [*forza*] than that of a single blast will be needed to stop it.' (p. 98.)
53. *Discourse on Bodies in Water*, trans. Thomas Salusbury, ed. Stillman Drake, (Urbana, Illinois, 1960), pp. 6–7, reprinted by permission of the University of Illinois Press.
54. *Opere*, **8**, 330. Cf. *Dialogue*: on a steelyard, Sagredo explained, a small weight counterbalances a great one because of the difference in their motions; 'the lesser weight moves the greater only when the latter moves very little, being weighed at a lesser distance, and the former moves quite a way, hanging at the greater distance. One must say, then, that the smaller weight overcomes the resistance of the greater by moving much when the other moves little.' To the exposition Salviati added that 'the speed of the less heavy body offsets the heaviness of the weightier and slower body. . . . Now fix it well in mind as a true and well-known principle that the resistance coming from the speed of motion compensates that which

depends upon the weight of another moving body, and consequently that a body weighing one pound and moving with a speed of one hundred units resists restraint as much as another of one hundred pounds whose speed is but a single unit. And two equal movable bodies will equally resist being moved if they are to be made to move with equal speed. But if one is to move faster than the other, it will make the greater resistance according as the greater speed is to be conferred upon it.' (pp. 214–15.) When he spoke of a body in motion resisting restraint and of a body resisting movement, Galileo attempted to derive dynamic conclusions from the statics of the simple machines. In both cases, he was applying the principle that the propensity of a body to move is measured by the force necessary to prevent the motion, with the virtual velocity relations of the steelyard added to the equation.

55. *Bodies in Water*, p. 16. Axiom I in the same treatise: 'Weights absolutely equall, moved with equall Velocity, are of equall Force and Moment in their operations.' Having introduced the concept of moment, he went on to define it: 'Moment, amongst Mechanicians, signifieth that Vertue, that Force, or that Efficacy, with which the Mover moves and the Moveable resists. Which Vertue dependes not only on the simple Gravity, but on the Velocity of the Motion, and on the diverse Inclinations of the Spaces along which the Motion is made; For a descending Weight makes a greater Impetus in a Space much declining than in one less declining.' (p. 6) Cf. *Mechanics*: 'it is not foreign to the arrangement of nature that the speed of the motion of the heavy body B should compensate the greater resistance of the weight A when this moves more weakly to D and the other descends more rapidly to E.' Hence, 'the speed of motion is capable of increasing moment in the movable body in the same proportion as that in which this speed of motion is increased.' (p. 156.)

56. Ibid., pp. 180–1.

57. Ibid., pp. 163–4. Cf. pp. 164, 168. The operation of the screw can be understood by the principle applicable to all mechanical instruments: 'whatever is gained in force by their means is lost in time and in speed.' To demonstrate this in the case of the screw (which is only an inclined plane wrapped around a cylinder), he imagined a weight E being drawn up an inclined plane by a line which passes over a pulley and which has a weight F hanging from its other end. As the weight F drops, it moves a distance equal to that which E moves along the plane, but the vertical displacement of E is less, of course, because of the incline. 'And since heavy bodies do not have any resistance to traverse motions except in proportion to their removal from the center of the earth, ... the travel of the force F has the same ratio to the travel of the force E as the line AC [displacement along the inclined plane] to the line CB [perpendicular displacement], or as the weight E has to the weight F.' (pp. 176–7.) After stating the principle in terms of speed, he carried out the analysis in terms of displacement.

58. 'Qui, primieramente, è manifesto che il peso delle dieci libbre, dovendo calare a perpendicolo, sarà bastante di far montare un peso di libbre cento sopra un piano inclinato tanto, che la sua lunghezza sia decupla della sua elevazione, per le cose dichiarate di sopra, e che tanta forza ci vuole in alzare a perpendicolo dieci libbre di peso, che nell'alzarne cento sopra un piano di lunghezza decupla alla sua per-

pendicolare elevazione; . . . adunque, caschi il peso di dieci libbre per qualsisia spazio perpendicolare, l'impeto suo acquistato, ed applicato al peso di cento libbre, lo caccerà per altrettanto spazio sopra il piano inclinato, a quale spazio risponde l'altezza perpendiculare grande quanto è la decima parte di esso spazio inclinato. E già si è concluso di sopra che la forza potente a cacciare un peso sopra un piana inclinato è bastante a cacciarlo anche nella perpendicolare che risponde all'elevazione di esso piana inclinato, la qual perpendicolare, nel presente caso, è la decima parte dello spazio passato sull'inclinata, il quale è eguale allo spazio della caduta del primo peso di dieci libbre; adunque è manifesto che la caduta del peso di dieci libbre fatta nel perpendicolare è bastante a sollevare il peso di cento libbre pur nella perpendicolare, ma solo per lo spazio della decima parte della scesa del cadente di dieci libbre.' *Opere*, **8**, 340–1.
59. Ibid., **8**, 338–9.
60. Ibid., **8**, 341.
61. *Discourses*, p. 181. Cf. the final topic of discussion in the *Discourses* in which the inevitable sag of a rope stretched horizontally is used to demonstrate that a projectile fired horizontally will not strike any target on the same horizontal line no matter how close it is. 'The curvature of the path of the shot fired horizontally appears to result from two *forze*, one (that of the weapon) drives it horizontally and the other (its own weight) draws it vertically downward. So in stretching the rope you have the *forze* which pull it horizontally and its own weight which acts downwards. The circumstances in these two cases are, therefore, very similar. If then you attribute to the weight of the rope a *possanza* and *energia* sufficient to oppose and overcome any stretching *forza*, no matter how great, why deny it to the weight [*peso*] of the bullet?' (p. 290). I have amended the translation in places.
62. The only passage I know of in which Galileo applied the concept of inertia to a motion other than horizontal occurs in the *Discourses*. By the super-position of motions, he demonstrated that the *momento* acquired by a body in descending along an inclined plane is just sufficient to carry it back to its original height. He imagined the body that had descended to be deflected along a plane that rises at the same angle. He calculated the distance that its inertial motion would carry it along that plane in a time equal to that of its descent, and from this subtracted the distance in the other direction the body would have travelled in the same time starting from rest (pp. 215–17). Possibly another example is to be found in the Sixth Day of the *Discourses*. Galileo imagined that the two bodies on the ends of the cord in his pulley arrangement are of equal weight. After the falling body snaps the cord, the system of two bodies is dynamically identical to a ball on a horizontal plane; the inclination of one body to move downward is exactly balanced by the repugnance of the other to rise. But the *momento* of the falling body before the cord snaps will put the other into motion; thus the system will move with a uniform motion, the one body rising and the other descending. Because of the equality of inclination and repugnance, Galileo decided that the velocity would be equal to that of the freely falling body at the moment the cord snaps. The whole arrangement is remarkably like Atwood's machine (*Opere*, **8**, 334–7). His recognition of terminal velocity is similar. Hence whenever the propensity to

motion of a body already moving is balanced by an equal force in the opposite direction, the body will continue to move with uniform velocity in Galileo's opinion. These isolated cases of uniform vertical motion did not play a significant role in his system, and it is not apparent that he saw the full implications of the two cases. Certainly, they do not modify the judgment that Galileo never considered free fall as a paradigm of the action of force. In both cases he merely found a repugnance to motion by yoking the falling body to another that was lifted – the other body on the cord, in the one case, the medium (in his interpretation of resistance) in the other.

63. *De motu*; ibid., pp. 64–5.
64. *Discourses*, p. 184. It should be noted that the passage was written by Viviani at Galileo's request and interpolated into the second edition. The same point was made in the demonstration of the equal time of descent along all chords of a circle; in this case, of course, the words were Galileo's own (pp. 189–90).
65. Ibid., p. 76.
66. Ibid., ebony is one thousand times as heavy as air, an inflated bladder four times as heavy. Therefore, air diminishes 'the inherent and natural speed' [*intrinseca e naturale velocità*] of ebony by one-thousandth and that of the bladder by one-fourth. Thus, in the time it takes a piece of ebony to fall from a tower of two hundred cubits, the bladder will fall only three-quarters of that distance. Lead is twelve times as heavy as water and ivory twice as heavy. Their speeds unhindered [*assolute velocità*] are equal; in water the speed of lead is diminished by one-twelfth, that of ivory by one half. Thus in the time it takes a piece of lead to fall through eleven cubits of water a piece of ivory will fall through six cubits. In both examples he has chosen to compare distances travelled during the same time. When the unit of time is the same, distances travelled starting from rest are proportional to the rate of acceleration. It is difficult to imagine that Galileo was merely lucky enough to state his problems in a way that was correct; despite his use of the term '*velocità*', he was clearly thinking of the natural acceleration being diminished by the buoyancy of the medium.
67. Ibid., p. 99.
68. Ibid., p. 74. Cf. the two bodies on a cord over a pulley, cited in note 62.
69. Pendulums might seem to offer a second case, but in fact they do not. Galileo's attempt to analyse the pendulum in detail, to show that the arc of a circle is the path of quickest descent, was a series of inequalities based on his demonstration that the times of descent along all chords are equal. It did not attempt to evaluate the acceleration at different points on the arc. Moreover, if it is relevant, Galileo fooled himself into thinking that he had demonstrated that the arc of a circle is the path of quickest descent, which is of course an error (ibid., pp. 237–40).
70. Ibid., p. 75.
71. Ibid., p. 81.
72. *Bodies in Water*, p. 67.
73. Ibid., p. 41.
74. *Dialogue*, p. 213. Cf. the discussion of the *Discourses* of the size of a vibrating string in relation to its pitch. For strings made of the same material, the pitch goes down as the size increases. When strings of different materials are compared, however,

the weight of the string is the factor to which its pitch must be compared; a gut string can have a higher pitch than a brass string of smaller diameter. The density of gold is twice that of brass; hence if two spinets were strung with wires of identical size and tension, the pitch of the gold wires would be lower by a fifth than that of the brass (because the size or weight must quadruple to lower the pitch an octave). 'And here it is to be noted that it is the weight rather than the size of a moving body which offers resistance to change of motion [*velocità del moto* – the word "change" does not appear in the Italian] . . .' (p. 103).

75. *Dialogue*, p. 149.
76. *Discourses*, p. 181.
77. On a horizontal plane, the body 'farà quello che piacerà a noi, cioè, se ve lo mettereme in quiete, in quiete si conserverà, e dandogli impeto verso qualche parte, verso quella si moverà, conservando sempre l'istessa velocità che dalla nostra mano averà ricevuta, non avendo azione nè di accrescerla nè di scemarla . . .' *Opere*, **8**, 336.
78. *Dialogue*, p. 215.
79. *Opere*, **8**, 323.
80. '. . . di tal percossa non si possa in veruna maniera assegnare una determinata misura.' Ibid., **8**, 337.
81. Ibid., **8**, 343.
82. Ibid., **8**, 344–5.
83. Ibid., **8**, 345–6. In the *Assayer*, Galileo gave a somewhat similar, though not identical, analysis of how a current of water sets a ship in motion. (Drake and O'Malley, *The Controversy on the Comets of 1618*, p. 281.)
84. *Discourses*, pp. 252–3.
85. *Dialogue*, p. 32.
86. In the *Discourses*, Galileo had Sagredo reply to the demonstration that the trajectory of a projectile is parabolic in the following way: 'One cannot deny that the argument is new, subtle, and conclusive, resting as it does upon this hypothesis, namely, that the horizontal motion remains uniform, that the verticle motion continues to be accelerated downwards in proportion to the square of the time, and that such motions and velocities as these combine without altering, disturbing, or hindering each other, so that as the motion proceeds the path of the projectile does not change into a different curve: but this, in my opinion, is impossible.' Sagredo and Simplicio went on to object further that on a round earth all perpendiculars to the horizon are not parallel and that the resistance of the medium will alter the motions. In his reply, Salviati dealt only with the latter two and said nothing about the question of motions compounding without interfering with each other (pp. 250–1).
87. *Dialogue*, p. 139. Galileo proceeded almost immediately, however, to apply the analysis to the motion of a stone dropped from the mast of a moving ship, a case in which the horizontal motion is not the natural motion of the earth. Later, in demonstrating that a cannon ball fired straight up will stay directly above the cannon even though the earth is turning, he compounded the natural horizontal motion with a violent motion in the vertical direction (pp. 176–7).
88. *Discourses*, p. 244.

89. Descartes to Mersenne, 11 October 1638; *Oeuvres de Descartes*, ed. Charles Adam et Paul Tannery, 12 vols. (Paris, 1897–1910), **2**, 380.
90. *Dialogue*, p. 166.
91. Ibid., pp. 33, 34.
92. Ibid., p. 37.
93. Conversely, the idea that like repels unlike also played a role in Galileo's thought. In the *Discourses*, Salviati discussed how water on leaves forms up in drops. It cannot be due to any internal tenacity acting between the particles. Such a quality should reveal itself the more when the water is surrounded by wine in which it is less heavy than it is in air, whereas in fact the drop collapses when surrounded by wine. The formation of the drop is due rather to the pressure of the air with which water has some incompatibility which he did not understand. Simplicio broke in at this point. Salviati made him laugh by his efforts to avoid the word antipathy. Salviati in turn replied ironically, 'All right, if it please Simplicio, let this word antipathy be the solution of our difficulty.' Here Galileo let it drop, but the passage illustrates beautifully the contrast between the drive for mechanical explanations on the one hand (external air pressure, not internal attraction) and the animism he rejected. His scornful phrase, 'the *word* antipathy', underlines his judgment of the empty quality of the prevailing philosophy of nature, words divorced from the reality of nature. The very fact that Galileo put the initial jibe into Simplicio's mouth is significant; manifestly Galileo found a world of difference between an 'antipathy' and 'a certain incompatibility which I do not understand . . .' (pp. 70–1).
94. Ibid., p. 166–7. I have altered the translation somewhat. Cf. the passage in the *Dialogue* in which Salviati responded to Simplicio's assertion that gravity is what causes heavy bodies to move downward. 'You are wrong, Simplicio; what you ought to say is that everyone knows that it is called "gravity". What I am asking you for is not the name of the thing, but its essence, of which essence you know not a bit more than you know about the essence of whatever moves the stars around. I except the name which has been attached to it and which has been made a familiar household word by the continual experience that we have of it daily. But we do not really understand what principle or what force it is that moves stones downward, any more than we understand what moves them upward after they leave the thrower's hand, or what moves the moon around.' (p. 234.)
95. Ibid., pp. 445, 462, cf. pp. 410, 419–20.

Chapter Two

Descartes and the Mechanical Philosophy

IF Galileo was the primary fountainhead of the new science of mechanics, Descartes' contribution to it was scarcely less important. It was Descartes who extracted the new conception of motion from the philosophic morass in which Galileo had left it mired. It was Descartes who consciously formulated a new conception of nature consciously purged of organic features. Once and for all, the ordered cosmos of Greek thought, in which natural motions were conceivable, gave way to the mechanical universe, which was inconceivable without the new idea of motion. The indifference of matter to motion, which in Galileo's thought was associated with horizontal motions, was extended by Descartes to all motions whatever, and beside the uniform matter of which Galileo conceived all bodies in the universe to be composed, Descartes placed the second cornerstone of the mechanical philosophy, a uniform conception of the motion by which all bodies in the universe move. 'I do not accept any difference between violent and natural motions.'[1]

Scion of an established family of the minor French nobility, René Descartes was born in 1596, a full generation later than Galileo. After completing his education at the Jesuit school, La Flèche, and the University of Poitiers, he took service for a time as a professional soldier and eventually chose to establish himself permanently in the Netherlands where he found a climate of relative intellectual freedom. There he devoted himself to mathematics and philosophy, and there he was exposed to the new ideas of motion put forward by his friend Isaac Beeckman. Although his reputation spread, Descartes did not publish anything before 1637 when his *Discours de la méthode* (*Discourse on Method*) introduced three accompanying essays on geometry, dioptrics, and meteorology. During the following seven years he published two other works, *Meditationes de prima philosophia* (*Meditations on First Philosophy*, 1641) and *Principia philosophiae* (*Principles of Philosophy*, 1644) which, with the *Discourse*, expounded a new departure in philosophy, including natural philosophy. More than any other one man, Descartes was responsible for installing the mechanical philosophy as the prevailing conception of nature, and as part of the mechanical philosophy, a conception of motion similar to Galileo's. He died in 1650 in Sweden, where he had gone at the behest of Queen Christina to instruct her in philosophy, a testimony in itself to the speed with which his influence spread.

Nothing testifies more strikingly to the analytic clarity of Descartes' thought than his discussion of motion. In the very midst of the process of formulating a new idea of motion, and lacking entirely the perspective of time, he was still able to recognise the central features of the new conception and to state them in terms acceptable to historians of science today. Not surprisingly, he found the Aristotelian definition of motion – *Motus est actus entis in potentia, prout in potentia est* (motion is the act of being in potency in so far as it is in potency) – so obscure as to be meaningless. In contrast, his idea was so clear, he said, that geometers, those who conceive things most clearly, had been accustomed to using it when they described a line as the motion of a point.[2] By motion, Descartes referred explicitly to local motion – 'for I can conceive no other kind, and do not consider that we ought to conceive any other in nature . . .'[3] In the vulgar sense, motion was held to be nothing more than the action by which a body passes from one place to another. The word 'action' rendered this definition unsuitable in Descartes' eyes. A prejudice, arising from our experience of the effort needed to move a body, convinces us that motion requires action whereas rest seems to require none. The prejudice is mistaken; we can deliver ourselves from it by reflecting that as much effort is needed to stop a moving body as to put it in motion. A satisfactory definition states that motion is 'the transference of one part of matter or one body from the vicinity of those bodies that are in immediate contact with it, and which we regard as in repose, into the vicinity of others.'[4] In fact this definition, published in the *Principles of Philosophy* in 1644, and amending an earlier definition found in the manuscript treatise *Le monde*, which Descartes withheld from publication after Galileo's brush with the Inquisition, had been deliberately constructed by Descartes after Galileo's condemnation so that he could assert that the earth does not move. According to his system, the earth does not move away from the bodies in immediate contact with it (the atmosphere and the subtle matter filling its pores) and hence, according to the definition, the earth is at rest. If the definition neither fooled nor mollified the Church, it was cleverly enough constructed that it did not betray anything essential in Descartes' conception of motion. It expressed clearly two interrelated principles basic to his conception, the relative nature of motion and the ontological identity of motion and rest. If A moves with respect to B, then, in Descartes' conception, B moves with equal velocity in respect to A. From a philosophic point of view, one is in motion as much as the other. Descartes stressed his use of the word 'transference' in the definition. Motion is not an action or force which transfers. It is the transference itself. It is nothing outside the body moved. Motion means only 'that a body is otherwise disposed, when it is transferred, than when it is not, so that motion and rest are only two different modes of it.'[5]

Against the background of this discussion, Descartes was ready to state what he called the first law of nature.

> That each thing remains in the state in which it is so long as nothing changes it.[6]

If a body is square, it remains square unless something changes it. If it is at rest, it remains at rest and does not begin to move itself. Why should motion alone among the states in which a body can be contain the seed of its own destruction? It is another prejudice from our youth that bodies in motion tend toward rest. On the contrary, once a body begins to move, there is no reason why it should not continue to move with the same velocity as long as it meets nothing that stops it. Here, more clearly than Galileo ever succeeded in stating it, was the kernel of the principle of inertia. As Descartes expressed it to Constantijn Huygens, 'I consider the nature of motion to be such that when a body has begun to move, that suffices to make it continue with the same speed and in the same straight line until it is stopped or turned aside by some other cause.'[7]

As the reference to motion in a straight line suggests, Descartes also went beyond Galileo in asserting, not merely a concept of the conservation of motion which approaches our principle of inertia, but the concept of the conservation of rectilinear motion which is indistinguishable from our principle of inertia. To the first law of motion he added a second:

> That every body which moves tends to continue its motion in a straight line.[8]

As the first law depends on the immutability of God, who, acting always in the same manner, conserves each body in its state unless it be changed by contact with another, so the second law derives from the same immutability of God, who conserves each body as it is at any moment. Rectilinear motion differs from circular motion in being more simple. Whereas two successive instants are necessary to define a circular motion, one instant can define a rectilinear one, giving both its location and the direction in which it is tending to move. Hence the Cartesian reduction of all motions to a single type had at least two facets. On the one hand, he denied any distinction between natural and violent motion. On the other hand, he rejected the idea of a primary circular motion. Of itself, every body in motion tends in a straight line. If in fact it moves in a curve, some external cause must be operating to turn it.

Descartes' systematic analysis of motion established the foundation on which the modern science of dynamics has been constructed. By means of the first law of motion, he stated, 'we are freed of the difficulty in which the Schoolmen find themselves when they seek to explain why a stone continues to move for a while after leaving the hand of the thrower. For we should rather be asked why it does not continue to move forever.'[9] Here, applied to a specific problem, was the central question to which the principle of inertia has directed dynamics – not what causes motion, since the state of motion is held to endure without the continued operation of a cause, but what causes changes of motion.

How liberating the new point of view could be is seen in his correspondence with Mersenne on the acceleration that cannon balls were held to receive after they leave the cannon. Experienced gunners were all agreed that such an acceleration takes place. A cannon does not exercise its greatest effect at point blank range; the ball requires a certain distance to gain its maximum force. Mersenne attempted to explain the supposed phenomenon by means of the impetus or impetuosity impressed on the ball by the shot. In contrast, Descartes threw experienced opinion to the wind and flatly refused to believe the pretended fact. The impetus impressed on the ball is merely its motion, which is greatest at the moment the ball leaves the gun. From there on, it can only slow down as the air resists its passage.[10]

Not only did Descartes define the central problem of dynamics, but he took the further step of reducing most changes of motion to a single causal agency. The matter of Descartes' universe was wholly inert – something, he said, 'which can be moved in many ways, not, in truth, by itself, but by something which is foreign to it, by which it is touched: for to have the power of self-movement, as also of feeling or of thinking, I do not consider to appertain to the nature of body . . .'[11] The 'something which is foreign' to a body could be God, and indeed Descartes held that God is the ultimate cause of all motion in the universe. God's action was confined to the origin of motion, however; since then He has limited Himself to maintaining the original quantity. The human soul, in its unique union with the body, also exercises a power of moving it. In the vast majority of natural phenomena, however, in the cases of all purely physical phenomena, only the impact of another body can operate to change a body's motion. The 'force with which a body acts upon another or resists its action consists in that alone, that each thing endeavours as much as it can to remain in the same state in which it finds itself, conformably to the first law . . .'[12] Because the world is a plenum, a body in motion must continually strike others, and the whole of nature's phenomena result from the changes in motion which thus occur. The quantity of motion in the universe remains constant because of the immutability of God, but God has constructed the universe such that motion can pass from one body to another in the process of impact. With the exception of acts derived from a human will, then, all changes of motion in Descartes' universe, after the creation, were reduced to a single causal agency. On such a philosophic foundation the science of dynamics could rest securely.

*

And yet a science of dynamics cannot be found in Descartes' writings. In part, its absence is a result of the fundamental thrust of his philosophy, which concerned itself less with exact quantitative mechanics than with demonstrating the possibility of the mechanical causation of phenomena. Nevertheless, a consistent set of dynamic conceptions would have enriched his philosophy, even though his insistence on impact as the universal causal agency would have led undoubtedly to a dynamics different from the one we employ. Such a set of dynamic conceptions are not to be found in his writings. Much of the interest that Descartes holds for the history of dynamics lies in the fact that he did not succeed in formulating a workable dynamics even though he had cleared away the philosophic obstacles that hindered Galileo. His example demonstrates how far from obvious the concepts were, and how intricate the problems to be solved. Among the chaos of terms used with dynamical signification by Descartes, the verb '*agir*' and three other words etymologically related to it (*agité, agitation,* and *action*) played a central role. As he employed the word 'motion' in a far more confined sense than it had held before the seventeenth century, so he applied the old and broader concept of 'action' to changes of local motion. A body acts on (*agit contre*) another when it changes (or, occasionally, endeavours to change) its state of motion. Descartes might not have admitted that one body can ever act, in the true sense of the word, on another. God is the only causal agent in the universe. In impact, one body does not act on another; rather God maintains the same quantity of motion in the universe. In practice, Descartes made no effort to maintain this ultimate metaphysical point of view, and he spoke of one body acting on another when it strikes it. Closely associated with the verb '*agir*' were the participle '*agité*' and the noun '*agitation*'. Although in Descartes' own terms logical necessity did not demand that a body be in motion in order to act on another, in fact he applied '*agir*' almost exclusively to bodies that are '*agités*' or have '*agitation*'. His insistence on the relativity of motion and rest could justify the usage, although in fact Descartes' discussions of motion frequently neglected his own principle of relativity. A body that is '*agité*' or has '*agitation*' can '*agir*', and the more strongly it is '*agité*' the more strongly it can '*agir*'. To '*agir*' is to communicate '*agitation*', and the body which '*agit*' loses as much '*agitation*' as it communicates to another.[13] A basic ambiguity in the use of the word '*agitation*' made an exact quantitative meaning of the word almost impossible. On the one hand, '*agitation*' frequently referred to a tumultuous random motion of particles. Thus heat was an '*agitation*' of the parts of a body.[14] In other contexts, however, '*agitation*' can only be rendered by 'velocity' or 'momentum'. In reference to a comet travelling through a vortex, he spoke of 'the force . . . which it acquires to continue to be transported, or rather to move, which is what I call its

FORCE IN NEWTON'S PHYSICS

agitation . . .'[15] On occasion, Descartes attempted to relate the effect of a body not only to its velocity or '*agitation*', but also to the time during which it acts. He argued that a man can break a bone more easily when he holds it in his hand than when he lays it on an anvil because the hammer remains in contact with the bone for a longer time.[16] For the most part, however, Descartes implicitly set a body's ability to '*agir*' proportional to its motion or '*agitation*'.

The word '*action*' was not limited to motions. The signification of '*action*', Descartes argued, 'is general, and includes, not only the power or the inclination to move, but also movement itself.'[17] Especially in optical contexts he liked to maintain the distinction – light is an '*action*'

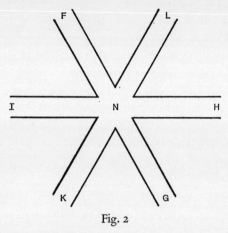

Fig. 2

transmitted through a medium, an '*action*' that obeys the same laws as motion although particles of matter do not actually move in the direction of the action.[18] The distinction allowed Descartes to explain how rays of light can cross without interfering with each other; although a body can move in only one direction at a time, it can transmit several actions at once. Imagine, for example, that six pipes are joined together at *N* and that six men blow into them (see Fig. 2). Even though the particles of air at *N* can only move in one direction at a time, 'they do not cease to be able to serve to transfer all the *actions* which they receive, and it can be said that the *action* which comes from *F* passes in a straight line to *G*, the fact notwithstanding that perhaps none of the particles of air which come from *F* and arrive at *N* are not turned from there toward *I* or toward *L*; for in doing so they transfer the *action* which determined them toward *G* to other particles of air which come from

H and from *K* and which tend toward *G* just as though they came from the point *F* . . .'¹⁹ In at least one place, Descartes referred to the product of force multiplied by time as '*action*'. Moreover, his analysis of simple machines employed the product of force times distance, what we call 'work', but what he frequently called '*action*'. For the most part, however, the word fell into the same category as '*agir*' and '*agitation*', and he did not use it with controlled precision.

Descartes' use of '*agir*' and its cognates illustrates how the main thrust of his philosophy directed him away from the formulation of a dynamics such as we know. The attention of his mechanical philosophy of nature was focused on the causes of phenomena, and his principal dynamic concepts were directed, not toward the changes of motion that a body undergoes, but toward bodies that '*agit*' on others to change their motions. What '*action*' signified to him was the operation of one body on another to change its state of motion or rest. Because of the inertness of matter, a body cannot change its own state of motion. To '*agir*' then, a body must '*agir*' on another. All '*action*' must come from a source external to the body that is acted upon. Beyond this generalised meaning, consonant with the basic tenets of his mechanical philosophy, Descartes made no effort to define '*agir*' and '*action*' precisely.

To one dynamical concept he did implicitly assign a quantitative meaning – the 'force of a body to act.'²⁰ I say 'meaning' rather than 'definition'; to call it a 'definition' would give the impression of greater consistency than he demonstrated in its employment. The '*force*' of a body to act, or the '*force*' of its motion, or its '*force*' to continue to move, as Descartes used the terms, are roughly equivalent to our concept of momentum. They are only roughly equivalent, however. Descartes' idea of size or extension differs from the idea of mass that enters into the definition of momentum, although the quantity of the third element, as he sometimes put it, is closer to mass. Moreover, the surface area of a body figured in its '*force*' to act, although he was not consistent in referring to its role. Again, the concept was attached to the active rather than the passive body, suggesting a measurement, not of the change of motion a body undergoes in defined conditions, but of the power of a body to act on others. As in the case of Galileo, the primary dynamic concept of the man who formulated the principle of rectilinear inertia expressed the force associated with a body in motion, a concept ultimately derived from the medieval idea of impetus. To us, as post-Newtonians, the principle of inertia suggests free fall as the model of force, an external action that changes a body's inertial state. In contrast, both Descartes and Galileo thought of force in terms of the model of impact. Force is the capacity of a body in motion to act.

The force of a body was not taken to be a measure of the act it performs, however, for the action depends as well on the other body. For example, a body rebounds more from a hard obstacle than from a soft one because its motion is retarded, not in proportion to the resistance it meets, but in proportion to the resistance it overcomes; and the other body, in being moved, receives that part of the force of motion which the first one loses.[21] In relating the force of motion lost by one body to the force of motion gained by the other, Descartes took a first step toward both the principle of the conservation of momentum and Newton's third law of motion. In the development of quantitative mechanics, this aspect of his thought exercised more influence than any other. The concept of the force of a body's motion, however, his one quantified or potentially quantifiable dynamical concept, did not in itself lead to a quantitative mechanics. The force of a body referred to something a body has, not to something that acts on it. It is akin to our idea of kinetic energy (as Leibniz's later alteration of the idea showed more clearly), a concept of minimal use in mechanics until it is yoked with another by which changes in kinetic energy can be measured. Although he formulated a conception of motion that made changes in motion the objects of explanation, Descartes' quantitative dynamical concept focused, not on changes, but on the steady motion which, by his own words, requires no explanation. As such, the idea of the force of a body in motion became more of an obstacle to a quantitative mechanics than an aid.

The application of the concept was complicated by a distinction, on which Descartes insisted, between the force of a body's motion and its determination. It is necessary to note, he said, 'that motion is different from the determination bodies have to move in one direction rather than another . . . and that properly speaking force is needed only to move bodies and not to determine the direction in which they ought to move; for that determination does not depend so much on the force of the mover as on the position both of the mover and of the other surrounding bodies.'[22] The velocity of a given body's motion is distinct from the direction of its motion. Even though force that is not determined [*vis sine determinatione*] cannot exist, he said, 'nevertheless the

DESCARTES AND THE MECHANICAL PHILOSOPHY

same determination can be joined with a larger or smaller force, and the same force can remain although the determination is changed in any way whatever.'[23]

Descartes' derivation of the laws of reflection and refraction embodied one of the most important applications of the distinction that he made. Although he held that light is an action transmitted instantaneously through a medium, he contended that an action, as a tendency to motion, obeys the same laws as motion itself; and without further ado, he set out to derive the laws of reflection and refraction from the analogy of a tennis ball. Let a ball be struck so that it moves from A to B, where it hits the earth (see Fig. 3). To avoid extraneous difficulties,

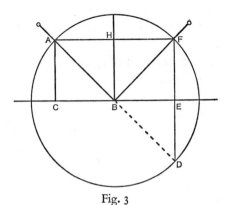

Fig. 3

he assumed that the earth was perfectly hard and smooth, and he abstracted from considerations of weight, size and figure. The determination of the motion can be separated into components parallel to the earth (AH) and perpendicular to it (AC). The encounter with the earth does nothing to obstruct the parallel component; the perpendicular one, obviously, cannot continue. To find where the ball will go, describe a circle around the centre B with the radius AB. Since the force of the ball's motion has not been altered in any way, in a time equal to that it took to cover AB the ball will travel an equal distance and find itself somewhere on the circumference of the circle. The exact point on the circumference is established by the undisturbed component of determination parallel to the surface. Draw a perpendicular FE as far removed from B as AC is. In the time that the ball moves to the circumference it will also be carried parallel to the surface as far as the line FE, and hence it will arrive at F where the line cuts the circle. The equality of the angles of incidence and reflection is a simple consequence of the geometry involved.

Essentially the same construction served for the law of refraction with the earth replaced by a piece of cloth through which the ball must break. The piece of cloth represents the refracting surface, and Descartes derived the change in determination from the different velocities assumed for light in the two media. Using the example of a ball rebounding from a hard and a soft surface, he argued that light travels more easily through a dense medium than through a rare one; and in the analogy of the tennis ball, 'more easily' was translated into 'more swiftly'. In such a case, when light passes from one medium into another through which it travels more easily, he imagined the ball to be struck again at the point B, increasing the force of its motion by the required amount. Not only was the utilisation of different velocities irreconcilable with his conception of light, which held its velocity to be infinite, and not only was there no conceivable mechanism in Descartes' nature to correspond to the second stroke of the racquet as the ball enters the denser medium, but the further assumption that all the change takes place at the surface, so that once past the surface the ball travels inertially through a resisting medium, was incompatible with his laws of motion. Be that as it may, assume that the ball is struck a second time as it passes through the cloth and its force increased so that in two seconds it traverses a space equal to AB, which it traversed, above the surface, in three seconds. As in reflection, the cloth does nothing to alter the component of determination parallel to it. Hence in two seconds that component will carry it to a line FE removed from B a distance two-thirds the distance of AC (see Fig. 4). Draw the circle with radius AB, and the light will arrive at I, where the line FE cuts the circle. A similar construction serves for refraction away from the perpendicular, when the force is diminished; and when the perpendicular FE falls outside the circle, refraction is impossible and the surface reflects.[24] The geometry of the construction entails the sine law of refraction, which Descartes was the first to publish.

The primary consequence of the distinction of force and determination was to remove changes of direction from the realm of dynamics. As he said in the case of reflection, it is not impossible that the determination of the ball be altered 'without there being in consequence any change in the force of its motion . . .'[25] That is to say, a change of determination is a change in which no action is exercised. He insisted indeed that

DESCARTES AND THE MECHANICAL PHILOSOPHY

the ball never stops. If it stops, if the force of its motion is interrupted, no cause which will set it in motion again can be found. The same force of motion then continues uninterrupted; neither the body in motion nor any other experiences a change in its state of motion or rest; and a simple reflection like that of the ball entails no dynamic action whatever. It is not true, he argued, that the efficient cause of a motion is the efficient cause of its determination. 'For example, I throw a ball against a wall; the wall determines the ball to return toward me, but it is not the cause of its motion.'[26] Descartes did not understand such a reflection to be elastic in any way. The presence of the wall, or the earth in his demonstration of the law of reflection, causes the reflection in the

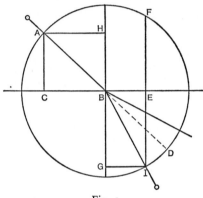

Fig. 4

sense that its presence is required for the reflection to take place, but the wall does not act upon the ball in any way to change the force of its motion.

From the point of view of mechanics alone, it is impossible to understand why Descartes embraced the distinction. For his dynamics, the distinction was disastrous. His first two laws of motion, taken together, asserted the vectorial nature of inertial motion. A body will continue to move in a straight line until something external acts upon it. As far as motion is concerned, a change of direction is as significant as a change of speed; and once the idea of rectilinear inertia is assumed, a workable dynamics will require that both changes be reduced to the same quantitative measurement of action. Descartes' distinction denied that the two changes have anything in common, and left changes of direction in the anomalous status of changes that are not changes, changes that involve no act. Only when we look beyond mechanics to his philosophy of nature, can we begin to comprehend the purpose of the distinction. In Descartes' kinetic mechanical philosophy, the ultimate causal agent was the moving particle, the power of which to act was summarised with helpful ambiguity in the 'force' of its motion. Velocity measured the force of a given body's motion, of course; and by eliminating from dynamics all changes except changes of velocity, Descartes protected a body's force from fruitless wastage and adulteration. In impact with another body it cannot move, a body's force alone continues to cause its motion. No extraneous forces connected with elasticity enter the picture to complicate and obscure it. The force of the body remains intact and unaltered, conserved entire for the impact in which it can act. Fire is a collection of rapidly moving particles, for example. Descartes noted explicitly that he had not said in what direction they move, 'for if you consider that the power to move and that which determines in what direction the movement ought to be made are two things entirely different and able to be separated from one another,' then you will realise that each particle in the flame will move in the direction 'which is made less difficult for it by the disposition of the bodies which surround it...'[27] In a flame, then, the particles move in all directions, but the flame generally tends upward because the bodies around it offer more resistance in every other direction. As we shall see, Descartes' analysis of impact concluded that a smaller body can never move a larger one at rest. As a result, the distinction of force and determination being given, the particles of subtle matter move almost always with the greatest velocity, as his philosophy of nature demanded, able to receive force from larger bodies, but suffering no reciprocal loss of force. Above all, perhaps, changes of direction are necessarily instantaneous since the force of the body's motion continues to operate without interruption. As instantaneous, they furnish the prototype of the Cartesian conception even of dynamic change, in which the conservation of the quantity of motion from one instant to the next is in fact derived from the immutability of God, not from the act of one body on another. In the distinction of force and determination, Descartes' mechanical philosophy dictated the substance of his dynamics. In that sense, the distinction summarised the whole of Descartes' mechanics. The demands of his mechanical philosophy of nature directed his dynamics into a channel we scarcely expect once the principle of inertia has been pronounced.

*

Beyond its distinction from determination, Descartes' concept of the force of a body involved another difficulty connected with the role of matter. In this case, the difficulty was not of his own making, being a problem at the very heart of the seventeenth century's conception of matter, and one we have encountered already in Galileo. It was not for that less troublesome to Descartes' mechanics. In common with the century as a whole, he held that matter is wholly inert, lacking in any internal force and dominated in its motion by external actions upon it. In Galileo's classic phrase, matter is indifferent to motion or rest. To Henry More, Descartes denied that motion causes violence to matter. The word 'violence' is relative to the human will which suffers violence when it undergoes something repugnant to it. 'In nature, however, nothing is violent, but to bodies it is equally natural to impell and push each other when they touch and to be at rest.'[28] It follows that bodies have no force to resist motion. The idea that matter contains such resistance is a prejudice, 'founded on our preoccupation with our senses, which derives from the fact that having tried since our infancy to move only bodies that are hard and heavy, and having in that always encountered difficulty, we have been persuaded since then that the difficulty proceeds from matter and is consequently common to all bodies; that has been easier for us to suppose than to realise that it was only the weight in bodies we tried to move which prevented us from lifting them and their hardness with the unevenness of their parts which prevented us from pulling them, and hence it does not follow that the same thing must happen with bodies which have neither hardness nor weight.'[29] One consequence of the indifference of matter to motion was Descartes' conviction that bodies start to move with a finite velocity.[30] Motion and rest (or different degrees of motion) are discontinuous, and a body starting to move does not pass through all the degrees of motion as Galileo maintained.

Over against the indifference and inertness of matter was the fact, acknowledged by Descartes and implicit in his idea of the force of motion, that 'size is always opposed to the velocity of motion.'[31] The word 'size' in itself offered further difficulties because of Descartes' equation of extension with matter, a difficulty sometimes but not consistently solved by the suggestion that the size of a body is measured by the quantity of the third element or tangible matter, it contains, so that the intangible matter filling its pores, the first and second forms into which the extended matter common to Descartes' entire universe is divided by mechanical necessity, is excluded from consideration. Furthermore, he asserted frequently that the quantity of surface on which other bodies can impinge modifies the force of a body to continue its motion. In the light of both problems, it is not too much to say that in Descartes' philosophy the size of a body was a concept so obscure as to render a quantitative mechanics impossible. Even if one assumes that the size of a body is a sharply defined idea, its relation to the indifference of matter to motion remains an acute problem. If matter is wholly inert, how can size oppose velocity?

One solution of the problem that Descartes proposed was to define a concept of 'natural inertia' in purely kinematic terms. I have been applying the word 'inertia' to Descartes' conception of motion since it is the term current in the twentieth century. Until the time of Newton, however, the word was not used in this way, and Descartes never spoke of 'inertial motion'. By 'inertia' he referred, as Kepler had done, to a supposed resistance on the part of matter to motion itself. The problem was to reconcile such an inertia to the inertness of matter.

> And since, if two unequal bodies both receive the same amount of motion, that equal quantity of motion does not give as much speed to the larger as it gives to the smaller, we can say in this sense that the more matter a body contains the more *natural inertia* it has; to which we can add that a body which is large can transfer its motion to other bodies more easily than a small one, and that it can be moved by them less easily. So that there is a sort of *inertia* which depends on the quantity of matter, and another which depends on the extension of its surface.[32]

That is, inertia or resistance to motion is an illusion. The reality behind it is the role that quantity of matter plays in quantity of motion, such that the larger the body the smaller the velocity for a given quantity of motion. This solution of the problem ignored dynamic considerations and confined attention to the kinematic concept of quantity of motion.

Descartes did not succeed always in confining his attention to kinematics, however. Especially in the discussion of impact, dynamic considerations slipped in unawares and with them a resistance to motion that cannot be reconciled to the inertness of matter. His refusal to admit a force in impact deriving from elasticity led Descartes to a doctrine of inelastic impact for most cases. Whenever a body B strikes a body C at rest and sets it in motion, the two bodies move together with a common velocity such that the quantity of motion of the two equals the quantity of motion that B had before impact. If B is unable to move C, it rebounds, its force of motion unchanged and its determination reversed, according to the analysis we have seen. The relative size of the two bodies determines whether *B* can move *C*; if *C* is larger by any amount whatever, *B* will not be able to move it. At this point, the natural thought occurred that *C* at rest remains a constant factor, whereas the force of *B* can increase indefinitely as its velocity increases. Surely at some point it will be able to move *C*. The more he examined the problem, however, the more Descartes was convinced that *B* can never move *C*, whatever its velocity.

> For in so far as *B* cannot push *C* without making it go as swiftly as *B* itself would go afterwards, it is certain that *C* must resist so much the more as *B* comes toward it more swiftly, and that its resistance must prevail over the action of *B* because it is greater than *B*. Thus, for example, if *C* is twice as large as *B* and *B* has three degrees of motion, it can only push *C*, which is at rest, if it transfers two degrees of the motion to it, that is, one for each of its halves, and if it retains only a third degree for itself, because it is no larger than each of the halves of *C* and after the impact it cannot move more swiftly than they. In the same way, if *B* has thirty degrees of speed [*sic*], it will have to give twenty of them to *C*; if it has three hundred, it will have to give two hundred, and thus always twice what it will retain for itself. But since *C* is at rest, it resists the reception of twenty degrees ten times more than two, and two hundred a hundred times more; so that the more speed *B* has the more resistance it finds in *C*. And because each half of *C* has as much force to remain at rest as *B* has to push it, and both halves resist *B* at the same time, it is evident that they must prevail in forcing *B* to rebound. With the result that whatever the speed with which *B* strikes *C*, which is at rest and larger, it can never have the force to move it.[33]

As with nearly every passage he published, Descartes found himself discussing and justifying the assertion in his correspondence. To Clerselier he explained that

> it is a law of nature that a body which moves another must have more force to move it than the other has to resist. But this excess can only depend on its size; for that body which is without motion has as many degrees of resistance as the other, which moves it, has of speed. Of which the reason is that if it is moved by a body which moves twice as fast as another, it must receive twice as much motion from it; but it resists twice the motion twice as much.[34]

From the point of view of the science of mechanics that has come to prevail, there is much to be objected to in this analysis. By treating all impacts in which a body at rest is put in motion as inelastic impacts, he imposed unnecessary constraints on his conclusions. Since he treated the change of motion as instantaneous, the resistance he attributed to matter had to be a resistance to motion itself and not to change of motion. From the point of view of his own philosophy, the resistance to motion that he admitted could not be reconciled to inertness which was a consequence of the essential nature of matter. The defects of the analysis should not blind us, however, to its brilliance. It constituted the first attempt to bring the fact of experience, that more effort is required to give a large velocity to a body than a small one, within the domain of rational mechanics. If the idea of a resistance in matter was in conflict with the inertness of matter, mechanics completed itself in the end only by pursuing the analysis Descartes initiated and embracing two irreconcilable notions in one conception of mass.

*

Descartes' major contribution to the science of mechanics in general and the science of dynamics in particular was located in the realm of philosophic analysis – philosophic analysis of the mechanical conception of nature, philosophic analysis of the new conception of motion and divers of its implications. Unlike Galileo's contribution, it was not directed primarily toward the detailed examination of individual problems within mechanics. To three such questions, however, he did devote considerable attention, and in all three he influenced the science of mechanics that followed him. Typically, the contribution involved the clarification of problems and the resolution of ambiguities. As I have indicated, Descartes' philosophy of nature encouraged the choice of a model of force that proved to be fruitless. On the other hand, his analytic gifts might almost have been created to deal with the other two sets of conceptual difficulties that confronted mechanics.

One of the three questions he examined at length was the treatment of simple machines, in a letter to Constantijn Huygens in 1637, and in a number of other letters consequent to it. As far as the basic ratios between force and resistance (the accepted terms, which Descartes also tended to employ with the simple machines) were concerned, they had been established long before, and there was nothing for him to add. Descartes was well aware of the situation and regarded what he wrote on the subject as an explanation of the basic principle from which the ratios were derived. No one, he contended, had ever done it so well.[35]

The fundamental principle held that 'the effect must always be proportional to the action that is necessary to produce it . . .'[36] The use of the word 'action' was a promising advance. In this context, he defined it with precision; and had he employed it with consistency, the term could have added to the clarity of discussion in mechanics. Unfortunately, Descartes himself usually employed the word *'force'*, which was wholly lacking in precision. He assured Mersenne that by 'force' he did not mean anything like the power [*puissance*] which is called the strength [*force*] of a man when you say that one man has more strength [*force*] than another. When you say 'that less *force* must be employed for one effect than for another, that is not to say that less power [*puissance*] is needed, for even if there is more, it will not matter; but only that less *action* is needed. And in that piece [his short essay on simple machines] I did not consider the power [*puissance*] that is called the strength [*force*] of a man, but only the *action* that is called the *force* by which a weight can be raised, whether the *action* comes from a man or a spring or another weight, etc.'[37]

In the context of the simple machines, '*action*' or '*force*' meant simply the product of weight times its vertical displacement.

> The contrivance of all these machines is based on one principle alone, which is that the same *force* which can raise a weight, for example, of a hundred pounds to a height of two feet can also raise one of 200 pounds to the height of one foot or one of 400 to the height of a half foot, and so with others, provided it is applied to the weight.[38]

Although he understood the principle of virtual motion, and hence of virtual work,[39] Descartes' discussion of simple machines dealt mostly with real motions and real work as the example above illustrates. The practice led him into confusion in the case of the lever, which he imagined to turn through a full angle of 180°. The force, always applied perpendicularly to its end of the lever, travels through a semicircle, while the load on the other end, in travelling through a semicircle, is only raised a vertical distance equal to the diameter of its circle. Hence a factor of π showed up in Descartes' analysis of the lever.[40] As far as the problem he defined is concerned, his solution is entirely correct and demonstrates how fully he understood his basic principle, even though the problem and its solution do not happen to be very useful in understanding the lever. In the other simple machines similar extraneous considerations did not confuse the analysis.

In the consideration of simple machines, velocity and displacement had been used interchangeably, by Galileo, for example, whose work on the subject Descartes knew. In contrast, Descartes adamantly insisted that displacement alone can serve. Reference to velocity in this context is an error the more dangerous as it is more difficult to recognize, 'for it is not the difference of velocity which determines that one of these weights must be double the other, *but the difference of displacement,*

> as would appear, for example, from the fact that to raise the weight F by hand to G it is not necessary, if you wish to raise it twice as swiftly, to employ a force which is exactly double that otherwise necessary; but it is necessary to employ a force that is more or less than twice as large according to the varying proportion that velocity can have with the factors that resist it; *whereas to raise it with the same speed twice as high, that is, to H, a force that is exactly double is necessary,* I say *that is exactly double, just as one and one are exactly two: for a certain quantity of that force must be employed to raise the weight from F to G and then as much again of the same force to raise it from G to H.*[41]

He did not, he added, deny that the same ratio holds for velocities and displacements in the simple machines. The ratio of velocities is a fact. Only the ratio of displacements explains why the proportion of force and resistance varies as it does. Before you can discuss velocity intelligently, you must understand weight, and to understand weight you must understand the whole system of nature. Hence he had chosen to discuss the simple machines in terms of a principle so simple as to be evident by itself. Mersenne contended that if a force can raise a weight a certain distance in one moment of time, then twice the force can raise the weight twice as far in one moment. 'I do not see the reason for it in any way,' Descartes replied.

> And I believe that you can easily demonstrate the contrary by experiment if, having a balance in equilibrium, you place in it the least weight that can make it turn; for then it will turn very slowly, whereas if you put twice as much weight in it, it will turn more than twice as fast. And, in contrast, if you take a fan in your hand, without your having to employ any force except that necessary to support it, you can raise and lower it with the same speed that it would fall by itself in the air if you let it go; but to raise and lower it twice as fast, you have to employ some force, which will be more than double the other since that one was nil.[42]

In his insistence on displacement, Descartes seized on the crucial factor that could be separated from the proportions of the lever in use as velocity could not. The product of size times velocity, derived from the law of the lever, bedevilled mechanical discussions throughout the century as scientists attempted to apply to real motions a quantity valid for virtual velocities. Descartes also employed the concept extensively as the force of a body's motion, and through that concept the statics of simple machines continued to dominate dynamics. In the case of the simple machines, however, he saw that the product of force times displacement was a quantity not confined to the virtual motions of a lever, one therefore that could be used as an independent principle. Later in the century, Leibniz was to use it – and with it Descartes' other principle, that an equal action must produce an equal effect – to demonstrate at Descartes' expense that the force of a body's motion cannot be proportional to its velocity.

It is essential to note that in this case again the careful analysis of vertical motion at the surface of the earth produced a major step forward in conceptual clarity. Like Galileo's description of uniformly accelerated motion, Descartes' definition of work (to employ the word that is now in use) was nearly confined to such conditions. It was measured precisely by the product of a weight times the vertical distance through which it is lifted. To a certain extent the simple machines allowed Descartes to generalise. Instead of lifting a weight vertically, a smaller force could pull it up an inclined plane, and the work performed would be measured by the product of that force times the distance it moved (which was equal of course to the product of the weight times its vertical displacement). As Galileo had done, Descartes employed a concept of static force that was measured by weight.[43] Hence a weight on a line over a pulley could replace any force applied to a simple machine, and equally a weight on a line could measure any force exerted uniformly, such as that of a man, an animal, or a spring. The success of generalisation depended on the extent to which a weight moving vertically could be imagined to replace another force, and the weight moving vertically was seen less as an example of a force than as its measure. The analysis focused on the vertical motion of heavy bodies and confined itself to forces that imagination could readily replace by such.

Descartes' analysis of the simple machines led to an important clarification of the concept of force. As I have already noted, he usually

FORCE IN NEWTON'S PHYSICS

employed the word '*force*' to express his concept of '*action*', that is, our concept of work. Inevitably the ambiguity of the word '*force*' led to misunderstanding, and nowhere does the analytic power of Descartes' mind appear more vividly than in his response to the questions raised.

> Above all it must be realized that I spoke of the *force* which is employed to raise a weight to some height, *which force always has two dimensions,* and I did not speak of the force which is employed to support it at each point, which always has only one dimension, so that these two *forces* differ from each other *as much as a surface differs from a line.* For the same *force* which a nail needs to support 100 pounds one moment of time suffices also, provided it does *not* decrease, to support it throughout a year. *But the same single quantity of that force which is employed to raise that weight to the height of one foot does not suffice to raise it to the height of two feet, and it is not more clear that two and two are four than it is clear that twice the force must be used.*[44]

Suppose someone objects to his treatment of the pulley that the nail to which one end of the line is tied supports half the weight, and that therefore only half the force must be applied to the other end to raise the load hanging from the pulley. In one sense, the objection is correct, and yet it is not the weight supported by the nail that allows the force to be cut in half. If the line, for example, is passed over another pulley hanging from the ceiling, the nail from which it hangs also supports weight, but the force that must be applied to the end of the line does not decrease for that.

> Hence, in order not to be deceived by the fact that the nail *A* supports half the weight *B*, nothing should be concluded except that by this application [of the pulley] one of the dimensions of the *force* which must be at *C* [the free end of the line] to raise that weight diminishes by half and consequently the other is doubled. So that if the line *FG* represents the *force* necessary to support the weight *B* at some point without the aid of any machine, and the rectangle *GH* that which is necessary to raise it to the height of one foot, the support of the nail *A* decreases the dimension which is represented by the line *FG* by one half and the doubling of the cord *ABC* makes the other dimension, which is represented by the line *FH*, double; and thus the *force* which must be at *C* to raise the weight *B* to the height of one foot is represented by the rectangle *IK* [See Fig. 5]. And as it is known in Geometry that one line added to or taken from a surface neither increases nor decreases it at all, thus it must be noted here that the *force* with which the nail *A* supports

the weight *B*, having but one dimension alone, cannot bring it about that the *force* at *C*, considered under its two dimensions, must be less to raise the weight in this manner than to raise it without the pulley.⁴⁵

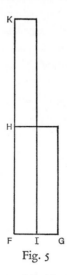

Fig. 5

In his original paper, he continued, he had not said simply that the force which can raise fifty pounds four feet can raise two hundred pounds one foot, but he had said it can do it 'provided it is applied to the weight'.

> Now the fact is that it is not possible to apply it to the weight except by means of some machine or other device which brings it about that the weight rises only one foot while the *force* acts through the whole length of four feet, and thus the device transforms the rectangle, by which the *force* needed to raise the weight of 200 pounds to the height of one foot is represented, into another which is equal and like that which represents the *force* needed to raise a weight of 50 pounds to the height of 4 feet.⁴⁶

As I suggested before, the consistent use of the word 'action' to designate what he analysed as two dimensional force would have contributed further to the clarity of discourse. Nevertheless, the passages above do not cease to be remarkable for that. Before anyone else, Descartes realised what ambiguities lay hidden in the word 'force'. Had his analysis been heeded and extended, one of the major conceptual difficulties that the science of dynamics had to surmount would have dissolved away into thin air. As it was, the ambiguous meaning of 'force', ironically dependent primarily on the example of the simple machines, continued to plague mechanics into the nineteenth century.

The mechanics of circular motion constituted a second problem on which Descartes made a major contribution to the science of dynamics. According to his second law of motion, every body that moves tends to move in a straight line, and if in fact it moves in a curve, the presence of other bodies must have obstructed its rectilinear course. Elsewhere, as I have discussed, Descartes distinguished the force of a body's motion from its determination and argued that no action whatever is required to change the direction of a body's motion. Had he applied his own distinction, the question of a dynamics of circular motion could not have arisen. What one part of his system denied, however, another demanded, and the dynamics of circular motion, the tendencies or forces set up when bodies are constrained to move in circles, played a leading role in his explanation of some of the basic phenomena in nature. On the one hand, the plenum made circular motion necessary. In a plenum, one body can move only if an entire circuit of matter, like the rim of a wheel, moves with it. Every motion, then, implies a closed circuit, and circular motion, in a loose, non-geometrical sense of the word 'circular', is necessary. On the other hand, circular motion is unnatural. By the second law of nature, a body tends to move in a straight line. To move in a circle, it must be externally constrained; in resisting the constraint, it strives to move away from the centre. Whirl a stone in a sling, and you can feel it pull on the string as it strives to move away from its centre of revolution. The effort or *conatus* away from the centre was fundamental both to Descartes' conception of light and to his explanation of the phenomenon of weight.

Although Descartes was convinced, in this context, that curvilinear motion entails a dynamic process, he did not find the process easy to analyse. To get a grip on the problem, he imagined the rectilinear motion along the tangent, the instantaneous tendency of every body in circular motion, to be composed of two other tendencies, one along the circle and the other radially away from the centre. We frequently find that divers causes acting on the same body prevent each other's effects, he explained, so that it is possible to say that a body tends in various directions at the same time. When a stone in a sling moving in the circle AB is at A, it tends toward C, along the tangent, if we consider its motion alone (see Fig. 6). If we consider its motion as restrained by the sling, it tends along the circle toward B. And finally, if we exclude from our consideration that part of the motion the effect of which is not impeded, the stone makes an effort to move radially outward toward E. Every effort, Descartes asserted, presupposes resistance. The effort of the stone away from the centre represents that part of its tendency along the tangent which the sling resists.[47]

In the *Principles of Philosophy*, he repeated the basic elements of the analysis above, which he composed for *Le monde*, the first exposition of his mechanical system of nature, written but not published in the early 1630's, and added two new examples by way of illustration. In one of

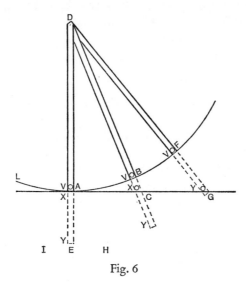

Fig. 6

them, a ruler turns about one end while an ant crawls, or strives to crawl, outward along the ruler.[48] The second example placed a ball in a hollow tube; as the tube rotates around one end, the ball moves lengthwise in it away from the centre of rotation.[49] Neither example was entirely apt since in both of them motion away from the centre takes place. It takes place, however, because there is an effort or tendency to move in that direction, and the whole purpose of his analysis was to demonstrate that such a tendency necessarily arises whenever a body is constrained to move in a curvilinear path.

Beyond the demonstration that such a tendency exists Descartes did not go. The tendency itself was all his system seemed to require; he did not see that a quantitative treatment of it would contribute to his aims. Perhaps he was also unsure of how to proceed mathematically toward a quantitative treatment. Nevertheless certain aspects of his analysis were to be suggestive to others. His diagrams showed two successive positions of the sling beyond the point of contact with the tangent, manifestly revealing that the distance between the circle and the tangent, the length of the radius extended beyond the circle to the tangent, increases at a rate that is not uniform. His discussion of the ant's motion, in the example introduced in the *Principles*, seemed to refer to that geometrical fact.

> I do not doubt that the motion of this ant must be very slow at the beginning and that its effort cannot seem very great if it is measured solely by that first motion; but it must also be said that it is not entirely nil, and since it increases in proportion to the effect it produces, the speed it causes quickly becomes rather large.[50]

With the ball in the tube he was more explicit yet.

> At the first moment that the tube is moved around the center E, this ball will proceed only slowly toward Y [the outside end of the tube]; but it will proceed a little more quickly in the second because in addition to the force communicated to it in the first instant, which it will have retained, it will acquire another new one from the new effort it will make to move away from the center since this effort continues as long as the circular motion lasts and renews itself at each moment.[51]

The tendency away from the centre apparently shared the characteristics of the tendency of heavy bodies to fall to the earth. The language Descartes used to describe the motion caused by the one was virtually identical to that for the other. Christiaan Huygens was to be impressed by the similarity, which was heightened by Descartes' derivation of the tendency of bodies called heavy to move toward the centre from the tendency of other bodies to recede. Huygens realised as well that the geometry of the circle, implied in Descartes' diagrams, could be employed further to demonstrate the similarity of the two tendencies and to reduce the tendency to recede to quantitative treatment.

From the point of view of the later science of mechanics, one of the peculiarities of Descartes' analysis of circular motion was his sole attention to the tendency of a body to recede from the centre. According to his own laws of nature, a body continues to move in a rectilinear path, and is diverted from it only by the action of some other body. His analysis of a body moving in a circle always assumed the existence of an external constraint. Nevertheless, he focused his discussion, not on the external constraint or force necessary to divert a body from its inertial path, but on the internal effort or force exerted by the body thus constrained. What makes the issue more interesting is the fact that every significant investigation of circular motion in the generation after Descartes began with a consideration of the force away from the centre. Following Newton (who himself had started initially from the Cartesian stance), we say now that a centripetal force must be applied to constrain a body in circular motion. Investigators in the seventeenth century said that a body constrained in circular motion exerts a centrifugal force (a term coined by Huygens). Although the two are quantitatively identical, the idea of centrifugal force became an obstacle to the clarity of thought in dynamics and had finally to be rejected. Probably Descartes' example played a major role in determining the approach to circular motion, for his was the pioneering effort on which the others built. It established a pattern which itself became a constraint. More than a mere example was involved, however. His approach appeared so natural that no one thought at first to question it.

In Descartes' case, two factors that reinforced each other seem to have determined his approach. One was the mechanical philosophy with its attention on the moving particle of matter as the ultimate causal agent in nature. The quantitative science of dynamics we know is built on a conception of force which is measured by the change of motion it causes. The mechanical philosophy, in contrast, focused its attention on the force of a moving particle to act, that is, in the case of circular motion, the force by which a particle strives to move away from its centre of revolution. Transmitted through the vortex as a pressure, this tendency was seen as the physical reality of light. In displacing bodies with a lesser tendency away from the centre, it was held to cause heaviness.

Reinforcing the demands of the mechanical philosophy was the enduring hold that an established conception of circular motion exercised over the minds even of those who were engaged in replacing it. Descartes asserted that inertial motion is rectilinear and that a body must be constrained to move in a circle. In analysing the dynamics of circular motion, however, he employed a conceptual scheme which treated the tangential path as the resultant of a circular motion and a radial tendency away from the centre. When he referred to the circular component as that part of the body's tendency 'the effect of which is not impeded,'[52] he returned unconsciously virtually to embrace the idea of a natural circular motion. I say 'virtually to embrace', for the conception of circular motion implicit in the analysis was closer to that held by Galileo, a conception in which circular motion appeared as a state of equilibrium between whatever constraint exists to turn bodies into a circular path and the force bodies exert against it. Such a state of equilibrium was compatible with the eternal circularity as fundamental in its own way to Descartes' universe as it was to Galileo's.[53] Such a state of equilibrium appeared equally self-evident to the men who followed Descartes' lead in the treatment of circular motion. The motion that appeared natural in a finite cosmos had become an enigma in the infinite universe. A riddle to the seventeenth century, it transformed itself in their groping efforts to treat it into a riddle for us. A generation of talented men, cozened by the enduring fascination with the circle's apparent perfection, avoided a confrontation with the enigma and refused to recognise that their own principle of inertia demands that uniform circular motion be uniformly accelerated motion.

The third problem in mechanics to the solution of which Descartes contributed was impact. Since the impact of one body on another was the sole means of action acceptable to the mechanical philosophy, it was inevitable that Descartes should consider it in detail. The new statement that he gave the problem set it on the path to ultimate solution.

What made Descartes' statement of the problem significant for dynamics was his effort to remove it from dynamics and to convert it to a kinematic problem based on the principle that God in His immutability maintains eternally the same quantity of motion in the universe. We can find much to object to in the principle. The idea of a given quantity of motion in the universe has meaning only if there is an absolute frame of reference. If motion is relative, as Descartes insisted, the idea of an absolute quantity of motion is meaningless. Applied to the problem of impact, however, it offered a means of avoiding the extraordinarily complicated issue embodied in the phrase, 'the force of percussion'. By ignoring the dynamic question and proceeding on the principle that the quantity of motion of two bodies after impact must equal the quantity before, Descartes stated the problem in a new form susceptible of solution. In so far as he remained faithful to his insight, he helped to find the solution.

Of the seven individual cases of impact that he considered, his solutions of two alone appear correct to us, and of them, paradoxically, one under conditions of perfectly elastic and the other of perfectly inelastic impact. The apparent failure stemmed directly from his lack of fidelity to the kinematic approach and his readmission of dynamic considerations. In the statement of his third law of nature, the kinematic and dynamic elements clash in irreconcilable disharmony.

> The 3rd, that if a body which moves strikes another stronger than itself, it loses none of its motion, and if it strikes a weaker one which it can move, it loses as much of its motion as it gives the other.[54]

The final clause states the kinematic principle of conservation of the total quantity of motion, but what can the idea of a body stronger than another mean in the context of Cartesian philosophy? In practice it meant size primarily, and its introduction into the problem meant that the needs of the mechanical philosophy as a system of nature were to determine the details of the science of mechanics. Less than five years before he published the rules of impact, Descartes agreed that a smaller body in motion can set a larger body at rest in motion.[55] In the system of nature published in the *Principles*, considerable use was made of the proposition that a large body can easily transfer its motion to a small one but a small one normally cannot transfer its motion to a large one. With its attention concentrated always on the particle of matter as a causal agent, the mechanical philosophy spoke of its perseverance in its state of rest or motion. Its size became the measure of its force of perseverance, and beside the kinematic principle of the conservation of motion Descartes admitted the dynamic principle that a body with more force 'must always produce its effect and prevent that of the other.'[56]

The first case of impact presented two equal bodies with equal motions in contrary directions. Since neither is stronger than the other, the conservation of motion dictates that both reverse their determinations with their motions unchanged. In the second case, the speeds are equal and opposite as before, but B is larger than C. Since B is stronger, C cannot affect it. B can force C to reverse its determination, however, and after impact they move together with their original speeds. In case three, B and C are the same size, but B moves more swiftly than C. In this case velocity determines strength; and by the principle already given, the stronger, B, forces C to reverse its determination. Obviously something else must change as well since B now moves in the same direction as C, following it with a speed which is greater unless there is a transfer of motion. *B* must give *C* half its extra motion so that they move together after impact with the same speed. The other cases follow similar principles. When *C* at rest is larger than *B*, *B* rebounds with its motion intact. When *C* at rest is smaller, *B* transfers enough of its motion to *C* that they move together after impact with equal speed. When *C* at rest is exactly equal in size to *B*, there is no reason why one should achieve its effect more than the other, with the result that each achieves half its effect. *B* rebounds, not transferring half its motion to *C*, as it would do if it achieved its whole effect, and not retaining its motion intact, as it would do if *C* attained its whole effect, but transferring one quarter of its motion to *C*. Finally, Descartes considered *B* and *C* to be moving in the same direction with *B* overtaking *C*. Strength in this case is measured by the product of size times velocity, and when *B* is the stronger, the amount of motion it transfers is determined by the conservation of motion.[57]

Descartes concluded the discussion with the assertion that his demonstration of the rules of impact was so certain 'that even though experience should seem to teach us the contrary, we would nevertheless be obligated to place more faith in our reason than in our senses.'[58] Forthwith, he proceeded to show that in a fluid everything happens otherwise. Whereas a smaller body would not have the force to move a larger one at rest in a vacuum, any the least force can move a body suspended in equilibrium in a fluid.[59] In Descartes' universe, of course, bodies are always suspended in fluids, and vacuums do not exist. Thus reason restored what reason had denied, and small bodies were allowed to move larger ones after all. We are left wondering what the entire exercise was about. If we are wondering, the seventeenth century was not. Everyone agreed that Descartes' rules were faulty. Those who corrected them, however, did so by applying his principles more rigorously than he had done, thus avoiding the intractable force of percussion.

*

Wherever elements of a science of mechanics appeared in Descartes, their shape had been moulded to the demands of the mechanical philosophy of nature. At every point, the tentative effort toward a quantitative mechanics touched the framework of the system, requiring that an adjustment be made between them. By no means did the adjustment always work to the disadvantage of the science of mechanics. As I have argued, it operated to purge the conception of motion of inconsistent

DESCARTES AND THE MECHANICAL PHILOSOPHY

features, reducing all motions and changes of motion to a common pattern. But the demands of the system of nature also led the science of mechanics on occasion in directions it would not likely have taken by itself. The mechanical philosophy exercised its influence less in the treatment of particular problems or phenomena than in shaping the questions that the science of mechanics asked and in determining the conceptual language that it employed. Descartes was not the sole architect of the mechanical philosophy, and its role in mechanics was not limited to his case. Nevertheless, his voice exerted the greatest influence in spreading the mechanical philosophy, and its role in his mechanics was typical of much that followed.

Basic to the mechanical philosophy, with its radical alteration of the accepted picture of nature, was the denial that 'the ideas which are in me are similar or conformable to the things which are outside me.'[60] After he had doubted them away, Descartes demonstrated to his own satisfaction that physical bodies do in fact exist. 'However, they are perhaps not exactly what we perceive by the senses, since this comprehension by the senses is in many instances very obscure and confused.'[61] Since his epistemology rested on the principle that God, who does not deceive, would not have created us with faculties which deceive, Descartes was forced to explain away the deficiencies of the senses. Sense perception was given us by God for our preservation; by it we are told what is beneficial and warned of what is harmful. We misuse the instrument when we think that it teaches us the nature of the external world.

> ... I see in this, as in other similar things, that I have been in the habit of perverting the order of nature, because these perceptions of sense having been placed within me by nature merely for the purpose of signifying to my mind what things are beneficial or hurtful to the composite whole of which it forms a part, and being up to that point sufficiently clear and distinct, I yet avail myself of them as though they were absolute rules by which I might immediately determine the essence of the bodies which are outside of me, as to which, in fact, they can teach me nothing but what is most obscure and confused.[62]

In Descartes' philosophy, the conviction that physical reality is utterly different from the world our senses picture applied primarily to the question of qualities. We have sensations of colour, for example, but our sensations in no wise prove that colours exist in bodies.[63] The mechanical philosophy could explain how particles of matter in motion cause sensations of colour, and in explaining the sensations of colour, it explained away its reality. If the reality of colour did not concern the science of mechanics, the reality of weight did, and equally the reality of various apparent attractions and repulsions, such as magnetic action. The prevailing philosophy of nature referred such attractions to divers occult qualities, qualities which no sense perceives directly as the eye perceives colour, but qualities which manifest themselves by the phenomena of attraction they cause. The word 'occult' was a word of anathema to mechanical philosophers, and their denial was a matter at once of conviction and passion.

Making common cause with the assertion that physical reality is wholly different from the world of appearance was a second belief, held with equal conviction, that nature contains nothing mysterious and beyond human knowledge. Already in Descartes' early *Rules for the Direction of the Mind*, composed about 1628, the precursor, though in no sense the first draught, of the *Discourse on Method*, the discussion of magnetism expressed an attitude that remained unchanged throughout his life. When people consider magnetism, he said, they expect complexity, and they start by dismissing from their minds everything that is known and seizing on what is most difficult, in the vague hope that they will find something new. His own approach was otherwise. If the cause of magnetism lay wholly outside the realm of our experience, we should need to be equipped with a new sense in order to grasp it. In fact, we know all that man can know about it when we demonstrate how the entities and processes with which we are familiar can produce the known magnetic phenomena.[64] Descartes did not believe that entities beyond the power of the human mind to grasp do exist. Deliberately, he had purged his universe of the occult. Apparent attractions were to be explained away by particles of matter in motion, just as colours and other qualities had been. In the *Principles of Philosophy*, he did just that and remarked that

> no qualities are known which are so occult and no effects of sympathy and antipathy so marvellous and strange, and finally nothing else in nature so rare (provided it proceeds entirely from purely material causes lacking in thought or free will) for which the reason cannot be given by means of these same principles.[65]

For the science of mechanics, the exclusion of the occult meant the rejection of the concept of attraction. Descartes' reaction to Roberval's

Aristarchus, which employed two distinct kinds of attraction in explaining the solar system, established a pattern for similar reactions to similar theories through the rest of the century. To suppose, as Roberval supposed, that all particles of matter in the universe attract each other, and that in addition parts of the earth attract each other with a separate property, he remarked acidly, we must suppose as well, not only that the particles of matter are animated by at least two souls, 'but also that those souls are intelligent and indeed divine, since they are able to know, without an intermediary, what is happening in places far removed from them and even to exercise their forces there.'[66]

In place of notions that were unpalatable, Descartes substituted invisible mechanisms, justified by the argument he applied to the magnet, that the causes of all phenomena must be similar in kind. A great deal of mechanistic discussion appeared in such passages, but causal explanation by imaginary mechanisms is not to be confused with the science of mechanics. Conscious as he was of the new concept of nature which he proposed and of its assertion that physical reality differs widely from appearance, Descartes' primary purpose was to show that every phenomenon can be produced mechanically. Explanation and causation – these were the foremost concerns, and they had little to do with the exact description of forces and motions with which the science of mechanics occupies itself. Quite the contrary, his discussions of phenomena are filled instead with micro-mechanisms justified by nothing more than their apparent plausability as mechanical and therefore acceptable explanations of phenomena.

Descartes formed the habit, he said, 'of always employing the aid of my imagination when I think of corporeal things . . .'[67] Having asserted that physical reality differs from the world of experience, he proceeded to affirm implicitly that it corresponds exactly to one aspect of that world. The microscopic level consists of nothing but miniature reproductions of the mechanisms we know on the macroscopic level. The first requirement of a mechanical explanation was a picturable image. As it turned out, nothing was easier than imagining particles of whatever shape and motion seemed appropriate to the phenomena in question. The properties of water, especially the ease with which it is driven out of bodies it has penetrated by heat or wind, shows that its particles must be long and flexible like eels.[68] The sharp acid called spirit of salt can only be obtained by means of hot fire applied to pure salt or to salt mixed with other dry and fixed bodies. Obviously then, its particles are particles of salt but such as are unable to rise in an alembic and become volatile until the fire has beaten them against each other and made them flexible. In the process their shape is altered, the rod-like particles of salt being flattened out like the leaves of an iris. Their sharp taste derives from their shape, for on the tongue they cut like knives.[69] Descartes argued that *ad hoc* explanations of individual phenomena can easily be constructed, but that principles which prove adequate to explain all phenomena have to be true.[70] In fact, the rest of the century demonstrated that the game of imagining invisible mechanisms recognised no limits; there was no phenomenon whatever for which an imaginary mechanism could not be constructed. The mechanical philosophy as such contained no criteria to judge the validity of pretended phenomena, and in his eagerness to explain everything in mechanical terms, Descartes was not disposed to play the sceptic. The mechanism devised to explain attractions in general explained as well why the wounds of a murdered man bleed when the murderer approaches.[71] Lightning is caused by a certain exhalation in the air. Other exhalations are equally possible, and clouds can press them together to compose matter that looks like blood, milk, flesh, or stones, matter that can even engender small animals when it decays. Thus even reported prodigies, in which blood, milk, flesh, stones, and animals rain from the air, found their explanations in the mechanical philosophy.[72] Explanation in terms of entities deemed ultimate, that is, particles of matter in motion, was and continued to be the basic concern of the mechanical philosophy. Constantly, it diverted attention away from the detailed examination of forces and motions which was the business of the science of mechanics.

Equally, the mechanical philosophy sometimes functioned directly to obstruct the science of mechanics. The question of attraction summarises the problem. On the one hand, the mechanical philosophy hung the label 'occult' on the very word 'attraction' and refused to admit the concept into natural philosophy. On the other hand, the idea of attraction offered to mechanics the handiest vehicle to a consistent and simple conception of force, especially in the possible model of free fall. The dilemma that Galileo had faced, which made it impossible for him to see free fall as the paradigm case of dynamics, was generalised for the entire seventeenth century by the mechanical philosophy.

Although 'my entire physics is nothing but mechanics,' Descartes wrote to de Beaune, 'nevertheless I have never examined questions which depend on measurement of velocity in detail.'[73] Here was posed in vivid contrast the distinction between a mechanics conceived as causal mechanisms and a mechanics conceived as a quantitative science of motion. Descartes liked to stress as well the central role of mathematics in his philosophy, but he referred to the geometrization of matter inherent in the idea of extension and not to mathematical mechanics such as Galileo had pursued.[74] The most important step he took in the direction of a mathematical mechanics was his analysis of impact. Whatever its defects, it posed the issue in a new way which set mechanics on the path toward its solution. Descartes concluded the passage with a remark that, for obvious reasons, experience seems to refute his rules. The rules supposed that the bodies in question are perfectly hard and isolated from the action of others, two conditions that are impossible. We need then to know not only how two bodies act on each other when they meet, 'but beyond that we must consider how all the other bodies that surround them can increase or decrease their action.'[75] In view of the plenum, the statement was tantamount to abandoning the effort to treat impact in precise terms.

To Galileo's treatment of free fall he opposed much the same objection.

> All that he says about the velocity of bodies that fall in a vacuum, etc., is built without foundation; for he ought first to have determined what weight is; and if he knew the truth, he would know that in a vacuum it is nil.[76]

The inhibiting effect that the mechanical philosophy's obsession with causation had on the method of idealisation, which proved so fruitful in mechanics, could hardly be stated more clearly. If bodies accelerate uniformly, he acknowledged, Galileo's demonstrations are correct. Manifestly, however, bodies cannot accelerate uniformly, and Descartes was correspondingly uninterested in the problem. The resistance of the air, from which Galileo abstracted, was not the major problem. The trouble lay rather with the mechanical cause of weight. As a body gathers speed, the impulses of subtle matter that drive it down must decrease in force, and hence acceleration can at best approach uniformity and that only at the beginning of fall.[77] Analogous considerations led him to doubt that all bodies fall with the same acceleration, as Galileo maintained,[78] and that a uniform horizontal motion can compound without interference with a uniform acceleration downwards, as Galileo's demonstration of the parabolic trajectory assumed.[79] Beyond Galileo's treatment of free fall, Kepler's laws of planetary motion were the other leading example of mathematical kinematics, and Descartes' mechanical philosophy brushed them aside as summarily as it dismissed Galileo.[80]

Such was the dilemma of mechanics in the seventeenth century. Implicated as it was in the two leading currents of scientific thought, the mathematisation of nature on the one hand, the mechanical philosophy on the other, it could only be torn when the two currents diverged. Until a means of reconciliation between the conflicting motives was found, a consistent dynamics could not emerge.

NOTES

1. Descartes to Mersenne, 11 March 1640; *Oeuvres*, **3**, 39. Cf. his letter to More in August 1649 in which he denied that motion causes violence to matter and asserted that it is as natural to bodies as rest (ibid., **5**, 404).
2. *Le monde*; ibid., **11**, 38-9.
3. *Principles*, II, 24; *The Philosophical Works of Descartes*, trans. Elizabeth S. Haldane and G. R. T. Ross, 2 vols., (New York, 1955), **1**, 265-6, reprinted through permission of Dover Publications, Inc., New York. I have used the Haldane and Ross translation for all that it includes; everything cited from the *Oeuvres* is in my own translation.
4. *Principles*, II, 25; ibid., **1**, 226.
5. *Principles*, II, 27; *Oeuvres*, **9**, 78.
6. *Principles*, II, 37; ibid., **9**, 84.
7. Descartes to Huygens, 18 February 1643; ibid., **3**, 619.
8. *Principles*, II, 39; ibid., **9**, 85.
9. *Le monde*; ibid., **11**, 41.
10. Descartes to Mersenne, 15 September 1640 and 11 November 1640; ibid., **3**, 180, 234.
11. *Meditations*; *Philosophical Works*, **1**, 151.
12. *Principles*, II, 43; *Oeuvres*, **9**, 88.
13. *Le monde*; ibid., **11**, 54, 129. *Principles*, III, 88; IV, 95-106; ibid., **9**, 153, 253-8. Cf. *Principles*, III, 76: although the first element has both a rectilinear motion and a circular one, 'it employs the greatest part of its *agitation* in moving itself in all of the other ways that are necessary continually to change the shapes of its little particles...' Therefore 'its *force* is weaker, being thus divided...' (ibid., **9**, 145).
14. *Principles*, IV, 29, 60, 87, 108; ibid., **9**, 215, 234, 247, 258-9.
15. *Principles*, III, 121; ibid., **9**, 174. Cf. *Le monde*; ibid., **11**, 75, 85. Also his letter to Mersenne, 2 November 1646, in which he discussed the concept of a centre of agitation and its geometrical relation to the centre of gravity. Weight and *agita-*

tion 'are two powers which concur in making bodies fall in a straight line when they are free . . . (which appears from the fact that it is the *agitation* which makes them fall more swiftly at the end than at the beginning).' (Ibid., **4**, 546.)

16. Descartes to Mersenne, 11 June 1640; ibid., **3**, 74–5. Descartes' conclusion to the discussion suggests that he did not regard the product of force by time as a significant quantity in itself: 'But the proportion that must obtain between the force of the blow and its duration in order to make the action greater varies as the parts of the body struck need more or less time to come apart.' For other discussions of the role of time in the effectiveness of a blow see Descartes to Mersenne, 11 March 1640 and 4 March 1641; ibid., **3**, 41, 327.
17. Descartes to Morin, 13 July 1638; ibid., **2**, 204.
18. *Dioptrique*; ibid., **6**, 88.
19. Descartes to de Beaune, 20 February 1639; ibid., **2**, 518–19. Cf. his letter to Morin of 12 September 1638 in which again he maintained that the word '*action*' is more general than the word 'motion', including in its meaning an inclination to motion. If two blind men hold the two ends of a stick, for example, and push it equally or pull it equally so that the stick does not move, then from the immobility of the stick each man can conclude that the other is pushing or pulling with the same force as he himself. What he feels in the stick from its lack of motion can be called 'the various *actions* which are impressed on it by the various efforts of the other blind man.' (Ibid., **2**, 363.)
20. See Appendix B.
21. *Le monde*; ibid., **11**, 42.
22. Descartes to Mersenne, 11 June 1640; ibid., **3**, 75.
23. Descartes to Mersenne for P. Bourdin, 29 July 1640; ibid., **3**, 113. A. I. Sabra has recently denied that Descartes' concept of determination is to be understood simply as direction (*Theories of Light from Descartes to Newton*, (London, 1967), pp. 116–21). He argues that we should understand the distinction of force and determination as the distinction between a scalar and a vector quantity, so that determination, in his view, would be similar to our concept of momentum. As we shall see, this interpretation has much to recommend it if we attend solely to Descartes' *Dioptrique*. It appears incapable, however, of reconciliation with his statements elsewhere. To Mersenne he wrote, for example, that 'the force of motion and the direction in which it is made [*le costé vers lequel il se fait*] are things wholly distinct as I said in my Dioptrique . . .' (11 March 1640; *Oeuvres*, **3**, 37). Obviously, his words here, where he refers to the distinction without mentioning 'determination', limit the meaning of the term to direction. Individual passages aside, dynamic considerations make it imperative to understand 'determination' as a synonym for 'direction'. The point of the distinction, I shall argue, lay in the contention that a change in determination entails no dynamic act by another body. Such could hardly obtain if 'determination' is equivalent to our 'momentum'.
24. *Dioptrique*; ibid., **6**, 93–101. Both in the demonstration of the law of reflection and in that of the law of refraction, Descartes implicitly extended the meaning of determination so that it became virtually equivalent to momentum, as Sabra has argued. To say as much is not to concede that he intended the distinction in this sense or that this sense is compatible with its dynamic purpose. Scholarship has

been at work on Descartes too long for us to affect incredulity at the thought of an inconsistency in his writing.

25. *Dioptrique*; ibid., **6**, 94.
26. Descartes to Mersenne for Hobbes, 21 April 1641; ibid., **3**, 355.
27. *Le monde*; ibid., **11**, 8–9.
28. August 1649; ibid., **5**, 404.
29. Descartes to Morin, 13 July 1638; ibid., **2**, 212–13.
30. 'I see nothing better, to convince those who maintain that a body passes through all the degrees of velocity when it begins to move, than to propose for their consideration two bodies extremely hard, the one very large which moves by the force that was impressed on it when it was impelled, so that the cause which set it in motion no longer acts, like a cannon ball which flies through the air after having been driven by the powder; and the other very small which is suspended in the air in the line along which the larger one moves; and to ask them if they think that the larger body, for example the cannon ball *A*, being impelled with great violence toward *B*, must drive the body *B*, which does not hold onto anything that prevents it from moving (*qui ne tient à rien qui l'empeche de se mouvoir*), before it. For if they say that the cannon ball must stop next to *B* or rebound in the other direction, since I suppose that the two bodies are extremely hard, they will make themselves ridiculous because there is no reason why their hardness should prevent the larger from impelling the smaller; and if they assert that *A* must impel *B*, by the same token they must assert that from the first moment that it is impelled it moves with the same speed as *A*, and thus it does not pass through several degrees of velocity. For if they say that it must move very slowly at the first moment that it is impelled, it will be necessary that *A*, which will be joined to it, moves as slowly; for being both of them very hard and touching each other, the one that follows cannot go more slowly than the one ahead of it. But if that which follows goes very slowly for one moment alone, there will be no cause which makes it then regain its original speed since the powder in the cannon, which impelled it, is no longer acting; and it is the same for a body to be without motion or to move very slowly for a moment as for a very long time.' (Descartes to Mersenne, 17 November 1642; ibid., **3**, 592–3. Cf. Descartes to Mersenne, 7 December 1642; ibid., **3**, 601.) The reluctance to consider an elastic force and the consequent tendency to use the idea of perfectly hard bodies stood directly athwart a concept of force proportional to acceleration. Descartes was convinced that bodies start to fall with a finite velocity; in terms of dynamics, their motion downward was held to be caused by impacts analogous to the one described above, with the sizes of the bodies reversed.
31. *Le monde*; ibid., **11**, 51.
32. Descartes to de Beaune, 30 April 1639; ibid., **2**, 543–4. Cf. Descartes to Silhon (?) March or April 1648; ibid., **5**, 136.
33. Principles, II, 49; ibid., **9**, 90–1.
34. 17 February 1645; ibid., **4**, 184.
35. Descartes to Mersenne, 12 September 1638; ibid., **2**, 358.
36. Descartes to Mersenne, 13 July 1638; ibid., **2**, 228.
37. Descartes to Mersenne, 15 November 1638; ibid., **2**, 432–3. Cf. Descartes to

Mersenne, 13 July 1638: Along an inclined plane, the length of which is twice its vertical rise, the *'puissance'* that can support a given weight will perform the same *'action'* in moving it the length of the plane as one twice as great would perform in lifting it vertically to the same height (ibid., **2**, 232).
38. Descartes to Huygens, 5 October 1637; ibid., **1**, 435–6. Cf. the same letter, pp. 437, 439.
39. Descartes to Mersenne, 13 July 1638; ibid., **2**, 233–4.
40. Descartes to Huygens, 5 October 1637; ibid., **1**, 443–7.
41. Descartes to Mersenne, 12 September 1638; ibid., **2**, 354.
42. Descartes to Mersenne, 2 February 1643; ibid., **3**, 614.
43. He did obscure the concept by suggesting that lifting a weight with a simple machine (a pulley in this case) is different from holding it in equilibrium. 'It must also be noted that a little more force is always necessary to lift a weight than to support it; which is the reason why I have spoken of the one here separately from the other.' (Descartes to Huygens, 5 October 1637; ibid., **1**, 438.) This is an unfortunate lapse on the part of a man who insisted on the conceptions both of rectilinear inertial motion and of the relativity of motion. It is equally unfortunate in its implication of a resistance to motion as such, something that Descartes denied.
44. Descartes to Mersenne, 12 September 1638; ibid., **2**, 352–3.
45. Descartes to Mersenne, 12 September 1638; ibid., **2**, 356–7.
46. Descartes to Mersenne, 12 September 1638; ibid., **2**, 357. Further on in the same letter, he applied the analysis to the inclined plane. Let BA be an inclined plane twice as long as its vertical rise, AC, and let a weight D be pulled up its entire length. FG represents the one dimensional force and GH ($=BA$) the distance through which it acts. Now let NO, another one dimensional force raise another weight L through a vertical distance LM ($=OP$) equal to BA. Then the force represented by the rectangle NP will equal the force represented by the rectangle FH. But the motion of D up the plane BA can be considered as the sum of two other motions, one along the horizontal plane BC and the other up the vertical plane CA. The motion along BC requires no force whatever; hence the entire force ($=FH$) is employed in raising it vertically through the distance CA. By definition, $CA = \frac{1}{2}BA$. Since the force to raise it equals the force NP which raises L through the distance LM, then $D = 2 \times L$ (ibid., **2**, 358–60).
47. *Le monde*; ibid., **11**, 84–6.
48. Shortly before the *Principles* came out, Descartes hit on the figure of an ant on a ruler to illustrate the composition of motions (Descartes to Huygens, 18 February 1643; Descartes to Mersenne, 23 March 1643; ibid., **3**, 628–9, 640–1). In the case of circular motion, however, he was attempting to compound an actual motion (along the circle) with a tendency to motion (radially outward) to obtain as a resultant the rectilinear tangential path the body would have followed were it not confined to the circle.
49. *Principles*, III, 57–9; ibid., **9**, 131–3. John Herivel (*The Background to Newton's 'Principia'*, (Oxford, 1965), pp. 47, 54) has recently argued that Descartes' analysis of circular motion assumes two independent tendencies in the body so moving, the one along the tangent and the other, independent of the tangential tendency,

radially outward. I am convinced that Herivel has mistaken Descartes' perplexity in treating the dynamics of circular motion for a conscious intention to assert an independent, and in Herivel's terms, uncaused tendency. After all, the analysis was a pioneering effort, and we should not be surprised that he found it difficult to grasp the central factors and to hold them firmly at the focus of his attention. Moreover, the point of the analysis was to demonstrate that a tendency outward from the centre is *caused* when matter, which left to itself moves rectilinearly, is constrained to move in a circle. When you whirl a stone in a sling, he said in *Le monde*, it presses the sling and pulls on the cord, 'demonstrating clearly thereby that it always inclines to move in a straight line and that it only follows the circle by constraint' (ibid., **11**, 44). In the discussion of his second law, which asserted the rectilinear nature of motion, he considered a stone in a sling moving in the circle LAB. At the moment when it is at A, it is determined to move in the line AC which is tangent to the circle at A. 'But it cannot be imagined that it is determined to move circularly, because, although it has arrived at A from L following a curved line, we do not conceive that any part of the curvature is in the stone when it is at the point A; and we are assured of the same by experience because the stone moves rectilinearly toward C if it leaves the sling, and does not tend in any way to move toward B. Which makes us see manifestly that every body moved in a circle tends constantly away from the circle it describes. And we can even feel it with our hand while we make the stone turn in the sling; for it pulls and stretches the cord to move directly away from our hand. This consideration is of such importance, and will serve in so many places following, that I shall explain it at still greater length when there is time for it.' (Ibid., **9**, 86.) The last sentence indicates the role that the tendency away from the centre, caused by the circumstance that circular motion, though necessary in a plenum, is unnatural, plays in Descartes' system. We should read his detailed analysis in the light of the system as a whole and not be misled by the examples of the ant and the ball in the tube, which are merely attempts to illustrate the outward tendency. Even in the detailed analysis, he stated again that when the stone is at point A, it tends along the tangent toward C if we consider the force of its motion alone. 'Finally, if instead of considering all the force of its agitation, we take account solely of that part of it the effect of which is prevented by the sling and which we distinguish from the other part the effect of which is not thus prevented, we shall say that the stone, when it is at the point A, tends solely toward D [radially outward], or rather that it makes an effort solely to move away from the centre E along the straight line EAD.' (Ibid., **9**, 131.) But this is to repeat the assertion that effort implies resistance. That component of the tangential motion which follows along the circle is not obstructed; only that component which departs from the circle is, and hence constraint to follow a circle sets up against itself a radial effort away from the centre. If the constraint to follow the circle does not cause the tendency away from the centre, whence does it derive? Descartes' matter was deliberately stripped of internal tendencies apart from motion itself.

50. *Principles*, III, 59; ibid., **9**, 132.
51. *Principles*, III, 59; ibid., **9**, 132–3.
52. *Principles*, III, 57; ibid., **9**, 131.

53. Cf. Descartes to Mersenne, 13 July 1638 (ibid., **2**, 232–3) in which, in a passage that appears downright Jesuitical to the reader from the twentieth century, he argues that bodies weigh less the closer they are to the centre of the earth. The argument hinges on the fact that the earth is round and perpendiculars to its surface can therefore not be parallels. As one descends along an inclined plane, for example, the component of gravity along the plane continually decreases until it reaches zero at the point where the plane cuts a radius at right angles. The fact that Descartes should have injected this consideration into his analysis of simple machines illustrates how conscious he was of the circularity of the world. In regard to his treatment of circular motion, it is perhaps relevant that the tendency or force away from the centre corresponds to the force necessary to move a weight away from the centre with one of the simple machines. The correspondence could have played some role in his choice of the force away from the centre, rather than the force toward the centre, as the crucial dynamic element.
54. *Principles*, II, 40; ibid., **9**, 86.
55. Descartes to Mersenne, 25 December 1639, 28 October 1640; ibid., **2**, 627; **3**, 210–11.
56. *Principles*, II, 45; ibid., **9**, 89.
57. *Principles*, II, 44–52; ibid., **9**, 88–93.
58. *Principles*, II, 52; ibid., **9**, 93.
59. *Principles*, II, 61; ibid., **9**, 99–100.
60. *Meditations*; *Philosophical Works*, **1**, 160.
61. *Meditations*; ibid., **1**, 191. *Le monde* begins with the following statement: 'Since I propose to discuss light in this treatise, the first thing I want to make clear to you is that there can be a difference between the sensation of it that we have, that is the idea of it which is formed in our imagination when we use our eyes, and what it is in objects which produces this sensation in us, that is to say what it is in flame or in the sun to which we give the name light. For even though each one of us is generally persuaded that the ideas we have in our thought wholly resemble the objects from which they proceed, in no way do I see any reason which assures us that such is the case; but on the contrary I observe a number of experiences which ought to make us doubt it.' (*Oeuvres*, **11**, 3–4.)
62. *Meditations*; *Philosophical Works*, **1**, 194.
63. 'It is ... evident when we say that we perceive colours in objects, that it is the same as though we said that we perceive something in the objects of whose nature we were ignorant, but which yet caused a very clear and vivid sensation in us, and which is termed the sensation of colours. But there is a great deal of difference in our manner of judging, for, so long as we believe that there is something in objects of which we have no knowledge (that is in things, such as they are, from which sensation comes to us), so far are we from falling into error that, on the contrary, we rather provide against it, for we are less likely to judge rashly of a thing which we have been forewarned we do not know. But when we think we perceive a certain colour in objects although we have no real knowledge of what the name colour signifies, and we can find no intelligible resemblance between the colour which we suppose to exist in objects and what we are conscious of in our senses, yet, because we do not observe this, or remark in these objects certain

other qualities like magnitude, figure, number, etc., which clearly we know are or may be in objects, as our senses or understanding show us, it is easy to allow ourselves to fall into the error of holding that what we call colour in objects is something entirely resembling the colour we perceive, and then supposing that we have a clear perception of what we do not perceive at all.' (*Principles*, I, 70; ibid., **1**, 249.)

64. *Regulae*; ibid., **1**, 47–55.
65. *Principles*, IV, 187; *Oeuvres*, **9**, 309.
66. Descartes to Mersenne, 20 April 1646; ibid., **4**, 401.
67. *Meditations*; *Philosophical Works*, **1**, 186.
68. *Météores*; *Oeuvres*, **6**, 249.
69. *Météores*; ibid., **6**, 263–4.
70. Descartes to Mersenne, 28 October 1640; ibid., **3**, 212.
71. *Principles*, IV, 187; ibid., **9**, 309.
72. *Météores*; ibid., **6**, 321.
73. 30 April 1639; ibid., **2**, 542.
74. Descartes was not wholly uninterested in the sort of mathematical kinematics in which Galileo engaged, and on several occasions he produced demonstrations of the relations between time, distance, and velocity, sometimes botching them dreadfully and other times not. In a letter to Mersenne of 13 November 1629, he included one such demonstration, based on an implicit dynamics of fall. He began

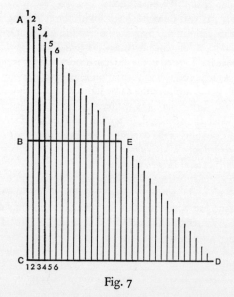

Fig. 7

with the assumption 'that the motion that is once impressed on some body remains in it perpetually if it is not destroyed by some other cause; that is to say that the body which once begins to move in a vacuum continues to move forever with a uniform speed.' Suppose that a body at *A* is impelled 'by its gravity' [*gravitas*, that is, weight] toward C. If its gravity leaves it at the moment it begins to move, it will continue in its motion to C, traversing the distance AB in the same time as it traverses the equal distance BC (see Fig. 7). In fact it does not move in this way, 'but its gravity remains with it and presses it downward and in successive moments adds new forces urging it to fall...' As a result the second half of the path, BC, is covered much more quickly than the first since the body retains all of the impetus with which it traversed AB while its gravity continues to increase its impetus still more. The triangular diagram shows how the velocity increases. The first vertical line represents the 'force of the speed' impressed in the first moment, the second line beside it the 'force impressed' in the second, and so on. The triangle ACD thus represents the 'increase of the speed of motion' during the descent from A to C, the smaller triangle ABE on the upper half of the path represents the increase during that half of the fall, and the trapezoid BCDE the increase during the second half. Because the trapezoid is three times as large as the triangle, it follows that the body will traverse BC in one-third the time it takes to traverse AB (ibid., **1**, 71–3). In another letter to Mersenne of 14 August 1634, he showed again how misplaced his condescension to Galileo was and how easy it was to confuse the relations. He agreed with what Galileo had to say about the fall of bodies, he stated – 'that the spaces heavy bodies traverse when they fall are to each other as the squares of the times of fall, that is to say that if a ball takes three moments to fall from A to B, it will only take one to continue from B to C, etc....' (ibid., **1**, 304). Nine years later, however, in a letter to Huygens, he gave a flawless version of the kinematic relations (ibid., **3**, 619–20).

75. *Principles*, II, 53; ibid., **9**, 93.
76. Descartes to Mersenne, 11 October 1638; ibid., **2**, 385.
77. Descartes to Mersenne, October 1631, 15 November 1638, 11 March 1640, 11 June 1640, and Descartes to de Beaune, 30 April 1639; ibid., **1**, 221–2; **2**, 442–3; **3**, 37–8, 79; **2**, 544.
78. Descartes to Mersenne, 18 December 1629 and November or December 1632; ibid., **1**, 95–6, 261.
79. Descartes to Mersenne, 23 March 1643; ibid., **3**, 644.
80. Cf. his letter to an unknown correspondent in 1648 or 49: In his *Principles*, he declared, he had not bothered to describe all of the movements of the planets in detail, but he had supposed in general all that observers had noted and had tried to explain the causes. In so far as all the planets have one thing in common, 'that they depart irregularly from the regular circle which it is imagined they ought to describe,' a fact which has led astronomers to assign apogees or aphelions, perigees or perihelions to them, he has given the reason for such. He has also given the reason why the aberrations are greater in the quadrants than in syzygies, to wit that the sky has an elliptical shape, and other causes can affect the motions of the planets more when they are in the ends of the ellipses where the sky is most expanded (ibid., **5**, 259). In a similar way, he held that the period of a pendulum could not be established by theory but only by experiment, and hence he had never bothered with it (Descartes to Cavendish, 2 November 1646; ibid., **4**, 560). To Mersenne's queries about the effect of air resistance on pendulums, he answered that 'it is impossible to reply to the question because it cannot be

handled scientifically (*et sub scientiam non cadit*); for if it is warm, if it is cold, if it is dry, if it is damp, if it is clear, if it is cloudy, and a thousand other circumstances can change the resistance of the air; and beyond that, if the bob is lead, iron, or wood, if it is round, if it is square or of some other shape, and a thousand other things can change the proportion, which can be said generally of all questions concerning the resistance of the air.' (13 November 1629; ibid., **I**, 73-4.) Cf. Descartes to Mersenne, 22 June 1637 (ibid., **I**, 392) where the same question is discussed in the same way.

Chapter Three

Mechanics in the Mid-Century

OF the two dominant themes in mechanics in the seventeenth century, that represented by Galileo and that represented by Descartes, neither can be taken solely as the influence of the man in question. To be sure, Galileo's definition of uniformly accelerated motion and his geometric derivation of its consequences played a major role in the science of motion as it developed. More important in Galileo's legacy, however, was the new conception of motion, to which Descartes contributed at least as much and others significant amounts as well, and the ideal of a mathematical mechanics, which derived from Archimedes who exercised a potent influence independently of Galileo. Indeed, we can perhaps best refer to this theme as the Pythagorean tradition. As for the other theme, if Descartes presented the mechanical philosophy of nature more effectively than anyone else, he cannot be mistaken for its originator, and his voice was only the most persuasive among a considerable number advocating mechanical conceptions of nature. Here again, Hellenic antecedents existed, and we can speak intelligibly of the Democritean tradition. Galileo and Descartes did not determine the course of mechanics; rather they embodied two currents which, to a considerable extent, both in their positive influence and in their covert antagonism, did determine its course.

Descartes' mechanical philosophy of nature competed with the atomism of Pierre Gassendi, another French philosopher of the same generation, who returned directly to the ancient sources of the Democritean tradition, the surviving works of Epicurus and Lucretius. A priest in the Catholic Church who lived from 1592 to 1655, and a voluminous writer, Gassendi was the primary agent in the revival of ancient atomistic thought. A lifetime's interest in atomism culminated in the *Syntagma philosophicum*, an immense work which appeared posthumously in 1658. Among other things, it contains Gassendi's final position on the nature of motion. From the point of view of the history of philosophy, important issues separated Descartes and Gassendi – on the one hand insistence on the plenum, on the other acceptance of the void; on the one hand infinite divisibility of matter, on the other ultimate undivided atoms; on the one hand rationalism, on the other empiricism. From the point of view of mechanics, however, the philosophical differences fade into insignificance. Gassendi's mechanical picture of nature accepted a general conceptual scheme which scarcely differed from Descartes', and the influence of both men, though not their influence alone, helped to establish the framework of concepts deemed acceptable within which the science of mechanics developed.

In approved mechanical style, Gassendi asserted that all of the properties of bodies arise from the various contextures that the particles composing them assume. As the letters of the alphabet are the units from which syllables, words, sentences, and books are composed, so are atoms the units of corpuscles and larger bodies. As A and O, differing in shape, denote different sounds, so atoms of different shapes cause different sensations when they strike a nerve. As the same three lines in different positions form the letters N and Z, so atoms in different positions cause distinct sensations. Finally, as a few letters suffice to compose the innumerable words in which all the books in the world are written, so atoms, innumerable in their different shapes, can join in various contextures to produce the almost infinite variety of properties that nature affords.[1] Atoms, in turn, must move. Constant change is the rule of nature, and if God is the first cause of every phenomenon, the proximate cause is motion. Gassendi quoted with approval the opinion of Epicurus that all change is a species of the genus motion.

> Certainly he held that generation or corruption, growth or decrease, alteration, that is, heating or cooling, becoming white or becoming black, etc., that all these changes are nothing else but certain local motions by which atoms come, go, collide, assemble, mix, separate, displace each other and change their place or location in a body.[2]

To the picture of nature as particles of matter in motion, he added the condition that one atom can act on another only by direct contact. It is inconceivable that a body act at a distance, at a place where it is not. We can, it is true, move an object some distance away if we have a stick in our hand; we can injure a man farther away by throwing a stone. To account for analogous cases, Gassendi employed a concept of instruments and a distinction between conjunctive and disjunctive instruments.

> Hence, obviously, you understand that if a magnet acts on iron, if fire acts on you, if the sun acts on the earth across an intervening distance, for the same reason they must transmit from themselves instruments, whether conjoined or separated, that is corpuscles, which with their surfaces touch the surfaces of the things they act on.[3]

For all practical purposes, the ontology of Gassendi's physical universe was identical to the ontology of Descartes'. It is not surprising that he shared Descartes' fastidious refusal to hear of occult qualities. As Walter Charleton, Gassendi's English translator and epitomiser summarised the issue, to assign such phenomena as magnetic action and the attraction of amber for straws to occult qualities is equivalent to a confession of ignorance. The same is true when we refer them to secret sympathies and antipathies, 'forasmuch as those Windy Terms are no less a Refuge for the Idle and Ignorant, than that of Occult Properties ... For, no sooner do we betake ourselves to Either, but we openly confess, that, all our Learning is at a stand, and our Reason wholly vanquisht, and beaten out of the field by the Difficulty proposed.'[4] In the style mechanical philosophers liked to affect, Gassendi imagined that 'from the body of the earth imperceptible corpuscles emerge, by the action of which bodies called heavy are pulled down.'[5] In other places, he specified that the imperceptible corpuscles are joined into slender threads, one end of which remains attached to the earth. Similar threads cause magnetic attractions. By such invisible mechanisms, the standard devices of the age, he preserved the dictum that one body acts on another only by direct contact.

Gassendi's atomic, mechanistic universe provided the context in which he was able to formulate the concept of inertia. Imagine a body isolated in an immense void.

> We have said that if anyone would push it, it would move in whatever direction the push was made, uniformly and according to the slowness or speed of the impulse, and indeed perpetually in the same line, since there is no cause which would accelerate or retard it by diverting its motion.[6]

Someone may object that a violent motion cannot be perpetual. In the case that he set, however, the motion would not be violent because the body would not have any repugnance to it, but would be wholly indifferent since in the imagined void there would be no centre toward which the body would be heavy. With projectiles on the earth the case is otherwise; the lines of gravitational action continually deflect a projectile from its course until it falls to earth and comes to rest. What must be sought is the cause of the decay of motion, and that cause is the gravitational attraction of the earth.

> Whence also we conclude that every motion once impressed is by its nature indelible, and does not diminish or cease unless through an external cause which checks it.[7]

Gassendi summed up the problem in an aphorism happier than his usual prose: the question is not 'what motive Virtue that is, which makes the Persevering motion, but what hath made the motion, that is to persever.'[8]

In abstracting to the conditions of an immense void, Gassendi succeeded not only in imagining the absence of friction, as Galileo had done, but also the absence of gravity, and in so doing apparently liberated inertial motion from the circularity Galileo applied to it. I say 'apparently', because it is not wholly clear that Gassendi attached the same significance to the phrase 'in the same line' that we attach to it today. To illustrate inertial motion, he used the example both of planetary motion and of a ball rolling on a perfectly smooth earth, Galileo's examples, both circular motions. With Galileo again, he distinguished horizontal motion, which is uniform, from vertical motion, which is accelerated or retarded. He tried to introduce the dichotomy of natural and violent, calling that motion which is according to nature and without repugnance natural, and that which is beyond nature and with repugnance violent.[9] The essence of the distinction lay in the fact of heaviness – to a heavy body any motion upward is repugnant. Gassendi argued that the motion of a ball rolling on a horizontal plane is natural because there is no repugnance. If it is thrown upward, the motion is violent because there is repugnance. Thus the crucial factor in imagining the body isolated in an immense void was the absence of gravitational lines which 'would accelerate or retard it by diverting its motion.' To Gassendi, the important factor was not the change of direction as such, but the acceleration or retardation, caused by heaviness, that follow when a body is diverted from a horizontal path into a vertical one. Nevertheless, whether he understood its importance or not, in eliminating gravitational attraction, he eliminated circularity as well and stated the principle of rectilinear inertia. Even in the hands of a relatively unsystematic thinker, the mechanical philosophy of nature contributed to clarifying the new idea of motion.

To changes of direction as such, Gassendi devoted no attention, but he understood the conservation of horizontal motion clearly enough to apply it to the problem of a stone dropped on a moving ship, as Galileo had done. It was Gassendi who actually performed the experiment after it had been debated more than half a century, and he used its outcome (as Galileo had used its imagined outcome) to destroy the case against the rotation of the earth.[10]

Despite passages such as those cited above, a serious case cannot be made for Gassendi as an important source of the new conception of motion. Gassendi held more than one conception of motion and made no attempt to reconcile them. As I have already indicated, circular motion embodied one problem; he treated the orbital motion of planets as inertial motion without recognising that such an idea was in conflict with the rectilinear nature of inertial motion, which he also stated. A more important conflict was embedded in his concept of gravity or weight. Gravity or weight was one of the properties of atoms, in addition to figure and magnitude. By the concept, Gassendi did not refer to weight in the ordinary sense of the term, the tendency of bodies toward the centre of the earth which he consistently referred to an external action and never treated as an internal tendency. The gravity of atoms – not of bodies we can perceive, but of atoms – was a concept derived from Epicurus and the ancient atomists. In the words of Walter Charleton, the gravity of atoms is 'the principle of their Motion,' 'a certain special Faculty, or Virtue' which is the 'Cause' of their motion. When God created atoms, He 'invigorated or impraegnated them with an Internal Energy, or Faculty Motive . . .' which is the first cause of all the motion in the world.[11] The same Gassendi who stated the principle of inertia also stated that '*materiam non inertem, sed actuosam esse,*' that matter is not inert, but active.[12] The same Gassendi who denied that force is necessary to conserve a motion, referred to the gravity of atoms as their 'native force' and 'motive force.'[13] Because of the gravity of atoms, nothing is absolutely at rest. The atoms that compose a body continually strive to free themselves, causing a perpetual agitation. 'Hence it results, not only that motion is universally more natural than rest, but also that every notion is natural because of its source, in so far as it derives from atoms which by nature move spontaneously. If some motions are in fact violent, they are so secondarily or because of the nature of concrete bodies according as they move with repugnance.'[14]

Undoubtedly the contradiction we face here derived in part from the character of Gassendi as a philosopher. Unlike Descartes, who attempted to sweep his mind clean of the past and to build a new structure of knowledge on a consistent plan, Gassendi was an eclectic in the ultimate sense of the word, a scholar consciously combing the tradition to isolate every strand of value. From Isaac Beeckman and from Galileo, he appropriated a new conception of motion, which appeared to have validity and appeared to fit into his mechanical conception of nature. From Epicurus and Lucretius, the primary sources of his atomism, he inherited the gravity of atoms, and he never stopped to ask if the one idea were compatible with the other. Motion is not the only example of eclecticism in Gassendi's philosophy; for mechanics, certainly, it is the most important.

Nevertheless, eclecticism does not appear to explain the inconsistency entirely. The gravity of atoms, an internal principle of motion, a motive force, recalls Descartes' force of a body's motion and suggests the compulsions built into a mechanical philosophy of nature to deal in a concept that expresses the efficacy of the causal particle. In a universe in which all action derives from the impact of one body on another, the moving particle embodies force. Undoubtedly the mechanical philosophy of nature accounts for a major part of the appeal that the model of impact held for seventeenth-century dynamics. The gravity of atoms, with its intuitive extension to composite bodies, provided the central dynamic concept in Gassendi's philosophy.[15]

There is no question of finding a mathematical dynamics in Gassendi. No quantitative genius urged him in that direction. Rather he deployed a purely verbal dynamics utilising the terms intuitively present to the mechanical philosopher. On the one hand, to the extent that he was a disciple of the new conception of motion, Gassendi denied that a mover impresses a 'force' on a projectile. Active force is confined to the projector; in the projectile there is only a passive force, sometimes called impetus, which differs in no way from the motion itself.[16] The very concept of a passive force suggests how difficult it was to manipulate a mechanical philosophy in purely kinematic terms. Walter Charleton expressed the dilemma when he said that 'every thing in Nature is judged to have just so much of Efficacy, or Activity, as it hath of Capacity to move either it self, or any other thing.' The motions of things are the same as their actions. Passivity is the privation of active power, and a passive body 'is compelled to obey the Energy of another.' Charleton also asserted that when all the atoms of a body move in the same direction, their 'united force' is determined in that direction, and the motive virtue of a compound body derives from the 'innate and co-essential Mobility' of its atoms.[17] The tendency, inherent in a mechanical philosophy, to think of the force of a body's motion found a rationale in the atomic concept of gravity.

However strongly Gassendi affirmed that motion alone is impressed, he found it easier to discuss motion in terms of forces. Consider his solution of Tycho's conundrum of the cannons, a solution that repeated Galileo's insight in Gassendi's terms. Imagine two cannons at points A and F on a line running due east and west; the cannon at A fires to F; that at F fires to A. Now imagine that the earth turns so that A moves toward F with a motion equal to that of the cannon balls. The ball fired from A toward F has its 'force' doubled by the addition of the earth's motion. Since the point F moves a distance FK equal to AF while the ball is in the air, the ball still strikes the same place on the surface of the earth. As for the cannon fired from F toward A, it is carried by the 'force impressed' by the cannon as far as it is carried in the opposite direction by the 'force impressed' by the earth. In respect to absolute space it remains immobile, but while it is in the air, of course, the point A moves to the position where F was originally, and the ball strikes the same place on the surface of the earth.[18] Nothing in the problem is changed if we accept Gassendi's assertions elsewhere that the word 'force' in such contexts refers to nothing but motion itself. Nevertheless, Gassendi habitually found himself more comfortable employing the dynamic term.

The same was true in his discussion of impact. When a body in motion strikes another in its path, 'because the second body has mass [*moles*] and obstructs the first, it is impelled by that force itself with which the body moves [*ea ipsa vi, qua corpus movetur*] and to which it is compelled to give way if it is not fixed or if it is unable to maintain its fixed position against the impetus, and it is compelled to move in the same direction in which the body that strikes it moves.'[19] Again, Charleton's rendition of Gassendi's thought emphasized the dynamic element. Motion is impressed upon a body only if it 'hath less force of Resistance, than the movent hath of Impulsion: so that the movent forcing it self into the place of the moveable, compels it to recede, or give way, and go into another place.'[20] Charleton indeed saw the concept of the inherent mobility of atoms as the foundation of Descartes' law that the quantity of motion remains constant – 'that just so much motive Force is now, and ever will be in the World, while it is a world, as was in the first moment of its Creation.'

> That, because look how much one Atom, being impacted against another, doth impel it, just so much is it reciprocally impelled by it; and so the Force of motion doth neither increase, nor decrease, but, in respect of the Compensation made, remains always the very same, while it is executed through a free space, or without resistence: therefore, when Concoctions, likewise mutually occurring, do reciprocally impel each other; they are to be conceived, to act upon, or suffer from each other, so, as that, if they encounter with equal forces, they retain equal motions on each side, and if they encounter with unequal forces, such a Compensation of the tardity of one, is made by the supervelocity of the other, as that accepting both their motions together, or conjunctly, the motion still continues the same.[21]

The kinematic element, which was the principal advance in Descartes' treatment of impact, was transformed wholly into dynamics, with an implicit equation between the force with which one body acts and the change in the force of the other body's motion. In Charleton's passage, if we can penetrate the pomposity of his Latinate prose, Descartes' third law, setting the change in one body's motion equal to that of the other, furnished an obscure prototype of Newton's third law. As far as Charleton and Gassendi are concerned, the passage above adequately represents their highest level of quantitative precision. We should not mistake a purely verbal statement, based on a vague and intuitive concept of force, for more than it is. In the case of a body striking an immovable object, they forgot the mutuality of action and reaction and held that the rebound is due to the original force of the body's motion.[22]

Gassendi's concern with impact was less with the exact description of motions than with the mechanical details of the impact itself. He tried to explain the equality of the angles of incidence and reflection in terms of fibres running through the body. When the body strikes perpendicularly so that its centre of gravity is along the line of fibres, it reflects straight back. When it strikes obliquely, however, the fibres on one side of the centre of gravity touch first. A disequilibrium of force exists, and the centre of gravity, still urging forward, causes the body to roll upon the surface. Like a pendulum, it passes the point of equilibrium, rolls an equal distance beyond and separating then from the surface reflects at an angle equal to the angle of incidence.[23]

Implicit in the idea of the force of a body's motion was the further idea that it is equal to the force which causes the motion. Gassendi's discussion of cannons firing east and west employed that equation. Such an idea shaded imperceptibly into a concept of force as that which causes motion, whether the force is applied with a single impulse, as in the model of impact, or uniformly over a period of time. The best example of the latter is gravity, considered as the cause of heaviness in bodies, which Gassendi consistently treated as an external force acting on bodies to draw them toward the earth. He considered it to be either identical or similar to magnetic attraction; both phenomena are caused by effluvial lines that emanate and draw bodies back to their source.[24] Despite his primary concern with the force of motion and the model of impact, then, Gassendi also explored the model of free fall in specifically dynamic terms.

His treatment of free fall went through a development which displays Gassendi's inadequacy as a mathematician. When he first considered the problem, in his letters on projectile motion published in 1642, he was convinced that two forces must act on a falling body. When it departs from equilibrium, the attraction of gravity acts on the body in the first instant, imparting one unit of velocity by which it traverses one unit of space. If gravity alone acted on it in the second instant, it would acquire a second degree of velocity, and with two degrees of velocity would cover two units of space. In the third instant, with a third degree of velocity, it would cover three units, and so on. But Galileo had demonstrated that the spaces traversed are, not as the series of numbers 1, 2, 3 . . ., but as the series of odd numbers 1, 3, 5 . . . Gassendi decided therefore that a second force must come into operation once the body is in motion. The air, rushing in behind the moving body, delivers an impulse. Hence, in the second instant, two degrees of velocity are added to the initial one; and with three degrees of velocity, the body covers three units of space. Two more in the third instant give it five degrees and so on, in the series Galileo had demonstrated.[25] In 1645, Gassendi recognised his mistake; and eliminating the role of air, he accepted gravity as the sole cause of the uniformly accelerated motion.[26]

To explain uniform acceleration, Gassendi imagined again a body placed in an immense void and given a push in some direction. Because it has no repugnance to motion and experiences no resistance, it will move with a uniform motion. Now let a second push equal to the first impel it. To the original velocity, it will add a second degree; a third impulse will add a third. Motion once gained is not lost, and the uniform attraction of gravity, adding a new degree of velocity each instant, produces a uniform acceleration.[27] By explicitly examining the dynamics of free fall, Gassendi arrived at a verbal statement of Newton's second law.

Before we hail him as the father of modern dynamics, however, we might recall that the model of free fall never replaced the model of impact as the embodiment of the prince of eclectics' central dynamic concept. A passage from Charleton describing the motion of a body thrown upward reveals the problem. Because of the attraction of gravity, he stated, a body can be moved away from the earth only by a force that is greater than its gravity.

> And hence it is, that by how much the greater force is imprest upon a stone, at its projection upward; by so many more degrees of excess doth that imprest force transcend the force of the Retentive Magnetique lines, and consequently to so much a greater Altitude is the stone mounted up in the Aer: and *e contra*. Which is also the Reason, why the Imprest Force, being most vigorous in the first degree of the stones ascent, doth carry it the most vehemently in the beginning; because it is not then Refracted: but afterward the stone moves slower and slower, because in every degree of ascention, it looseth a degree of the Imprest Force, until at length the same be so diminished, as to come to an Aequipondium with the Contrary force of the magnetique Rays of the Earth detracting it Downward.[28]

If Gassendi's ultimate position on free fall revealed how immediately the central equation of dynamics emerged from its dynamic treatment, the passage from Charleton recalls anew the conceptual morass in which dynamics was still bogged down. What is the model of force, and what is its measure? Three mutually incompatible measures are used by Charleton – impulse (Δmv), momentum (mv), and acceleration (ma). As we have had ample occasion to see already, and as we shall see again, the problem was not confined to Gassendi and Charleton. The ambiguity of the concept 'force' was a major conceptual problem of seventeenth-century mechanics, and a workable dynamics could not be constructed until the concept was clarified. Gassendi's intuitive dynamics contributed nothing toward that end.

Two other aspects of Gassendi's treatment of fall invite comment. The image of threads or lines emanating from the earth and drawing bodies down suggested that gravitational attraction cannot be constant over large distances. As the lines reach out from the earth they spread through a larger space and become less dense, and thus 'less vigorous' in their attraction.[29] The same image suggested an equation between the density and rarity of bodies on the one hand and their heaviness and lightness on the other. The more atoms a body contains in a given space, the more lines could attach to pull it down. The concept of mass (*moles*), referring to the quantity of solid matter or atoms in a body, further implied the reason for the equal fall of all bodies whatever their weight. One attracting line could attach itself to each particle, and the equal acceleration of gravity that all bodies experience would merely express the fact that bodies are accumulations of atoms, all of which accelerate equally – Galileo's idea done over in specifically mechanistic terms.[30] How familiar the ideas look. How easily their significance can be over-estimated. There was less here than meets the eye. Like his attribution of the uniform acceleration of fall to the uniform force of gravity, what mattered most to Gassendi was, not the mathematical consequences that we see in his words, but the mechanical image itself. If we treat the image as merely suggestive, we can move immediately, with the knowledge of hindsight, to the basic formula of dynamics. The image was the most important factor, however; indeed it was everything. Gassendi was merely providing a mechanism to explain Galileo's earlier conclusions. As we explore the image, moreover, the mathematical formula appears as a mere approximation. Can attractive lines from the earth produce a uniform acceleration? Given the innate differences of atoms, can his mechanism lead to an exact proportion of mass and weight? Again, the mechanical image bars the road toward a mathematical mechanics.

★

Pierre Gassendi was not the only philosopher who published a mechanical system of nature to rival Descartes'. Thomas Hobbes of England produced a third important mechanistic system, expounded primarily in his work *De corpore* published in 1655. Since Hobbes exercised less influence as a philosopher of nature than Descartes or Gassendi, and since his system was presented with far less detailed treatment of particular phenomena, a work devoted to the history of mechanics has no occasion to examine it at length. Suffice it to say that he presented a typically mechanical picture of nature in which all change was traced to particles of matter in motion.

> To act and to suffer, moreover, is to move and to be moved; and nothing is moved but by that which toucheth it and is also moved . . .[31]

Since the motion of the agent constitutes the efficient cause of every phenomenon, 'all active power consists in motion . . .'[32] Power is not an entity that differs from action. It is an act itself, motion – an act which is called 'power' because it produces another act. Hypotheses such as the above, the hypotheses on which his system rested, he said at the conclusion of *De corpore*, are both possible in themselves and easy to comprehend. Someone might be able to demonstrate the same conclusions from other hypotheses; and if the hypotheses were conceivable, Hobbes was prepared to grant the equal validity of that system.

> For as for those that say anything may be moved or produced by *itself*, by *species*, by *its own power*, by *substantial forms*, by *incorporeal substances*, by *instinct*, by *antiperistasis*, by *antipathy, sympathy, occult quality*, and other empty words of schoolmen, their saying so is to no purpose.[33]

The principal interest that Hobbes holds for the science of mechanics attaches to his concept of 'endeavour' [*conatus*].

> I define endeavour *to be motion made in less space and time than can be given*; that is, *less than can be determined or assigned by exposition or number*; that is, *motion made through the length of a point, and in an instant or point of time*.[34]

At first glance, Hobbes seems to have been defining instantaneous velocity. In the light both of what follows and of the use to which he put the concept, however, the definition appears to have had other ends in view. Immediately following it, he defined impetus or quickness of motion as '*the swiftness or velocity of the body moved, but considered in the several points of that time in which it is moved. In which sense* impetus *is nothing else but the quantity or velocity of endeavour.*'[35] Clearly, 'impetus' meant instantaneous velocity, and since Hobbes added that a lead ball falls with greater endeavour than a ball of wool, it is equally clear that 'endeavour' referred to more than velocity.

For one thing, endeavour had a vectorial nature, which did not appear in his discussion of impetus.[36] More important was his use of 'endeavour' as a concept applicable to static contexts as well as dynamic.

Hobbes insisted that motion alone can be an active agent. 'It is therefore manifest, that rest does nothing at all, nor is of any efficacy; and that nothing but motion gives motion to such things as be at rest, and takes it from things moved.'[37] It follows that a concept similar to static force as we use it was impossible in Hobbes' system. Every pressure is in fact an implicit or virtual motion. 'For to endeavour is plainly the same as to go.'[38] An endeavour may be resisted; resistance is merely the endeavour of another body in the opposite direction.[39] Equilibrium occurs when the endeavour of one resists the endeavour of the other so that neither is moved. Weight is simply the aggregate of all the endeavours downward by the separate parts of a body.[40] Hobbes' discussion of the concept of moment, using implicit motions, suggests again how the balance and lever influenced the science of dynamics in the seventeenth century.[41] In static situations, Hobbes' 'endeavour' was identical to the virtual motion of a simple machine. By means of implicit motion, the concept served as a bridge uniting statics with dynamics. Statics in fact ceased to exist in Hobbe's philosophy, and the whole of mechanics became the interplay of endeavours.

In the context of his mechanical philosophy of nature, moreover, the concept of endeavour, which operated as Hobbes' central dynamic concept, was peculiarly shorn of dynamical content, with the result that dynamics itself was in turn reduced to kinematics. Hobbes' treatment of the restitution of strained bodies illustrates how he restricted dynamic problems to a consideration of motions alone. Things that are removed from their places by forcible compression or extension [*per compressionem vel tensionem vi factam* – a generalised and vague use of the word 'force' with unmistakable connotations of violence] and restore themselves to their original situation when the force is removed 'have the beginning of their restitution within themselves, namely, a certain motion in their internal parts,' which is present when they are compressed or extended. 'For that restitution is motion, and that which is at rest cannot be moved, but by a moved and contiguous movent.' Merely removing the straining force cannot cause the restitution, for the removal of impediments does not have the efficacy of a cause. Therefore the cause of restitution must be a motion, either in the ambient medium or in the body strained. It cannot be the former. 'It remains therefore that from the time of their compression or extension there be left some endeavour or motion, by which, the impediment being removed, every part resumes its former place.'[42]

Hobbes did define a concept of force.

> I define force *to be the* impetus *or quickness of motion multiplied either into itself, or into the magnitude of the movent, by means whereof the said movent works more or less upon the body that resists it.*[43]

He did not explore the meaning of multiplying quickness into itself. Multiplied into the magnitude of the movent, however, it was obviously equivalent to Descartes' 'force of a body's motion'. Thus Hobbes said that when two movents are equal in magnitude, the swifter works with greater 'force' on a body that resists. When they are equal in velocity, the larger works with greater 'force.'[44]

The concept of force did not play a significant role in Hobbes' philosophy, however, while the concept of endeavour did. Endeavour applied to a larger range of phenomena. Not only could it be used in static problems, but it could be applied to 'actions' that were not obviously motions at all. In a plenum, every endeavour proceeds to infinity.

> For whatsoever endeavoureth is moved, and therefore whatsoever standeth in its way it maketh it yield, at least a little, namely, so far as the movent itself is moved forwards. But that which yieldeth is also moved, and consequently maketh that to yield which is in its way, and so on successively as long as the medium is full; that is to say, infinitely, if the full medium be infinite ... Now although endeavor thus perpetually propagated do not always appear to the senses as motion, yet it appears as action, or as the efficient cause of some mutation.[45]

Light was an obvious example of such an action. A grain of sand, visible at a short distance, continues its action on the eye even when it is removed so far that it cannot be seen. However far it is, enough others can be added to it to make the aggregate visible, which would be impossible unless an action proceeded from each part alone. Following the example of Descartes, though to his own damage not the details of Descartes' demonstration, Hobbes derived the laws of reflection and refraction from the motions of bodies. 'But if we suppose,' he added, 'that not a body be moved, but some endeavour only be propagated ..., the demonstration will nevertheless be the same. For all endeavour is motion ...'[46] On the basis of a specious argument similar to Descartes' contention that tendencies to motion follow the same laws as motion itself, endeavour could be applied to problems in which the force of motion could not.

One might expect that the concept of endeavour, with its implicit equation of statics and dynamics, would have led Hobbes to deal with the 'force of percussion'. In fact, his one reference to it was to deny that it can be measured by a static weight. The effect of percussion is so different from that of traction or pulsion that it is scarcely possible to compare their forces. Given an effect of percussion, such as a pile driven a certain distance into ground of a certain tenacity, it is difficult if not impossible to define what weight in what time would drive it without a blow an equal distance into the same ground.

> The cause of which difficulty is this, that the velocity of the percutient is to be compared with the magnitude of the ponderant. Now velocity, seeing it is computed by the length of space transmitted is to be accounted but as one dimension; but weight is as a solid thing, being measured by the dimension of the whole body. And there is no comparison to be made of a solid body with a length, that is, with a line.[47]

While we applaud the conclusion, we are apt to find the rationale disconcerting. In another form, it expressed Hobbes' tendency to reduce dynamic problems to kinematic terms.

One other aspect of Hobbes' mechanics should be mentioned. The concept of 'action', which played a prominent role in Descartes' mechanics, was present also in his. Unlike 'endeavour', 'action' did not receive a formal definition. It was an obvious translation into the mechanical universe of a concept too well established to require definition. In the paired terms, 'action' and 'passion', 'active' and 'passive', 'agent' and 'patient', the idea had been present in the western philosophic tradition from its origin. Thus it is not wholly surprising to find suddenly, without any introduction, the following statement by Hobbes:

> *Action* and *reaction* proceed in the same line, but from opposite terms. For seeing reaction is nothing but endeavour in the patient to restore itself to that situation from which it was forced by the agent; the endeavour or motion both of the agent and patient or reagent will be propagated between the same terms; yet so, as that in action the term, *from which*, is in reaction the term *to which*.[48]

The passage suggests the extent to which Newton's third law only made precise a general notion that had, almost of necessity, been present in the tradition of the mechanical philosophy.

In the development of the science of mechanics, Hobbes did not play a role of any significance. Despite his mathematical pretensions, his attempts to deal in precise quantitative mechanics were dispiriting. He seized on unlikely concepts, such as the velocity of a whole motion, which was defined as the sum of the instantaneous impetus, or the product of the mean impetus and the time of motion.[49] He had an unhappy knack of setting problems which were unrealistic or worse.[50] He was sadly deficient in physical intuition. His interest for the history of mechanics lies primarily in his concept of endeavour, another concept which appeared irrelevant to his successors, but one which is revealing for the relations of the mechanical philosophy and dynamics. Related to the Cartesian 'force of a body's motion', though different from it, the concept of endeavour suggests how the mechanical philosophy of nature inherently pursued a concept of force attached to the model of impact while it tended to ignore the model of free fall. Force as we understand it today had no place in a mechanical ontology which contained only particles of matter in motion. The concept of endeavour, on the other hand, was composed from exactly those entities the mechanical universe did contain, and if it proved to be fruitless in the history of mechanics, it did give expression to the philosophic climate within which mechanics developed.

*

Within that philosophic climate, no influence on the science of mechanics was more powerful than Galileo's. Above all, his treatment of free fall brought a whole new range of phenomena within the scope of scientific analysis, and there was no student of mechanics who was not impressed by the achievement. Galileo's *Discourses* were published in 1638 through the agency of Mersenne. Already in the years immediately following, a number of works took up and pursued its strands of thought. Common to them all was a dynamic point of view. I have argued that an implicit dynamics underlay Galileo's kinematics; nevertheless, the heart of his enduring contribution to mechanics lay in pursuing the kinematics of uniformly accelerated motion. In the seventeenth century, one man alone, Christiaan Huygens, fully savoured the possibilities of a kinematical mechanics. Nearly everyone else devoted himself to exploring a dynamics from which the kinematics might be derived.

De motu naturali gravium solidorum, by Giovanni Battista Baliani, appeared in 1638. A work of no particular distinction, *De motu naturali* did make explicit a dynamic foundation of the kinematics of vertical motion similar to that we can understand as implicit in Galileo. The equal fall of all heavy bodies, Baliani argued, expresses the relation of gravity to matter. In fall, gravity operates as the agent, and matter serves as the patient on which it acts. Heavy bodies fall at a rate determined by the proportion of gravity to matter, and where there is no resistance they fall at an equal rate because gravity and matter vary in proportion to each other. Where resistance is present, motion is proportional to the excess of the active virtue over the passive impediment, which excess is called momentum. 'As the momentum of a heavy solid is to momentum, so is velocity to velocity.'[51] Since Baliani proposed the formula in order to handle motion down an inclined plane, the choice of terminology was unfortunate. Whereas in free fall the passive factor was placed in the denominator of the ratio, it was apparently moved to the numerator in motion down inclined planes. Whatever the deficiencies of his terminology, Baliani's treatment of the problem surmounted them. By resistance he referred, not to the friction of the plane, which he ignored, but to the decrease in the effective component of gravity. As a 'supposition,' he suggested that the momentum of a body on an inclined plane is to its gravity as the perpendicular is to the length of the plane. Hence if two bodies descend, one perpendicularly and the other on an incline, the proportion of velocities is reciprocally as the proportion of distances.[52] It is significant that Baliani, who employed Galileo's analysis of accelerated motion, expressed himself in such a context in terms of velocities, and we must supply the factor of unit time to the two motions in order to understand his conclusion. Galileo's kinematics of free fall was a revolutionary innovation, and only with difficulty did men accustomed to think of velocity as distance covered in unit time learn to think rather of acceleration. Despite his hesitation with the kinematics, Baliani proposed a dynamics for motion on an inclined plane virtually identical to that which Viviani added at Galileo's behest to the second edition of the *Discourses*. Surpassing Viviani's passage was his implicit concept of mass.

Book Two of Father Mersenne's *Harmonie universelle* contains miscellaneous observations on motion, mostly inspired by Galileo, even though they were published two years before the *Discourses* appeared. Mersenne, like Gassendi a cleric in the Church, was a focal point of scientific correspondence in the first half of the seventeenth century, encouraging the work of such diverse figures as Pascal, Huygens, and Descartes. Familiar with Galileo's mechanics, he was responsible for the publication of the *Discourses* in the Netherlands. Miscellaneous observations were as close as Mersenne came to a systematic mechanics. Their interest lies in his persistent inclination to read Galileo's science of motion in dynamic terms, and further, to draw on the law of the lever for the intuitive dynamics he employed. His discussion of a ball rolling down an inclined plane resembles Baliani's; but instead of the term 'resistance', he invoked the balance by referring to a counterweight – 'the part of the ball supported by the plane is a sort of counterweight which, as much as it can, impedes the descent of the ball, which rolls more slowly in proportion as it is more supported.'[53] To motion in general Mersenne applied a dynamic vocabulary suggesting an equivalence of motion and force. In falling, a heavy body acquires 'an impetuosity capable of carrying it up again.'[54] When he examined Galileo's idea that the planets gained their velocity by falling toward the sun, he spoke of 'the force of the rectilinear motion' as the source of the velocity in orbit.[55] If cannon balls rotated on their axes as they move through the air, they would have a greater effect 'because they would join the force of the drill or the vise to their impetuosity.'[56] With such a view of motion, Mersenne found it logical to measure impact by static weight. It should be possible, for example, to determine whether bodies falling in air reach a terminal velocity. If a piece of lead falling twelve feet into the pan of a balance can raise a weight of one pound in the other pan, then by Galileo's analysis of free fall it should be able to raise two pounds when it falls from forty-eight feet. If it fails to raise the calculated weight when it falls from a given height, the air must have obstructed its motion and prevented the velocity from attaining its anticipated value.[57]

Similar concepts appeared in Mersenne's approach to free fall. Although he agreed with Galileo that all bodies fall at the same rate, when a resisting medium such as air is present the principle is valid only for dense bodies. When a body is so light 'that it scarcely has the force to overcome the resistance of the air,' a different rate of fall results.[58] The resistance of a body to motion derives apparently from two sources. On the one hand, it stems from its natural inclination to descend, 'for the inclination it has to move down is equal to the resistance it has to move up . . .' Weight is not the sole factor, however; velocity enters as well. The same weight 'has more force in proportion as it is more removed from the pivot of a balance, because it moves a great deal in the same time that the other moves only a little, so that the speed of the smaller compensates precisely for the weight of the larger . . .' With Galileo, he concluded 'that it is as difficult to resist a body of one pound that moves with one hundred degrees of speed as another of a hundred pounds of which the speed is only one degree.'[59] An utterly unstructured discussion based on an implicit equation of force and motion, with weight considered commensurable, drew on the example of the balance to arrive at a general, if exceedingly vague, principle of dynamics. Despite the influence of Galileo, the balance helped to steer him toward impact as the model of dynamic action.

Unlike the miscellaneous comments of Mersenne, *De proportione motus*, by Johannes Marcus Marci, a Czech doctor who contributed significantly to medicine and optics as well as mechanics, attempted to present a system of dynamics. Published in 1639, the work bore the obvious influence of Galileo, a number of whose conclusions were embodied in it. In accepting certain conclusions from Galileo, however, Marci did not accept the basis on which they rested in their author's work, the inertial conception of motion as a state which maintains itself without the active exertion of a cause. Marci's dynamics derived from an older conception of motion. Its historical value lies in the example it presents of traditional dynamic concepts applied to the new kinematics.

Basic to the entire system was the concept of impulse.

> Impulse is a locomotive virtue or quality, which moves in a finite time and through a finite distance.[60]

Impulse itself is also finite; it will not move any body whatever but only those for which it exceeds the resistance holding them in place.

> Impulse is a necessary agent and begets a motion equal to itself.[61]

Unlike some natural agents, such as heat, that do not, or at least not at once, produce effects equal to themselves, impulse immediately causes its maximum motion. It is a transient quality, however, which gradually decays, the velocity it causes decreasing in equal proportion to its force.

Manifestly, Marci's 'impulse' was a new rendition of the medieval concept of impetus. A force internal to a body, it causes motion equal to it. Applied to uniform motion, the concept of impulse led to a pair of propositions analogous to propositions in Galileo's *Discourses*. 'An equal impetus moves through an equal distance in the same or equal time.' 'A greater impulse moves through a greater distance in the same or equal time, or through an equal distance in a shorter time.'[62] In the light of Marci's assertion that impulse decays spontaneously, the full signification of these propositions remains dubious at best.

One thing they clearly indicated, however, was the proportionality of impetus to velocity. After proposition six, Marci defined what he meant by a 'greater impulse'. He meant one greater, not extensively, but intensively, such that velocity of motion varies in proportion to it. Later in the work, the apparent distinction of intensive and extensive magnitude disappeared, however, and he suggested that the impulse that moves a given body with some speed will move one half as heavy with twice the speed.[63] To this proportion he continued to couple the arbitrary condition that a given body can be moved only by an impulse of sufficient size. An impulse must be proportioned to the body moved, 'for the impulse by which a wooden ball is set in motion, will certainly not move an iron ball of the same size or a larger one.'[64] The condition, which repeats the Aristotelian dictum that a motive force generates a motion only when it is larger than the resistance, contradicts the ratios of impulse to velocity and size that Marci asserted. With the exception of that condition, Marci's 'impulse' bore a close resemblance to Descartes' 'force of a body's motion' despite its connection with a radically different idea of motion. It suggests at once the historical roots and the inherent plausibility of a dynamic concept which fit so readily into the scheme of the mechanical philosophy. Given a quantitative definition by the law of the lever, 'impulse', or the 'force of a body's motion,' expressed the dynamic perception that underlay the Aristotelian conception of motion. Even though the mechanical philosophy rested on a different conception of motion, it found the dynamic perception of the force in motion too compelling and too congenial to be denied. Either as Marci's 'impulse' or as Descartes' 'force of a body's motion', it evoked the image of impact as the model of dynamic action and obstructed the ready exploration of the model of free fall.

Marci had been impressed by Galileo's kinematics of free fall, however, and he attempted to supply its foundation from his own dynamics. To this end, he identified gravity as an impulse.

> The more an impulse is impeded by another impulse the less it moves; gravity however is an impulse moving downward or toward the center of the earth; in a perpendicular direction, therefore, because it is not

impeded by any impulse and is a necessary agent, it will beget a motion equal to itself, and the velocity of motion will be equal to gravity.[65]

In many ways, Marci's concept of impulse was similar to Hobbes' concept of endeavour, though in Marci's case its dynamic intent was made explicit. As with Hobbes, it obscured the difference of static and dynamic contexts, effectively equating rest with motion. 'From the beginning of motion in a perpendicular direction,' he asserted, 'the velocity is equal to gravity...'[66] In so far as all motions were traced to an impulse, Marci's analysis denied the traditional distinction of natural and violent motions. Because he had to explain the acceleration of free fall, whereas his identification of gravity as an impulse implied that bodies should fall at a uniform rate, he reintroduced the distinction in a modified form. In violent motions, the velocity steadily decreases, 'the impulse as it were growing old.' Because the source of the motion is external and the body separates from it, the impulse cannot be renewed. 'By its own nature' and by the contrary action of gravity, the impulse dies. In natural motions, on the other hand, the source of motion is internal; 'from gravity an impulse is produced which will produce a motion equal to itself because it is a necessary agent, and before it will complete this motion, another impulse and yet another from the same source continually reproducing the velocity of motion will increase it by a continual increment.'[67] When a heavy body is at rest, the resistance on which it rests destroys the impulses as fast as gravity generates them; only in motion can an impulse preserve itself. Hence the acceleration of free fall.

Marci's treatment of free fall presents an interesting comparison to Galileo's discussion in *De motu*. Whereas Galileo proposed to make impressed force commensurate with weight by considering it as artificial lightness, Marci took the opposite course of considering weight as an impulse. The advantages of Marci's position as the starting point for a general science of dynamics should be obvious. By applying a dynamic interpretation to Galileo's mature analysis of free fall, he arrived at once at something recognisably similar to the basic formula of Newtonian dynamics. As he said at the end of the discussion of fall, the result is the same whether gravity is an internal principle or an external traction such as magnetism.[68] His reproduction of Galileo's triangular diagram was explicitly dynamic; the horizontal coordinate, which to Galileo represented the increasing velocity, represented the increasing impulse as well to Marci, and he suggested that the size of the angle at the vertex of the triangle represents the size of the effective motive principle (see Fig. 8). In free fall, where no resistance opposes gravity, the angle is wide and the impulse increases rapidly. On an inclined plane, the angle is narrower as the incline is less, and the impulse grows at a lesser rate.[69]

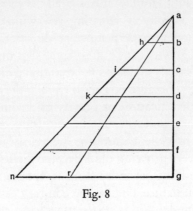

Fig. 8

As far as internal consistency is concerned, his discussion of free fall was hardly a contribution to his dynamics. The self-expending nature of impulse was one problem. If impulse naturally decays, as Marci asserted, his treatment of free fall could not arrive at a uniform acceleration, although Marci intended uniform acceleration as its conclusion. His distinction of natural and violent was hardly to the point. Whatever the perseverance of impulse, moreover, if it always produces a motion equal to itself, dynamics could never escape from the entangling necessity of a finite velocity at the first instant of motion. Although the uniform operation of gravity, whether it is an internal principle or an external force, produces a uniformly accelerated motion, that motion begins from a finite rather than a null velocity.[70] Above all, the analysis required a covert redirection of the ultimate thrust of his dynamics. Although he began by calling gravity an impulse, a constant force that produces a constant motion could hardly account for acceleration, and he had to shift his ground and treat gravity as a source of impulse.[71] The deficiencies of the analysis were legion, but it remains of interest as a demonstration that the basic formula of dynamics pre-emptorily demanded to emerge from a dynamic analysis of free fall. Even an analysis dominated by concepts ultimately incompatible with Galileo's kinematics subtly transformed itself into a fair approximation of the second law of motion, which assumes Galileo's kinematics.

MECHANICS IN THE MID-CENTURY

Meanwhile, the uniform acceleration of free fall was hardly the central issue of Marci's dynamics. Other problems suggest that other intellectual factors and influences were of more crucial significance in shaping his mechanics. The simple machines offered a more compelling image. In applying the general proposition, that a contrary prevents the effect of an equal contrary, to his conception of impulse, Marci seized on the balance as the obvious instrument for the demonstration.

> For let a weight *a* of 8 pounds, which presses the arm down with an impulse of 8 pounds, be placed in the balance abc, and its impulse will only be restrained by the impulse of an equal weight of just as many pounds.

To illustrate the principle, Marci imagined that a line attached to the other end of the balance runs over a pulley so that a weight *e* on the

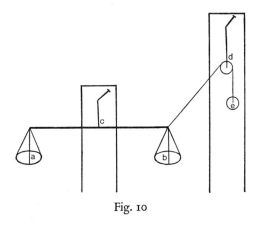

Fig. 10

end of the line pulls that arm of the balance up (see Fig. 10). He then proceeded to get the effect exactly reversed although the numerical values reveal his intention clearly enough.

> If a weight *e* of 5 pounds draws the same arm up [in his diagram it draws the other arm up, and indeed at an angle not perpendicular to the length of the arm], the impulse at *a* will be 3 pounds; therefore the weight or impulse at *e* contrary to the impulse at *a* takes a part from *a* equal to itself. In the same way, if two spheres of equal effort, meeting each other in the same line of motion, collide, no motion will follow from their contact; indeed a greater impulse will reflect a smaller one, and the greater will move after the contact with a velocity smaller in

proportion as the resistance of the smaller impulse is larger; because a smaller impulse, at the same time it expires, undoubtedly takes a part equal to itself from a larger one; therefore the excess of the larger will be the principle of motion after contact, and since it is a necessary agent it produces a motion equal to itself.[72]

Here was a principle far more in harmony with the concept of impulse and the model of impact it recalled than ever a dynamics drawn from free fall could be.

Marci employed the proposition on contrary impulses to examine, not only impact as the passage above suggests, but also motion down an inclined plane. He confined his attention to rolling spheres; and in an analysis similar to Mersenne's, he called the point of contact the fulcrum. The increase of velocity down an inclined plane, let us say the acceleration, is less than the acceleration in free fall because part of the sphere, lying on the other side of the fulcrum, functions as a weight in the opposite pan of a balance. It furnishes a contrary impulse which subtracts a quantity equal to itself, and the acceleration decreases in proportion as the impulse causing it is less.[73] In practice, Marci made no effort to compute the sizes of the two parts of the sphere. He solved the problem rather by means of the moment of the centre of gravity. It is typical of his level of conceptual development that he simply extended the concept of impulse to embrace moment. 'The impulse of gravity increases in proportion to the distance of the centre of gravity from the fulcrum.'[74] By construction, he set dc equal to the distance of the centre of gravity from the fulcrum when a ball moves on an inclined plane, whereas df (the radius) represents that distance when the plane is vertical (see Fig. 11). 'Therefore because the impulse increases in proportion to the distance of the centre from the fulcrum, by Axiom 6, and produces a motion equal to itself by Proposition 2, velocity of motion is in proportion to the distances; however, by Proposition 7, as the greater impulse fd is to the lesser impulse dc, so will the motion on ba be to the motion on bf.'[75] In the statement of the conclusion, Marci slipped back into the language of uniform motions. Since he had set up the problem such that elapsed times on ba and bf are equal, and since he was explicitly discussing the two motions as accelerated, we can restate the conclusion in terms of the velocities attained. Nevertheless, it is revealing that he did slip into the other terminology, for it was in harmony with his basic dynamic conceptions as uniformly accelerated motion never was.

MECHANICS IN THE MID-CENTURY

More than the inclined plane, the problem of impact was the ultimate goal of Marci's dynamics. He introduced it with a general proposition:

> No motion results from a contrary and equal impulse, but the motion from an impulse that is contrary and unequal equals the excess of the larger.[76]

Marci left the meaning of 'impulse' in problems of impact somewhat ambiguous. He spoke of a blow producing an impulse, and he even tried hesitantly to examine the dynamics of the blow as such.[77] In an equal number of cases, however, 'impulse' referred to the motive virtue

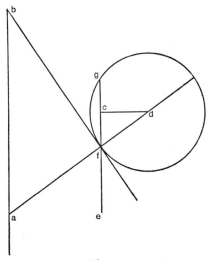

Fig. 11

of the body in question, the product of its weight times its velocity. Unlike Galileo's attempt to measure the force of percussion, Marci's treatment of impact resembled Descartes'. It tried to define the motions of two bodies after impact in terms of their motions before impact and their sizes. As with Descartes, the more he attended to dynamic considerations, the more perplexed his discussion became. Since Marci did not assert the conservation of motion, the results at which he arrived differed widely from Descartes'.

With Descartes he agreed that a body in motion is not able to put every other body it may strike into motion. An impulse must be adequate to a body in order to move it. When a body strikes another it cannot move, if the second body is soft or fragile, the impulse will simply be absorbed. If the second body is hard, the first body will rebound with a velocity approaching its original one as the hardness approaches perfect hardness. The body in motion is initially brought to rest by the blow, but the obstacle it strikes strikes in return, and the new impulse causes it to rebound. Despite what he had asserted earlier, Marci now decided that when two bodies of equal impulse moving in opposite directions strike each other, each one acts as an immovable obstacle to the other. The contrary impulses destroy each other, but the blows generate a new impulse in both, and they rebound. The nub of the issue lies in the succession of the two impulses. If the second blow comes after the first, the two impulses do not mix and destroy each other, and the body then rebounds.[78]

When the two impulses are not equal, the problem becomes more complicated. If the two bodies in question are equal in size, their velocities are proportional to their impulses. The one moving more swiftly strikes the other harder than it is struck in return.

> From which manifestly the velocity of those which strike each other when they are in motion is caused to be exchanged by the percussion; for those which strike harder rebound less impetuously, and those which strike less rebound more impetuously.[79]

When neither the bodies nor the impulses are equal, Marci employed two criteria to determine if a given impulse will cause a body to rebound. The impulse able to move a larger body can move a smaller body in all circumstances. When a larger body strikes a smaller one that is unable to make it rebound, the larger gives the smaller a velocity equal to the larger's original one, and it continues to move in the same direction with that part of the impulse remaining to it.

> And thus it happens that they are borne from that unequal blow with unequal velocity, the larger indeed with the smaller velocity because of the forces hacked off and mutilated by the percussion, but the smaller with the larger velocity because of the same forces acquired unimpaired.[80]

The impulse capable of moving a smaller body, however, can move a larger one only if the proportion of the impulse to the body it moves is larger than the proportion of the other body to that body. Two elements contributed to Marci's formulation of this unlikely comparison. First, he had specified that gravity is an impulse, and hence the weight of a body can have a proportion to its motion. In this case, he pointed out (incorrectly it would seem, since the motions are not vertical) that gravity is a contrary impulse, and hence the impulse or motion of the one body must exceed the gravity of the other in order to move it.[81] Second, the omnipresent image of the balance, with its proportions of weights and virtual velocities, also suggested the comparison. References to the 'ratio of the fulcrum' fill the discussion of impact.[82] A body on one end of the balance can raise a larger one at the other end only if the product of its weight times its virtual velocity exceeds that of the larger body. When such is the case in impact, the more powerful agent prevails, and the larger body is forced to rebound. The smaller body rebounds as well.

> For since the impulse of the smaller body is larger than the gravity of the larger body, because of the smaller proportion of the latter than of the former, if the smaller body strikes the larger one, the larger will be moved by the blow; moreover the smaller body also rebounds from the larger because the smaller is moved by any blow whatever from it.[83]

It was at once the irony and the dilemma of seventeenth-century dynamics that the phenomena of impact, which furnished the model of dynamic action most readily apparent to the age, were themselves most amenable to kinematic analysis. As Marci illustrates anew, nearly every attempt at a dynamic analysis of impact issued forthwith in a trackless morass.

★

More incisively than Marcus Marci, Galileo's great Italian disciple, Evangelista Torricelli, explored the dynamic significance of his kinematics. Under the picture of the author which serves as frontispiece to Torricelli's *Lezioni accademiche*, when they were finally published in 1715, was the Latin phrase '*En virescit Galilaeus alter*,' an anagram for Evangelista Torricellius which can bear the translation 'Behold, another Galileo flourishes.'[84] In its statement of the relationship of Torricelli's work in mechanics to Galileo's, such that the younger man consistently took the achievement of the older as his starting point, the motto expresses an historical truth. It expresses only half the truth, however, for if Torricelli took Galileo as his starting point, his mechanics departed significantly from the master's by explicitly emphasising dynamics. The set of lectures on percussion that form part of the heterogeneous collection labelled *Academic Lectures* were delivered in Florence about 1642, apparently as Torricelli's inaugural lectures when he succeeded Galileo, whose final months of life he had tended, as mathematician and philosopher to the Grand Duke. In addition to the Lectures, which were not known during the seventeenth century, he also produced *De motu gravium naturaliter descendentium et proiectorum libri duo*, published in 1644, an extension of Galileo's kinematics into dynamics, which was known and influential.

Indeed the dynamic point of view pervaded Torricelli's mechanics to the extent of determining a conception of motion fundamentally different from Galileo's. What is it that carries a projectile, Torricelli asked. Galileo had demonstrated that it cannot be the medium. It must then be 'the impetus impressed in the grossness and corpulence of the matter.' What impetus is, he did not know, but that it is, he did. As long as an arm pushes a projectile, the fact that it moves arouses no wonder;

> but after it is released and separated from the hand that sped it along, I should remain astonished that it continues to move through a great distance if I could not imagine some assisting virtue impressed in that body and able to carry it through the air. For a cannon ball to fly through the air, impeded by the ambient medium and not helped by some power that accompanies it, would be an effect without a cause, that is an absurdity in nature. Hence of necessity some virtue (whatever it is), able to cause a motion and a velocity more or less great as the impressed virtue is greater or less, impresses itself in the moving body...[85]

Torricelli's favourite term was not 'impetus' but 'moment' [*momento*], usually in the plural [*momenti*]. Velocity, he insisted is merely an effect; the reality of which it is a manifestation are the intrinsic *momenti* in the body, *momenti* which can subsist by themselves, without the accompaniment of a velocity, though a velocity cannot exist without them. Weights placed at different distances on the arm of a balance or on planes of different inclinations have different *momenti*, whereas their velocities are only potentially different. 'But velocity by itself cannot indeed subsist without the internal *momenti*.'[86]

The passages above are quoted from the *Academic Lectures*, which date from the early 1640's. Torricelli's position in *Two Books Concerning the Motion of Bodies Naturally Descending and concerning Projectiles* (*1644*) frequently appears different.

> If a projectile which follows the parabola *ABC* were deprived of all its gravity at some point *B* of the parabola, then without doubt it would continue its translation with a motion always uniform in the straight

line *BD* tangent to the parabola. For any cause which might inflect or accelerate or retard the motion has been taken from it. Hence it is manifest that the impetus [*impetum*] of the body will continue always to be the same in any part whatever of the tangent *BD* as it was at the point *B*.[87]

In this case, impetus has been shorn of dynamic features, and the passage stands as one of the early statements of the principle of inertia. Elsewhere in *De motu gravium*, however, the dynamic meaning reappeared. Thus he stated that the *momentum* compounded from the two *impetus* at any point along a parabolic trajectory must be a power [*potentia*] equal to them.[88] The apparent contradiction serves to recall, as Galileo's ambiguous language frequently did as well, how nearly impossible it is at times to distinguish the concept of a permanent impetus from the concept of inertia. In Torricelli's case, it seems clear that he understood impetus (or *momenti*) primarily in an active sense. It was more than a mere verbal equivalent of motion; it was the dynamic cause of motion. As such, it was similar to Marci's concept of impulse.

Perhaps the relation of Torricelli to Galileo is nowhere more clear than in *De motu gravium*. The full title (*De motu gravium naturaliter descendentium*) is reminiscent of Galileo, and Torricelli followed the master in regarding the movement of heavy bodies downward as a unique phenomenon of nature. Already in the introduction, however, Torricelli announced his own approach to the subject. Galileo's treatment of accelerated motion, he stated, rested on the principle, announced but not demonstrated, that the degrees of velocity acquired by the same heavy body in descending different inclined planes are equal for equal vertical displacements. Torricelli undertook to supply a demonstration from a principle of mechanics:

> The moments of equal heavy bodies on planes unequally inclined are in the same proportion as the perpendiculars of equal parts of the same planes.[89]

To be sure, the principle is identical to that used by Viviani in the demonstration he added to the second edition of the *Discourses*, and referred to by Galileo even in the first edition in his theorems on motion along the chords of circles. In Torricelli's case, the statement of purpose was equivalent to announcing an explicit dynamics of vertical motion.

In order to carry the demonstration through, Toricelli proposed the following assumption.

> Two bodies joined together are not able to put themselves in motion unless their common centre of gravity descends.
>
> For whenever two bodies will have been connected to each other in such a way that the motion of one will entail the motion of the other, whether it is done by a balance or a pulley or some other mechanical device, those two bodies will be as one composed of two; moreover a body of this sort will never move unless its center of gravity descends. Indeed when it will have been arranged in such a way that its center of gravity cannot of itself descend, the body will rest entirely in its place, for if it will move otherwise, that is, with horizontal motion that in no way tends down, it will move in vain.[90]

The assumption expressed a new principle of immense importance for the science of mechanics. Under properly defined conditions, two or more bodies can be treated as a single body concentrated at their centre of gravity, and the conclusions established for single bodies can be applied as well to the motion of their centre of gravity. Torricelli himself applied it only in a distinctively Galilean way, being concerned with the natural, that is vertical, motion of two bodies so joined as to constitute a single heavy body. Others were to apply it more broadly with still more significant results.

In order to demonstrate Galileo's proposition about motion on an inclined plane, Torricelli imagined two planes of unequal inclination descending from a single point to the same plane. A body rests on each, and a line which passes over a pulley at the vertex, connects them so that they can move only as a unit. The conclusion follows immediately from the construction of the problem – the two bodies will have 'equal *momentum*' when their proportion to each other is as the proportion of the lengths of the planes. Using the phrase 'total *momentum*' to refer to the weight of a body, Torricelli summarised his conclusion in the following Lemma:

> The total *momentum* of a body is to its momentum on an inclined plane as the length of the inclined plane to its elevation.[91]

In the language of trigonometry, the *momentum*, the effective moving force along the plane, varies as the sine of the angle of inclination. As he proceeded toward his conclusion, treating *momentum* as the cause of

MECHANICS IN THE MID-CENTURY

motion down an inclined plane, Torricelli stated the basic formula of dynamics as a matter of course.

> With Galileo we assume here that velocities on planes of different inclinations are as the *momenta* when the mass [*moles*] is constant.[92]

In the context, which treats accelerated motion along the chords of a circle, as in Galileo's problem, 'velocities' refers to the velocities at the end of unit time, and the statement says in effect that for a constant mass acceleration is proportional to the moving force. The translation of '*moles*' by 'mass' is a gratuitous modernisation of a word difficult to render otherwise. Torricelli did not distinguish mass from weight, and perhaps we might better translate '*eadem moles*' by 'the same body'.[93] Nevertheless, the statement is a remarkable indication of what a frankly dynamic interpretation of the kinematics of uniform acceleration could produce.

Already, in the earlier *Academic Lectures*, he had demonstrated the same thing in an even more remarkable way. The problem there was not motion down an inclined plane but percussion, and again Galileo supplied the starting point. Torricelli recited a fascinating set of experiments Galileo had performed in Padua, experiments that are not mentioned in Galileo's own writings on the force of percussion. Fastening a bow firmly in a vice, he suspended a lead ball weighing two ounces from the middle of the bowstring on a chord one *braccio* (roughly two feet) in length. When the ball was dropped from the level of the bowstring, its force of percussion bent the bow, and by means of a vessel that would ring when struck Galileo determined how far the ball pulled the chord. Say it was four inches. He then determined what static weight hanging from the chord would pull it the same distance. Say it was ten pounds. Now with a stronger bow he performed the same experiment and found that more than twenty pounds of dead weight were needed to pull the chord as far as the ball of lead did. With a third bow stronger yet still more was needed; and as the bow became stronger and stronger, he needed a more and more heavy weight to equal the 'force' [*forza*] of the same ball of lead falling the same distance. He concluded that with a very strong bow a thousand pounds would be required. But a bow a thousand times as strong can be imagined, and with that bow the ball of lead would cause an effect equivalent to a million pounds – 'a perfectly clear indication that the force of that little weight and of that two feet of fall is infinite.'[94]

In order to analyse the infinite force of percussion, Torricelli defined a problem of his own. Imagine a marble table that can be broken by a weight of a thousand pounds resting on it. If we place a weight of a hundred pounds on it, the weight will do nothing as far as breaking the table is concerned. To be sure, the weight presses down in each instant with a '*forza*' or '*momento*' of one hundred pounds, but the table resists in each instant with a *momento* of a thousand pounds. If we put ten balls of a hundred pounds each on the table, 'or if we were able to concentrate the virtue and all the activity of the ten in one alone' we would have a '*forza*' sufficient to overcome the resistance of the table. When we put ten balls on it, we multiply the matter. If instead we were able to multiply the time, finding some way to conserve the *momento* with which one ball presses down in each instant, we could achieve the same effect. And this we can do very easily – by letting the ball drop.[95] Already it is clear how Torricelli's concept of impetus or *momento* will fit into an analysis of the force of percussion. It was no accident that he referred frequently to the *forza* of projectiles when he discussed them.[96]

Suppose we need a hundred bottles of water from a fountain that flows at a rate of only one bottle an hour. Obviously, we fill a bottle every hour for a hundred hours.

> Gravity in natural bodies is a fountain from which *momenti* continually flow. The body we are dealing with produces a *forza* of a hundred pounds in every instant of time; hence in ten instants, or better ten of the shortest units of time, it will produce ten of those *forze* of a hundred pounds each, if only they can be conserved. But so long as it rests on the body that supports it, it will never be possible to have the *forze* that we desire collected all together, because at the moment when the second *forza* or *momento* appears, the preceding one has already disappeared or, so to speak, been extinguished by the resisting opposition of the support which murders all the *momenti*, one after another, the very moment they are born. But without further tedious prolixity [*sic*!], the very definition of naturally acceleratd motion that Galileo provides suffices to unveil this mystery of Nature concerning the *forza* of percussion. Let the fountain of gravity be opened. Let the heavy ball be lifted high enough that it remains in the air for ten instants of time when it falls back down and consequently generates ten of its *momenti*. I say that those *momenti* will be conserved and will be collected together. That is manifest from the constant experience of bodies falling and of accelerated motion, since bodies have greater *forza* after a fall than they had at rest. But reason also convinces us of the conclusion, since if that obstacle supporting the body

extinguished all the *momenti* previously mentioned with the constant resistance of its hateful touch, now that the obstacle is taken away the effect ought to be removed with the cause. Hence when the body hits after the fall, it will no longer employ the mere *forza* of a hundred pounds, offspring of a single instant, as it did before, but multiplied *forze*, offspring of ten instants, which will be equivalent to a thousand pounds, exactly as many, united and employed together, as are needed to leave the marble broken and conquered.[97]

In accelerated motion, then, the increasing velocity of the body expresses the more fundamental reality, the *momenti* accumulating in it. The *forza* of percussion is only another expression of the same dynamic reality. The *forza* of percussion is not merely the measure of the *momenti*; it is the *momenti*, the *momenti* which manifest themselves equally in the force of a blow and in the velocity of motion. The *forza* of a projectile, Torricelli asserted, 'cannot be anything except the virtue impressed on it by the machine that set it in motion, and it is numerically precisely the same virtue that flowed from the same machine.'[98]

Obviously, velocity alone is not the measure of *momenti*; a huge body strikes with immensely more force than a small one moving at the same rate. What role can matter play in percussion? Torricelli was convinced that matter as such can play no role at all.

> This much is certain, that matter in itself is dead and serves only to obstruct and resist the operating virtue. Matter is none other than a bewitched vase of Circe which serves as a receptacle of *forza* and of the *momenti* of *impeto*. *Forza*, then, and *impeti* are abstractions so subtle, quintessences so spiritual, that they cannot be contained by any other jar except the innermost materiality of natural bodies. Such at least is my opinion.[99]

When no obstruction destroys them, *momenti* accumulate in a body, manifesting themselves as accelerated motion. Hence the force of a body in percussion is greater than its force at rest. If we neglect the ambiguity involved in the word '*forza*' for a moment, we can say that Torricelli's conclusion, in the case of free fall, is equivalent to the formula

$$mv = \Sigma \, F\Delta t,$$

where F represents the weight of the body falling.

In fact, Torricelli advanced to the very brink of recognising the ambiguity in his use of *forza* and *momento*. The initial point of his inquiry was the infinity of the force of percussion. Nothing said so far indicates that it should be infinite. In the case discussed above it is merely the product of a body's weight times the number of instants of fall. Much depends on Torricelli's understanding of time. Time, in his conception, is granular – a succession of discrete segments, which of course he called instants. Though discrete, instants are infinitesimal in duration, and the crux of the issue in regard to the force of percussion lay in the opinion that an infinite number of instants compose every finite segment of time. Since a single *momento* flows from the spring of gravity each instant, the force of percussion after a finite fall must be infinite.[100]

At first glance, it appears that Torricelli might have proved too much. If the analysis above is correct, any pebble falling from any height ought to break our marble table; such, fortunately, is not the case.

> To this the response is as follows [Torricelli continued]. If the percussion were instantaneous, then an infinite effect would follow every percussion, however small – that is, if the percussant could apply all that accumulation of *momenti*, which are indeed infinite in number, that it has collected together in itself and impose them all on the resistance in one instant of time alone. But if in applying them, it applies them over some space of time, it is no longer necessary that the effect be infinite...[101]

It is well known that the motion which a body acquires in falling can carry it back straight up to its original height in a time equal to its fall. A small repugnance to moving upward, equal to the weight of the body and constant during the interval of time, suffices to extinguish the infinity of *forze* accumulated in the fall. In a percussion, the immensely greater resistance achieves the same result in a shorter time. 'Times reciprocally proportional to resistances are equivalent for extinguishing the same *impeto*.'[102]

> Let a body fall and multiply its *momento* say a hundred times. If it applies all its multiple *forze* in one instant alone in percussion, the resistance will feel a violence as of a hundred, exactly as much as the *forza*. But if it applies it and distributes it, for example, over ten instants, the resistance will never feel a hundred *momenti* of *forza* all together, but rather ten at a time.[103]

One striking figure that Torricelli used to illustrate his point was an exhausted peasant sleeping for an hour on the stones under some gate. He is uncomfortable to be sure, but the pressure is endurable because it is spread out over an hour. Now imagine him asleep for an hour as he falls from the moon. No feather bed could be softer than the air, but when he hits the earth, all the pressure distributed before over an hour is concentrated into the brief duration of the blow.[104] If you tried to break a diamond placed on a wooden table, the table would give with the blow, spreading its force out over a finite time, and the *momenti* in the hammer could not all act together. But if you used a steel hammer on a steel anvil, the diamond 'would be forced to receive the *momenti* of the blow all together, and I am certain that it would be reduced to dust, any degree of hardness whatever notwithstanding.'[105]

In fact, in Torricelli's opinion, even the impact of a steel hammer on a diamond requires some time because every impact is elastic.[106] He stood as it were poised to announce a major dynamic generalisation – that every change of motion requires the action of a force and is equal to the product of the force multiplied by the time. Stated in this way, the generalisation would have been limited to uniform forces. Without a sophisticated concept of elasticity available to him, Torricelli assumed that the resistance in impact remains constant through the time of impact. Despite his considerable mathematical gifts, he might well have been unable to deal with the problem even had he known of a relation such as Hooke's law. Important as it is, this remains a technical detail the solution of which would have been found in due time. The conceptual problem was the more fundamental, and close as he was, Torricelli did not take the final step. He never applied the same terminology to the destruction of motion that he applied to its generation. Neither the word *'momento'* nor the word *'forza'* appeared in the context of elasticity. He spoke rather of the effect of a blow, of its violence, of the resistance of the body struck. Perhaps the closest he came to a generalised point of view was his use of a body rising to illustrate the destruction of *momento*. Even in this case, however, he employed a different terminology. Bodies have a repugnance to moving upwards, not a negative *momento*. Indeed the concept of a negative *momento* could have had no place in Torricelli's thought. *Momento* is a cause of motion, a motive virtue. Something other than *momento* must destroy it.

Another problem inherent in Torricelli's dynamics, related to the one above since it involves the concept of force, was his effort to equate statics and dynamics and to measure the force of percussion in terms of static force. Again, he seemed at times on the verge of defining the distinction of statics and dynamics. To his conclusion that the accumulated *momenti* of a falling body constitute an infinite *forza*, he raised the possible objection that the body's velocity ought to be infinite also. To answer the objection, Torricelli drew a distinction between absolute quantity and increase. The *momenti* of a falling body increase infinitely. Since the body starts from rest, its velocity also increases infinitely although it remains finite in quantity.[107] In Torricelli's own terms, the answer is unsatisfactory. His concept of *momento* was such that only an effective denial of a static situation, the acceptance of an initial velocity of fall such as Marci and Hobbes asserted, was compatible with it. Perhaps we should see his answer as a revision, not fully conscious and hence incomplete, of his concept of *momento*. When he concluded that the *momenti* of a falling body are infinite, he implicitly concluded that its *forza* is incommensurable with its static *forza* or weight. To use terminology more common in the twentieth century, the force of a body's motion is incommensurable with static force. A dynamics cannot be built merely by extending the principles of statics. Torricelli stood on the verge of this realisation. Had he coined a different term, so that the two would be recognised as distinct concepts, he would have defined a workable distinction of statics and dynamics. He did not, and perhaps the greatest achievement of his mechanics remained incomplete.

Another facet of the same problem reveals itself in his treatment of elastic rebound. Again Torricelli appears to have been ready to state a generalised conception of force embracing all the actions that generate motion. Since in rebound he was dealing, not with the destruction of motion, but with its generation, there was no obstacle to his seeing it in analogy to gravity. Inflated balls have this property, he asserted,

> that when their surface is compressed by any violence, it has the *forza* to return to its original state, and that with more or less *impeto* in proportion as the *forza* which compressed the imprisoned air was more or less.[108]

The air inside, compressed beyond normal and seeking to return to its original condition

> presses the floor with a great *forza*, like a boatman standing in a skiff who pushes the bank, not to make the bank, but rather the boat move. At that furious thrust of the enclosed air, the ball rises a distance equal to its compression and returns to its original state in an insensible time, that is, with great rapidity, and therefore with an *impeto* which conserves itself for a time, once it is conceived, and causes the rebound.[109]

A body rebounds from the skin of a drum, he repeated, 'not indeed because any part of the *impeto* of projection remains in it, but only because a new *impeto* is generated in it by the *forza* of the skin.'[110] Although the usage is somewhat ambiguous, in the passages above Torricelli appears to have meant by *forza*, not the instantaneous pressure at any moment during the rebound, but the sum of instantaneous pressures equal to the total impetus or motion generated. *Forza* in this sense would be equal to the force of the body's motion, or its *momenti*, but would not be commensurable with its static weight.

The crucial factor in Torricelli's analysis of the force of percussion was his exploitation of the other model of force, the model of free fall. Using his concept of *momento*, he extracted the dynamic consequences inherent in the kinematics. The extent of his insight is revealed in his ability to apply the analysis to totally different dynamic situations in which another motive source replaced the action of gravity. Why is it, he asked, that a huge galleon shakes the entire pier when it is pulled in by a man, whereas a small skiff, pulled the same distance by the same man with the same *forza*, shakes nothing at all, even though it hits the pier with a greater speed. The role of matter and its faculty of absorbing force is central to the problem of course. There is a recalcitrance in matter which, as it is greater, prolongs the action of the force and allows a greater accumulation to be impressed on a body.[111]

> Hence the *forza* of the man pulling [the rope] is what acts and what strikes [the pier]. I do not say the *forza* that he exerts in that instant of time when the boat arrives to give the blow, but all that which he previously exerted from the beginning until the end of the motion. If we should ask how long he expended himself when he was pulling the galleon, he would answer that perhaps a half hour of time and constant exertion was required to move that huge machine through a distance of twenty feet. But no more than four beats of music were employed to pull the tiny skiff. Hence the *forza* which flowed from the arms and sinews of that laborer, as from a lively fountain, through the space of half an hour did not merely disappear in smoke or fly off through the air. It would have vanished had the galleon not been able to move at all, and all of it would have been destroyed by the stone or anchor that prevented its movement. Rather it was deeply impressed in the bowels of those timbers and bolts from which the ship is made and with which it is loaded; and there inside it continually conserved itself and grew, apart from the small amount that the resistance of the water can have taken away. What marvel will it be then if that blow which carries with it

the *momenti* accumulated during half an hour causes a much greater effect than that which carries with it only the *forze* and *momenti* accumulated during four beats of music.[112]

For better or worse, we have stifled the poetic fancy in mechanics, but our analysis of the momentum of the two ships is essentially the same. In a context other than free fall, (though still entangled in the ambiguities of the word '*forza*') Torricelli asserted again a conclusion equivalent to the formula

$$mv = \Sigma\, F\Delta t.$$

Torricelli's *Lezioni accademiche*, unlike his *De motu gravium*, dropped completely out of sight after their delivery in the early 1640's. Later in the century, G. A. Borelli reported his failure to locate a copy of them after he had heard of their delivery.[113] Only in 1715 were they published, and by then the science of mechanics had progressed to the point that they were of interest only as an historical document. Inevitably the question arises whether the history of mechanics might have been different had they become part of the available record and played a role in the discussion. No other work on mechanics in the first half of the century approached the *Lectures* in the power and consistency with which they explored the dynamic model of free fall. I have argued that his dynamics contained an inherent source of confusion in his attempt to measure the force of percussion in terms of static force. The confusion on this point, however, was the confusion of the entire century in its jumbling together of various irreconcilable concepts under the heading of 'force'. Torricelli's dynamics only shows again what the conceptual difficulties in the development of dynamics were, and with the exception of Descartes, no one before Newton progressed so far toward unravelling them. Fascinated as he was with the notion of an infinite force – 'perhaps the most recondite and the most abstruse of all the arcana of Nature'[114] – he stopped short of drawing what appears to us as the obvious conclusion when he recognised the incommensurable nature of static force and momentum. With the possible exception of Leibniz, no one in the seventeenth century analysed the dynamics of percussion with equal insight. Surely, one is tempted to judge, the publication of Torricelli's *Academic Lectures* would have advanced the science of dynamics by nearly half a century.

One is only tempted so to judge, however. Other considerations suggest instead that his own century might have found Torricelli's dyna-

MECHANICS IN THE MID-CENTURY

mics incomprehensible. Not only did he hold a conception of motion which was already largely displaced among the leaders of European science, but behind the conception of motion lay a conception of nature wholly at odds with the mechanical philosophy which was rapidly establishing its dominance over scientific thought. Quantitatively, Torricelli's *forza* of percussion or *momenti* may have been identical to Descartes' force of a body's motion. Conceptually the two were worlds apart. *Forza* or *momento* had ontological status in Torricelli's universe. Perhaps it is best understood as a quantitative rendition of a Paracelsian active principle – the animating reality that activates the dull and inactive corpulence of matter.

> I would perhaps incline to believe that if it were possible to confine or to cram all that *forza* and exertion that were produced by pulling our imagined vessel during half an hour inside the most abject, though unbreakable, shell of a walnut, I say I would believe that perhaps that extremely light shell would cause the same effect when it strikes that the huge mass of the ship causes.[115]

He went on to say that what he imagined here is impossible. Nevertheless, the very proposal suggests an idea of force that was anathema to mechanical philosophers. The mechanical philosophy of nature was formulated, indeed, primarily for the purpose of abolishing such incorporeal agents from natural philosophy. Torricelli's achievement in dynamics was directly dependent on his willingness to employ concepts that were unacceptable to most of the leading students of mechanics during the century. Perhaps it was not wholly by accident that he failed to publish his *Lezioni accademiche* and expressed himself far more circumspectly in *De motu gravium*. While the science of mechanics pursued its course within a different conceptual framework, Torricelli's *Lezioni* stand as a permanent reminder that no inherent necessity compelled dynamics to develop along the lines it finally took. In a different philosophic climate quite another dynamics might very well have taken shape; and judging from Torricelli's achievement, we have no reason to believe that its technical power would have been less. As far as Torricelli's dynamics is concerned, it was confined to problems of rectilinear motion. The dynamics of circular motion was one of the central problems of seventeenth-century mechanics, however, and to the solution of its riddle Torricelli offered nothing. To us, it is apt to appear that he was unable to offer anything; in circular motion the continuous action of a force does not increase the force of a body's motion. It is only worth recalling that we hardly expected a quantitative dynamics of rectilinear motion in his terms.

NOTES

1. *Syntagma philosophicum*; *Opera omnia*, 6 vols. (Lyons, 1658), **1**, 367.
2. *Syntagma philosophicum*; ibid., **1**, 363.
3. *Syntagma philosophicum*; ibid., **1**, 364–5.
4. Walter Charleton, *Physiologia Epicuro-Gassendo-Charltoniana: or a Fabrick of Science Natural upon the Hypothesis of Atoms*, (London, 1654), p. 343. Cf. Gassendi, *Syntagma philosophicum*; *Opera*, **1**, 449.
5. *Syntagma philosophicum*; ibid., **1**, 205.
6. *Syntagma philosophicum*; ibid., **1**, 354. Gassendi's original statement of the position is found in *Epistolae tres. De motu impresso a motore translato*: What would happen to a stone placed in empty space and put in motion by some force? 'Respondeo probabile esse, fore, ut aequabiliter, indefinenterque moveretur; & lente quidem, celeriterve, prout semel parvus, aut magnus impressus foret impetus.' A ball placed in motion on a horizontal plane is never accelerated or retarded because it experiences no perpendicular motion; 'adeo, ut quia in illis spatiis nulla esset perpendicularis admistio, in quamcumque partem foret motus inceptus, horizontalis instar esset, & neque accelararetur, retardareturve, neque proinde unquam desineret.' (Ibid., **3**, 495.)
7. *Syntagma philosophicum*; ibid., **1**, 355.
8. I quote Charleton's translation. (*Physiologia*, p. 465.) Gassendi's Latin is as follows: '& vis motrix quaerenda sit non quae motum perseverantem faciat, sed quae fecerit perseveraturum.' (*Syntagma philosophicum*; *Opera*, **1**, 354.)
9. *Syntagma philosophicum*; ibid., **1**, 343.
10. *Epistolae tres. De motu impresso*; ibid., **3**, 478–83. The second of the three letters (pp. 500–20) is devoted to the motion of the earth.
11. Charleton, *Physiologia*, pp. 112, 126.
12. *Syntagma philosophicum*; ibid., **1**, 335.
13. His words are 'ipsa nativa Atomorum vis' and 'motrix sua vis.' (*Syntagma philosophicum*; ibid., **1**, 343.)
14. *Syntagma philosophicum*; ibid., **1**, 343.
15. See Appendix C.
16. *Epistolae tres. De motu impresso*; ibid., **3**, 498–9.
17. *Physiologia*, pp. 271, 269.
18. *Epistolae tres. De motu impresso*; *Opera*, **3**, 503–4.
19. *Syntagma philosophicum*; ibid., **1**, 385.
20. *Physiologia*, p. 465.
21. Ibid., p. 445.
22. *Syntagma philosophicum*; ibid., **1**, 358. Cf. Charleton: The cause of the restoration of a flexible body is the same force that originally distorts it, just as the rebound

of a ball from an obstacle is due to the original force with which it strikes. If, for example, you lay a stick over the edge of a table and strike the projecting end, the other end flies up. If some obstacle is put above the other end, it hits it, bounces down, bounces up again, and continues thus 'till the force of resistance in the 2 beams hath wholly overcome that of the first percussion or impulse.' All of the rebounds are caused by the force originally impressed on the stick. If now the stick is set upright in a hole and pulled to one side, it vibrates back and forth, as the stick in the first case did, 'till the resistance of the hole hath wholly overcome the force thereupon imprest, by your hand.' The self-restoring motion in flexible bodies is similar to the rebounding of the stick, and both are caused by the original impressed force. (*Physiologia*, p. 332.)

23. *Syntagma philosophicum*; *Opera.*, **1**, 354, 360–1. Cf. Charleton, *Physiologia*, pp. 471–4.
24. *Epistolae tres. De motu impresso*; *Opera*, **3**, 491–5. *Syntagma philosophicum*, ibid., **1**, 346.
25. *Epistolae tres. De motu impresso*; ibid., **3**, 497.
26. *Epistolae tres de proportione qua gravia decidentia accelerantur*; ibid., **3**, 622–3.
27. *Syntagma philosophicum*; ibid., **1**, 349–50.
28. *Physiologia*, p. 285.
29. I quote from Charleton, ibid., p. 283. Cf. *Epistolae tres. De motu impresso*; *Opera*, **3**, 494; *Syntagma philosophicum*; ibid., **1**, 353.
30. *Epistolae tres. De motu impresso*; ibid., **3**, 495.
31. *Concerning Body*; *The English Works of Thomas Hobbes of Malmesbury*, ed. Sir William Molesworth, 11 vols. (London, 1839–45), **1**, 334. I have altered the translation slightly. The Latin is as follows: 'Agere autem et pati est movere et moveri; et quicquid movetur, a moto et contigue movetur . . .' *Thomae Hobbes Malmesburiensis opera philosophica quae latine scripsit ommia*, ed. Sir William Molesworth, 5 vols. (London, 1839–45), **1**, 272.
32. *Concerning Body*; *English Works*, **1**, 131.
33. *Concerning Body*; ibid., **1**, 531.
34. *Concerning Body*; ibid., **1**, 206.
35. *Concerning Body*; ibid., **1**, 207.
36. *Concerning Body*; ibid., **1**, 215–16.
37. *Concerning Body*; ibid., **1**, 213. There can be no cause of motion, Hobbes asserted, except a body contiguous and moved. By the same reason, whatever is moved will continue to be moved in the same direction with the same motion unless it is hindered by some other body that is contiguous and moved, 'and consequently . . . no bodies, either when they are at rest, or when there is an interposition of vacuum, can generate or extinguish or lessen motion in other bodies. There is one [Descartes] that has written that things moved are more resisted by things at rest, than by things contrarily moved; for this reason, that he conceived motion not to be so contrary to motion as rest. That which deceived him was, that the words *rest* and *motion* are but contradictory names; whereas motion, indeed, is not resisted by rest, but by contrary motion.' (*Concerning Body*; ibid., **1**, 125.) 'Mutation therefore is motion, namely, of the parts either of the agent or of the patient. . . . And to this it is consequent, that rest cannot be

the cause of anything, nor can any action proceed from it; seeing neither motion nor mutation can be caused by it.' (*Concerning Body*; ibid., **I**, 126.)

38. 'Nam conari simpliciter idem est quod ire.' *De corpore*; *Opera*, **I**, 271. Author's translation.
39. *Concerning Body*; *English Works*, **I**, 211.
40. *Concerning Body*; ibid., **I**, 351.
41. *Concerning Body*; ibid., **I**, 353-4.
42. *Concerning Body*; ibid., **I**, 344-5.
43. *Concerning Body*; ibid., **I**, 212.
44. *Concerning Body*; ibid., **I**, 217.
45. *Concerning Body*; ibid., **I**, 342.
46. *Concerning Body*; ibid., **I**, 385.
47. *Concerning Body*; ibid., **I**, 346-7.
48. *Concerning Body*; ibid., **I**, 348.
49. 'The velocity of any body, in whatsoever time it be moved, has its quantity determined by the sum of all the several quicknesses or impetus, which it hath in the several points of the time of the body's motion. For seeing velocity . . . is that power by which a body can in a certain time pass through a certain length; and quickness of motion or impetus . . . is velocity taken in one point of time only, all the impetus, together taken in all the points of time, will be the same thing with the mean impetus multiplied into the whole time, or which is all one, will be the velocity of the whole motion.' (*Concerning Body*; ibid., **I**, 218-19.) He saw that distance traversed is proportional to the whole velocity, and he went on to try to compute the distance traversed when impetus increases as the square, the cube, the fourth power, and the fifth power of time (ibid., **I**, 224-7).
50. Cf. the following problem: 'Any length being given, which is passed through in a given time with uniform motion, to find out what length shall be passed through in the same time with motion uniformly accelerated, that is, with such motion that the proportions of the lengths passed through be continually duplicate to that of their times, and that the line of the impetus last acquired be equal to the line of the whole time of the motion.' (*Concerning Body*; ibid., **I**, 237.)
51. Giovanni Battista Baliani, *De motu naturali gravium solidorum*, (Genoa, 1638), p. 9.
52. Ibid., pp. 8, 23.
53. Marin Mersenne, *Harmonie universelle*, (Paris, 1636), p. 124.
54. Ibid., p. 108.
55. Ibid., p. 103.
56. Ibid., p. 120.
57. Ibid., p. 130.
58. Ibid., p. 142.
59. Ibid., p. 147.
60. 'Impulsus est virtus seu qualitas locomotiva, quae non nisi in tempore, & per spatium movet finitum.' Johannes Marcus Marci, *De proportione motus seu regula sphygmica ad celeritatem et tarditatem pulsuum ex illius motu ponderibus geometricis librato absque errore metiendam*, (Prague, 1639), Propositio I. The work is not paginated.

61. Ibid., Propositio II.
62. Ibid., Propositiones V and VI.
63. Cf. ibid., Propositio XXXVII and the fourth Problema near the end.
64. For a body to move not just any impulse suffices 'sed [impulsum] proportionatum illi mobili: impulsus enim, quo globus ligneus ad motum concitatur, haud quaquam loco movebit pilam ferream ejusdem molis aut maiorem . . .' Ibid., Propositio XXXVII.
65. Ibid., Propositio VIII.
66. 'Velocitas a principio motus per lineam perpendicularem est aequalis gravitati, minor vero per lineam inclinatam.' Ibid., Propositio VIII.
67. Ibid., Propositio IX.
68. Ibid., Propositio IX.
69. Ibid., Propositio XII. It is true that Marci rather bungled the demonstration. He stated the proposition in the following manner: 'Incrementa velocitatis [sic] rationem habent quam temporum quadrata.' The demonstration then proceeded to confuse velocity (or impulse) with distance traversed: 'Quia virtus locomotiva eo modo augetur, quo triangulum sibi simile manens . . . propterea quod hujus augmentum sit perfectio intensiva; cum ex illo puncto quietis veluti latescit, angulum constituit sui augmenti, majorem minoremve pro cuiusque perfectione, quam obtinet in principio motus, sive ex natura sua, sive ex impedimento: majori enim perfectioni maior angulus debetur.' Setting the angle at 45° for the demonstration, he divided the time into equal periods or minutes, *ab*, *bc*, *cd*, etc.; 'velocitas ergo motus augetur impulsu augescente in primo quidem minuto in *hb*, in 2. in *ic*, in 3. in *kd*, atque ita consequenter aequata area illius trianguli rectanguli, cujus longitudo numerus minutorum, basis vero terminus augmenti. Quia vero eadem est ratio motus & virtutis impulsivae, virtus quidem dupla in eodem aut aequali tempore movebit per spatium duplum,' therefore if the virtue latent in the point *a*, with which the velocity increases equally, increases to *hb* in the first minute, to *ic* in the second minute, and to *kd* in the third, the distance traversed in two minutes will be to that traversed in one as the triangle *iac* to the triangle *hab*. Therefore the motion of two minutes will be to the motion of one minute as the square of the side *ic* to the square of the side *hb*. 'Itaque si quadratum lateris *ab*, hoc est primi minuti, subtrahas a quadrato *ac* secundi minuti, numerus reliquus dabit velocitatem motus in eodem minuto . . .'
70. Marci's most perceptive discussion of acceleration, a largely qualitative discussion, occurs in Propositio XX, on the motion of a pendulum. By considerations of symmetry he showed that to the increment of impulse (as measured by the inclination of the tangent) at each point in the pendulum's descent there is a corresponding equal decrement of impulse during its ascent, and hence it stops after an ascent equal to its descent. In Propositio XXIII, which also discusses pendulums, Marci revealed the limitations of his conceptual scheme again. The proposition states that a pendulum moves through equal arcs of the same circle in unequal times, the times being longer the closer the arcs are to the perpendicular. Let *a* be the centre around which the pendulum oscillates, *ab* the perpendicular, and *bd* and *df* equal arcs (see Fig. 9). The motion in *df* will be swifter than the motion in *bd*. 'Quia enim motus per arcus ejusdem circuli rationem habent,

quam sinus, ... est autem sinus *bg* major sinu *bt*, erit velocior motus in *f* quam in *d*: & quia arcus *bd*. *df* sunt aequales, minori tempore movebitur in arcu *df* remotiore, quam in arcu *bd* stationi [i.e., the perpendicular] propiore ...' To the objection that the motions are not uniform, he replied that the velocity collected from *f* to *d* is greater than the velocity collected from *b* to *d*. 'Quia enim velocitatis ex *d* in *b* continuo quoque minora fiunt incrementa; velocitas inde collecta erit minor velocitate ab aequalibus ipsi *d* incrementis collecta: at vero

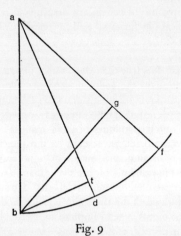

Fig. 9

velocitas in *f* majora ex *f* in *d* sumit incrementa, quam ut aequalia sint velocitati in *d*: velocitas ergo ex *f* in *d* collecta est multo major velocitate ex *d* in *b* collecta, ac proinde minori tempore illos arcus perambulat aequales.' The discussion continually confused increments in velocity at individual points with the velocity itself. If we ignore his problems of expression, indicative of the difficulties that a whole generation experienced as it learned to discuss accelerated motion, we can interpret Marci's proposition in a sense acceptable to us by taking the arc *bd*, not as the second half of the greater arc *fb*, but as a separate arc in which the pendulum swings, starting from rest at *d*. In this sense the proposition is trivial, of course; since all swings are taken to be isochronous, the first half of the larger swing is bound to take less time than the shorter swing.

71. Propositio XI states: 'Impulsus in quolibet motu seu recto [i.e., vertical], seu inclinato est major gravitate.' The discussion includes the following: 'Motum in quolibet puncto lineae perpendicularis esse majorem sua gravitate nullum est dubium: nam cum velocitas cum ipso motu incipiat augeri, sicuti a principio est aequalis gravitati, ita in progressu erit major gravitate.' With the premises he accepted, his demonstration that every motion on an inclined plane is also greater than gravity was, of course, fallacious.
72. Ibid., Positio [sic] II.
73. Ibid., Propositio X.
74. Ibid., Positio VI.

MECHANICS IN THE MID-CENTURY

75. Ibid., Propositio XIII. In Propositio XXXIX he attempted to use the analysis to demonstrate that the angle of reflection is equal to the angle of incidence when a ball bounces against a plane. The inclined plane serves as the plane against which the ball bounces, and the angle of incidence is the angle between the plane and the vertical. He found the line of reflected motion by compounding a motion perpendicular to the plane with one parallel; he determined the size of the latter from his analysis of the inclined plane.
76. Ibid., Propositio XXX.
77. Ibid., Propositio XXXVII. Since the blow does not derive from the mere contact of two bodies but rather 'ex irruptione violenta' by which the percussant as it were penetrates the body struck, the blow requires motion and motion requires time. In Porismae II and III on percussion, Marci extended this vague doctrine slightly. Percussion takes place through time; as the blow builds up, it moves the body originally at rest until the velocities are equal, at which time, of course, the percussion ends. When the percussant is the larger body, its impulse can impart an equal velocity to the other body without being entirely exhausted. When the percussant is smaller, however, the other body cannot reach the percussant's original velocity since the impulse is entirely exhausted in moving the larger body with a lesser velocity.
78. Ibid., Propositio XXXVII.
79. 'Ex quo fit manifestum illorum velocitatem, quae in motu se percutiunt, a percussione permutari: quae enim magis percutiunt, minus; & quae minus percutiunt, magis impetuose reflectunt.' Ibid., Porisma V.
80. 'Itaque fit ut ex illa inaequali plaga, velocitate ferantur inaequali, minori quidem majus ob vires a percussione accisas & mutilates, majori vero minor ob easdem vires de integro acquisitas.' Ibid., Propositio XXXVII.
81. Ibid., Porismae III and IV.
82. When two bodies that are not equal in violence strike each other, the larger is not brought to rest 'propterea quod minus non habeat rationem hypomochlij ad majus . . .' Ibid., Propositio XXXVII.
83. 'Quia enim major est impulsus minoris gravitate majoris, ob minorem hujus quam illius rationem, si minor percutiat majorem, movebitur ex illa plaga major: reflectit autem & minor a majori, propterea quod a quacunq; hujus plaga movetur minor.' Ibid., Porisma VI.
84. Evangelista Torricelli, *Lezioni accademiche*, (Firenze, 1715). Cf. the opening paragraph of *De motu gravium*: 'Scientiam de motu . . . a pluribus quidem tractatam, ab unico (quod ego sciam) Galileo Geometrice demonstratam, aggredi libet. Fateor, quod ille totam hanc segetem tamquam falce demessuit, nec aliud superest nobis, nisi ut tam seduli messoris vestigia subsequentes, spicas colligamus, si quae ab ipsa vel relictae fuerint, vel abiectae: sin minus, ligustra saltem, et humi nascentes violas decerpamus; sed fortasse et ex floribus coronam contexemus non contemnendam.' (*Opere*, ed. Gino Loria and Guiseppe Vassura, 4 vols. (Faenza, 1919-44), **2**, 103, cited with permission of the Comune di Faenza.)
85. *Lezioni accademiche*; ibid., **2**, 30.
86. *Lezioni accademiche*; ibid., **2**, 22.
87. *De motu proiectorum*; ibid., **2**, 159.

143

88. *De motu gravium*; ibid., **2**, 125.
89. *De motu gravium*; ibid., **2**, 104.
90. 'Praemittimus: Duo gravia simul coniuncta ex se moveri non posse, nisi centrum commune gravitatis ipsorum descendat.

    'Quando enim duo gravia ita inter se coniuncta fuerint, ut ad motum unius motus etiam alterius consequatur, erunt duo illa gravia tanquam grave unum ex duobus compositum, sive id libra fiat, sive trochlea, sive qualibet alia Mechanica ratione, grave autem huiusmodi non movebitur unquam, nisi centrum gravitatis ipsius descendat. Quando vero ita constitutum fuerit ut nullo modo commun ipsius centrum gravitatis descendere possit, grave penitus in sua positione quiescet: alias enim frustra movebitur; horizontali, scilicet latione, quae nequaquam deorsum tendit.' *De motu gravium*; ibid., **2**, 105.
91. *De motu gravium*; ibid., **2**, 106.
92. *De motu gravium*; ibid., **2**, 108.
93. For example, in a miscellaneous note, Torricelli spoke of two heavy bodies 'di diversa gravità in if specie ma di mole eguali . . .' *De motu ac momentis varia*; ibid., **2**, 238. '*Mole*' can only be translated by volume here.
94. *Lezioni accademiche*; ibid., **2**, 22–3.
95. *Lezioni accademiche*; ibid., **2**, 6–8.
96. Cf. *De motu proiectorum*; ibid., **2**, 228–30.
97. *Lezioni accademiche*; ibid., **2**, 8–9. At the beginning of the third lecture, Torricelli summarised what he had asserted in the earlier lectures – 'che la gravità ne i corpi naturali è una fontana continuamente aperta, la quale ad ogni istante di tempo, o (se non piacciono gli istanti) ad ogni brevissimo tempo, produce un momento eguale al peso assoluto di detti corpi. È ben vero, che quando i gravi stanno quiescenti, tutti gl'impeti prodotti se ne trascorrono via, venendo o ricevuti, o annichillati dal corpo sottoposto, il quale col contrasto dell'indiscreta repugnanza va continuamente estinguendo tutti quei generati momenti. Ma quando i medesimi gravi cadono per l'aria, quegl'impeti non s'estinguono più, ma si conservano là dentro, e vi si moltiplicano: a però quando i gravi velocitati arrivano a percuotere, la forza, o virtù loro deve essere infinitamente accresciuta.' (*Lezioni accademiche*; ibid., **2**, 15.)
98. *Lezioni accademiche*; ibid., **2**, 30.
99. *Lezioni accademiche*; ibid., **2**, 27.
100. *Lezioni accademiche*; ibid., **2**, 10.
101. *Lezioni accademiche*; ibid., **2**, 10–11.
102. *Lezioni accademiche*; ibid., **2**, 11.
103. *Lezioni accademiche*; ibid., **2**, 12.
104. *Lezioni accademiche*; ibid., **2**, 29.
105. *Lezioni accademiche*; ibid., **2**, 12.
106. *Lezioni accademiche*; ibid., **2**, 16–17.
107. *Lezioni accademiche*; ibid., **2**, 15–16.
108. *Lezioni accademiche*; ibid., **2**, 18.
109. *Lezioni accademiche*; ibid., **2**, 18.
110. *Lezioni accademiche*; ibid., **2**, 19.
111. *Lezioni accademiche*; ibid., **2**, 30–2.

112. *Lezioni accademiche*; ibid., **2**, 27–8. Torricelli briefly introduced one other example. Suppose that in order to knock down a wall someone were to push on it all day. We would certainly laugh at this latter day Samson. But if '*le forze*' produced by him could be conserved and concentrated in some receptacle, applied to the wall they would knock it down like those of Jericho. (*Lezioni accademiche*; ibid., **2**, 28–9.)
113. Torricelli, *Lezioni accademiche*, xlvii–xlviii.
114. *Lezioni accademiche*; *Opere*, **2**, 5.
115. *Lezioni accademiche*; ibid., **2**, 28.

Chapter Four

Christiaan Huygens' Kinematics

THE name of Christiaan Huygens cannot be omitted from any list of the prime contributors to the science of mechanics in the seventeenth century. Between Galileo and Descartes at the beginning of the century, on the one hand, and on the other, Leibniz and Newton at its end, Huygens stood in the middle as the indispensable link whereby the endeavours of the age acquired continuity and retained coherence. Dutch by birth, son of a prominent father who was widely acquainted throughout the intellectual circles of Europe, Huygens was exposed to the achievements of Descartes and Galileo while he was still a boy, and he went on to become the greatest scientist of the later seventeenth century before the advent of the Newtonian age. When Colbert founded the *Académie royale des sciences* in 1666, Huygens, already established in Paris as a leading scientist, was automatically appointed as a member despite his nationality. As the heir of Galileo and Descartes, he assimilated their achievements. As a scientist too brilliant to be merely a student, he advanced their work to a new level of sophistication. As the predecessor of Leibniz and Newton, he prepared the way for their creation of a quantitative dynamics. Much as he contributed to it, Huygens did not himself create a dynamics. The peculiar nature of his combined legacy from Galileo and Descartes both enabled him to proceed to the point where the principles of dynamics lay open before him, and restrained him at that point from advancing further to make them his own. The ultimate key to Huygens' mechanics is the mutual interplay of the influence of Galileo and Descartes, of the Pythagorean and Democritean traditions in seventeenth-century science.

The first problem in mechanics to which Huygens devoted extensive study was impact. A passage written in 1656 and intended as part of a preface to the completed work stated the rationale for the study in terms which set the tone for his entire scientific career. Impact is more than a problem in mechanics; it is central to the whole of natural philosophy.

> For if the whole of nature consists of certain particles, from the motion of which all the diversity of things arises, and by the extremely rapid impulse of which light is propagated and spreads through the immense spaces of the heavens in a moment of time, as many philosophers deem probable, this examination [of nature] will seem to be helped no small amount if the true laws by which motion is transferred from body to body be made known.[1]

Here spoke the mechanical philosopher, the son of Descartes' friend Constantijn Huygens, and through his father the acquaintance and disciple of Descartes. In a world which consists solely of particles of matter in motion, the sole means of action is the impact of particle on particle, and this Huygens set out to investigate.

The relation of Descartes to Huygens' investigation of impact was not limited to his influential role in establishing the mechanical philosophy of nature. The specific origin of the study was Descartes' own attempt to establish the laws of impact, and the particular temper of the study reflects the fact that from its beginning it was devoted to the refutation of Descartes' rules. First he found that the rules did not conform to experience. Next he found that they were inconsistent among themselves.[2] In fact, nothing solid was known about the nature of percussion – 'nothing ... is more frequent in use and more powerful in effect, but of no inquiry are the principles equally unknown.'[3] The ancients had done nothing at all with the question. Galileo, who had explained so much about motion, had published nothing about percussion beyond a few remarks on its power. After Galileo, a few desultory efforts had resulted in nothing that was consistent with experience or demonstrated with rigour. Almost alone, Descartes had proposed a complete set of rules, and his rules were incorrect.

In the intellectual circle in which Huygens travelled, the refutation of Descartes was a task not lightly undertaken. His mentor, van Schooten, urged him to abandon the study; and though he persisted to its conclusion, he did not venture to publish more than brief résumés of his conclusions, and not even those until a decade and more had elapsed.[4] Nevertheless, Huygens was fully convinced that he was right and Descartes wrong. No doubt his conviction derived in large measure from his procedure. Although he confronted Descartes' rules with experimental results, and regarded the agreement of his own with experiments as a considerable argument in their favour, he had not derived his rules from experiments and did not consider them to rest ultimately on such a foundation. What Huygens had done primarily was to confront Descartes with Descartes and to demonstrate his error from his own principles. To state it thus, however, is to tell only half the story. It was not entirely by accident that Huygens seized, as the subject of his first major effort in mechanics, the one point at which Descartes had attempted to reduce the motions of bodies to exact mathematical description. Nor was it entirely by chance that Huygens indicated his familiarity with Galileo's published discussion of impact. If Huygens convicted Descartes with his own principles, they were equally the principles of Galileo, and what Huygens undertook was the incorporation of impact into the body of rational mechanics represented in the seventeenth century by the work of Galileo. Above all, his study of percussion was the confrontation of the Democritean tradition by the Pythagorean. As such, it supplied the initial strand to the central thread of his career.

The first version of his treatise on impact, composed in 1652, began with the statement of a premise taken directly from Descartes.

> If two equal bodies that are perfectly hard and are moved with equal speed in opposite directions collide with each other, each one will be reflected in the direction from which it came having lost no part of its speed.[5]

How is the assumption to be interpreted? In the final treatise, substantially the same assumption was posited on an implicit appeal to symmetry. Descartes had justified it on dynamic grounds, however, and the overwhelming indication is that Huygens proposed it initially on the same grounds. Equality in the bodies' sizes and speeds expresses itself in the equilibrium of their forces of motion, and the equilibrium of forces leads to their equal reflections. The second case, for example, involved two bodies, A and B, also moved with equal and opposite speeds but differing in size such that $B = 2A$. If half of B hit A, he reasoned, it would be reflected with its original speed. The other half of B would try to continue in the initial direction with the same speed. Hence, the two halves cancel each other's motion and B comes to rest.

Since $\frac{1}{2}B$ would suffice to reflect A with its original speed, and since, in fact, it is impelled by twice as much, A is reflected with twice its original speed.[6] The appeal to dynamic considerations in this case cannot be mistaken. Although Huygens crossed the passage out and replaced it with a different problem, in which A at rest is equal to B which strikes it, he also referred the new case to dynamic factors. The 'force of collision,' he said, is the same whether A is at rest and B moves with a given speed, or A moves to the right and B to the left with half that speed.[7] Because of the relativity of motion, case two reduces to case one, and the equal forces of the two bodies cause equal changes in each other's motion. He also added an axiom which virtually defined a concept of force from the model of impact.

> The same force which gives a certain velocity to a body at rest is able to confer half that speed on a body that is twice the other.[8]

Years later, Huygens' initial reaction to Leibniz' concept of *vis viva* evoked the image of his original attack on impact.

> Leibniz asserts that it is necessary to say that motive forces are in a ratio composed, not of the bodies and of the speeds as such [en générale], but of the bodies and of the heights which produce speed, that is, of the bodies and of the squares of the speeds. But one can ask him why in that case, when two bodies, which according to him have equal motive forces, meet and strike each other squarely, each does not retain its original speed in the rebound. For example, if the body B is four times A, let the speed of A be double that of the body B. Here their motive forces are equal according to M. Leibniz. And therefore it seems that when they meet at C, one of these forces ought not to prevail over the other, but that each body ought to turn back with the speed that it had. Which, however, is not the case. But that result occurs when the speed of A is four times that of B. It appears then that one should rather say that the motive forces are equal in this last case, and not when one prevails over the other.[9]

In the second version of the tract, composed as the first in 1652, Huygens employed the same reasoning to extend his treatment from equal bodies to unequal ones.

> ax. 3. If a larger body A strikes a smaller body B, but the velocity of B is to the velocity of A reciprocally as the magnitude A to B [sic], then each will rebound with the same speed with which it came.
>
> If this is granted, everything can be demonstrated. Descartes is forced to grant it however.[10]

On what grounds was Descartes forced to grant it? Surely on the grounds of its implicit appeal to the equilibrium of forces, calculated in the very way that Descartes had calculated such forces. Already the aborted second case of the earlier draft had taught Huygens how deceptive the appeal to forces could be; and to the present case he added the further note, 'But it must be seen whether it can be demonstrated from principles that are known better.' At the same time, he crossed out the label 'ax. 3.'

What principles were known better? As axiom one of the second version, he had initially stated the conclusion he had earlier drawn as the only valid part of the original second case of version one – 'The speed of separation after the impact of two bodies is the same as the speed of approach was.'[11] This axiom he likewise cancelled, with the note that it too could be demonstrated from another axiom more basic. In turn, that axiom, which remained as one of the five hypotheses of the ultimate treatise, made its initial appearance further down the same sheet, circled with a wavy line to indicate its importance.

> Axiom. If two bodies coming from opposite directions strike each other, and one of them is carried back with the same speed with which it approached, losing none of its motion, the other one also will rebound with that speed with which it approached.[12]

If it is possible to reconstruct the progress of Huygens' thought, it appears to have followed an ever broadening realisation of the import of the principle he had employed with the amended second case of version one. To establish the identity of all cases in which the two bodies are equal, he had imagined that the space in which they are moving also moves with a uniform velocity, and by adjusting the quantity of that velocity, he could adjust any impact of equal bodies to appear to an observer outside the moving space as the impact of equal bodies with equal and opposite velocities. In the first version, he had even crossed out the word 'space' and replaced it with the word 'boat', adopting the device that continued through the ultimate version of the treatise. Every impact occurs on a boat coasting smoothly along a Dutch canal, its speed adjusted to the experiment in question, and every impact is observed by two men, one on the boat and one on the shore. Both observe the same event. Nothing more than adding and subtracting uniform velocities is required to transpose one and the same event from one frame of reference to the other. Here, of course, was the principle of the relativity of motion, Descartes' own principle turned against his own conclusions. In the ultimate version of the treatise, the principle of relativity was made explicit as one of the five hypotheses. Equally the new axiom of version two could be traced to Descartes. It asserted, in effect, that in impact, and in impact alone, is motion transferred from one body to another; and hence if one of two bodies conserves its original motion, the other must do so as well. As throughout his consideration of impact, Huygens was dealing with what he called perfectly hard bodies, which by definition yield the same results in impact as the other ideal bodies we call perfectly elastic, although in Huygens' view hard bodies were not elastic in any way. What the principle of relativity suddenly revealed was that every impact can be observed from a frame of reference in which one body (and hence the other as well) rebounds with its original speed. As he indicated, the equality of the speeds of approach and separation follows then immediately from the new axiom. He saw intuitively as well that the frame of reference in relation to which each body rebounds with its original speed is that in which the ratio of speeds is reciprocally as the ratio of sizes, although he later realised that this conclusion does not follow necessarily from the axiom given and requires a separate proof based on additional principles. Meanwhile, if the speed of separation always equals the speed of approach, not only are individual rules of impact as given by Descartes wrong, but his ultimate principle of the conservation of motion is wrong. Let a smaller body strike a larger body at rest. Since their speed of separation equals their speed of approach, the sum of their two speeds must equal the initial speed of the smaller body; but since part of the motion now belongs to a larger body, the total quantity of motion has increased. In one frame of reference, of course, the quantity of motion remains constant. But this is only to state in more general terms, as Huygens realised, the ultimate incompatibility of Descartes' two principles, the conservation of motion and the relativity of motion. For any two bodies, the quantity of motion varies with the frame of reference; if motion is in fact relative, quantity of motion cannot be an absolute value.

The effect of the changes introduced was to alter the tone of the treatise as well as its logical structure. With each successive draft, the kinematic content grew at the expense of the dynamic, emphasising that aspect of Descartes' treatment which had been most promising and minimising that element which had compromised it. Reference to a '*vis collisionis*' dropped out of the text, and Huygens chose to speak instead only of the transfer of motion, even of the 'apparent' transfer of motion since the transfer itself was relative to the frame of reference. Thus, the draft of 1654 explained the impact of equal bodies E and D. It is agreed that when their speeds are equal, they rebound equally. Huygens now rested this assertion, which had to remain an hypothesis rather than a demonstration in the final treatise, not on an appeal to dynamic factors, but, as I have suggested, on implicit considerations of symmetry. Now let the boat move with the same speed as the two bodies, so that E appears to a man on the shore to be at rest. 'Hence it appears that if the body E is at rest in respect to the man G and is impelled by the equal body D, it will receive all its motion from it, and the body D itself will remain motionless in the place E.'[13] The temper of the prefaces changed in a similar manner. Whereas the preface of 1654 discussed the multiplication of forces, the infinite power of percussion, and the importance of comprehending such action, the preface of 1656 gave the force of percussion one short phrase and devoted itself instead to the shortcomings to Descartes' rules.[14] Huygens was unable completely to eliminate dynamics from the work. His final demonstration that the speed of separation equals the speed of approach slipped in the assumption that a smaller body in motion striking a larger one at rest gives it a speed less than the original speed of the smaller body.[15] Nevertheless, Huygens converted the work into essentially a kinematic treatment of impact. Significantly, the title he gave it was not the standard *De vi percussionis* of similar seventeenth-century treatises, but a uniquely kinematic name, *De motu corporum ex percussione*.

As he proceeded, moreover, that which had been implicit from the beginning became ever more explicit and ever more clear. If Descartes' principle of the conservation of motion was incorrect, as he had demonstrated it to be, what was the ultimate principle on which the treatment of impact must rest? The principle of relativity and the principle that both bodies conserve their motions if one of them does were, to be sure, basic, but there was in fact a principle even more basic on which they rested as well. From the beginning, Huygens had employed the concept of inertia, and in the ultimate treatise he stated it as Hypothesis I. What he realised with increasing clarity was that two bodies in impact can be considered as one concentrated at their centre of gravity. Torricelli's principle, confined by Torricelli to heavy bodies constrained by a simple machine to move together vertically, was generalised by Huygens to the motion of bodies not constrained together in any way, beyond their mutual constraint to the principle of inertia. The frame of reference in relation to which each body rebounds with its original speed – what is it but the centre of gravity's frame of reference? And if the bodies recede from the centre of gravity with the same speeds by which they approach, the centre of gravity itself undergoes no change in motion.

> It must be noted, however, that according to what I have said, the center of gravity of bodies taken together continues always with a uniform motion in the same direction and is not disturbed by any impact of the bodies.[16]

In realising that two bodies in impact can be considered as one concentrated at their centre of gravity, Huygens converted the study of the force of percussion into an extension of Galileo's kinematics of uniform motion.

It is necessary to be explicit on this point. If *De motu ex percussione* stemmed initially from Descartes' rules of impact, the model for the treatise was Galileo's *Discorsi* and its effort to reduce the phenomena of motion to geometrical description. Like the final two days of the *Discorsi*, *De motu ex percussione* constructed its argument on the Archimedean pattern of axioms, lemmas, and propositions. Like the *Discorsi*, it sought to express the properties and motions of bodies quantitatively. Velocity emerged here as a physical quantity more clearly than it had in Galileo's writings.

> Therefore we measure the ratio of velocities by the ratio of distances which are traversed in the same or equal times. So that when body *A* is said to move with velocity *AC* and *B* at the same time with velocity *BC*, it is to be understood that in the same interval of time *A* would traverse the distance *AC* and *B* the distance *BC*, the velocities maintaining that ratio between themselves which is the ratio of the lines *AC* and *BC*.[17]

Whereas Galileo's diagrams presented the paths a body follows and only incidentally its velocities, Huygens' presented the velocities and only incidentally the paths that it follows. The change marked a significant step forward in the construction of a quantitative mechanics.

In its ultimate form, essentially completed in 1656 although the definitive text dates from a later period, *De motu corporum ex percussione* began with three hypotheses – the principle of inertia, the symmetric case of equal bodies with equal and opposite velocities, and the principle of relativity.[18] With these three, he solved all possible cases of equal bodies in impact. To deal with unequal bodies, he had to postulate two more hypotheses – that a larger body striking a smaller body at rest gives it some motion and consequently loses some of its own, and that when one body in impact conserves all its motion the other one also neither gains nor loses motion.[19] By applying the principle of relativity to the first of the two additional hypotheses, Huygens reversed another conclusion of Descartes. 'A body however large is moved in impact by a body however small that has any speed whatever.'[20] When the frame of reference is reversed, the motion lost by the larger body when it sets the smaller one in motion becomes the motion caused by the impact of the smaller body on it. Proposition III, as it was numbered, reaffirmed the inert nature of matter fundamental to the mechanical philosophy. Galileo had asserted that any force can move any body however large, but his conceptual equipment had been limited to the treatment of vertical motions alone. What Huygens' proposition added was the means of calculating the motion generated in horizontal planes, to be sure not when a force sets a body at rest in motion, but when a body given in size and velocity strikes another at rest.

Using the fifth hypothesis, Huygens attacked the central case in the impact of unequal bodies, the case in which velocities are inversely proportional to the bodies themselves. In the early drafts, he had assumed that the hypothesis stated the conditions of this case. Now he realised that either he must broaden the scope of the hypothesis, thereby weakening its claim to intuitive acceptance and admitting covert dynamic assumptions, or he must demonstrate that when velocities are in inverse ratio to magnitudes the conditions of Hypothesis V obtain. In order to demonstrate it, he appealed to yet another assumption, introduced as 'the most certain axiom in mechanics' – that 'the common centre of gravity of bodies cannot be raised by that motion of the bodies which is generated by the gravity of those bodies themselves.'[21] To apply the axiom to his problem, Huygens employed Galileo's analysis of the uniform acceleration of free fall. Imagine that the speeds before impact, inversely proportional to the bodies, were generated by falls through distances which, from Galileo's kinematics, are proportional to the squares of the speeds. In rebound, assume any possible pair of speeds whatever, other than the original ones, and it can be demonstrated that the common centre of gravity of the two bodies, placed at the heights to which those speeds could raise them, would be higher than their common centre of gravity when they are in the positions from which they were imagined initially to fall.[22] Unless the bodies rebound with their original speeds, a perpetual increase in motion is possible, and a perpetual increase in motion is absurd. Huygens had called again on the insight that an isolated system of two bodies can be treated as one concentrated at their centre of gravity. As Galileo had shown in the case of a single body, such a system gains sufficient motion in falling to raise it again to its original height.

By imagining the centre of gravity in vertical motion, Huygens had substantiated his intuition that the inertial motion of the centre of gravity in a horizontal plane is not disturbed by the mutual impact of bodies. He considered the frame of reference of the centre of gravity to be the ultimate frame of reference for any impact; and from its point of view, nothing whatever happens. Impact is not a dynamic action. It is purely kinematic. Bodies are not brought to rest and set in motion anew by a force of elasticity of some sort. This had been the point of view of Torricelli, and others, such as Wallis and Mariotte, would adopt it. Huygens saw that it required the presence of a dynamic action, and the entire development of his analysis had been directed toward the elimination of such considerations. Such was the point of his insistence on perfectly hard bodies rather than perfectly elastic ones. When soft bodies meet in impact, they are gradually brought to rest through a short period of time; they do not rebound because there is nothing to give them a new motion.

> But with hard bodies the situation is other, for their speed continues always without being interrupted or diminished, and therefore it is not surprising that they rebound.[23]

The impact of hard bodies is instantaneous.[24] When bodies with speeds inversely proportional to their sizes meet in impact, each conserves its original motion. Huygens was careful always to put it in these terms – not that they rebound with speeds equal to their original speeds, but that they conserve the motions they had. Their motions are never interrupted; their directions are merely changed. But every impact is between bodies with speeds inversely as the bodies – from the point of view of the centre of gravity.

In his eagerness to emphasize his correction of Descartes, Huygens may have failed to realise the extent of his continuing debt. Implicit in his view of impact was the Cartesian distinction between the speed and the direction of a body's motion, and the further conviction that a mere change of direction involves no dynamic action whatever. Although he stated the principle of inertia in unexceptionable terms – that a body continues its uniform motion in a right line unless something opposes it – he did not draw the conclusion which seems so obvious to us, that a change of direction, as an alteration of the inertial state, is dynamically identical to a change of speed. For that matter, he sought to treat changes of motion in purely kinematic terms as well, speaking of the transfer of motion from one body to another but not of its dynamic destruction and generation. Involved in this view also was the Cartesian conviction that motion is always a positive quantity. In fact, Huygens realised that the conservation of motion could be salvaged as a principle of impact if one added the condition that the motion conserved is motion in a given direction.[25] In effect, this idea would have converted the concept of quantity of motion into the concept of momentum, a vectorial quantity capable of taking on a negative value. The concept did not please him greatly, however, and he continued to speak formally of quantity of motion in the Cartesian sense, a scalar quantity which is always positive in value. Motion in this sense of the word suffers no change in impact, from the point of view of the centre of gravity. The effortless movement from one frame of reference to another, with its consequent change in the apparent loss, gain, or conservation of motion, merely expresses the ultimately kinematic nature of impact in which the transfer of motion is accomplished without dynamic action. In the rigorously geometric world of the mechanical philosophy, the ultimate particles of which Huygens took to be his perfectly hard bodies, matter in motion can accomplish everything without the intrusion of an alien principle of force. With a final twist of irony, the mechanical philosophy, which constantly encouraged impact as the model of dynamic action, suddenly revealed that it could do without that one as well.

Even though it was valid in the frame of reference of the centre of gravity, and in other frames in relation to which neither body reverses direction, Huygens was wholly unwilling to grant Descartes' principle of the conservation of motion. The relativity of motion, which makes it possible to view every impact from the centre of gravity, makes it equally possible to view every impact from other frames as well, and motion is not conserved in every frame. As Huygens realised, however, another quantity is.

> When two bodies strike each other, the quantity which is obtained by adding the products of the magnitudes of the individual bodies multiplied by the squares of their velocities [*id quod efficitur ducendo singulorum magnitudines in velocitatum suarum quadrata, simul additum*], is found to be the same before and after the impact of the bodies.[26]

The exit of the Cartesian 'quantity of motion', that is, was only the cue for the entrance of a new quantity (mv^2) onto the stage of rational mechanics. (A minor anachronism is involved in the use of the symbol *m*, of course. As I shall later discuss, Huygens did not arrive at a wholly clear conception of mass.) In some sense, the quantity had been implicit already in Galileo's kinematics of free fall, in the demonstration that the velocity of a body falling from rest is proportional to the square root of the distance it has fallen (or $v^2 = 2as$ in our terms). Perhaps Huygens' derivation of the new equation of conservation was connected to his perception of the relation of free fall to the impact of bodies inversely proportional to their speeds. His demonstration of that case, it will be recalled, employed the kinematics of vertical motion. The source of his insight is purely speculative, however. What is certain is the fact that Huygens first made explicit a quantitative relation destined to play a central role in the future history of mechanics.

What was the advantage of mv^2? Unlike the Cartesian quantity, it did not conflict with the relativity of motion. It was valid in every frame of reference. If we examine it from this point of view, we can understand its significance in Huygens' eyes. Obviously, the sum of the products of magnitudes and velocities squared is not a constant for all frames of reference. What is constant is the sum before and after impact in each frame of reference. From Huygens' point of view, mv^2 did not replace quantity of motion as the measure of a body's force. Dynamic concepts had dropped out of his treatment of impact once and for all. Impact did not present itself to him as a model of force. As a measure of force, the new quantity is no more reconcilable with the relativity of motion than the old; every frame of reference gives its own value. Instead of a measure of force, Huygens believed he had discovered a new kinematic formula, one that was valid for every frame of reference as Descartes' was not. The sum of the products of magnitudes and velocities squared was merely a number, however, a number obtained by a given arithmetical operation, a number wholly lacking in metaphysical significance. Toward the end of his life, under the tutelage of Leibniz to whom he himself had taught the conservation of mv^2, Huygens appears to have granted the number a somewhat broader significance. When he encountered Leibniz's conservation of *vis viva* for the first time, however, Huygens' response had all the enthusiasm of a dry cough. 'But he cannot pretend that this principle of the conservation of motive forces will be granted as though it had no need of proof.'[27]

*

If Huygens rejected the model of impact as the measure of force, it was not to embrace the alternative model of free fall, for his approach to free fall was equally kinematic. His study of pendulums entailed considerable attention to the properties and consequences of uniformly accelerated motion. In November, 1659, the investigation of circular motion had led him to an understanding of the conical pendulum and the variation of its period with the square root of the vertical height of the cone. Already, as he well knew, Galileo had established the kinematic significance of the square root of vertical distance. Not only does the time of fall from rest vary with the square root of the vertical distance, but the period of the ordinary simple pendulum also varies as the square root of its length. Even though the latter conclusion was correct, Galileo's demonstration of it was erroneous, as Huygens recognised. Nevertheless, he had clearly had it in mind when he had posited the identity of minimal oscillations of conical pendulums with minimal oscillations of ordinary pendulums. Via that identity, he would be able to proceed from his ability to state the period of the conical pendulum as a function of the acceleration of free fall to the period of the ordinary pendulum as a function of the same value. The periods of both are equal for minimal oscillations. The periods of both vary as the square root of vertical height. The period of an ordinary pendulum, then, is equal to the period of a conical pendulum the vertical height of which is equal to that of the ordinary pendulum. So far, the argument rested on the intuition that minimal oscillations are identical, and the demand for rigorous demonstration, revealed in his treatment of impact, was not likely to be satisfied by mere intuition. It is not surprising, then, that a manuscript dated December, 1659, which contains his discovery that the cycloid, rather than the circle, is the curve of isochronous oscillation, also includes a further statement of purpose.

> It is inquired what ratio the time of the minimal oscillation of a pendulum has to the time of perpendicular fall from the height of the pendulum.[28]

Huygens' investigation of pendulums was an extension of Galileo's kinematics of free fall which culminated both in the rigorous demonstration that the cycloid is the curve of isochronous oscillation and in the derivation of the period of oscillation as a function of the pendulum's length and the acceleration of gravity. Like the investigation of impact, the study of pendulums reveals the basic motivations of Huygens' mechanics.

An early exposition of uniformly accelerated motion in free fall presents an interesting comparison with his initial approach to impact. Although it dates from 1659 and therefore several years after the definitive triumph of the kinematic treatment of impact, it indicates that Huygens initially treated accelerated motion in frankly dynamic terms. Because of the principle of inertia, his starting point again, a heavy body falling retains the velocity it has gained at any moment; its acceleration consists in the fact that new velocity is continually being added to that which the body already has. Should it come to a place 'where no action of gravity is felt,' it would continue to fall with a uniform velocity. There is no such place, of course, 'but the uniform force of gravity is present everywhere.' Hence the motion of free fall is uniformly accelerated.[29] As to the origin of gravity, the young mechanical philosopher had no doubt that it is due to the impact of particles that drive bodies called heavy downwards.[30] Such a view neither modified the essentially dynamic approach of the paper, nor troubled its author in the use of a dynamic vocabulary that was not dependent on his view of the source of gravitational action.

Although the paper frequently adopted a purely kinematic attitude, dynamic considerations reappeared intermittently throughout it. The fourth paragraph, which set out to demonstrate that a perfectly hard body that falls perpendicularly on a surface to which it communicates none of its motion will rebound, in a time equal to that of its fall, to the height from which it fell, carried the demonstration out in dynamic terms. Let the body fall from A. It arrives at B with a velocity by which it could traverse a distance equal to $2AB$ in the time of its fall, and since it loses none of its motion in impact it would rise to a height $2AB$ 'if the force of gravity did not continually endeavour to drive it in a direction that is the very contrary,' and by its contrary impulse cause it to rise only to the height AB. To show that this is so, he imagined a ball C initially at rest on the horizontal table FD 'to be drawn by the force of a spring or by the force of a weight E' attached to the end of a line that passes over a pulley. C moves to the left through the distance FD, reaching D with a velocity by which it could traverse a distance $2FD$ in an equal period of time. Now imagine that the table is on a boat moving to the right with exactly the velocity that C has when it reaches D – a typically Huygensian device. The motion of the boat represents the motion of the original body upward after its impact, of course, and 'the motion by which in fact the ball meanwhile is drawn to the left through FD represents the contrary motion downward by which the heavy body is drawn by the force of its weight.' Moreover, the motion represented by that to the left produced by the spring or weight can never be omitted since 'we have established that the attraction of gravity always acts in the same way and therefore also produces the same effect' whatever the other motion of a body. It follows immediately and obviously that the body rebounds to position A.[31] Similarly, he later demonstrated that 'the distances traversed in the same time by the same body on different inclined planes are to each other as the forces [*potentiae*] by which they can be sustained on each of the planes.'[32] The demonstration spoke explicitly of static equilibrium, it is true, but its utilisation of Galileo's proposition on motion along the chords of a circle implicitly set the force accelerating bodies along inclined planes equal to that which can hold them in equilibrium. Both in these demonstrations and elsewhere in the paper, Huygens simply applied a dynamic interpretation to Galileo's kinematics of uniform acceleration. When he employed the example of the body on a horizontal plane, and when he further replaced the weight on the line with a spring, he took two steps in the direction of generalising the dynamics. Since he did not trouble himself to define any dynamic terms, we must assume that the dynamics he employed appeared obvious.

In 1673, a revised and extended version of the paper above appeared as Part II of Huygens' master work, *Horologium oscillatorium*. It now carried a title reminiscent of that finally attached to his work on impact – *De descensu gravium & motu eorum in cycloide*, Concerning the Descent of Heavy Bodies and their Motion in a Cycloid. As the title implies, the process of revision had been accompanied by a systematic reduction of the dynamic content. Total elimination of dynamic considerations was impossible since the whole depended on the premise, accepted in effect as an axiom, that heavy bodies move downward. With dynamics reduced to that minimal content, he attempted to proceed on exclusively kinematic grounds.

Part II began with three hypotheses.

> I. If gravity did not exist and the air did not affect the motion of bodies, each body having once been set in motion would continue in a right line with a uniform velocity.
>
> II. Now in fact, by the action of gravity, from whatever source gravity may arise, it happens that they move with a motion compounded from their uniform motion in this or that direction and from the motion downwards due to gravity.
>
> III. And each of these can be considered separately and neither obstructs the other.[33]

Proposition I, which established the relations of velocity and time in free fall, was more explicitly dynamic than the corresponding passage in Galileo's *Discourses*. Whereas Galileo had presented uniformly accelerated motion as a definition, Huygens treated it, following the hypotheses, as the continuous alteration of inertial motion by the action of gravity, and showed that if the action of gravity is uniform, velocity increases in proportion to time, and the spaces traversed in equal successive increments of time form a series in which the difference of the terms is constant.[34] In the proposition, he employed phraseology that is implicitly or explicitly dynamical – '*motus à gravitate productus*,' '*actio gravitatis*,' even '*vis gravitatis*.' In the following propositions, he proceeded in purely kinematic terms to derive the relations of distance traversed to velocity and time in uniformly accelerated motion, and then to demonstrate dynamically that with the velocity gained in falling a body will rise again to its initial height.

Compared to the paper of 1659, the dynamic content of the propositions is considerably muted, and a paragraph inserted at this point marked a change of tone which expressed Huygens' uneasiness with even a muted dynamics. The demonstration that a body will rise to the height from which it fell depended, he said, on the proposition that there is a determinate relation between the spaces traversed in equal successive increments of time. This of course was the conclusion from the dynamics of Proposition I, and Huygens did not apologise for the conclusion as such. If it be denied, 'it must be admitted that the investigation of the proportion of these spaces is hopeless.' Nevertheless, he continued, it is possible to demonstrate the spaces traversed in uniformly accelerated motion, without employing the specific conclusion of Proposition I as a premise, by following Galileo's method. Since the first step, the demonstration that the space traversed in any given time by a body falling from rest is equal to half the space it could traverse in the same time with a uniform velocity equal to that acquired at the end of the time in question, had been imperfectly formulated by Galileo, he proposed to replace it with a rigorous one.[35] What followed was a purely kinematic exercise which proceeded on Galileo's definition of uniformly accelerated motion without regard to the dynamics behind it. He divided the time into a succession of equal increments in each of which the velocity is uniform and demonstrated in effect that when $\Delta t \to 0$, $s = \int v \, dt$.[36]

Huygens did not derive anew the ability of bodies to rise to the height from which they fell, although the paragraph quoted above implied that the spaces traversed in uniformly accelerated motion, now demonstrated in kinematic terms, be superimposed, according to the dictates of the third hypothesis, on a uniform vertical motion to arrive at that conclusion. Accepting the conclusion, he proceeded to another proposition crucial to the kinematics of descending bodies, that whatever the inclination of the plane by which it descends, a body acquires the same velocity for the same vertical drop. Again he complained of Galileo's demonstration, this time more vigorously, calling it 'rather insubstantial' [*parum firma*].[37] What he referred to, of course, was the explicitly dynamical demonstration inserted in the second edition of the *Discorsi* in which the acceleration of the body was set proportional to the component of its weight parallel to the plane. Huygens replaced it with one which utilised the demonstrated ability of a body to retrace its path and rise to its original height, and he further assumed that it cannot rise above its original height. Huygens considered the latter assumption, which is equivalent to denying the possibility of perpetual motion, so obvious that he called its negation absurd. Nevertheless, it did introduce a further modicum of dynamic content into the treatise. Suppose, he argued, that a body descending along one plane acquires a greater velocity for the same vertical displacement than along another. If that were the case, it could acquire the lesser velocity in some lesser drop and then by ascending the first plane, which the prior demonstration showed it could do, reach a height above that from which it fell.[38] Only the acquisition of equal velocities for equal vertical displacements is compatible with the earlier conclusion that a body, with the velocity it acquires in descent, can rise again to its original height.

What Huygens had done was to trim the dynamic content implicit in Galileo's kinematics of descending bodies to the bone itself. Despite his conviction that gravity arises from the impact of subtle particles, indeed it would appear because of his commitment to the mechanical philosophy from which that conviction derived, Huygens held the very concept of force under suspicion for the occult tendencies he felt to be implicit in it. More even than Galileo, he tried to base the kinematics of heavy bodies on the fact, which was to be accepted as empirically given, that heavy bodies descend with a uniformly accelerated motion. In the context of the *Horologium*, gravity was not considered as an external force acting on bodies, but only as the empirical fact of their weight. Hence '*gravitas*' and '*pondus*' were interchangeable words, and if he referred on occasion to '*vis gravitatis*,' he tended rather to say '*vis gravitatis suae*,' the force of a given body's weight, the quantitative measure of the fact that bodies are heavy. His usage was intimately connected with the long tradition of the simple machines, in which the 'force' applied to one end of a lever measured the weight it held in equilibrium on the other end. In Part IV of the *Horologium*, for example, in discussing the proposition that the centre of gravity of a system of bodies cannot raise itself, Huygens evoked the established connotations of the word 'force'. A number of unconnected bodies have a centre of gravity the height of which must be considered as the height of the composite system since all the bodies can be moved to that level 'without the introduction of any other power [*nulla alia accersita potentia*] than that which is in the weights (*ponderibus*) themselves . . .' They have only to be connected by bars and turned about the centre of gravity 'for which no measurable force or power [*nulla vi neque potentia determinata*] is required.' As all the bodies together, brought to that plane, cannot raise themselves 'by the force of their gravity' [*vi gravitatis suae*], so also the centre of gravity of the system cannot rise of itself when they are otherwise disposed.[39] Repeated here precisely was the Galilean point of view, itself the product of the tradition of the simple machines, that the tendency of heavy bodies to descend can be measured by, and can replace, the force applied to the other end of a lever or balance to keep them in equilibrium. It was wholly in keeping with that tradition that the man who demonstrated the isochronism of the cycloidal pendulum and developed the first precision clock should have proposed to derive a universal standard of length from the tendency of heavy bodies to descend – one-third the length of a seconds pendulum, which at that time he thought was a constant, and which he called the *pes horarius*.⁴⁰ As Galileo's utilisation of the downward

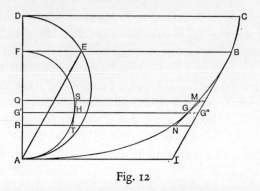

Fig. 12

tendency of heavy bodies formed an implicit dynamics behind his kinematics, such was present in Huygens' work as well, a ready model for a generalised concept of force. If anything, however, Huygens' kinematics of descending bodies moved further away from rendering the dynamics explicit than Galileo's had been.

From the basic demonstrations about descent and ascent on vertical and inclined planes, Huygens drew conclusions which exceeded Galileo's in generality. Since a curve can be considered as an infinite number of planes, the change of velocity for a given vertical displacement is constant whatever the path followed, and 'the body . . . will always have the same velocity, in ascending and descending, at points equally high.'⁴¹ This latter proposition became, in turn, the basis of his demonstration of the isochronism of the cycloid. Treating the cycloid as the limit of a set of planes, (see Fig. 12) he compared the time necessary for a body to traverse any segment of the cycloid, after it had descended from some arbitrary starting point *B*, with the time necessary to traverse a corresponding segment of the line *BI*, tangent to the cycloid, with a uniform velocity equal to one half that which the body would acquire in descending through the entire length *BI* of the tangent. By a summation of segments he was able to show that the time for descent from the starting point to the lowest point on the cycloid is to the time to traverse the tangent with a uniform velocity equal to one half that which the body would acquire in descending the tangent as the circumference of a semicircle is to its diameter, that is, as the ratio $\pi/2$. But by Galileo's theorem, the time of descent along any chord is equal to the time of descent along the vertical diameter DA.[42] The length of a pendulum swinging between cycloidal cheeks, so as to trace the cycloid, is twice the diameter of the generating circle. Hence, Huygens had not only demonstrated the isochronism of the cycloidal pendulum, but he had also derived the formula for its period,

$$T = 2\pi\sqrt{l/g}.$$

By measuring length directly and time by the rotation of the earth, he used the formula to establish the value of g with a degree of accuracy not even approached before.[43] Meanwhile, the distinguishing mark of the derivation, what made it wholly different from the analysis to which we are accustomed, was its independence of dynamic considerations. It depended entirely on Huygens' imaginative extension of the kinematics of uniformly accelerated motion, and only employed the minimal dynamic assumptions implicit in that kinematics.

In what was perhaps the supreme achievement of his mechanics, Huygens extended the analysis of the simple pendulum to embrace the physical pendulum, without the introduction of further dynamic considerations. Basic to his procedure was the insight first enunciated by Torricelli and extended by Huygens with considerable imagination to a variety of problems, that a system of bodies can be considered as a single body concentrated at the centre of gravity. Thus a physical pendulum can be resolved in thought into its constituent parts, which are then imagined to move independently. What amounts to the same thing, he imagined a pendulum that has swung to its lowest position to strike a set of separate bodies each one of which is equal in weight to the constituent part of the pendulum that hits it (see Fig. 13). From his study of impact, he concluded that the pendulum will come to rest and the separate bodies will be set in motion, each with a velocity equal to that of the corresponding part of the pendulum. The motion of the separate bodies can in turn be imagined to exhaust itself in raising them, and the centre of gravity of the system will reach exactly the height from which the centre of gravity of the physical pendulum initially descended. Huygens called this last principle, which had already played a role both in his analysis of impact and in his derivation of the formula for the simple pendulum, 'the chief principle of mechanics.'[44] Together with the insight that an isolated system of bodies is equivalent to a single body concentrated at their centre of gravity, it furnished the starting point for his assault on the physical pendulum.

When it is part of a physical pendulum, the velocity of the bob and the amount of its descent are related to those of other parts of the

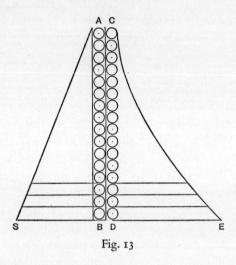

Fig. 13

pendulum in a simple ratio set by their distances from the axis of rotation. When they are separated, the potential ascent of each part is proportional, not to its velocity and distance from the axis, but to the square of its velocity. Hence he could write an equation in terms of weights and distances from the axis which set the actual descent of the physical pendulum equal to the potential ascent of its constituent parts with the velocities gained in descent. Now let an isochronous simple pendulum swing with an equal arc beside the physical pendulum. The ratio of its length to that of the physical pendulum provides the ratio of the velocities of their bobs, and with the simple pendulum the potential ascent of the bob is by definition equal to the actual descent. By substituting the ratios of the two pendulums, the simple and the physical, into his original equation, he was able to state the length of the isochronous simple pendulum as a function of the length of the physical pendulum and the relative weights of its parts.[45] The analysis gave Huygens the means to turn his understanding of the simple pendulum, which corresponds to the idealised motion of a body on a frictionless plane, into the actualised instrument which is the precision clock. Among other things, it showed him how to adjust the clock. A small weight added at the centre of oscillation does not alter the period, but placed below the centre, it retards the pendulum, and placed above, accelerates it.[46]

At the heart of the analysis of the physical pendulum stood the same quantity that had emerged from the analysis of percussion – mv^2.[47] The influence of Leibniz, whom Huygens himself had instructed in these mysteries, ultimately taught him to think of the quantity as force.

> In all movements of bodies whatsoever, no force is lost or disappears without producing a subsequent effect for the production of which the same amount of force is needed as that which has been lost. By force I mean the power of raising a weight. Thus, a double force is that which is capable of raising the same weight twice as high.[48]

No similar comments are to be found either in the *Horologium* or in the earlier manuscripts in which the analysis was originally pursued to a successful conclusion. There, the physical pendulum was treated within the domain of the kinematics of heavy bodies, and only references to the impossibility of perpetual motion even hinted at the later idea of the conservation of force. Force in this meaning of the word, moreover, had nothing to do with Newton's second law. Though its derivation depended on the analysis of free fall, conceptually it expressed the model of impact – not an external force that acts on a body to change its motion, but the force that a body in motion has, Descartes' 'force of a body's motion' given a new quantitative expression.

*

De vi centrifuga – the title of at least one major investigation testifies that Huygens was unable wholly to eliminate a concept of force from his mechanics. As I have argued, circular motion was one of the riddles that mechanics in the seventeenth century confronted. Huygens' treatise on centrifugal force, of which only a skeletal set of propositions, stripped of the flesh of demonstration, appeared during his lifetime, advanced the understanding of the problem to a new level of sophistication, replacing Descartes' qualitative discussion with a quantitative one worthy of the man who had reduced impact to exact description. In its very triumph, however, his treatise helps to reveal the full depth of the enigma that circular motion posed to seventeenth-century mechanics, for the man who derived the quantitative expression of centrifugal force did not succeed in fully solving the riddle. In his treatment, circular motion remained ultimately unreconciled to the principle of inertia, the foundation of his conception of motion. In its use of the word 'force', his analysis also suggests that point at which Galileo's kinematics of heavy bodies became inadequate to the problems that mechanics faced in the seventeenth century.

The completed treatise started indeed with an exposition of the central facts in the descent of heavy bodies, stated for his purposes here in terms more frankly dynamic than in any other finished work. 'Gravity is the endeavour to descend.'[49] In virtue of that endeavour or *conatus*, bodies fall with an accelerated motion such that the distances traversed are proportional to the squares of the time. Experiments confirm that heavy bodies do in fact fall with such a motion, and if the resistance of the air interferes with the ideal proportion, the perturbation does not invalidate the law of uniformly accelerated motion. 'Hence when a heavy body is suspended from a string, it pulls the string since it endeavours to move away in the direction of the string with an accelerated motion of this sort.'[50] If a ball tied to a string rests on an inclined plane, it still pulls on the string as it endeavours to move, but the endeavour is smaller in proportion as the space that it would travel in a given time is smaller. The same situation obtains in every case in which such an endeavour exists – whatever the cause of the *conatus*, it can manifest itself as a pull on a string. Moreover, only the first instant of the potential motion is significant in measuring the *conatus*; when a ball hangs vertically, for example, above a surface that would make it follow a curved path if released, only the initial endeavour at the first instant of motion is relevant in determining the 'force of the endeavour' (*vim conatus*).

> Hence whenever two bodies of equal weight are each held by a string, if they have endeavors to the same accelerated motion by which equal spaces in the directions in which the strings are extended would be traversed in the same time: we assert that equal tensions in those strings are also felt, whether they are pulled down or up or in any direction whatever.[51]

Huygens' use of 'weight' as the measure of a body reveals one of the unsolved problems in his mechanics, as I have indicated before, and detracts a modicum from his insight. Nevertheless, the statement contains the first generalised conception of force to be announced by the science of mechanics in the seventeenth century.

It is worth pausing to note as well the relation of statics to dynamics that the discussion of *conatus* implied. Huygens did not devote any serious attention to statics. He probably regarded it as a subject already adequately treated. Be that as it may, the passage above embodies the first wholly satisfactory statement of the relation of statics and dynamics. Statics is merely a special case of dynamics, in which the endeavour, or force, which would otherwise produce an accelerated motion proportional to it in a given body, is held in equilibrium. There was in Huygens' mechanics no attempt whatever to derive a dynamics from the ratios of the simple machines. Although he did not discuss them at length, the reference above to inclined planes makes it evident that he saw the simple machines in terms of equilibrium, and not as a paradigm for the understanding of force in dynamics. Equally, he had seized, as no one before him had, the full significance of Galileo's analysis of free fall as the paradigm of non-equilibrium. A generalised dynamics that exploited the model and defined force in its terms lay open before him had he chosen to follow that path.

Perhaps the least satisfactory application of the concept of *conatus*, at least from our point of view, was the one for which he specifically developed it, the *vis centrifuga* of circular motion. Imagine that a man holds a body on a string as he stands on a turning platform. He feels a tension in the string arising from the 'force of rotation' [*vis vertiginis*]. It can be proved by the geometry of circles, moreover, that the body, in trying to follow the tangent, attempts to separate itself from the circle by distances which would increase as the square of the time.

> Hence it is certain that this endeavour will be wholly similar to that which is felt when a ball is held suspended from a string, since in this case also it endeavours to move away in the direction of the string with a motion similarly accelerated.[52]

By exploiting the geometrical properties of circles, he was able to derive exact quantitative expressions of centrifugal force in terms of the velocity of the body and the diameter of the circle in which it moves. When two equal bodies move with equal angular velocities on unequal circles, the distance the tangent diverges from the larger circle for a given angle, in comparison to the corresponding distance from the smaller circle, is proportional to the diameters (see Fig. 14). The given angle is the measure of time, and the distance of the tangent from the circle is of course the measure of the *conatus* for unit time. Hence for equal angular velocities, the centrifugal force varies as the diameter of

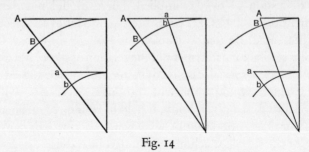

Fig. 14

the circle. By a similar procedure, he showed that for equal circles, centrifugal force is proportional to velocity squared; in the case of equal bodies with equal linear velocities on unequal circles, when their centrifugal forces are equal, their periods vary as the square root of the diameter.[53] If we allow the symbol *m* once more and assume units that

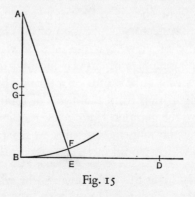

Fig. 15

set the constants of proportionality equal to one, Huygens' results can be expressed by the formulae we still employ for circular motion:

$$F = mv^2/r = mr\omega^2.$$

For Huygens, of course, the question of units was more complicated. To reduce the proportionalities of centrifugal force to accepted units, he set out to determine the conditions under which the centrifugal force of a body is equal to its weight. Suppose that a body falls through a distance *CB* equal to half the radius of a circle (see Fig. 15). With the velocity it acquires, it can traverse a distance twice as great, *BD*, in an equal time. Now by construction set $CG/CB = BE^2/BD^2$. *BE* is then proportional to the time in which the body will fall from rest through *CG*, and in the same time it will move the distance *BE* with the velocity acquired in the fall through *CB*. When the angle is very small, $BE = BF$, and the body will move the distance *BF* along the circle with a uniform speed.

> Hence we say that there is in it an endeavour to recede from the point *B* through the distance *FE* with a motion naturally accelerated (for such it has been shown to be) in the same time in which it would traverse the distance *BE* with a uniform motion by the speed of its revolution, that is, in that time in which it would traverse the distance *CG* from rest with an accelerated motion. Therefore, if it will have been shown that the distances *CG* and *FE* are equal, it will be established that the endeavour of the suspended body to descend with an accelerated motion is exactly equal to the endeavour of the same body, when it moves in the circumference, by which it attempts to recede in the direction of its string with a motion similarly accelerated: since the endeavour to accelerated motions is obviously the same when equal distances would be traversed by those motions in equal times.[54]

The geometry of circles easily establishes the equality of *CG* and *FE* for small angles, and hence the conclusion emerges that a body describing the circumference of a circle with the velocity that would be acquired in falling through half its radius has a centrifugal force equal to its weight. By this conclusion the units are established; and from Galileo's formulae for free fall, it is a simple matter to show that Huygens' proposition is equivalent to the formula

$$F = mv^2/r.$$

Once he had proceeded this far, Huygens inevitably applied his conclusion to the problem set by Copernican astronomy – if the earth turns on its axis, why are bodies not thrown off its surface? Galileo had addressed himself to the problem without an adequate understanding of circular motion. Huygens was now in a position definitively to solve it. In fact, when he first took it up, he was not quite yet in that position. In effect, he wished to compare the centrifugal acceleration calculated from the size of the earth and its speed of rotation with the measured acceleration of gravity. Not only was the accepted radius of the earth inaccurate, but the best value of *g* available, measured directly from freely falling bodies, was more inaccurate. When Huygens first carried out the comparison, he calculated the radius of the sphere on which a given body would have a centrifugal force equal to its weight when the sphere turns at the earth's rate of rotation. For a given angular velocity, centrifugal force varies directly with the radius, and he found that the earth would have to be 265 times larger than it is, the acceleration of gravity remaining unchanged, in order to project bodies.[55] The figure was too small, but soon thereafter his study of the pendulum gave him the means to remove the major source of error by correcting the value of g.[56] A fully satisfactory comparison had to wait two decades for more accurate measurement of the earth. Meanwhile, the formula for centrifugal force, even when applied to inaccurate data, had effectively demolished one objection to the heliocentric system – if indeed the objection was ever taken seriously by anyone except unregenerate geocentrists.

Despite the fact that his analysis of circular motion was quantitatively correct – perhaps indeed because it was quantitatively correct – it poses a problem. Like Descartes before him, Huygens stated the concept of rectilinear inertia explicitly. As we have seen, it provided the starting point for his treatment both of impact and of pendulums. To us, it appears obvious that the principle of rectilinear inertia requires that changes of direction be equated to changes of speed. The action of an external force is required in both cases to alter the inertial state. We call the force that alters the direction of a body's motion 'centripetal', a force directed toward or seeking the centre around which the body turns, and Huygens' results were quantitatively identical to our formula for centripetal force. Like Descartes, however, Huygens saw the dynamics of circular motion in exactly the opposite terms – not a force seeking the centre which diverts a body from its rectilinear path, but a force by which a body constrained in a curvilinear path endeavours to flee the centre, a centrifugal force, Huygens' own term coined to replace clumsy phrases about tendencies to recede.

To explain why he should have approached circular motion in terms that seem to us not only incorrect but contradictory to his own principles, we must undoubtedly start where he started, with Descartes. As the first man who attempted to analyse the mechanical elements of circular motion, Descartes promulgated a given conceptualisation of the problem which was bound to influence those who followed him. Certainly the fact that Descartes spoke of a tendency to recede from the centre of circular motion was one reason why Huygens thought in terms of centrifugal force.

Nevertheless, Descartes' statement of the problem is not, by itself, a convincing explanation of Huygens' conceptualisation. In fact, Descartes' influence extended far beyond the specific passage in which he attempted to analyse the dynamic factors in circular motion. To Descartes, circular motion was not just a problem in mechanics; it was central to his vision of a mechanical cosmos. If Huygens did not follow every detail of Descartes' system, he accepted enough that the approach to circular motion appeared to fit naturally into a broader setting. Above all, Huygens' view of circular motion was conditioned by his understanding of the cause of gravity, a Cartesian vortical mechanism. His treatise began with an extended comparison of centrifugal force to weight. He saw their relation, not merely as analogues, but as cause and effect. In the early autumn of 1659, even before he had arrived at the quantitative expression for centrifugal force, Huygens had been convinced that it had to be similar to weight.

> The weight of a body is equal to the endeavor of an equal quantity of matter, moved very swiftly, to recede from the center. Whoever holds a body suspended prevents that matter from receding; whoever allows the body to fall thereby offers to the same matter the opportunity to recede from the center along the radius; since however at the start it recedes from the center according to the series of odd numbers beginning with one, it is impossible that it not force the heavy body to approach the center with a motion similarly accelerated, with the result that these motions – the recession of matter from the center, and the approach of the falling body toward the center, are necessarily equal at the start. And therefore, having found how much the body descends in a given time, for example, $\frac{3}{5}$ of a line in 1 second second, we will also know how much that matter ascends from the center, which of course will also be $\frac{3}{5}$ of a line in the time of 1 second second.[57]

Huygens never renounced the central idea behind this view, which appeared in his discourse on gravity published with the *Treatise of Light* at the end of his career. The concept of centrifugal force was no less essential to his mechanical system than it had been to Descartes'.

The mechanical philosophy of nature also influenced Huygens' conceptualization of circular motion in a more subtle way. Its basic conviction that the moving particle is the ultimate causal agent tended to focus attention on the particle and its capacity to act, not on forces external to the particle that alter its motion. It was basic to Huygens' view that a body moves in a circle only if it is constrained to do so, but his attention focused on the body's capacity to act and not on the constraint. Let a body on a horizontal plane revolve in a circle with a period equal to the time required for it to fall from a height of $\pi^2 D$. When it revolves in such a circle, 'in which it is held by a line tied to a peg, the length of which line is half the said diameter, then the line is pulled by a force, due to the endeavour of the body to recede from the centre, as much as it would be pulled by the same body if it were suspended from the line, to wit, by its proper weight.'[58] Similarly, in Huygens' terminology a body moving in a circle 'has' a given centrifugal force.[59] The concept of centrifugal force repeated the model of impact which the mechanical philosophy of nature invariably encouraged.

Another influence was also at work, an influence older than the mechanical philosophy by far and ultimately irreconcilable to it, the concept of perfect circular motion. As Galileo had reminded seventeenth-century scientists, circular motion alone is compatible with an ordered cosmos. The ordered cosmos had dissolved into the mechanical universe, but not without leaving behind the conviction that circular motion maintains the relations of bodies unchanged. A sphere rotating on its axis does indeed present analogies to a body at rest. Orbital motion is more complicated than axial rotation, but a considerable effort of the imagination is required to see as uniformly accelerated a motion that is constant in speed. It is small wonder that the generation engaged in digesting the kinematics of uniformly accelerated linear motion found it difficult to subsume circular motion under the same categories. To Huygens, for all his derivation of the quantitative expression for centrifugal force, circular motion presented the analogy, not of uniformly accelerated motion, but of uniform inertial motion or of rest, which were of course identical in his eyes. 'We see two sorts of motion in the world, rectilinear and circular ...' he stated.[60]

> In regard to bodies considered simply, without that quality called heaviness, their movement is naturally rectilinear or circular. The first pertains to them when they move without impediment; the other when they are retained around some centre or when they turn on their own centres.[61]

On more than one occasion, he referred to the *conatus* of bodies to recede from the centre 'with a naturally accelerated motion...'[62] The phrase repeated the idea underlying his comparison of the endeavour of bodies to descend and the endeavour of bodies in circular motion to recede from the centre. The accelerated motion, be it noted, was not the circular motion but the potential motion away from the circle which, by definition, could be realised only through the destruction of the circular motion whence the endeavour derived.

The analogy of weight and centrifugal force involved another unexpected dimension as well. A body hanging on a string endeavours to

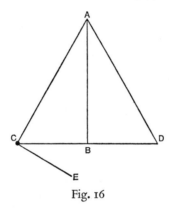

Fig. 16

descend, its weight held in equilibrium by the opposite pull of the string. A body moves in a circle only when it is constrained to do so; its centrifugal force, like the endeavour of the heavy body, stands in equilibrium with the constraining force. The analogy of circular motion with inertial motion was equally the analogy with static equilibrium. The conical pendulum, which stood near the origin of his investigation of centrifugal force, presented itself to him, for example, as a problem of equilibrium on an inclined plane. Consider the conical pendulum *AC*, to which the line *CE* is perpendicular by construction (see Fig. 16). It is obvious that there must be a centrifugal force in the bob *C*, which holds the string extended in that position;

> and since the body *C* has the same endeavour to descend, due to its gravity, as though it rested on a plane *CE*, but the centrifugal force by which it attempts to recede from the axis *AB* along the line *BC* opposes that endeavour of gravity, it is necessary that the centrifugal force mentioned be equal to the power exerted in the line *BC* parallel to the horizon by which the body *C* can be sustained on the inclined plane *CE*.[63]

FORCE IN NEWTON'S PHYSICS

He went on to examine the force with which the bob pulls the string in comparison to the pull of its weight alone. Again he imagined a plane (*BE*) perpendicular to the string, (see Fig. 17), but he replaced the centrifugal force with another string *BG*, which passes over a pulley and pulls the bob in the direction of *BG* by a weight which is to the weight of the bob as *BF*:*BE* – 'to which weight the centrifugal force is

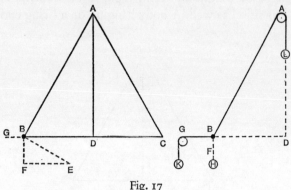

Fig. 17

then equivalent.' But the bob not only pulls away from the axis along *BG* by its centrifugal force; it also pulls straight down along *BF* with all its weight.

> Hence the situation is obviously the same as though the end of the line, *B*, is tied to two others, *BG*, *BF*, of which *BF* has hanging from it a weight equal to that of the body *B*, and the other *BG* is pulled by a weight which is to the weight of *B* as *BF* to *FE*.[64]

To find how much the combined load pulls on the line, he replaced the point of suspension *A* with another pulley and found the weight *L* which must be attached to the line to hold the other two in equilibrium. The conical pendulum had been transformed into Stevin's triangle of static forces. No small part of Huygens' spontaneous choice of centrifugal force as the quantity to analyse derived from his perception of uniform circular motion as a state of equilibrium.

The phrase 'centrifugal force' testifies to Huygens' ultimate inability to state mechanics entirely as kinematics. The context in which he viewed it also testifies, however, to the distance which separated his centrifugal force from a generalised dynamics. As I have argued, the ultimate analogy of circular motion, in his eyes, was not a dynamic action at all but static equilibrium. His very choice of the word 'force' (or *vis*) from the welter of available terms was undoubtedly determined by its established usage in connection with simple machines. It is true that static equilibrium, which he saw as a tendency to descend held in check, fit into a potentially dynamic context. The tendency to descend, when free to act, generates the uniformly accelerated motion of Galileo's kinematics, and Huygens insisted as well that the same model be seen in its application to centrifugal force. This is only to say, however, that Huygens possessed virtually all the conceptual equipment needed for a generalised dynamics. With the concept of 'centrifugal force', he took a step toward such a dynamics which was as small as the problems he considered allowed.

★

Part of the fascination in Huygens' mechanics lies in the fact that despite his constant effort to eliminate dynamic concepts and to treat mechanics as kinematics, he stood constantly face to face with the basic concepts of a generalised dynamics. His very concern with the kinematics of uniformly accelerated motion meant that he repeatedly confronted one of the models of dynamic action, a model, as we have seen, which he recognised readily enough, even though his notion of scientific propriety led him to suppress its expression. In a manuscript composed in the middle 1670's, more than a decade after the effective completion of his investigations of impact, pendulums, and circular motion, he explored more fully the concepts sketched in the introduction to *De vi centrifuga* and laid out explicitly the foundations of a consistent dynamics.

The paper began with the statement of three axioms which are virtually laws of motion:

> A body which has acquired a certain speed of movement continues to move with that same speed if there is nothing that acts to diminish its movement and nothing which incites it anew.
> If something acts continually to decrease the movement of a body which is moving, it will gradually lose its speed.
> And if, on the contrary, something acts continually on a body, pushing it in the direction in which it is already moving, its movement will be continually accelerated.

The first axiom employed the verb '*inciter*', and Huygens went on to define its substantive, '*incitation*', as the force which acts on a body to put it in motion or to change its speed. The 'incitation' of the manuscript is conceptually identical to the *conatus* of *De vi centrifuga*. It provided the central element in the dynamics that Huygens proceeded to elaborate.

The quantity of the incitation 'is measured by the force that must be employed, in the place where the body is and in the direction it can take [*dans la direction qu'il a*], to prevent it from beginning to move.' Consider a ball on a curved surface, for example. Its incitation at any point on the surface is equal to the force that must be exerted on a cord parallel to the tangent at that point to hold it at rest. The incitation is the same whether the ball is at rest or already in motion since the weight of the ball does not change. The point of the concept of incitation was not confined to a dynamic interpretation of descent, however; rather, it sought to exploit the model of free fall in the derivation of a generalised dynamics. If a spring pushes a body, 'its incitation at each point in its path is also measured by the force that is necessary at that place to prevent it from beginning to move in the direction that the spring pushes it.'

> It is manifest [he continued] from what has been said so far that the incitations of a body can be equal to each other although they are caused by different causes, as weight, elasticity, wind, attraction of a magnet, or something else.

Despite its explicit abstraction from the cause of incitation, the paper was unable wholly to forget the context of mechanical philosophy in which Huygens viewed every question of force. He defined 'perfect incitation' as that in which the speed of the cause is infinitely great in comparison to that of the body moved, a definition which implicitly evokes an ultimate mechanism. Perfect incitation applied particularly to gravity. By it, Huygens intended to avoid the problem that had led Descartes to neglect mathematical kinematics – because gravity causes a perfect incitation, the incitation operating on a falling body remains constant despite its increase in speed, and a perfect incitation is the dynamic source of a uniformly accelerated motion. Although he could not bring himself to ignore such issues, Huygens was primarily concerned in the paper to abstract from causation to a quantitative dynamics valid for all causes that operate to alter a body's state of rest or motion. An hypothesis stated the point again.

> If two bodies are taken equal, and if they traverse equal lines with incitations that are equal to each other at each pair of points correspond-

ingly placed on the two lines, the two lines will be traversed in equal times although the incitations of the bodies derive from different causes.[65]

With the hypothesis, the paper came to an end. Immediately before it, three brief phrases had suggested problems that he intended to solve by the dynamics of incitation – the movement of equal bodies with different uniform incitations, the times required to traverse equal distances (implicitly with different uniform incitations), and the movement of unequal bodies with equal incitations. Although the solutions were not given, they were obvious enough from what had gone before. By specifically treating the descent of heavy bodies as a model of dynamic action, making more explicit the concept expounded as *conatus* in his work on centrifugal force, the paper carried dynamics to a higher level of generality than the science had hitherto achieved in the seventeenth century. Huygens' repeated insistence that incitation be in the direction of motion suggests that his generalisation stopped short of realising that changes of direction are dynamically equivalent to changes of speed in inertial mechanics. As he presented it, incitation was concerned exclusively with changes in linear velocity – another reflection of his understanding of circular motion. Despite these limitations, the paper exploited the kinematics of free fall to achieve a generalised conception of force.

Why did Huygens seize on the term 'incitation', which was unknown in the literature of dynamics in the seventeenth century beyond his own manuscripts, instead of a term in current use, such as 'virtue', 'power', or even 'force'? Since he did not explain himself, we can only speculate, but the very choice of a word foreign to the prevailing literature is surely suggestive. Half the problem of dynamics was the ambiguity of words. What was the definition and measure of force? It had at least four distinct and conflicting uses, and the ambiguity was spread equally over all the words in common use. By coining a new dynamic term, to which he gave a single and precise definition, Huygens was trying apparently to pull himself free from the morass of words and to lay a solid footing on which a secure structure could rest. To define 'incitation', he found it necessary to employ the word 'force', but when he stated that 'incitation' is measured by the 'force' necessary to prevent a body from moving, he seized on the most precise and most generally accepted use of the word. In the tradition of the simple machines, 'force' was applied at one point, say the end of a lever, to move a load attached to another point. By the seventeenth century, it was understood almost universally that the ratios of the simple machines are the conditions of equilibrium; and when Huygens stated that the incitation of a body on an inclined plane is measured by the force that will prevent its motion, he was defining 'incitation' in terms of a usage generally accepted.[66]

As with his concept of *conatus*, his definition of 'incitation' implied an unexceptionable statement of the relation of statics to dynamics. Statics is concerned with forces in equilibrium, dynamics with unbalanced forces, the products of which are accelerated motions. The same perception of statics in relation to dynamics expressed itself in his investigation of the vertical motion of bodies in resisting media. First, Huygens found the resultant force acting on the body by adding the resistance (seen as a function of velocity) to the weight when the motion is upward, or by subtracting it when the motion is down, and then he set the resultant force proportional to the acceleration.[67] Huygens had fully digested the principle of inertia and understood that it had ruptured, once and for all, the bond that had tied dynamics to the simple machines. There was in his work, manuscript or published, no effort to derive a dynamics from the law of the lever. He asserted that the force of percussion is infinite, that is, incommensurable with static force.[68] With the concept of incitation, he attempted instead to exploit the dynamic principle inherent in the kinematics of free fall.

The word '*incitation*', applied to the concept of force in the paper cited above, had appeared a few years earlier (as the Latin *incitatio*) in a paper devoted to the vibration of strings. Undefined there, it was applied intuitively to an interesting attempt to analyse the dynamics of simple harmonic motion. The paper suggests the breadth of the dynamics that lay open before Huygens. It began with a statement of the dynamic conditions of pendular motion, a conclusion neglected by his earlier kinematic examination of pendulums. For any two arbitrary points A and B on a cycloid, 'the gravity of a weight in A will be to the gravity of the same weight placed in B as the arc AC to the arc BC.' That is, the component of weight parallel to the path at any point on a cycloid is proportional to the displacement from the lowest point measured along the cycloid. The structure of the paper suggests that this relation reminded him of the conical pendulum and of his analysis of centrifugal force which showed that it is proportional to the radius when angular velocity is constant. To apply the dynamics of the pendulum, Huygens imagined a horizontal string stretched by a weight hanging over a pulley. All of the matter of the string was initially imagined to be concentrated at the centre, and its weight was ignored. For small circular vibrations, the constant tension of the string corresponds to the constant weight of the bob of a conical pendulum. The 'incitation' toward the centre of small circles is proportional to the displacement, and in Huygens' view of circular motion the centrifugal force must equal it if circular motion is maintained. Hence the centrifugal force is proportional to the radius, and all (small) circular vibrations of whatever radius are equal in period. As earlier, in the case of pendulums, he perceived the ultimate identity of small circular vibrations with vibrations in one plane when the amplitude approaches zero, and to analyse vibrations in one plane further he applied the dynamic relation stated above for the pendulum. Because the weight on the line maintains a constant tension, the 'incitation' urging the body toward the centre when it has been pulled to one side is proportional, for small vibrations, to the displacement. By setting the body concentrated at the centre of the string equal to the weight stretching the string, he was able to equate the vibration of the string to the oscillation of a pendulum equal to it in length. From this analysis he could proceed to other tensions in the string, showing that the period is inversely proportional to the square root of the tension. The results depended on the assumption that all of the matter to be moved by the 'incitation' is concentrated at the centre of the string, corresponding to the simple pendulum, but the theory of the physical pendulum allowed him to transpose the results to the case of physical strings.[69] As in the case of the later paper discussed above, the concept of 'incitation' used in the analysis of vibrating strings can only be understood as a generalised concept of force. Not only does it show how far the kinematics of descent had led Huygens toward a workable dynamics, but also how useful dynamics could be in transposing the results of kinematics to other problems.

Closely associated with his ventures into dynamics was Huygens' groping toward a concept of mass. Anyone who applied dynamic concepts to the kinematics of descent and reflected on the equal acceleration of all bodies in free fall faced an implicit distinction between weight and mass. It is not surprising, then, to find the distinction becoming explicit in papers from 1668 that considered the cause of weight. One paper, a set of questions – or perhaps propositions – tentatively explored the relation between the specific gravity of bodies and their density, in a context which assumed that gravity is an epiphenomenon caused by the impact of particles subtle enough to penetrate the pores of bodies. He asked whether metals can be compressed and made heavier by hammering them, and whether they weigh the same when they are hot (and by implication expanded). In the midst of these questions appears one more pregnant.

> Whether the force of a hard body is exactly proportional to the gravity of the same body. For it would then seem to follow that the gravity is proportional to the quantity of matter present in any body whatever.[70]

Whether or not he had already carried out the experiments called for or at least satisfied himself as to their outcome, he had done so before he composed another paper from the same year. Descartes had maintained that the weights of bodies are not proportional to their quantities of matter, he noted.

> As for me, I assert that the weight of each body is proportional to the quantity of matter that composes it and is at rest, or can be taken to be at rest, in respect to the infinitely swift motion of the matter that passes through it [causing its weight]. Such is evident from the effect of impulsion which is exactly proportional to the ratio of the weight of the bodies.[71]

Huygens' recognition that impact offers a means independent of weight to measure the quantity of matter in a body is significant. In dynamics, mass is more than merely quantity of matter. By functioning as a resistance to changes in a body's inertial state, it establishes the ratio of a force to the acceleration it produces. In a preface written about 1689 for his treatise on impact, Huygens remarked that Descartes had set the resistance of a body proportional to the quantity of motion it must acquire and arrived at the conclusion that a smaller body can never set a larger one in motion. On the other hand, Pardies had realised correctly that a body at rest and not held in any way cannot resist motion as such, but he had gone on to conclude that any speed can be transmitted to it with equal ease. Hence a body at rest receives the speed of the body striking it – a conclusion more incredible than Descartes'.[72] Although Huygens did not say it, Pardies' assertion was the logical outcome of the seventeenth-century concept that matter is indifferent to motion. The extent of Huygens' dynamic insight is revealed in his refusal to draw that conclusion. Indifferent to motion in one sense of the word, as the concept of inertia in its rejection of natural motions demands, matter also resists, not motion itself, but being put in motion. Every body 'resists the velocity of the motion which one would give to it, in proportion to the quantity of matter which it contains and which is obliged to follow this motion.'[73] Thus, Huygens could assume, as his argument on impact required, that when a body B smaller than A strikes the body A at rest and sets it in motion, it gives A a smaller velocity than it would if A were only equal to B in size.[74] The resistance bodies make to being set in horizontal motion cannot be caused by their weight since a lateral movement does not remove them farther from the earth and in no way opposes the action of weight which pushes them toward the earth.

> There is then nothing but the initial quantity of matter that each body contains which causes this resistance, with the result that if two bodies both contain the same amount of it and strike each other with equal speeds, they will reflect each other equally, or both of them will come to rest, according as they are hard or soft.[75]

Present here were all the components that went to make up our concept of mass. It remains impossible, nevertheless, to maintain that Huygens held and used a concept of mass clearly distinguished from weight. The investigation of vibrating strings mentioned above, for example, depended at a crucial stage in the argument on the imaginary concentration of the mass of the string, represented by G, at its middle point, where it is moved by the tension caused by the weight K hanging on the end of the string. Huygens even made his intention clear by explicitly excluding the weight of G from consideration. Nevertheless, his statement of the problem readmitted all the ambiguity he had been at pains to exclude – '*Ponatur* pondus G *aequale* K.'[76] When he investigated the centre of oscillation of the physical pendulum, he imagined bodies attached to a rigid weightless bar. Where we would introduce the mass of the bodies into the analysis, he spoke instead of their '*gravitas*.'[77] No doubt part of the problem was the issue of nomenclature. Sometimes he used the word '*moles*' and sometimes the word '*magnitudo*' to refer to the size of bodies – both vague words, though perhaps susceptible of exact definition. Such a definition he did not provide, nor did he consistently use either word, or even the phrase 'quantity of matter.' He tended rather to fall back on *'gravitas'* or *'pondus'* to express the size of bodies, a usage no doubt made easier by his conviction that weight is proportional to quantity of matter. To say as much is only, of course, to repeat the proposition that Huygens did not hold and employ a distinct concept of mass. Ultimately the explanation of this fact must surely return to his distrust of dynamics and his preference for kinematics. In a dynamic context, the logic of his inquiry led Huygens directly to the distinction of mass and weight. A dynamic context was what his mechanics tried consistently to avoid, however, and the distinction hardly mattered to kinematics.

*

The final assessment of Huygens' role in seventeenth-century mechanics hinges on his distrust – which finally became almost instinctive – of dynamics. The concept of incitation was thrust aside. Not only did it fail to achieve publication by its author – a fate it shared, to be sure, with such major achievements of his mechanics as the discourses on percussion and centrifugal force – but unlike them it failed even to be deemed worthy of elaboration. After devoting a few short pages to sketching the concept in, Huygens simply forgot it. It is difficult to avoid the conclusion that his distrust of dynamics was rooted in his philosophy of nature, to which dynamic considerations too readily summoned up the images of concepts banished from the mechanical universe. Almost the first scientific demonstration that Huygens undertook, appropriately concerned with the kinematics of uniformly accelerated motion, assumed that there is no resistance from the air or from any other source, he told Mersenne, 'but only a uniform attraction downward, be it great or small.'[78] Over forty years later, when Fatio de Duillier had written him describing the contents of the forthcoming *Principia*, he replied that he looked forward to seeing it and did not care that Newton was not a Cartesian 'as long as he does not propose any suppositions like that of attraction to us.'[79] Attraction was indeed the heart of the problem. His youthful letter to Mersenne had indicated how readily a dynamic interpretation of the kinematics of descent summoned up the idea of attraction, and attraction was what, as a mechanical philosopher, he would in no way admit. Toward the end of his career, Huygens assessed the influence Descartes had exerted on him.

M. Descartes had found the way to have his conjectures and fictions taken for truths. And to those who read his Principles of Philosophy something happened like that which happens to those who read novels which please and make the same impression as true stories. The novelty of the images of his little particles and vortices are most agreeable. When I read the book of Principles the first time, it seemed to me that everything proceeded perfectly; and when I found some difficulty, I believed it was my fault in not fully understanding his thought. I was only fifteen or sixteen years old. But since then, having discovered in it from time to time things that are obviously false and others that are very improbable, I have rid myself entirely of the prepossession I had conceived, and I now find almost nothing in all his physics that I can accept as true, nor in his metaphysics and his meteorology.[80]

Huygens was deceiving himself grossly. If he had broken with many particulars of Cartesian natural philosophy – so many indeed that he might with justice argue that it had ceased to exist for him as a coherent system – it continued to exert the overwhelming influence on the way in which he approached natural philosophy. Under the tutelage of Descartes, he had become a mechanical philosopher, and a mechanical philosopher he remained. As a mechanical philosopher, he refused to deal in categories such as attractions. As a mechanical philosopher, he avoided the concepts of dynamics and sought to confine his mechanics to kinematics, a mechanics that measured itself by the ultimate reality of the mechanical universe, the particle of matter in motion.

All the obvious phenomena, which in Newton's hands became examples of the forces of attraction present in nature, continued to present themselves to Huygens in terms of mechanical images that were basically Cartesian in inspiration. This was so of magnetism, which he explained by special vortices of a special subtle matter.[81] It was equally so in the case of electricity.[82] Above all, it was so in the case of crucial importance to seventeenth-century mechanics, that of gravity. In a paper from 1668, Huygens put some pointed questions to Descartes' explanation of gravity. According to Descartes' explanation, there should be no gravity in deep mines, but in fact there is. His explanation requires that the subtle matter causing gravity rebound from the earth when it strikes it, but the cause of gravity must penetrate the most solid bodies to their centres. Although gold is twenty times heavier than water, Descartes said that it contains only four or five times as much solid matter because the internal motion of the particles of water gives it a tendency away from the centre. If that is true, Huygens noticed, ice should be heavier than water. Finally, contrary to Descartes' assertion, bodies do accelerate uniformly as they fall, and the cause of gravity continues to act on them as they descend, even when they have acquired considerable speed.[83] The last two issues were of crucial importance to rational mechanics in its relation to the mechanical philosophy. Galileo's kinematics of descent rested on two concepts that became the foundations of quantitative mechanics – uniformly accelerated motion, and the constancy of the acceleration of gravity for all bodies. How were they to be reconciled with mechanisms imagined as the causal explanations of gravity? In reflecting on these two issues, among a number of other questions, Newton was led to reject the causal mechanisms and to embrace an idea of attraction which transposed itself readily into the concept of force that dynamics required. Huygens, in contrast, was bound to the mechanical philosophy by ties that refused to break. Its domination of his vision of mechanics found a symbolic expression in a cryptic note from 1688 which greeted the *Principia* – 'Vortices destroyed by Newton. Vortices of spherical motion in their place.'[84]

Twenty years earlier, on the occasion of a formal discussion of gravity at the *Académie*, his own vortical theory of gravity had already made essentially the same response to his own questioning of Descartes' explanation. In August, 1669, Roberval, Frénicle, and Huygens presented their views on the nature of gravity before the *Académie*. Huygens' paper, the most elaborate of the three, furnished the principal substance of his *Discourse on the Cause of Gravity*, which was published in 1690 with the *Treatise on Light*. Both Roberval and Frenicle asserted that gravity, the heaviness of bodies whereby they seek to descend, is caused by an attraction between terrestrial bodies and the earth, a mutual attraction of like bodies for like which holds the earth together as a unified whole. When Huygens presented his discourse following the other two, his opening remarks constituted a mechanical philosopher's statement of faith, virtually a call to battle against the forces of obscurantism.

> To discover a cause of weight that is intelligible, it is necessary to investigate how weight can come about, while assuming the existence only of bodies made of one common matter in which one admits no quality or inclination to approach each other but solely different sizes, figures, and motions...[85]

An explanation that is intelligible, then, includes nothing but matter in motion. To attribute the descent of bodies toward the earth to an attractive quality in the earth or in the bodies, he told Roberval, was to say nothing at all.[86] Huygens' own intelligible explanation was a variant of Descartes' vortex. Its major change was to introduce movement of the subtle matter in all planes so that its centrifugal tendency is away from the centre and the opposite thrust on bodies is toward the centre. This was what he meant by a vortex of spherical motion. In contrast, the Cartesian vortex would generate motions perpendicular to the axis, as many had pointed out. To account for the fact that the acceleration of gravity is constant for all bodies (or that the ratio of weight to quantity of matter is constant), he contended that bodies are extremely porous, allowing the subtle matter to act upon every particle of matter in them. To account for the continued uniform acceleration of bodies as they fall, he calculated a velocity of subtle matter, in comparison to the velocity of a point on the earth's equator, great enough to account for continued acceleration through the range of velocities observed in natural phenomena. The solution to the first problem was patently inadequate. That to the second, not wholly unlike Descartes' reaction to Galileo's kinematics, treated uniformly accelerated motion as a mere approximation to actual fall, even before the resistance of the air was taken into consideration. In both cases, the effect of the causal mechanism was to suggest that an exact quantitative dynamics corresponding to Galileo's kinematics was a delusion not worth pursuing.

When Huygens finally published the *Discourse on Gravity*, Newton's *Principia* had appeared, and he added a passage that took account of it. It was impossible, he agreed, to withhold assent from the mathematical demonstration of so many phenomena, especially the demonstration of Kepler's laws. Newton must be right in his contention that gravity extends throughout the solar system and that it decreases in strength in proportion to the square of the distance. To proceed on to the assertion that gravity is an attraction was another matter, however, and one he could not accept. Gravity must be explained by motion.[87] He applied the concept of a spherical vortex, by which he explained terrestrial gravity, to the entire solar system, and calculated that the subtle matter at the surface of the sun must be moving forty-nine times as fast as the subtle matter that accounts for terrestrial gravitation.[88] Newton had argued that a subtle matter in space would obstruct the motion of the planets. Far from obstructing motion, Huygens replied, such matter causes the centripetal motion of bodies.

> It would be another question if one assumed that weight were a quality inherent in corporeal matter. But that is something which I do not believe Mr. Newton accepts because such an hypothesis separates us entirely from the principles of Mathematics and Mechanics.[89]

Here indeed was the nub of the matter. That a concept of attraction was inconsistent with the principles of mechanics as Huygens used the phrase is clear enough. Mathematics is another question, however. In the work of no one man in the seventeenth century was the tension between the mechanical philosophy of nature and the mathematical science of mechanics revealed more clearly than in Huygens. Had he chosen to pursue the dynamics he sketched, exploiting the model of free fall and expressing in effect the dynamic conditions of the kinematics of descent, it is reasonable at least to speculate that textbooks today would refer to Huygens' two laws of motion instead of to Newton's three, and we might discuss dynamics in terms of 'incitation' instead of 'force'. His concept of 'incitation' was not dependent on causal mechanisms, and it might appear in the abstract that he could have developed it whatever his philosophy of nature. Whenever Huygens thought of gravity or weight, however, he did not think of the abstract concept of incitation to which the acceleration was proportional. Quite the contrary, in his mind's eye he saw the all too concrete image of subtle particles in vortical motion, pushing bodies toward the centre as they strive themselves to recede from it. And with the image arose at once the problems it posed for a quantitative dynamics. Huygens possessed the conceptual equipment that such a dynamics required. His mechanical philosophy of nature effected its early abortion.

NOTES

1. Christiaan Huygens, *Oeuvres completes*, 22 vols. (La Haye, 1888–1950), **16**, 150, reprinted by permission of Martinus Nijhoff N.V.
2. Ibid., **16**, 139.
3. Ibid., **16**, 99–100.
4. 'Extrait d'une lettre de M. Hugens à l'auteur du Journal,' *Journal des sçavans*, 18 March 1669, **2** (1667–71), 531–6. (*Oeuvres*, **16**, 179–81.) 'A Summary Account of the Laws of Motion, Communicated by Mr. Christian Hugens in a Letter to the R. Society,' *Philosophical Transactions*, No. 46, 12 April 1669, **4** (1669), 925–8. (*Oeuvres*, **6**, 429–33.)

5. Ibid., **16**, 92.
6. Ibid., **16**, 92.
7. Ibid., **16**, 93.
8. Ibid., **16**, 94.
9. Ibid., **19**, 164–5. On a scrap that dates from about 1667, Huygens considered the impact of two soft bodies that are equal in size and that have equal and opposite velocities. By using the argument that their common centre of gravity cannot rise of its own accord, he showed that the two bodies must come to rest. Not long before, the French Jesuit, Claude-François de Chales had published an analysis of the same problem based on the implicitly dynamic argument that the two bodies must come to rest since they both have equal quantities of motion and therefore neither can prevail over the other. Huygens commented on de Chales' demonstration, indicating some of the reasons why he had come to prefer another approach to impact: 'Des Chales. There are some who think they have adequately demonstrated the truth of this Theorem from the fact that the quantity of motion in each body is equal to that in the other. The reasoning is probable, but it is not satisfactory. In the same manner it would be demonstrated that a body at rest that is struck by different bodies with the same quantity of motion would acquire the same speed in every case, which however is false.' (Ibid., **16**, 164.)
10. Ibid., **16**, 96.
11. Ibid., **16**, 94.
12. Ibid., **16**, 96.
13. Ibid., **16**, 110.
14. Ibid., **16**, 104–5 and 137–40.
15. Ibid., **16**, 43.
16. From the version of 1654; ibid., **16**, 132. In 1669, Huygens composed some anagrams for theorems that he chose not to include in his communication on impact to the Royal Society (though apparently in the end he did not send the anagrams either). The first of the anagrams concealed the following theorem, which did not appear in the final version of his treatise on impact: 'Before and after impact, the centre of gravity of two bodies continues with a uniform motion in the same direction.' (Ibid., **16**, 175 fn.) In his publication of seven rules of impact, in the *Journal des sçavans* in the same year, he referred to 'an admirable law of nature' which appeared to be generally valid for all impacts whatever. 'It is that the common centre of gravity of two or three or however many bodies you wish continues to move uniformly (*avance toûjours également*) in the same direction in a right line before and after their impact.' (Ibid., **16**, 181.) Whereas Huygens' completed treatise on impact confined itself to perfectly hard bodies, in 1667 he considered the impact both of soft bodies and of imperfectly hard ones and arrived at the conclusion that the centre of gravity continues in its inertial state in these cases as well, as his statement in the *Journal des sçavans* implies (ibid., **16**, 166–7).
17. From the version of 1656; ibid., **16**, 143.
18. Ibid., **16**, 31–3.
19. Ibid., **16**, 39–41.
20. Ibid., **16**, 39.
21. Ibid., **16**, 57.

22. Ibid., **16**, 53–65.
23. From an incomplete preface, composed about 1689, for a projected publication on impact and centrifugal force; ibid., **16**, 210.
24. In a paper from about 1667, Huygens considered three equal bodies, A, B, and C. A in motion strikes B and C which are at rest; how is it possible for B to transmit the shock from A to C without moving itself? In discussing the problem he brought in the time of impact. 'When a hard body strikes another equal body that is at rest, one can say that the one acquires motion in an instant and that the other loses it in an instant.' (Ibid., **16**, 160.) In another paper from the same period, he asked if perfectly hard bodies, which do not change their shapes, obey the same laws of impact as the elastic bodies with which one can actually experiment, bodies subject to flexion and restoration of their original figures. 'I do not see what prevents it,' he replied. 'For it does not seem impossible that motion be communicated in an instant or an indivisible moment of time (which is necessary), because the harder the bodies that we have are, that is, the less their figures are distorted and the less they are restored, that is, the shorter the time in which they communicate motion, the better they obey our laws of reflection.' (Ibid., **16**, 168.)
25. His paper of 1669 in the *Journal des sçavans* contained the following statement: 'The quantity of motion that two bodies have can be increased or decreased by their impact; but the quantity in the same direction, when the quantity of motion in the opposite direction is subtracted from it, always remains constant [*mais il y rest toûjours la mesme quantité vers le mesme costé, en soustrayant la quantité du mouvement contraire*].' (Ibid., **16**, 180.) In both the version of 1654 and that of 1656, he had stated the same principle in slightly different words. (Ibid., **16**, 130–1 and 146–7 fn.)
26. Ibid., **16**, 73. The conclusion had emerged already in the second version of the treatise in 1652. (Ibid., **16**, 95.)
27. Ibid., **19**, 164.
28. 'Quaeritur quam rationem habeat tempus minimae oscillationis penduli ad tempus casus perpendicularis ex penduli altitudine.' Ibid., **16**, 392.
29. Ibid., **17**, 125.
30. Ibid., **17**, 125–8.
31. Ibid., **17**, 130–1.
32. Ibid., **17**, 136–7.
33. Ibid., **18**, 125.
34. Ibid., **18**, 127–9.
35. Ibid., **18**, 137.
36. Ibid., **18**, 137–41.
37. Ibid., **18**, 141.
38. Ibid., **18**, 141–3.
39. Ibid., **18**, 249.
40. Ibid., **18**, 97. A short paper written in 1666 had asked if gravity decreases with distance from the centre of the earth. He had replied that the decrease, if it exists, is not noticeable in pendulums on the tops of mountains 3,000 feet high (ibid., **17**, 278). What did become noticeable two decades later was the variation of the

length of a seconds pendulum at different latitudes. An expedition in 1686–7 to the Cape of Good Hope and back established the variation beyond question. Huygens attributed the variation to the centrifugal effect of the earth's rotation. At any rate, it destroyed the project of the *pes horarius* (ibid., **18**, 636–43).

41. Ibid., **18**, 145–9.
42. Ibid., **18**, 171–83.
43. Cf. Alexandre Koyré, 'An Experiment in Measurement,' *Proceedings of the American Philosophical Society*, **97**, (1953), 222–37.
44. Huygens to De la Roque, 8 June 1684; *Oeuvres*, **8**, 499.
45. A considerable number of manuscripts chart the progress of Huygens' investigation; ibid., **16**, 385–439. The finished presentation appeared as Part IV of the *Horologium*; ibid., **18**, 243–359.
46. Ibid., **16**, 423.
47. Let me repeat that the use of the symbol *m* is mildly anachronistic in that the concept of mass did not emerge with full clarity in Huygens' mechanics. Nothing in his treatment of pendulums stimulated the clarification of the concept. Equally, nothing in the treatment was undermined by its absence.
48. Ibid., **18**, 477.
49. 'Gravitas est conatus descendendi.' Ibid., **16**, 255.
50. Ibid., **16**, 257.
51. 'Porro quoties duo corpora aequalis ponderis unum quodque filo retinetur, si conatum habeant eodem motu accelerato, et quo spatia aequalia eodem tempore peractura sint, secundum extensionem fili recedendi: Aequalem quoque attractionem istorum filorum sentiri ponimus, sive deorsum sive sursum sive quamcunque in partem trahantur.' Ibid., **16**, 259.
52. Ibid., **16**, 263.
53. Ibid., **16**, 267–73.
54. Ibid., **16**, 275–7.
55. Ibid., **16**, 304.
56. Proposition VI determines the diameter of a circle which a body will trace in one second when its centrifugal force equals its weight. It assumes a known value of g. When Huygens originally wrote the manuscript, he stated that 'experiment' [*experientia*] gives us the value of g. At some later time, he deleted the word '*experientia*' and replaced it with the word '*calculus*'. In the margin, he added a note of explanation: 'On the contrary, calculation as I found out later after the proposition [comparing] perpendicular fall to fall through a cycloid or the vibration of a pendulum was known.' (Ibid., **16**, 278 fn.) Late in 1665 or early in 1666, he again compared centrifugal force on the equator with weight. He still employed a figure for g that was too small, and he applied it to a diameter of the earth more than fifty per cent too large. Hence his result on this occasion—that weight is 165 times as great as the centrifugal force of a body on the equator—is less accurate than this earlier one (ibid., **16**, 323–4).
57. Ibid., **17**, 276–7. Later in 1667 apparently, Huygens wrote a brief paper on subtle matter in which he sketched in a brief statement of the cause of gravity similar to the one above (ibid., **19**, 553). In August 1669, he presented his completed explanation of gravity before the *Académie*. It had been objected to the Copernican

FORCE IN NEWTON'S PHYSICS

system, he said, that the rotation of the earth would hurl terrestrial objects into the air. He intended to demonstrate just the contrary, that this very effort away from the centre is the cause whereby other bodies are pushed toward the earth (ibid., **19**, 632).

58. Ibid., **16**, 303. Cf. Huygens' calculation of the force with which the bob pulls the string at B, the lowest point on its path, after it has swung downward through a full quadrant. In his words, 'when it will have arrived at B, it will pull the string AB more strongly than if it were suspended with its mere weight. But how much more? Three times as much.' (Ibid., **16**, 296 fn.)

59. For two examples from the innumerable ones that might be cited, cf. ibid., **16**, 271, 275.

60. His discourse on gravity before the *Académie* in 1669; ibid., **19**, 631.

61. *Discourse on the Cause of Gravity*; ibid., **21**, 451. In the published Discourse, these two sentences replaced the one cited immediately above (fn 60), which was in his presentation before the *Académie* in 1669 and in the MS. of 1687. The change was apparently made after he read the *Principia*, and its minimal nature indicates how deeply ingrained his view of circular motion was.

62. Ibid., **16**, 269. Cf. pp. 275-7, quoted above as note 54.

63. Ibid., **16**, 285. Cf. his examination of a similar problem, a ball B in (unstable) equilibrium in a tube AB inclined at an angle of 45° and rotating around the vertical axis AC. If the radius of the circle the ball traces is $9\frac{1}{2}$ inches and its period is one second, the ball will be held in equilibrium in the tube. 'For if it were placed on a wheel at that distance of $9\frac{1}{2}$ inches from the centre, it would be able to sustain an equal ball D tied to the line DCB and hanging down through the centre of the wheel. In the rotating tube, however, it attempts to move away from the centre along the right line CB. The same force acting in the line CB is required, however, to sustain the ball at B on the plane BA and to sustain the same ball hanging freely. Therefore the ball B will be able to sustain itself in the tube AB by the same endeavour to move away from the centre by which (when it is placed on a wheel) it is able to sustain the hanging body D equal to itself.' (Ibid., **16**, 306-7.) Note that when D and B are in equilibrium on the same line, Huygens visualised the problem, not in terms of D holding B in its circular path, but rather of B supporting D by means of its endeavour to recede. B rather than D was seen as the active body.

64. Ibid., **16**, 310.

65. Ibid., **18**, 496-8.

66. See Appendix D.

67. Ibid., **19**., 102-57.

68. Ibid., **19**, 24.

69. Ibid., **18**, 489-94. Cf. the similar but less fully developed dynamics on which Lord Brouncker demonstrated that the cycloid is the curve of isochronous oscillation: 'But first I must ask, that it be granted me, that the increase of the velocity of the same body descending is always in proportion to the power of the weight. As, that because the powers of the weight are in proportion to the perpendicular altitudes of the inclinations of the planes, as you may see in Stevinus, *Livre I, de la Statique*, prop. xix. cor. 2. therefore, AB and DE being of the same length, and EF to BC as 1 to a, the increase of the velocity . . . in descending BA is

to the increase of the velocity of the same bullet in descending ED as a to 1; and consequently, that the time it descends ED, is to the time it descends BA, as $\sqrt{a}$ to 1.' (Thomas Birch, *The History of the Royal Society of London for Improving of Natural Knowledge, From its First Rise*, 4 vols. (London, 1756–7), **1**, 70–1.)

70. 'An vis percutiendi in corpore duro sequatur precise gravitatem corporis ejusdem. Hinc enim sequi videtur gravitatem sequi quantitatem materiae cohaerentis in quolibet corpore.' *Oeuvres*, **19**, 625.
71. 'Moy je dis que chasque corps a de la pesanteur suivant la quantité de la matiere qui le compose et qui est en repos, ou peut estre prise pour estre en repos a l'egard du mouvement infiniment viste de la matiere qui le traverse. Cela paroit de l'effect de l'impulsion qui suit exactement la raison de la pesanteur des corps.' Ibid., **19**, 627. In his résumé of the rules of percussion presented to the *Académie* in 1669, he made a similar comment: 'Ie considere en tout cecy des corps d'une mesme matiere, ou bien j'entends que leur grandeur soit estimée par la poids.' (Ibid., **16**, 180.) In *De vi centrifuga*, when he sought to equate centrifugal tendency to weight, he initially used the word 'weight' to indicate the size of the body experiencing a centrifugal tendency. Realising the problem he had raised, he went on to say that the centrifugal forces of bodies moved uniformly in equal circles vary 'as the weights of the bodies, or their solid quantities' [*sicut mobilium gravitates, seu quantitates solidas*] (ibid., **16**, 267).
72. Ibid., **16**, 207.
73. Ibid., **19**, 482. I have used the translation by Silvanus P. Thompson, *Treatise on Light* (London, 1912), p. 30.
74. *Oeuvres*, **16**, 43.
75. From his discourse on gravity before the *Académie* in 1669; ibid., **19**, 638.
76. Ibid., **18**, 490. My italics.
77. Ibid., **16**, 417.
78. Huygens to Mersenne, 28 October 1646; ibid., **1**, 27.
79. Huygens to Fatio, 11 July 1687; ibid., **9**, 190.
80. A set of comments written in 1693 on Baillet's biography of Descartes; ibid., **10**, 403.
81. A number of papers were devoted to magnetism, mostly about 1680; ibid., **19**, 565–603.
82. Apparently in the 1690's, he wrote a number of papers on electricity; ibid., **19**, 611–16.
83. Ibid., **19**, 626–7.
84. 'Tourbillons detruits par Newton. Tourbillons de mouvement spherique a la place.' Ibid., **21**, 437.
85. 'Pour chercher une cause intelligible de la pesanteur il faut voir comment il se peut faire, en ne supposant dans la nature que des corps faicts d'une mesme matiere, dans lesquels on ne considere nulle qualité, ny inclination a s'approcher les unes des autres, mais seulement des differentes grandeurs, figures et mouvemens...' Ibid., **19**, 631.
86. Ibid., **19**, 642.
87. Ibid., **21**, 472.
88. Ibid., **21**, 478.
89. Ibid., **21**, 474.

Chapter Five

The Science of Mechanics in the Late Seventeenth Century

IN the hands of a master such as Huygens, the problems of mechanics aligned themselves with such transparent ease and the science so readily assumed the outlines of a familiar shape that we are prone to underestimate the difficulties. To remind ourselves of them anew, and to appreciate in its full dimensions the achievement involved in their solution, it is necessary to descend from the Olympian heights and to breathe the common atmosphere again. In fact, not the common atmosphere – the men I examine in this chapter were in no sense common men. They were leaders of the European scientific community in the second half of the seventeenth century, nearly every one of them the bearer of a name honoured in the history of science. If the science of dynamics was not obvious to them, if they bungled problems that even beginning students of today are expected to solve, we shall conclude of course, not that their reputations are ill-founded, but that the clarification of basic concepts was a work the difficulty of which is not readily perceived in the limpid simplicity of the finished product.

They differed from Huygens in another way as well. To a man, they preferred a frankly dynamic approach to mechanics. A concept of force, whatever the name to which it answered, appeared on virtually every page they wrote – and was mangled on virtually every page as well. Only after we have watched his peers struggle in vain to free themselves from the morass of dynamic terms in which they were hopelessly bogged down, can we fully appreciate Huygens' preference for kinematics. Only when we have watched his peers dealing unsuccessfully with analogous problems, can we fully appreciate the clarity of thought that created his concept of incitation. Although Huygens did not consider the paper worthy of publication, they would have welcomed it. Dynamics was exactly what they were about, and in the absence of a clearly established model, they sought to extract a dynamics from the disparate, not yet consolidated elements that the tradition of mechanics had delivered to them.

Father Ignace Pardies of the Society of Jesus, a quondam professor of rhetoric turned scientist who became the professor of mathematics at the College of Clermont in Paris, wrote two short works on mechanics shortly before his death in 1673, one entitled *Discours du mouvement local* and the other *La statique*. Both were concerned, if not with dynamics, at least with dynamic concepts. Despite its misleading name, the work on local motion devoted itself primarily to impact, a phenomenon it attempted to clarify with the shopworn idea of the 'force of percussion'. Pardies' analysis, connected at every point with the basic themes of seventeenth-century mechanics, illustrates how readily the model of impact misled dynamic thought.

The treatise began with an exposition of the principle of inertia which insisted on the indifference of bodies to motion or rest.[1] Imagine that a body at rest, 'wholly indifferent either to remain at rest or to take the motion that can be given to it . . .,' is struck by another. Since both are impenetrable, something has to happen. The body struck is free of any impediment and indifferent to motion. If it does not move, the other body cannot continue its motion. It is obvious, therefore, that it will start to move with the velocity of the body that strikes it.[2] The interesting feature of Pardies' analysis is his insistence that the size of bodies is irrelevant to their behaviour in impact if they are in fact indifferent to motion. Descartes had contended that bodies have the 'force to attach themselves in the place where they are at rest so that some effort is needed to tear them away . . .' Pardies found such an idea inconceivable. If a ship cannot hold itself immovable in the middle of the ocean or a stone in the air, how can a body in a void hold itself at rest when there is nothing firm to grasp?[3] In fact we cannot imagine any resistance in bodies except that connected with their weight, and by definition the analysis of impact is concerned with bodies abstracted from the property of weight. If the body struck is larger than the body that strikes in one case, and in another case equal, 'since . . . the larger is just as indifferent to rest or motion as the equal body, certainly the larger will not offer more resistance than the equal body since neither one offers any at all.'[4] Given the principle of indifference, which everyone accepts, 'more force is not required to move a large body than to move a small one, and . . . there will be no more effort in moving ten parts than in moving five since neither the five nor the ten offer any resistance.'[5]

In a passage that made no effort to conceal its scorn, Huygens expressed his amazement that anyone could have arrived at a conclusion so absurd.[6] Since Pardies explicitly contended that the presence of a fluid medium alters the conditions, such that in a medium large bodies are harder to move than small ones, Huygens' scorn was somewhat misdirected.[7] Perhaps it was pointed at the very idea of an analysis, like Descartes' earlier one, of conditions that do not in fact exist. Even so, it is impossible not to feel that at least half the scorn should have been directed at the concept of indifference itself. Pardies had drawn out its logical consequences in unexceptionable terms. If the results were absurd, the problem lay with the concept of indifference, not with Pardies' reasoning from it. Until dynamics recognised as much, it remained incomplete.

Perhaps also Huygens could have legitimately expressed more amazement that Pardies proceeded to a frankly dynamic theory of impact from a foundation which seemed to exclude dynamic factors. The dynamic element, excluded with the indifference of bodies to motion, entered with their impenetrability. When two bodies hit each other, there is a percussion, 'which is nothing else than the shock of two bodies which approach each other and impede each other by their impenetrability.' Because impenetrability is a property of both, the percussion is mutual. If we imagine the bodies to be sensitive, each will feel as much pain as the other. If instead we imagine nails to be fixed in both such that the two heads come together, each nail will be driven equally. 'Hence we can establish as a general maxim that *when two bodies strike each other, the percussion is mutual and equal for both*.'[8] Now return to the body at rest. If the body that strikes it moves with one degree of velocity, and if the body initially at rest is set in motion with that velocity, then the percussion of one degree ought equally to give the other body one degree of velocity in the opposite direction. That velocity, compounded with the body's initial motion, reduces it to rest. Now let both the bodies be in motion, approaching each other with velocities of one degree. The 'force of percussion' in this case is of two degrees, and each body will have its velocity reversed. Obviously the force of percussion is proportional, not to the velocity of either body, but to their relative velocity – *'percussions are always proportional to the relative velocities . . .'*[9]

If we agreed to overlook the problem associated with the concept of indifference, that is, in effect, the absence of mass from the measure of force, to what extent can we find a workable dynamic concept in Pardies' 'force of percussion'? At first glance, he appears to have set the change of velocity proportional to the force of percussion, and thus to have moved in the direction of a concept of force that measures changes in the inertial state. In fact, the appearances are deceiving. Like other students of impact, he let his attention be riveted by the apparent force associated with the moving body that strikes. The 'force of percussion' is the force of its motion, measured by its motion relative to the body it strikes and only incidentally by the change of velocity it produces. Indeed, the change of velocity is not always proportional to it, as his analysis of a body striking an immovable obstacle shows. To examine this case, Pardies first imagined that two bodies with velocities of one degree approach the same point of a sheet of some material from opposite sides. The presence of the sheet changes nothing in their impact; the sheet will remain unmoved, and they will clearly rebound as they do when it is absent. Now leave one of the bodies out and let the sheet become an immovable obstacle such as a wall. In the light of bodies' indifference to motion, Pardies might well have paused to reflect on the possibility of immovable obstacles. He did not, however, but rather affirmed that when the body meets such an obstacle, it will rebound as it did before. He also failed to note, or failed to consider it significant, that the force of percussion diminished by half while the change of velocity did not. At least he affirmed more than once that in such an impact the 'force of percussion' is proportional to the velocity of the body relative to the wall.[10] Pardies' concept of the 'force of percussion' was primarily an extension of an intuitive idea of force, found elsewhere in his mechanics, which was almost indistinguishable from 'strength'.[11] And such was the problem of dynamics in the seventeenth century – to give an intuitive and ambiguous concept a meaning at once precise and usable. Pardies' tendency to set the change of velocity proportional to the force of percussion was a step in that direction, but the perplexities involved in the model of impact contrived to limit the clarity and the utility of what he did.

Part of the problem that he faced in treating dynamics was the presence of another image of force, the lever, or more generally the simple machine. It is revealing that the full title of his second treatise on mechanics was *Statics or Moving Forces*, an equation of two terms that have since come to appear mutually exclusive. In the preface, he indicated his intention to construct 'a complete system of mechanics and to reduce all the science of motion to order.'[12] In the first of six projected volumes, he had shown how local motion is produced, conserved, and communicated. The present treatise would treat motions made 'with some violence in overcoming resistance which is present also.'[13] Others would examine the motion of heavy bodies, the motion of fluids, vibratory motions, and undulatory motions such as sound and light. In addition to the first two, which were published, the last one apparently reached some form of completion and exerted some influence on the wave conception of light. If the others were ever composed, they have disappeared. Meanwhile, Pardies had indicated where statics fits into the science of mechanics. Despite its name, statics concerns itself primarily with motion.

> It frequently happens that bodies are bound to each other in such a way that some cannot move unless others move as well; and sometimes even, when some attempt to move in a contrary direction to others, they mutually prevent each other's motion if their forces are equal; and if they are not, the stronger prevail and oblige the weaker to move against their own inclination.

In the balance, for example, one weight can only descend if another is raised contrary to its own inclination. As others had done, Pardies was confronting the central anomaly of the simple machines which created so many problems for seventeenth-century mechanics. Their analysis, which yielded proportions accepted by everyone as basic relations in mechanics, examined the conditions of equilibrium, but the simple machines existed to move loads, not to hold them in equilibrium. Pardies was perfectly clear about the purpose of statics.

> Hence it is these forces, necessary to move bodies despite the resistance of contrary forces which for their part act to prevent the motion; it is, I say, these forces that we must now treat, and it is this science that we call Statics.[14]

The intended division of topics, then, set on one side percussion, in which motion is communicated without resistance, and on the other statics, the study of motion against resistance. To a certain extent, the division of topics also set the motion of bodies abstracted from weight on one side, and the vertical motion of heavy bodies on the other. It would be a mistake, however, to imagine that Pardies had separated the questions that neatly or to suggest that the image of the simple machines did not influence his treatment of percussion. His conviction that the 'force of percussion' is proportional to relative velocity must have derived ultimately, as did all such equations, from the law of the lever. In at least one place, he appeared to derive his understanding of impact directly from the simple machines. In discussing the inclined plane, he showed how a body descending vertically not only can pull another body ten or a hundred times as great up a plane, but also 'in descending make the other weight . . . which rises move with an equal speed.'[15] Here, as he explicitly pointed out, was his model of impact.

Pardies' treatise did not attempt to reconcile his two ideas of force – the 'force of percussion', and the 'moving forces' of the simple machines. He probably did not recognise that they stood in need of reconciliation. Both were concerned with moving bodies; both were common-sense extensions of the intuitive idea of strength. Indeed, as its principal generalisation, his *Statics* arrived at a definition of force that differed from both – that the same amount of 'force' is required to raise a hundred pounds one foot and one pound a hundred feet.[16] Whatever the definition of force, he did suggest a generalisation of it. Imagine two weights in equilibrium on a balance. One weight can be replaced by a man pulling down with enough 'force' to maintain the other weight in equilibrium; the 'force of his hand will be equal to that of the weight.' Now replace the other weight with a second man pulling with the same 'force' as the weight; the equilibrium still remains. Statics, Pardies said, is concerned with the 'forces' necessary to move bodies despite the resistance of contrary 'forces'; it is not confined to the 'force' present in heavy bodies. 'It is true that because there is no force that cannot be expressed in some way by the force of weights, the example of heavy bodies is ordinarily used to make what is generally true of all sorts of tractive or motive forces comprehensible.'[17] Where Galileo had used weight to measure force but had not considered weight itself as a force, Pardies had advanced a step further to arrive at a generalised conception in which weight was only the most conveniently quantified example. There is no need to overstate the importance of the step. Others were arriving at the same idea independently. As in all his mechanics, so in this case, Pardies serves primarily as an example of the general level of mechanical thought in the latter half of the century.

The same can be said of Claude-François Milliet de Chales, another Jesuit whose *Cursus seu mundus mathematicus*, published in 1674, touched extensively on the science of mechanics. De Chales cannot be placed in the same category with Pardies. Where the latter was a recognised member of the scientific community, whose comments on Newton's theory of colours, for example, were neither to be ignored nor to be treated with disrespect, de Chales played the humbler role of the epitomiser and expositor. His *Cursus* frankly addressed itself to the beginner, although it did promise to conduct him to the highest pinnacles mathematics had reached. If the truth must be known, the sections, or as he preferred to call them, treatises on mechanics (treatise six is entitled '*Mechanice*' and treatise seven '*Statica seu de gravitate terrae*'[18]) ran out of breath well before they reached such altitudes. In their very limitations, however, they provide an excellent mirror of the general level of mechanical thought in the latter half of the century, and above all of the nature and source of its dynamic concepts.

Although he had digested the gross conclusions of Galileo's kinematics of free fall, de Chales was not prepared in any way to accept them in purely kinematic terms. His mechanical thought was entirely dynamic, and rested indeed on the peripatetic conviction that every motion implies a mover, a foundation on which a thorough assimilation of Galileo was not to be based. At the heart of de Chales' mechanical thought was a concept of '*impetus*', understood primarily as the cause of motion though not infrequently expressed as its consequence, and thoroughly confused with another concept of '*momentum*', which he wished apparently to distinguish from '*impetus*' despite his tendency to use the two identically. Further to confound the issue were the terms '*potentia*' and '*vis*'. For the most part, de Chales saw '*potentia*' as the source of '*impetus*'; thus it was vaguely analogous to our '*force*' in a context which makes all analogy somewhat misleading. In such passages, the distinction of '*potentia*' from '*impetus*' allowed him to account for the acceleration of free fall in a manner similar to Torricelli's.[19] '*Vis*' was sometimes used in a way that can only be translated as 'strength', but also sometimes in a way that makes it interchangeable with '*impetus*' and '*momentum*'.[20] Confusion of terms and of concepts was endemic in seventeenth-century mechanics, as I have been arguing, and there is no occasion to berate de Chales for the obscurity of a science struggling still toward clarity. Quite the contrary, he offers some insight into the factors that helped to maintain the confusion.

De Chales developed his concept of impetus in the treatise on mechanics, or the simple machines. Heavy bodies resist motion upward, and resist a greater motion more than a lesser, so that more impetus is required to raise a pound two feet than one. The resistance of a heavy body can be overcome by the descent of another which produces impetus in proportion to its descent. A body of one pound further removed from the fulcrum can raise another body of one pound that is closer. Hence de Chales arrived at the conclusion that 'more impetus is required to move the same weight through a greater space than through a less, whether the entire impetus is produced at once as in a projectile, or over a period of time as when a heavy body is pulled.'[21] Both the context and the examples suggest beyond the possibility of doubt that this measure of impetus by what we call work derived immediately from the simple machines. In seeking to separate it from the simple machines, as he did in the passage quoted, however, he wandered innocently down a path beset with every snare.

Moreover, as we have seen, the simple machines taught a perplexed lesson. If they revealed that one pound descending two feet can raise two pounds one foot, they also showed that one pound moving with a velocity of two is equal in strength to two pounds with a velocity of one. 'Those moving bodies are equal in force whose magnitudes and velocities are reciprocals.' De Chales supported the proposition by a direct appeal to the machines, and again went on to assert its general validity, whether bodies are connected by a machine or not. There is no reason why bodies 'are equal in virtue when they are connected on a machine and extension compensates intension, if this does not happen apart from a machine.'[22] To the product of size times velocity he also applied the term 'impetus'. It was possible to reconcile this measure of impetus with Galileo's kinematics of free fall, in which velocity increases as the square root of space,[23] but it was not possible to reconcile it with the measurement by work.

De Chales' dynamics was little more than a generalisation of the law of the lever. His investigation of percussion suggests what havoc the ratios of the lever could create in an inquiry to which they were inappropriate. A large weight placed on a wedge causes little to happen, whereas a blow by a hammer produces a considerable effect. The problem is to calculate 'how much more weight the motion adds'. Such was the standard definition of the problem, which implied the equivalence of percussion to static force and encouraged a solution by means of the simple machines. Without further ado, de Chales seized the ratios they offered. Suppose that a hammer of one pound, six feet long, is swung by a man swiftly enough that it would complete four full circles in one second, and with this velocity the hammer drives a nail one inch. From the circle, we can calculate the velocity as 144 feet per second. If the hammer fell 'by the force of heaviness' with the normal acceleration of heavy bodies, it would acquire such a velocity in the fifth second, after it had fallen 425 feet.[24] 'Hence we conclude that the height of 425 feet is the measure of the percussion...' Moreover, a weight of one pound falling 425 feet, 'acquires the force' to raise one pound 425 feet, and 'therefore this percussion seems to be in equilibrium with a weight of one pound moving through 425 feet.' But a weight of 5100 pounds that moves one inch is in equilibrium with one pound that moves 425 feet. Hence a dead weight of 5100 pounds will drive the nail one inch. Lest the perplexity not be sufficient, de Chales went on to insist that the acceleration and deceleration of bodies in vertical motion must be taken into account – though of course acceleration had appeared at one point already. Suppose that 100 pounds drives the nail one inch after a fall through ten feet. Given a fall of forty feet, what weight will also drive the nail one inch? Aspects of the earlier problem might lead one to answer twenty-five pounds. At this point, however, de Chales was concerned with the proportion of velocities and the answer was fifty pounds. 'The same effect will result if the quantity of motion is the same in both bases.' Beyond the insoluble dilemma in seeking to measure force of percussion by weight, de Chales compounded his difficulties by applying both the ratio of displacements and the ratio of velocities indiscriminately to unconstrained motions. 'Percussion,' he remarked, 'is extremely difficult...'[25] To which we might add, 'Not only percussion.'

Fully to comprehend de Chales procedure, we must probe his understanding of the simple machines. Like Pardies, he did not think of them in terms of equilibrium. The machines increase the strength of a force in a wonderful way, he said, 'so that what had been capable of overcoming a resistance say of a hundred pounds, is now made able to pull a thousand, or two thousand, or more.'[26] With a simple machine, a force 'overcomes' (*vincere* or *superare*) a resistance.[27] The machines exist to move loads, not to hold them in equilibrium, a view that encouraged one to apply their undoubted ratios to other problems and not to see that the displacements and velocities involved were merely virtual.

If de Chales did not play a major role in the history of mechanics, his use of the simple machines provides a key to decipher the enigmatic paper on impact of a more significant scientist, Christopher Wren, virtually his last contribution to natural science before architecture finally absorbed his creative talents. It is well known that the Royal Society received three important responses to its invitation of 1668 to investigate impact. Not only did Huygens submit the conclusions of his earlier treatise, which he had not published, but two English scientists sent in papers as well. John Wallis' was elaborated at some length in his *Mechanica, sive de motu* of 1672, which I shall examine later in this chapter. In contrast to both of the others, Wren's brief tract, entitled 'The Law of Nature Concerning the Collision of Bodies', is not illuminated by a longer work, and in its very brevity it has presented a puzzle. Although it contains a correct solution of all cases of impact of perfectly elastic bodies, it appears to offer no indication of how Wren derived it. The context in which he wrote, the prevailing level of mechanical thought, does.

The primary content of the paper is quickly repeated.

> The proper and most natural velocities of bodies are reciprocally proportional to the bodies.
>
> Hence when bodies R and S have proper velocities, they retain them also after impact.
>
> And when bodies R and S have improper velocities, they are restored to equilibrium by impact, that is, the [sum of the] amounts by which R exceeds and S falls short of proper velocity before impact is taken from R by impact and added to S, and conversely.
>
> Hence the collision of bodies that have proper velocities is equivalent to a balance oscillating on its centre of gravity.
>
> And the collision of bodies that have improper velocities is equivalent to a balance oscillating on two centres equally distant on one side and the other from the centre of gravity. When it is necessary, of course, the beam of the balance is extended.[28]

Wren's theory of impact rested on the assertion, proposed without further justification, that two bodies rebound with their velocities reversed but otherwise unchanged when the velocities are inversely as the bodies. Although his phrase, 'proper velocity', rings a bit unnervingly in the twentieth-century ear, it merely repeats a proposition accepted by everyone in the seventeenth century. Proper velocity is not an absolute quantity, of course; it is relative both to the given pair of bodies in impact and to the sum of their velocities. When one body exceeds its proper velocity, the other must of necessity fall short, and for any given pair of bodies with given velocities, the amount by which the one exceeds equals that by which the other falls short. To subtract the sum from the greater velocity is thus equivalent to subtracting twice the amount by which the body in question exceeds its proper velocity, and its velocity after impact falls short of the proper value by as much as it exceeds the proper value before. Such is the meaning of Wren's phrase about equilibrium. Let the line RS connecting the two bodies be divided at a, their centre of gravity (see Fig. 18). If the velocities are not in the proportion Ra and Sa, there is another point e on the

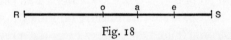

Fig. 18

line (not necessarily between R and S – as he said, the line can be extended) such that the velocities are in the ratio $Re:Se$. Now finds another point o on the other side of a (again the line can be extended) such that $ea = oa$. Then the velocities after impact will be in the ratio $oR:oS$. From this conclusion follows the line about oscillations on two centres. Since o and e are equally distant from the centre of gravity a, the line containing R, S, a, o, and e is like a balance poised, not on its centre of gravity a, but on two points e and o, equally removed from the centre.

Surely the key to this perplexing theory lies in its use of the balance, which it presents without further discussion as though the intended reference were wholly obvious. What alone could have been obvious to the seventeenth-century reader? – the dynamics understood to be implicit in the law of the lever. When $Rv_R = Sv_S$, $F_R = F_S$. Wren was proposing a dynamic theory of the impact of perfectly elastic bodies. When they have proper velocities (the phrase itself becomes intelligible in terms of the equilibrium of the balance), they have equal forces. Neither can prevail over the other, and they both rebound with their original velocities. At the time when Wren composed his paper, Huygens alone among seventeenth-century scientists had questioned the intuitive validity of this proposition, and Huygens himself had considered it highly probable.

In Wren's case, it is not quite true to say that each rebounds with its original velocity. As a dynamic approach to impact required, his theory contained an implicitly vectoral conception of velocity. Velocity Sa is not the same as velocity aS. Its quantity is identical, but its direction has been reversed. A dynamic action does take place in impact, causing a change in motion. What Wren added to the standard dynamic assumptions of the age was the recognition that every impact between two given bodies is dynamically equivalent to an impact with 'proper velocities'. It is necessary to recall again that 'proper velocity' is a relative term. It is relative first to the two bodies, R and S, and second, since $Ra:Sa$ is a ratio, to the sum of their initial velocities. In any impact between two bodies, R and S, each undergoes a change of velocity equal to twice its 'proper velocity'. Whatever the initial velocities Re and Se, subtract from each twice the 'proper velocity' of the body to find the final velocities oR and oS. In effect, this was Huygens' position set over into dynamical terms. Because of the principle of inertia, every impact can be viewed from the common centre of gravity where bodies approach each other with 'proper velocities'. To Huygens, with his kinematic approach, impact involves no dynamic action; bodies merely reverse their directions. To Wren, in contrast, body R with velocity Ra, as seen from the centre of gravity, rebounds with velocity aR, a change equal to $2Ra$. Since every impact is dynamically equivalent to an impact with 'proper velocities', in every case one subtracts $2Ra$ (or $2Sa$) from the initial velocity Re (or Se) to find the final velocity oR (or oS). The paper closed with rules of calculation that showed how to find Ra and Sa, given R and S and their initial velocities, and how to use Ra and Sa to calculate the resultant velocities.

Although Wren did not employ the word 'force' in his paper, the dynamics implicit in his theory involves serious ambiguities despite the fact that he was able to derive a correct solution of perfectly elastic impact. To what does 'force' refer? The use of the balance in equilibrium as the intuitively obvious case and the phrase 'proper velocity' associated with the balance both suggest that 'force' means the force of a body's motion – mv in our terms. When he applied the impact of bodies with proper velocities to impacts with improper velocities, however, the element he seized as crucial was the change of velocity – Δmv in our terms. His rules of calculation are essentially the rules by which first to calculate the change in velocity, and from it to calculate the final velocity. For every insight it suggested, the law of the lever as a foundation of dynamics presented at least one dilemma as well.

*

Although impact provided the prevailing model to seventeenth-century mechanicians, and the simple machines the means to quantify it, at least a few found their inspiration in Galileo's kinematics of free fall. Among them was Robert Hooke. Hooke is something of an enigma for the historian of science. He was recognised as a scientific leader in his own age. His apparently universal talent touched fruitfully on problems in a dozen fields, and he seemed destined to be remembered as one of the giants of the age of genius. Instead he has fallen into partial obscurity, and his name survives primarily through its connection with mechanics – on the one hand, through his claim that Newton had stolen the law of universal gravitation from him, a contention that was summarily crushed, and on the other hand, through his statement of Hooke's Law, a relatively trivial principle of elasticity. Although he was a practical mechanic of outstanding skill and a man of sporadically penetrating insight in theoretical mechanics, Hooke was not a systematic student of the science. He produced no treatise on mechanics. Instead, his works from the decades of the 1660's and '70's, the period of his most productive scientific activity, contain a number of brief passages devoted to mechanics in which Hooke, following his normal bent, announced dynamic principles of intended general significance. The passages appeared in widely scattered contexts, and the principles of dynamics they proclaimed were sometimes contradictory to each other. It should be obvious by now that Hooke was scarcely alone in this respect. A workable system of dynamics had yet to appear; different models offered different possible insights, and it was not immediately clear that the insights were mutually incompatible. For the most part, Hooke perceived that the uniformly accelerated motion of free fall was a more fruitful model for dynamics than the constrained motion of the simple machines, and his role in the history of mechanics is directly connected to that realisation.

If he looked beyond the simple machines for a model, he did accept part of the concept of force they encouraged. Among other things, force to Hooke was a property of a body in motion – what he referred to once as 'force in moving bodies...' Understood in this sense, force, he said in 1669 in a lecture to the Royal Society, is in duplicate proportion to velocity 'so that there is required a quadruple weight to double the velocity.' Two experiments supported this proposition. One employed a pendulum made, he said, after the manner of a fly counterpoised, which I take to refer to a mechanism similar to his spring-driven watches with weights on a line substituted for the spring that drove the balance wheel or fly. The effect would be equivalent to altering the acceleration of gravity for an ordinary pendulum. He found that doubling the number of oscillations required a quadrupling of the weight. The second experiment examined the relation between the depth of a fluid and the speed of efflux through a hole. An unfortunate leak in the vessel spoiled the demonstration, but the goal at which it aimed was clear enough – the velocity of flow varies as the square root of the depth.[29]

> Now this [he stated about the same proposition in a Cutlerian lecture eight years later] is exactly according to the General Rule of Mechanicks. Which is, that the proportion of the strength or power of moving any Body is always in a duplicate proportion of the Velocity it receives from it [sic]; that is, if any body whatsoever be moved with one degree of Velocity, by a determinate quantity of strength, that body will require four times that strength to be moved twice as fast, and nine times the strength to be moved thrice as fast, and sixteen times the strength to be moved four times as fast, and so forwards.

The rule holds good, he continued, in the motion of bullets shot from cannons and guns, from windguns, and from crossbows, in the motion of arrows shot from bows, of stones thrown by hand or by sling, of pendulums moved by the gravity of weights, of musical strings, of springs and all other vibrating bodies, of wheels and flies turned by weights or springs, of falling bodies – that is, the rule holds good for all mechanical motions.[30]

Implicit in the passage is an equation of external force exerted on a body with the force it generates in the body. In so far as external force, as he understood it, was equivalent to our concept of work, his principle approximates our work-energy equation. What he usually meant by external force was so equivalent. The implicit equation derived from Galileo's result that the square of velocity in free fall varies as the distance fallen. In a dynamic context, force in this usage referred to weight multiplied by distance, our concept of work. In an analysis of pendulums, for example, Hooke argued that the 'quantity of strength' moving the bob in a given swing is equivalent to the 'perpendicular lines of attraction' corresponding to the arc, that is, to the 'attraction' exerted through the vertical displacement *BC* (see Fig. 19), and he set the square of velocity proportional to the quantity of strength.[31] If we are willing

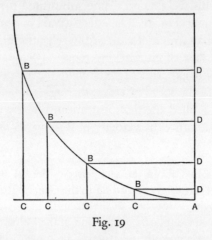

Fig. 19

to assume what possibly gives Hooke too much credit, that by velocity in this problem he meant the maximum velocity in a single swing, he had arrived at an important conclusion.

Nevertheless we cannot ignore the considerable confusion present despite the apparent clarity of the concept. It expresses itself in part through the multiplicity of terms. 'Strength', 'quantity of strength', 'force', 'force of a moving body', 'pressure', 'power', – Hooke employed them all interchangeably. On one occasion, in apparent despair at the chaos of terminology, he referred to 'force, pressure, indeavour, impetus, strength, gravity, power, motion, or whatever else you will call it . . .'[32] Behind the confusion of terms was the far more important confusion of concepts. In the analysis of the pendulum, he boldly seized a result from Galileo, relative to uniformly accelerated motion, which seemed to present a likely model, and applied it without any attempt at justification to a non-uniformly accelerated motion. In the case of the pendulum, Hooke was lucky, but luck was an unlikely foundation for a rigorous mechanics. If his use of 'force' (whatever the word) was usually similar to our 'work', at other times it was such that only our concept of force, which is something quite different from work, can be set equivalent. In one way or another, this confusion of concepts infected nearly everything Hooke did in mechanics.

Consider briefly three examples. In a paper that he read to the Royal Society in 1666, Hooke examined the mechanics of orbital motion. When left to itself, a body in motion continues to move in a straight

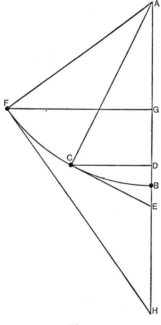

Fig. 20

line. If it follows a closed orbit, some cause must operate continually to deflect the motion. As one possible cause of the deflection, he suggested an attraction from the sun. At this point, he offered the conical pendulum as an illustration, but he noted that it was inexact in at least one respect. Gravity decreases with distance from the centre, but with a conical pendulum 'the degrees of conatus at several distances from the perpendicular' are proportional to the sines of the angle formed by the string and the perpendicular. When the bob is at C, for example (see Fig. 20), its conatus to descend is to that when it is at F as $\frac{CD}{FG}$, a proportion he went on to derive from the inclined plane. Hence 'the conatus of returning to the centre in a pendulum is greater and greater,

209

according as it is farther and farther removed from the centre, which seems to be otherwise in the attraction of the sun . . .'[33] This passage is significant at once for its clarity and for its confusion. On the one hand, it marked a major advance in the conceptualisation of orbital motion. From the principle of inertia he reasoned to the necessity of a deflecting force – a 'conatus of returning to the centre' instead of a conatus of fleeing from it. His dynamics was incapable of translating the conceptualisation into rigorous analysis, however. To handle the problem, he called upon the tradition of statics and compared the components of gravity perpendicular to the cord at two different positions of the bob.

The importance of Hooke's statement of the mechanical elements of orbital motion cannot be exaggerated. Following the pattern established by Descartes, every student of circular motion before him had spoken of the tendency of bodies in circular motion to recede from the centre. It was Hooke who broke the tyranny that pattern exercised and reconceptualised the problem. If the principle of inertia is given, the question is what constrains a body to follow a curved path and not the tendency to recede that it exhibits when so constrained. It is not too much to say that Hooke taught the science of mechanics this fundamental lesson and set it on the way to a satisfactory dynamics of circular motion. Having stated the proposition before the Royal Society in 1666, he restated it in his Cutlerian Lecture on the motion of the earth in 1674. The mechanics of the solar system that he intended to perfect depended on three suppositions, he said. The first held that all celestial bodies have a gravitating power toward their centres, the activity of which extends far enough that the sun, for example, can attract the earth and other planets.

> The second supposition is this, That all bodies whatsoever that are put into a direct and simple motion, will so continue to move forward in a streight line, till they are by some other effectual powers deflected and bent into a Motion, describing a Circle, Ellipsis, or some other more compounded Curve Line. The third supposition is, That these attractive powers are so much the more powerful in operating, by how much the nearer the body wrought upon is to their own Centers.[34]

The same statement of the problem appeared in his letter to Newton in 1679, when he spoke of 'compounding the celestiall motions of the planetts of a direct motion by the tangent & an attractive motion to-

SCIENCE OF MECHANICS IN THE LATE SEVENTEENTH CENTURY

wards the centrall body . . .,'[35] and via Newton Hooke reshaped the understanding of the mechanics of orbital motion.

To conceptualise the problem was one thing. To analyse its mechanical elements was another, and in this task, the second example of his mechanics that I want to examine, he quickly ran aground on the confusion inherent in his dynamics. As he said in 1674, the attractive power decreases with distance. In what proportion does it decrease? To answer the question, Hooke apparently applied Kepler's aborted 'law' of velocities, that the velocity of a planet at divers points in its orbit varies inversely as its distance from the sun, to his own dynamic formula, that force is proportional to the square of velocity. When Newton described the orbit of a body attracted by a uniform central force, Hooke replied that the curve was correct for the force assumed. 'But my supposition is that the Attraction always is in a duplicate proportion to the Distance from the Center Reciprocall, and Consequently that the Velocity will be in a subduplicate proportion to the Attraction and Consequently as Kepler Supposes Reciprocall to the Distance.'[36] On this basis, Hooke claimed the inverse square relation as his discovery, and even the publication of the *Principia* failed to disabuse him.

The third example is Hooke's major achievement in mechanics, his analysis of the vibration of springs. Everyone knows that Hooke's name is still attached to the law he announced, but it is less well known that the lecture containing Hooke's Law also contains an analysis of what we call simple harmonic motion. The analysis based itself on his law – that the 'power' of a spring is proportional to the degree of flexure. To each of the infinite degrees of strain there corresponds a proportional degree of power. 'And consequently all those powers beginning from nought, and ending at the last degree of tension or bending, added together into one sum, or aggregate, will be in duplicate proportion to the space bended or degree of flexure . . .' Thus, if the aggregate or sum of powers corresponding to a strain of one degree is one, it will be four when the degree of strain is two.

> The Spring therefore in returning from any degree of flexure, to which it hath been bent by any power receiveth at every point of the space returned an impulse equal to the power of the Spring in that point of Tension, and in returning the whole it receiveth the whole aggregate of all the forces belonging to the greatest degree of that Tension from which it returned; so a Spring bent two spaces in its return receiveth four degrees of impulse, that is three in the first space returning, and one

in the second; so bent three spaces it receiveth in its whole return nine degrees of impulse ... Now the comparative Velocities of any body moved are in subduplicate proportion to the aggregates or sums of the powers by which it is moved, therefore the Velocities of the whole spaces returned are always in the same proportion with those spaces, they being both subduplicate to the powers and consequently all the times shall be equal.[37]

The analysis rested ultimately on the brilliant intuition that a vibrating spring is dynamically equivalent to a pendulum. The 'aggregate of powers' in the one case is equivalent to 'quantity of strength' or 'perpendicular lines of attraction' in the other, and in both cases he set this sum or quantity proportional to the velocity squared. It is true that he treated two different kinds of sums as equivalent without bothering to justify the step – in one case a sum of forces that vary with distance, and in the other case a sum of forces that are constant through a vertical displacement. In both cases, nevertheless, we can recognise an interpretation of his words whereby they become equivalent to our work-energy equation. We would do well to pause before we impose that meaning on them. What he had done was to apply Galileo's result – $v^2 \propto s$ – derived for uniformly accelerated motion, to non-uniformly accelerated motion; and ignoring the complexities, he had drawn conclusions that were correct for reasons he did not understand. What is to be understood by 'velocity of the whole spaces returned', and how far could dynamics proceed with such concepts?

The three examples from Hooke nicely illustrate the problems of dynamics in the seventeenth century. Unlike most of his contemporaries, he seized on a dynamic action to quantify the concept of force – not the equilibrium of the balance but the uniformly accelerated motion of free fall. The problem still remained, however, to decide what factor in the action represented force. The one he chose, equivalent to what we call work, was not as independent of the simple machines as it might at first appear; such a measure of force, as in Descartes' example, stemmed directly from the law of the lever stated in terms of displacements. Hooke's measure of force was susceptible to systematic development, but the complexities baffled him. With the conical pendulum, he simply abandoned it; and in the case of orbital motion, he applied to instantaneous attraction exactly the words he applied to the aggregate of powers in the case of vibrating springs – that velocity is 'in subduplicate proportion' to force – in order to 'derive' the inverse square relation. Clearly more than a plausible model was required for a workable dynamics. The confusion of Hooke, who was not after all wholly lacking in ability, can be taken as a measure of the problem's difficulty.

★

Among the scientists who concerned themselves with mechanics during the period following Huygens' great work near the middle of the seventeenth century, three stood out both for the extent and for the quality of their work. Giovanni Alfonso Borelli in Italy, John Wallis in England, Edme Mariotte in France – before the definitive labours of Leibniz and Newton imposed an enduring pattern on dynamics, these three were its acknowledged masters.

In terms of sheer quantity, Borelli outdistanced the other two. Born early enough to have known Galileo, Borelli was the ultimate heir of the Italian tradition in mechanics in the seventeenth century. In extending the outlook of mechanics into every realm of scientific thought, he was equally the child of his age. He is best known today for *De motu animalium* (1680–1), perhaps the leading treatise in the tradition of iatromechanics. It was not the chief basis of his claim to a role in the history of mechanics, however, for Borelli was also the author of heavy tomes *De vi percussionis* (1667) and *De motionibus naturalibus, a gravitate pendentibus* (1670), standard exercises on which students of mechanics were expected to break their skulls. In *Theoricae mediceorum planetarum ex causis physicis deductae* (1666) he produced the first serious effort at a quantitative celestial dynamics since Kepler's *Astronomia nova*.

Although *De motu animalium* and *Theoricae planetarum* were not the principal fruits of his labour in mechanics, they were the works in which he expressed most clearly the thoroughly mechanist cast of his thought. Plato had proclaimed that geometry and arithmetic are the two wings on which we mount to heaven and unlock the secrets of astronomy. We can also affirm, Borelli added, 'that geometry and mechanics are the stairs by which we mount to the admirable science of the motion of animals.'[38] Animals are bodies; their vital operations are motions; bodies and motions are subject to mathematics. It follows then that the science that deals with their motion is geometrical as well, and is pursued by means of the same mechanical reasoning associated with balances and levers.[39] The muscles 'are instruments and machines, by which the motive faculty of the soul moves the members and parts of an animal.'[40] Borelli sounded much the same theme throughout his celestial dynamics. He intended to show how the motions of the heavenly bodies 'stem from their physical causes by the certain necessity of nature.'[41] As it requires more skill to build a machine that operates by itself than to build one that requires an operator, so also God, in constructing the universe, must have made it to run by itself.[42] Astronomers such as Bouillau had devised schemes in which the planetary orbits were generated by motions around empty points. To such systems Borelli was fundamentally opposed; physical force cannot inhere in an empty point. One can only imagine such schemes by positing the existence of intelligences to guide the planets in their paths, whereas he proposed to explain the generation of orbits 'without intelligences or angelic faculties but from the forces of nature alone.'[43] Here spoke the strident voice of the mechanical philosophy. Borelli distinguished himself from the common run of mechanical philosophers, however, by proposing to yoke their philosophical demand to a quantitatively articulated science of mechanics.

To that task he brought a thoroughly eclectic spirit. Deeply imbued with the outlook of the mechanical philosophy, which established the principal features of his idea of nature, he was influenced as well by Gilbert's view of magnetism which he accepted without effluvial mechanisms and extended to gravity. In the midst of his mechanical philosophising with its implicit acceptance of the conservation of motion, moreover, unsettlingly Scholastic passages unexpectedly assert that the end nature intends in the motion of heavy bodies is their rest at the centre of the earth.[44] Borelli had consumed and digested Galileo's kinematics – which he frequently expounded in the vocabulary of impetus mechanics. Equally, he had absorbed Kepler's celestial dynamics; and in that realm of activity, at least half his problems stemmed directly from his effort to harness Galileo and Kepler to the same wagon. Borelli was, in a word, a living repository of the mechanical tradition, and as such a striking example of how the tradition appeared most readily applicable to the problems now at hand.

Among those problems, some were essentially concerned with statics, and Borelli revealed a competent grasp of the subject. He employed easily a generalised conception of static force in which the elasticity of a spring or the pull of a hand is equivalent to a weight and measurable in terms of it. Statics applies to conditions of equilibrium; when the forces present are not in equilibrium, motion results. Borelli had taken to heart those passages in Galileo where he explained why velocity is not the direct dynamic expression of weight. Ten men can support a weight ten times that one can bear, but they cannot run ten times as fast carrying it. Ten dogs tied together cannot run faster than each one alone.[45] The weight of a heavy body held in equilibrium presses down as static force. When the body is free to move, a new situation obtains. The weight is now employed in generating motion, and there is no reason why the velocity should be proportional to the weight. In the case of a body in a fluid, the effective force by which the movement is caused is the difference between the body's weight and the weight of an equal volume of fluid. Although the insight was not original with him, in these passages Borelli recognised clearly the relation of dynamics to statics, the necessary starting point of a quantitative dynamics.

If dynamics takes up where static equilibrium breaks down, what are the key factors of the dynamic situation which must be placed in mutual relation to construct a quantitative dynamics? The received science of mechanics had conducted Borelli to an implicit posing of the question, but the received science gave no clear indication of the direction in which the answer lay. Perhaps we might better say the received science gave every indication that the answer lay in exactly the wrong direction. Faced with a dynamic problem, Borelli invariably fell back on the one undoubted equation of force, displacement, and velocity, the law of the lever. The most revealing aspect of his mechanics was his complete unreadiness to set force proportional to acceleration, despite his recognition of the vital elements in Galileo's work. He relied instead on a disparate set of dynamic concepts derived from the simple machines in one way or another – the most heterogeneous and bizarre dynamics that the century produced.[46] The net result confounded statics with dynamics anew and completely undid the results of his apparently clear distinction.

Borelli's dynamics rested on the concept of *vis motiva*, a protean concept capable of taking on a bewildering variety of forms but most frequently employed with a meaning implicitly identical to the Medieval concept of impetus. Imagine that a body A, moved with velocity DE by the 'motive virtue' R, strikes a body B that is indifferent to motion, and let

$$\frac{A+B}{B} = \frac{\text{velocity } DE}{\text{velocity } DF}.$$

> I say that DF is the retardation of the impelling body A after the impact ... Because one and the same motive force R acts always with the same endeavour, therefore the same energy with which the body A moved alone will impell it against the body B in the act of impact. Moreover, since B is indifferent to motion, it gives way to any impulse however small and therefore does not resist and does not diminish the motive force R, but gives way with complete freedom to the body A which is entering its place. The body B cannot leave its place unless it is moved, however, nor can it be moved unless it is propelled by some motive virtue, which cannot be anything other than R. Therefore, after impact, one and the same motive force R impells two bodies ... This motive force, however, moved only body A before. Therefore, two unequal corporeal masses, AB the greater and A the smaller, are moved by the same motive virtue R ...[47]

And so on in the same sparking style at continuingly excessive length to the desired conclusion.

At first glance, the concept presents a riddle. As the heir of Galileo and Descartes, Borelli commanded a principle of inertia, or a reasonable approximation thereto. Any body on which motion is impressed continues to move with the same velocity. If someone should ask what is impressed on the body, 'we see nothing else impressed ... beyond the motion itself newly added; this motion indeed is by its nature nothing but a migration from place to place, ... which would not be such if it were extinguished by itself without any contrary resistance.'[48] How is one to explain the blatant contradiction between the two passages? I suggest that the contradiction expressed the prevailing bewilderment of seventeenth-century mechanics as it faced dynamic phenomena. Motion alone was one thing; Borelli had seized the essence of Galileo's and Descartes' innovation. Impact, on the other hand, displayed a force inherent in motion such as mere translation did not. Impact was a dynamic action, and to deal with it he grasped blindly at available dynamic concepts whatever their import for his concept of motion.

How could one quantify the concept of motive force? The law of the lever was the one undoubted relation that expressed dynamic actions, and to it Borelli instinctively turned. In the example above, the two products set equal to the same motive force express the conditions of equilibrium of the two masses on a balance. What he did in a problem of impact, he did in virtually every problem.

> The simplest method of investigating the motive force of any power [he declared] derives from the recognition of its best known effect, which is the weight it supports. For the force of a power equals a resistance if they remain in equilibrium and one does not prevail over the other when they are immediately opposed or on a balance of equal arms. Or if there is a machine which reduces to a lever, then powers are reciprocally proportional to their distances or to the velocities by which they can be moved.[49]

In many of the problems to which his work on animal motion addressed itself, the analogy of the lever performed admirably. One of the basic concepts of the book was the proposition that muscles usually work at a considerable mechanical disadvantage. When a forearm holds a weight, the muscle that supports it is attached near the fulcrum of the elbow whereas the weight exercises the leverage of the entire forearm. At this point, Borelli confronted the paradox of the simple machines. He was interested in the motion of animals, not in their equilibrium, and he did not hesitate to extend the conditions of equilibrium to the explanation of motions. For example, Proposition 175 set out to demonstrate that the 'motive force' necessary for a man to leap into the air is 2900 times as great as his weight. In his established pattern, Borelli first calculated the conditions of equilibrium for the man in a crouched position by measuring the angles of the joints, and the mechanical disadvantage of the muscles that must support the weight in that position, and he concluded that the force the muscles exert is 420 times the weight of the man. As Borelli interpreted the figure, this was not the force that holds the man in equilibrium. It was rather the force that pulls him erect. Obviously, a much greater one is needed to project him into the air; and to compute how much greater, Borelli called on Galileo's analysis of free fall. The 'projectile force' that impels a body upward varies as the square root of the ascent. When a man pulls himself erect, he really produces a jump 'because the motion by which his centre of gravity is impelled upward is not generated without the impression of an impetus, which is not destroyed, and hence, when the extension of the limbs is completed, that active impetus produces its effect, raising the body of the man from contact with the floor; that is, it produces a leap.' Borelli called it an 'occult leap' and estimated its height at half an inch. An 'evident leap', on the other hand, raises the man two feet, and therefore requires a force seven times as great (*i.e.*, $\sqrt{48} \approx \sqrt{49}$). As a result, the 'motive force' necessary to produce a leap is more than 2900 times the weight of a man.[50]

Borelli's peculiarly dynamic conception of equilibrium underlay his treatment of motion. Consider a weight T supported on oblique lines by two forces R and S.

> And certainly, when the same weight T is held in equilibrium by the powers R and S pulling obliquely, the individual lines are drawn by opposed powers; and therefore, though it be granted that the powers are in equilibrium and are not in actuality transfered from one place to another, at least the proclivity to motion cannot be denied them; obviously, since that rest is not inert, but results from opposite pulls, they constitute a certain tonic motion which cannot be conceived as altogether identical to rest . . .[51]

Because equilibrium conceals a 'tonic motion' or an 'occult leap', the statics of simple machines extends directly into the dynamics of moving bodies. The dominant tradition of the simple machines had always approached them in such terms. The ultimate poverty of Borelli's dynamics stemmed from his failure to achieve a radical departure from the traditional perception of them.

In such a context, the central features of Borelli's treatment of circular motion become comprehensible. In his work on the satellites of Jupiter (the Medicean planets), he proposed a general celestial mechanics. As an Italian less than half a century removed from the condemnation of Galileo, Borelli feared openly to espouse the heliocentric universe; rather he implicitly adopted the Tychonic model, did not discuss the earth at all, but under the guise of explaining the motions of Jupiter's satellites proposed a dynamic solution both of planetary orbits around the sun and of the inequalities of the lunar motion around the earth. In all, it was an ambitious programme, the apparent audacity of which in the face of the Inquisition exceeded the grasp of Borelli's mechanics. Viewed as celestial dynamics alone, it presents a unique amalgam of Kepler, Galileo, and Descartes, which rests on Borelli's own view of dynamics as an extension of static equilibrium.

Basic to the theory is his understanding of circular motion. Borelli began from a Keplerian position. If the planets move at all around the sun, it is because the sun – 'as the heart, or vital spring of the planets' motive virtue'[52] – moves them as it turns on its axis. (Already the transparent disguise of the Tychonic system failed to hide the reality it pretended to cover. Borelli said nothing to justify the stability of the earth, which indeed he did not even mention, in the presence of the solar force, and nothing to justify a similar force in Jupiter, with which the work ostensibly dealt. If the Inquisition let him get away with this, it is probably incorrect to speak of his audacity. Apparently any pretence sufficed.) Borelli identified the solar force with light, the particles of which strike against a planet and move it in the direction the sun is turning. To explain how such a motion is possible, he drew upon Galileo's conception of the indifference of a body to motion on a horizontal (i.e., spherical) plane. No matter how small the force of the rays and how large the planets, the rays are able to move them somewhat.[53] After sufficient time, which elapsed long ago, the planets move with the rays' velocity – an unfortunate conclusion both for the variation of the velocity of a single planet and for the variation of velocities from planet to planet. Never mind – his celestial dynamics made no effort to encompass the latter problem, which was indeed beyond its capacity, and he simply forgot the explanation of planetary motion when the first arose. Meanwhile, the basic element of his celestial mechanics came directly from Kepler with a Galilean veneer applied for decoration.

Since Kepler's day, however, Descartes had attacked the problem of circular motion; and following Descartes, Borelli argued that some force or virtue must act to hold the planets near the sun and prevent them from straying. Heavy bodies on the earth have a 'natural instinct' to unite with the earth and iron bodies to unite with a magnet. It is not impossible that planets have a 'certain similar faculty', what he also called a 'natural appetite', to unite with the sun.[54] At this point, Borelli seemed well along the way to viewing circular motion as accelerated. In fact he was not. If the sole element of circular motion were a planet's appetite to join the sun, clearly it would do so. If the planet maintains a given distance from the sun, some other tendency must balance the first. Such a tendency Borelli found in the effort of bodies that move in a circle to recede from the centre. When the velocity of motion is sufficient, the tendency to recede balances the instinct to approach, and the body moves in a perfect circle – 'moreover this happens by necessity, because the virtue of the planet to approach cannot overcome the contrary repelling virtue, nor can it be conquered by the other...'[55] Circular motion presented itself to Borelli as an equilibrium. It was even more than an equilibrium – in a pregnant passage on the motion of a body thrown off a wheel, he asserted that 'the previous circular impetus degenerates into a rectilinear . . .'⁵⁶ In no sense did he see circular motion as the resultant of a constant accelerating force diverting a body from its rectilinear path. Borelli never comprehended the notion of a constant accelerating force. He predicated his dynamics on its continuity with static equilibrium, and he pictured orbital motion in such terms.

> Hence in the same manner, if in aethereal space we conceive a planet *I* which has a natural instinct to approach the sun *D* and is at the same time carried around the same solar centre with a velocity which suffices to cause the planet to recede exactly as much as it approaches the sun in each instant, clearly there can be no doubt that since these two contrary motions mutually compensate for each other, the star *I* will neither approach nor move away from the sun a distance greater than the semidiameter *DG* [of the circular orbit]; and therefore balanced and floating it will appear to be held by a substantial chain although it is placed in a perfectly fluid aether and neither rests on anything nor is supported by anything . . .⁵⁷

But planets do not move uniformly in circular paths. Both their velocities and their distances from the sun vary. The central feature of Borelli's celestial dynamics proposed a mechanical explanation of both variations, a theory ingenious enough to excite our admiration even

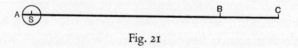

Fig. 21

while it instructs us in the dangers that beset the too facile application of analogical reasoning to complicated dynamic problems. As far as the variation in velocity was concerned, Borelli derived it directly from the law of the lever. Imagine the lever *ABC* turning on the fulcrum *S* (see Fig. 21). In his diagram, *S* represents the centre of the sun which is drawn such that *A*, one end of the lever, lies at the circumference of the sun. What he called the *potentia* at *A*, multiplied by its velocity, is the motive virtue of the sun which pushes the planets along as it turns on its axis. Borelli insisted that the motive virtue remains constant. If a planet varies in its distance from the sun, it will present a varying load on the lever. When a planet is at *B*, the sun can move it if its resistance multiplied by its velocity is a little less than the motive virtue of the sun. Now let the same planet with the same resistance move to *C*, further from

S. If the sun continues to move it, its velocity must decrease in the ratio BS/CS.[58] As I have noted more than once, the lever was understood as an instrument by which a force moves a resistance; Borelli manipulated it adroitly, not to say brazenly, to extract the ratio he sought. He made no effort to reconcile the passage with the earlier explanation of the planet's motion, thus sparing the reader an exercise in sophistry that must surely have exceeded the merely jesuitical. Meanwhile he had provided a new mechanical foundation to the Keplerian 'law' of velocities.

Some other mechanism must cause the distance to vary. Kepler had ascribed the action to magnetic attraction and repulsion, but Borelli explicitly rejected the explanation. Rather he suggested an analogy from the animal world – 'a pulsation in some way similar to the heart.'[59] As his mechanics of animal motion would suggest, the example he used to illustrate the pulsation was in no way animistic. Imagine a log half as dense as water floating in an upright position. When half the log is immersed, it will stand in equilibrium. If it is raised entirely out of the water, however, so that its lower end just touches the surface, and there released, it will bob up and down as first its weight and then the buoyant force of the water exceed each other. Compared to Hooke's analysis of a vibrating spring, Borelli's discussion of the dynamics of the bobbing log was crude. He did not realise that the cycles of the log's motion will be isochronous whatever the initial displacement – significantly his initial figure was a pulsing heart and not a pendulum. Since he was interested in proving that the two halves of a given cycle are symmetrical and equal in time, he examined the motion in successive finite instants, and thereby used 'force' in two incompatible senses, one equivalent to our sense of the word, and one equivalent to our concept of impulse, $\int F dt$. The passage is interesting nevertheless as another groping effort toward the analysis of what we call simple harmonic motion. If he bungled the details for lack both of a satisfactory conception of force and of a sophisticated mathematics, he recognised intuitively the central dynamic factors in an alternating motion.[60]

In the case of the planets, the forces never tire, as he said, and the perfect fluidity of the aether offers no resistance to oppose them. The instinct of the planet to approach the sun is analogous to the weight of the log, and the tendency to recede, arising from circular motion, corresponds to the buoyant force of the water. The tendency to recede varies as the velocity. If then, in the creation, God chose to make the planetary orbits eccentric, he had only to place a planet initially at a distance from the sun such that its instinct to approach is greater than its tendency to recede. (Carried away by the analogy of the log, Borelli mistakenly set the tendency to recede in this position at zero.[61]) As the planet approaches the sun, its oribital velocity increases; its tendency to recede increases equally, and the oscillating motion along the radius vector follows the alternating preponderance of the two forces. What could ever make the two periods, the circulation around the sun and the oscillation on the radius, equal, so that the line of apsides would be stable? Borelli discussed the problem, but in the end he could only cite the skill of God the mechanic. As celestial mechanics, the theory leaves much to be desired. As a revelation of the prevailing level of dynamic thought it is priceless.

Although the relation of force to resistance in the simple machines provided the quantitative elaboration of Borelli's dynamics, his primary model of dynamic action was percussion, the 'force and energy' of which, he said, are 'not inconsiderable but of immense power.'[62] To inquire into the subject, he composed a heavy tome with a familiar title, *De vi percussionis*. Roughly the final third of the book devoted itself to the various efforts that had been made to measure the force of percussion by static weight. From the beginning, he was sceptical of the idea. With some insight, he criticised experiments that had attempted to measure the force of percussion by the weight in one pan of a balance that is raised by a body dropping into the other pan, and he argued that the crucial factor is the height of the weight's ascent, and not the mere fact that it is budged. From such considerations, he concluded 'it is false that the energy of a single blow of a hammer is equivalent to any assignable weight...'[63] Such, in effect, had been Galileo's conclusion, that the force of percussion is infinite, but Borelli explicitly went a step further. Impetus, he said, in a context that employed the word as a synonym for 'energy of percussion,' is not a quantity directly comparable to weight; that is, the two 'are not contained in the same subordinate genus as they are commonly said to be.

> For quantities which can exceed one another when they are multiplied are said to be of the same genus, as lines are compared with each other, but because a line multiplied however much can never equal or exceed a surface or a body, so a weight multiplied however much will never exceed a force of impetus, because these quantities have no proportion to each other and are not to be considered of the same genus.[64]

Here was an insight of considerable potential importance for the science of dynamics. Since it appeared toward the end of the book on percussion, together with Borelli's criticism of his own earlier procedure, it is not entirely clear what bearing it had on the preceding discussion. At least relevant to the conclusion were divers passages scattered throughout the book in which, whether deliberately or unconsciously, he suggested that the force of percussion should be measured by the change of the quantity of motion. After his criticism of experiments that attempted to measure the force of percussion by dropping bodies into one pan of a balance, for example, he proposed quite another experiment in which bodies swinging in a vertical plane strike other bodies placed to receive a horizontal impulse; the impetus they receive, taken as the measure of the percussion, is determined by the distance they travel before they hit the floor.[65] The apparatus embodied an earlier conclusion, voiced in respect to a body which strikes another at rest, that 'the percussive action is measured by the impetus restrained and reflected, which is resisted and the progress of which is impeded, but not indeed by the persevering and active impetus which is not impeded . . .'[66] In briefer terms, 'the force and energy of percussion' is measured by the 'diminution of velocity' of the body which strikes another at rest and sets it in motion.[67] Borelli also insisted explicitly that the force can be measured in terms of either body. As much as the one acts the other resists, but there is only one impact seen from two different points of view.[68] In considering impact, moreover, Borelli arrived at a conception of *moles* or mass, the quantity of matter in a body considered without reference to weight, which he said is not operative in horizontal motions.[69] If we take such passages, isolated from the rest of the work and compared with each other, it appears that Borelli was approaching a dynamics rather different from that associated with the simple machines and the concept of motive force.

In fact, he never advanced beyond the level of isolated passages. The suggestions were never knitted together into a consistent and self-conscious statement. What is more, the very concept of motive force appears to have blinded Borelli to the significance of his own suggestions, for if he sometimes saw impact in terms of the change in motion it produced, for the most part it presented itself to him as a revelation of the force in a moving body. Such was the model of impact examined by the great majority of its students in the seventeenth century. The concept of the 'force of percussion' derived from this model, and

Borelli's choice of the well-worn phrase for his title indicates the basic direction of his work. Despite his tentative groping toward a different measure of force, he was mostly concerned in the book with the favourite diversion of seventeenth-century dynamics, the force of a body's motion, the measurement of which derived ultimately from the ratios of the simple machines. Far from opening a new vista to dynamics, then, his investigation of impact led him back to the continuity of statics and dynamics he had seemed ready to escape.

As a result of his approach to impact, Borelli's treatment of it was somewhat disjointed. Instead of developing a unified theory, he attacked different cases on an *ad hoc* basis, applying intuitive ideas of force as the spirit moved him. He began with perfectly hard bodies, explicitly denied any elastic property, and he investigated first of all the impact of one such body in motion on another at rest. In this case, force is obviously associated with the moving body. 'A projected body is carried by a virtue communicated and propagated by the projector.'[70] At exhaustive length, he explored the dimensions of the motive force and arrived at the less than novel conclusion that it varies in proportion both to the velocity and to the quantity of matter. As for the body at rest, it has no force of resistance. It is, in Galileo's phrase, 'altogether indifferent' to motion, so that it cannot stand against any force whatever 'but obeys it with complete freedom.'[71] Perhaps then the body struck begins to move, as Pardies argued, with the velocity of the body that strikes it? Not at all, in Borelli's opinion, for if that were the case, the same motive force would move a larger body at the same velocity, in violation of the entire analysis of motive forces. By its nature, a motive force not only moves the body in which it is located, 'but in addition [it] can be communicated and diffused in other mobile bodies that obstruct its progress...'[72] The solution of the problem is obvious. One and the same motive force successively moves two bodies – initially the percussant and then the composite body of percussant and percussed – and the velocities are inversely as the bulks. Borelli insisted that the communication of motion in this case takes place in an instant, a logical necessity with perfectly hard bodies.[73] He contended as well, and again with logical rigour on his side, that such impacts could never wholly destroy the motion of a body; by diffusing the motive force over larger and larger bulks, they could merely retard it indefinitely. Somewhat surprisingly, the tentative measurement of 'force of percussion' by the change (loss) of velocity emerged from this analysis despite the appearance that impact was being handled in terms primarily kinematic.

Now let the second body be in motion. Borelli wavered as he faced the problem. On the one hand, it appeared similar to the earlier problem. He suggested a solution by the summation of what he called partial percussions. For two bodies A and B, compute the percussion of A on B at rest and of B on A at rest. It is manifest 'that in the mutual impact of the bodies A and B the same contrary percussions will be performed . . .,' and the total percussion is the sum of the two partial percussions.[74] Analogous considerations obtain when the body struck moves more slowly in the same direction; obviously its motion in this case diminishes the force of the blow.

> Hence it is established that the force and energy of percussion depend, not on the impetus of the real motion of the percussant in absolute space, but on the relative motion by which one exceeds the other.[75]

Alas, other considerations upset the simplicity of the analogy. The solution of impact in the case of one body at rest depended on its indifference to motion, its total lack of force or resistance. Clearly the same does not obtain when it is in motion. Suppose that A and B meet with velocities inversely proportional to their bulks.

> Since the impulsive forces of the bodies A and B are equal to each other (because their sizes and velocities are inversely proportional to each other), both bodies therefore strike with equal energy, and both suffer from the other repulses of equal energy since the strength of resistance of body B exactly equals the impulse of A itself, and therefore the direct progress of the body A toward G is altogether impeded by the force of stability [*firmitudinis*] or resistance of B. Moreover, since the motive force of A is not destroyed but remains in vigour, it is necessary that it continue its motion in the direction in which it can; and hence at C it is reflected back toward A with the same active velocity DE. For the same reason, body B is reflected toward G with the same velocity H which it had . . .[76]

When A is greater than B and the motive force of A greater than that of B, B is reflected with its velocity increased by the larger force. When A is smaller than B though its motive force is larger, the situation is somewhat more complicated. As for B, clearly its forward progress must be altogether checked 'by the greater motive force,' and it is reflected with its original impetus intact and its velocity increased 'by the impulse of the stronger motive virtue of A . . .' With A the outcome varies. If its initial 'impetus' exceeds the 'velocity' of B so much that it is still larger than the velocity of B after impact, it can only proceed by moving in the opposite direction; that is, it is reflected. On the other hand, if its final velocity is less than B's, it proceeds in the same direction. 'The variety of reflections which the same bodies experience in oblique impacts,' he concluded with perhaps more nonchalance than was warranted, 'can be found by an easy application of the things demonstrated; hence, there is no need to pause any longer with these problems.'[77]

What had led Borelli to lose hold of the principle of relativity which he appeared initially to grasp? The structure of his argument suggests his concentration on the motive forces in impact. The presence of a contrary motive force made the outcome of an impact between moving bodies seem to differ from that when one of them is at rest and indifferent to motion. In the latter case, the two bodies move together after impact, since they are by definition devoid of elasticity. In the former case, however, two contrary forces strive against each other and the stronger prevails – again the image of the balance. The dissimilarity between the two cases extends beyond the fact that in one case the two bodies separate. As Borelli explicitly stated, the reflected body is not brought to rest and started anew in the opposite direction. Rather, its initial motive force, unaltered, continues to move it. Dynamically, reflection is no action at all, and there is no 'force of percussion' in the sense he defined above. On the one hand, the conclusion repeated Descartes'; on the other, it expressed as well Borelli's concept of the active equilibrium of forces in the balance.

In the impact of perfectly hard bodies, the equilibrium of forces results in the reflection of bodies, not in their coming to rest. Borelli was convinced that with perfectly hard bodies impact can never reduce a body to rest.

> It is clear therefore that the energy of a blow can create motion and impetus in a body hanging [at rest] but it is not possible for a blow to weaken or extinguish the same motion by another contrary percussion **once** it has been impressed . . .[78]

The world is not composed of perfectly hard bodies, however. All of those with which we deal are flexible and elastic to some extent, and with an elastic body the situation is different, because 'it will not have a complete indifference to motion but rather a positive resistance . . .'[79] What is the source of the resistance? As far as Borelli expounded it, the lapse of time was of central importance. The impact of perfectly hard bodies is instantaneous; the motive force, which is never interrrupted, continues in one direction or another, and the body never comes to rest. During an interval of time, however, a contrary impetus can be impressed which brings the original one into equilibrium and reduces the body to apparent rest.[80] He even used the example of a body in motion in one direction on a boat that moves in the opposite direction. Borelli never specified why an instantaneous impact cannot produce a motion equal and opposite to a given one whereas an action continued through time can. He did insist on it, however, and indicated the source of the conviction when he applied his analysis of the bobbing log to the action of an elastic body. With no attempt at quantitative treatment, he described how the elastic resistance, increasing with the deformation, like the buoyant force of water with greater immersion, gradually brings the motive force of the body into equilibrium and then impresses a new velocity equal to the original one as the elastic body restores itself. As usual, unfortunate complexities presented themselves. For example, if two perfectly hard bodies with opposite motions reflect each other, then an elastic body, such as a tennis racquet, ought to reflect one of them twice as fast if it hits it in motion.[81] Nevertheless, the analysis of a phenomenon treated satisfactorily at that time only by Torricelli, was not without insight.

Moreover, the analysis was significant for its further revelation of the apparently unlimited facility of the lever in the elucidation of dynamic action. The essence of elastic impact, whereby it differs from inelastic, he repeated several times, lies in the fact that an elastic body resists 'without percussion.'[82] The resistance arises in turn from the mutual intertangling of parts, 'such that they ought to be considered as levers which are pressed on one end by the projectile and on the other adhere with a resistance which resists the impulse of the projectile with a contrary effort . . .'[83] A prior discussion had already successfully obscured the distinction between the virtual velocity of a body on a balance in equilibrium and the real velocity of a body in motion. Borelli's resistance 'without percussion' was merely the equilibrium of a balance, and his concept of elasticity another application of the law of the lever.

Although Borelli continually sought to extract dynamic equations from the ratios of the simple machines, and continually drew on that source in his treatment of impact, he was aware of another potential model of dynamic action, the accelerated motion of free fall. Since Borelli's mechanics was infused with the idea that a constant motive force maintains a constant velocity, it is not clear that free fall should have presented itself to him as a dynamic model, and, in fact it did not do so explicitly. He treated it rather as a phenomenon to be explained by his dynamics but not as a paradigm case from which the dynamics might be drawn.

De vi percussionis divided all motions into two classes – those caused by external agents and those caused by internal. Projectile motion was an example of the first, animal motion and the descent of heavy bodies of the second. When he came to explain accelerated motion, however, he chose to examine its source in terms of external agents. Suppose that a body A 'with a constant persevering motive force and a constant impetus FC' continually pushes the body B during a period of time. In the first instant (read 'impact'), it impresses a degree of velocity FI on B. In the next instant (read 'impact'), it cannot press B with its whole velocity since B now moves in the same direction. Only the excess velocity IC acts and impresses an increment of velocity IK, smaller than FI, on B. In the third instant, it impresses still less, and so on until B reaches the velocity of A.[84] Obviously such is not a uniformly accelerated motion. Now suppose that body A with its motive force is carried along with B. When it gives B the velocity FI in the first instant, its own velocity increases by the same amount since it is carried along, and in the second instant, its velocity still being greater than B's by the amount FC, it imparts a second increment equal to the first. There was a certain ingenuity in the explanation – yoked to an utterly bizarre conception to be sure.

To explain what such a force carried with a body and continually striking it might be, he employed a number of figures, all exploiting the force of percussion. He compared it to a spring, unconnected to a boat, unbending in such a way that the boat feels the impulse in one direction but not the reaction in the other, to a hammer that strikes a boat to which it is not connected, to a bird that flies against one of the cross pieces of a boat. One evocative phrase referred to an 'intestine wind.'[85] Ultimately he decided against such a material representation on the grounds that rarer bodies, with larger pores to contain the intestine wind, ought to be heavier, and he concluded instead that either particles of matter are self-moving or they contain a self-moving spirit.

The image of successive impacts remained, however –

> the descent of heavy bodies is generated by the blows and impacts made by the internal motive virtue of the same heavy body . . .

The 'internal percussion' of gravity differs from ordinary percussions.

> For the force impelling a projected body creates a determinate degree of impetus in the body in a single instant, which impetus is not indivisible but finite, having a certain linear extension; but the virtue of gravity does not create a finite impetus in an instant but rather an indivisible one which afterwards, when multiplied by the inexpressible multitude of instants in the given time composes at length a finite and measurable velocity.[86]

It is easy for us to find in the passage an expression of the concept that a continuous force produces an acceleration, that force in this sense is incommensurable with impetus or motion, and that such a force acting over an interval of time produces an increment of motion ($Ft = \Delta mv$). Borelli was unable to express these relations with any corresponding degree of clarity, however. He was seeking rather to trace accelerated motion back to his own basic dynamic concepts. It became for him a series of tiny percussions, each measured in terms, not of a change in motion, but of a motion itself, the *vis motiva* on which his entire dynamics rested. He headed the chapter with the following title:

> The force of impetus of falling bodies is less than any impulsive force whatever impressed by a projected body.[87]

Again, the concept looks familiar to us, but Borelli saw it in terms of a crude mathematics of indivisibles in which the ultimate units are of the same kind as the whole. Although the impetus of gravity is extremely small, it is 'not altogether indivisible and without quantity . . .'[88] Despite his phrases about incommensurable quantities, such appears to have been the view most congenial to him.

Other figures reinforced the picture. He compared the acceleration of a falling body to the motion of a pendulum struck with a small blow on each swing. It is obvious that 'an extremely swift and vehement oscillation will at length be created in the great pendulum by all these tiny impulses multiplied . . .' Such is the case because 'any force however small and slow can impress a degree of impetus in a hanging and movable body . . .'[89] The adjective 'slow' in the passage is somewhat unsettling to us. It recalls the image of his 'intestine wind' which is carried along with the body. The chapter heading expressed a further dimension of the same idea which reveals the ultimate disparity between Borelli's approach to accelerated motion and ours.

> The smallest motive force in any body with a slow motion can impress in a vast body and make increase a velocity greater than that by which the impelling body is moved.[90]

It is relevant that in his book on natural motion, that is, the accelerated motion of descent, when he wished to argue, following Galileo, that the medium functions only to decrease the effective weight, he found himself constrained to speak in terms of velocities, not of accelerations. There was no question about the motion of heavy bodies; it is uniformly accelerated. His equation of motive force with a moving body, however, left him unable to state that the acceleration is proportional to the effective weight. He demonstrated, for example, that two heterogeneous bodies of equal weight enclosed by equal and similar figures descend with equal velocities in the same medium. In that medium, their effective weights are equal;

> but the motive virtues by which the bodies A and B are carried down are . . . nothing other than the energy of their weight; therefore bodies A and B have equal motive forces in the same fluid; furthermore these [forces] are equally impeded by the same fluid because of the similarity and equality of their figures; therefore the effects of the forces, that is the velocities by which the bodies are carried down, will also be equal to each other.[91]

His dynamics had carried him back to the equation of weight with velocity that he had initially denied.

In no sense can it be said that Borelli saw the accelerated motion of free fall as the model of dynamic action. In no sense, indeed, can it be said that he saw dynamic action as a problem distinct from statics. Rather he approached the whole of mechanics from the trusted relations of the simple machines; and via the concept of motive force, based on the model of percussion and measured by the law of the lever, he attempted to reduce all problems, static and dynamic, to the same ultimate terms.

★

John Wallis was one of the leaders of English science during the second half of the seventeenth century. A member of the group whose meetings in London in the 1640's were preliminary to the establishment of the Royal Society, he moved to Oxford with a number of them in 1649 after the Parliamentary victory and there continued to participate in the circle. Although Wallis remained in Oxford after the Restoration, he was one of the original members of the Royal Society; and as Savilian Professor of Geometry from 1649 until his death in 1703, he occupied a position of rare prestige in English science. He was more of a mathematician than a mechanician, and the attention he gave to mechanics, measured by his published works, was considerably less than Borelli devoted to it. His contribution was confined primarily to one book, *Mechanica: sive de motu*, published in 1670–1, and to exercises, such as his paper on impact submitted to the Royal Society, closely associated with it.[92] Even his *Mechanica* devoted more than half its pages to essentially mathematical problems of centres of gravity. *Mechanica* was a large and thick tome, however, and less than half its pages still constituted a weighty treatise. It was in fact the most important work on mechanics published in England before the *Principia*; and like Borelli's work, it reflected the highest level of prevailing mechanical understanding before Newton. Wallis was also like Borelli in his forthright repudiation of pure kinematics, so that mechanics in his treatment of the science was wholly dynamic.

Mechanica set out with a series of definitions.

> I call that which aids in producing motion *Momentum*, and that which prevents motion or impedes it *Impedimentum*.

Under the heading *momentum*, he included two quantities, motive force [*vis motrix*] and time, to each of which the motion generated is proportional. To *impedimentum* he also referred two quantities, resistance and distance, because motion is impeded the more as they increase. Motive force and resistance also required definition.

> I call a power that produces motion [*Potentiam efficiendi motum*] *Motive Force* or even simply *Force*....
>
> [I call] a power that is contrary to motion or resists motion *Resistance* or *Force that resists*. [*Vim resistendi*].[93]

It is impossible to mistake the conscious effort toward a generalised conception of force, especially in the use of the phrase 'force that resists'. Throughout the work, Wallis treated motive force and resistance as additive quantities, quantities of the same sort though their effects are contrary to each other. If the total motive force or *momentum* exceeds the total resistance or *impedimentum*, motion is generated or increased. If resistance or *impedimentum* is greater, motion is prevented or diminished. If the two are equal, motion is neither generated nor destroyed, and the existing state either of rest or of motion perseveres.[94]

Within such a context, gravity (i.e., *gravitas*, heaviness) appeared merely as one example of a broader class.

> Gravity is a motive force downward or toward the centre of the earth.... What is said about gravity in respect to the centre of the earth, moreover, can be understood in the same way about any other continuous motive force in respect to the terminus toward which it tends. Hence if that word [force], hitherto restricted to a particular usage in so far as it regards the center of the earth, is understood in a broader sense to apply to any continuous motive force moving directly toward its terminus, it will not be less accurately employed but rather more accurately, since general terms will be expressed generally. But since those things which are appropriate to all continuous motive forces are usually applied specially to gravity, I have accommodated myself so far to the common error that I must warn you that the things that will be expressed in a specific sense are true in a general sense.[95]

By explicitly refusing to discuss the cause of gravity, Wallis implicitly acknowledged force as an abstract concept relevant more to quantitative mechanics than to philosophy of nature. Gravity itself he distinguished from weight [*pondus*] which is its measure, and equally the measure of all motive forces.[96] I have already indicated that other mechanicians were moving in the direction of a generalised conception of force. This essential step in the construction of a quantitative mechanics found its clearest expression yet in Wallis' *Mechanica*.

In one important respect, Wallis carried this process further than any previous mechanician. In discussing the capstan, which is analogous to the windlass but is used primarily to draw loads horizontally, he introduced a distinction between 'contrary forces' (such as the weight a windlass lifts) and mere 'impediments' (such as the resistance due to the roughness of a surface over which a load is pulled, or the resistance of a nail being drawn). Unlike mere impediments, contrary forces have to be held at least in equilibrium, or they move of themselves in the other direction. When a load is pulled up an inclined plane, both an impediment and a contrary force must be overcome. When the contrary force strives to move down an inclined plane, however, the impediment opposes it, and may hold the load in equilibrium if the slope is gentle enough.[97] In effect, Wallis was distinguishing conservative from nonconservative forces and extending the generalisation of force by bringing friction within the domain of rational mechanics.

Toward the end of the book, Wallis applied his generalised conception of force to the model of free fall and without further ado arrived immediately at a statement which is indistinguishable, its verbosity aside, from the second law of motion.

> If a continuous application of motive force that is uniform be made, a motion continuously accelerated will be generated, and indeed accelerated in such a way that it will conceive equal increments of speed in equal times: which motion they call *Uniformly Accelerated*. If a similar application of an impeding force that is uniform be made, it will generate a similar retardation of motion, which motion they call *Uniformly Retarded*. . . .
>
> For let it be understood that any motive cause impresses one degree of speed on a body in one moment of time. This degree will persist even without a new cause unless it is destroyed by some impediment. . . .
>
> Indeed the same cause, acting in the same way and applied in the same manner during a second moment, will produce the same effect . . .
> Hence, since the first degree still persists, it will add a second to it. And similarly in a third moment, since those two degrees still persist, it will add a third to them, and so continually.[98]

Only a small additional step led on to the conclusion that motion on inclined planes, which is also uniformly accelerated, experiences a rate of acceleration that is proportional to the sine of the angle of inclination.[99] Wallis had demonstrated again how readily general dynamic equations emerged from the dynamic treatment of free fall.

We should be mistaken, however, to hail Wallis as the innovative precursor who made Newton's dynamics possible. The passage above on accelerated motion came near the end of a long book, and what had come before bore so little resemblance to it that the passage appears more as an unprepared assertion than as a logical culmination. A careful examination of the concepts of *momentum* and *impedimentum*, on which his entire mechanics rested, suggests how far removed it was from Newton's. As I have indicated, Wallis referred two factors to each of the two quantities. Both weight and distance contribute to *impedimentum*. At one point he remarked that what he said about distance was equally true of the density or tenacity of the medium,[100] but he never indicated how they might be substituted. He couched all of his discussion in terms of distance, and it is impossible to see how density or tenacity could function in a similar role. Wallis did not treat weight and distance as additive quantities, the sum of which constitutes *impedimentum*. His diagrams represent *impedimentum* by a rectangle, one side of which is weight and the other the distance it moves. Similarly, *momentum* is the product of force and time. 'In any motions whatever, compared together,' he concluded, 'the *momenta* are proportional to the *impedimenta*. . . . The product of weight and distance in one motion is in the same proportion to the product of weight and distance in another motion as the product of force and time in the first motion to the product of force and time in the latter motion, other things being equal.' What other things had to be equal he did not say, but he proceeded to express the proposition as an equation:

$$VT = PL,$$

in which V represents force [*Vis*], T time [*Tempus*], P weight [*Pondus*] and L distance [*Longitudo*]. 'Primarily on this proposition depends the comparison of motions with each other as to forces, times, weights, and distances.'[101]

The interpretation of this singular formula depends on the recognition of its two critical factors, force on one side of the equation and weight (or resistance) on the other. These were the traditional words in the terminology of the simple machines. Wallis' formula attempted to express the conditions of equilibrium. Rather I should say that like Borelli's dynamics, the formula attempted to extend the application of the simple machines. The basic analysis which I have been discussing appears in Chapter I, which bears the title 'General Principles of Motion' [*De motu generalia*]. Wallis understood himself to be moving from accepted and established principles toward a generalised dynamics. As the terms force and weight or resistance suggest, he too looked upon the simple machines for what in fact they were, devices to move loads, not to hold them in equilibrium.[102] He bowed ambiguously in the direction of the enigma by suggesting that beyond equilibrium a small additional increment of force is necessary for motion.[103] The title of a proposition at the end of the book adequately summarises the dilemma in which Wallis had caught himself – '*De motibus pure staticis*' 'Concerning Motions Entirely Static.'[104] Ironically, in attempting to generalise the law of the lever, he treated the motions as real and inserted both time and distance in the equation as independent variables, thus rendering the result unsatisfactory even for the simple machines.

To develop his equation one step further, Wallis divided both sides by time and arrived at the statement

$$V = PC \; [C \equiv Celeritas],$$

which is roughly equivalent, in our terms, to

$$F = mv.$$

Note that force in this equation represents motive force and not the force of a body's motion. By pursuing the analogy of the simple machines, Wallis had arrived at an essentially Aristotelian expression of motion despite his earlier statement of the principle of inertia.

> I assert that other things being equal, by whatever ratio the speed is to be increased or diminished, let the force applied be similarly increased or diminished in the same ratio, and the given weight will be moved with the speed desired.[105]

This proposition he employed to calculate, not merely virtual, but real velocities, as the statement of it suggests. Nevertheless, the image of the simple machines refused to be banished entirely, and like the ghost of Banquo continued to reappear without invitation. The problems he set assumed the machines – for example, 'To move a given weight with a given force.' Let the force V, by which a weight P can be moved with a speed C, be given, and it is required that the weight nP be moved by the same or equal force.

> I assert that other things being equal, if the situation is so arranged (by the application of a machine [*interposita Machina*]) that by whatever ratio the weight nP is greater or less than the weight P, by the same ratio inverted let the speed be decreased or increased, that is $\frac{1}{n} C$, and the same force V will move the given weight nP with the speed $\frac{1}{n} C$.

And he generalised the result in the following Scholium.

> Therefore mechanics is primarily concerned with the task of devising and placing in use a machine that adjusts the speed of the motions of the force and the weight applied to it, so that it compensates for the size of the weight by the slowness of its motion or the deficiency of the force by the length of time.[106]

The 'application of a machine,' 'devising a machine' – the image of the lever brooded over Wallis' discussion of the 'General Principles of Motion.'

As the title of chapter one promised to consider general principles of motion, the title of chapter two, 'Concerning the Descent of Heavy Bodies and the Inclination of Motions,' suggested an investigation of the rate of acceleration on inclined planes. As we have seen, Wallis understood the problem well enough and later in the work, in a particular rather than a general context, handled it unexceptionably. Chapter two, in contrast, despite its title, devoted itself to conditions of equilibrium on inclined planes. Wallis understood the critical factor in the inclined plane very well – that the vertical component of displacement alone is significant. Again, however, he wanted the motion to be real rather than virtual, and he wanted to employ the formula described above to calculate velocities.

> If further ... the ratio of speed is to be calculated, so that not just any motion occurs, but one with a given speed, the answer to this will be found from Propositions 29 and 30, Chapter I. That is, (in the three preceding propositions respectively) when the conditions under which motion at some speed can be made have been found, by whatever ratio that speed is to be increased, the length of the line *PO* found in Proposition 28 of this chapter will have to be increased by the same ratio; and the force found in Proposition 29, and the weight found in Proposition 30 will have to be decreased in the same ratio.[107]

The attempt to extract a generalised dynamics from the law of the lever could hardly be more explicit.

Generalising the concept of force was one thing; freeing it from the ambiguities inherent in the simple machines and the confusion of statics with dynamics was something else. Once Wallis had committed himself to the programme of extracting a dynamics from the law of the lever, he was hopelessly entangled in the full range of ambiguities that beset the concept of force throughout the seventeenth century. What was the measure of force? All four of the measures common in seventeenth-century dynamics appeared in Wallis' *Mechanica*, two and sometimes three of them (and not always the same ones by any means) frequently in the same sentence. As should be evident by now, in purely static problems Wallis commanded an adequate conception of force which he applied flawlessly, not merely to conditions of equilibrium on the simple machines, but equally to more complicated questions of air pressure and hydrostatics. The ambiguities arose when he attempted intuitively to extend the conception of static force to other problems. Even then, when he applied the same conception of force, referred to as the 'continuous application of motive force,' to the model of free fall, in the passage cited above, he arrived at consistent conclusions. The dynamic analysis of free fall appeared as Proposition 2 of Chapter X, 'Concerning Composite, Accelerated, and Retarded Motions and the Motions of Projectiles,' a subject to which, symptomatically, he devoted one third as much space as he gave to inclined planes. Meanwhile, Proposition 1 of the same chapter had indicated the pitfalls in the path of dynamics when it approached such problems.

> If a new force or new impetus in the same direction be added to a body placed in motion, an acceleration of motion occurs. If an impediment or contrary force [be added], a retardation is caused. And in both cases, [the acceleration or retardation] is proportional to that new impetus, or impediment or contrary force. Hence if the impediment or contrary force is less than the given force [*Vi posita*], the motion will continue in the same direction with a diminished speed. If it is equal, or also if the impediment is greater, the motion will be destroyed. But if the contrary force is greater, it will even generate a motion in the opposite direction.[108]

Two different measures of force, both of them different from that in Proposition 2 which set force proportional to acceleration, are present here. As the demonstration explicitly stated, the 'force or impetus' by which the body is moved, what he called the 'given force' in the statement of the proposition, is equal to the body's weight multiplied by its speed. ('$V = PC$;' by prop. 27, Chap. I.') This force (mv) is compared directly to the new force, impediment, or contrary force (Δmv) that generates an increment of motion.

In Proposition 3, after the analysis of uniformly accelerated motion in Proposition 2, he added that measure of force as well (see Fig. 22).

> Similarly if a heavy body in *A* is understood to be projected upward with that force [Δmv] which would impress on it a speed such as *Aa*, it would traverse a distance represented by the parallelogram *ABaa* if the speed continued for a time *AB*. In fact, the speed will continually decrease because of gravity which acts as a contrary force [ma] and diminishes the *momentum*: when the lines *ab*, *ab* which compose the triangle *aaB* are subtracted, leaving the lines $b\underline{b}$ which represent the

remaining degree of speed, as ab increases and becomes aB, the ascent will cease at the point of time B. But as the impetus caused by gravity increases further, not only will the body not be carried higher, it will begin to descend (the force [mv] downward now prevailing in power), and indeed with a uniformly accelerated motion (because of the impetus increasing uniformly as the lines cb, cb, which compose the triangle BCb); until, when the triangle BbC has been made equal to AaB, it will have descended as far as it ascended earlier; and if nothing stops it, it will continue to descend.[109]

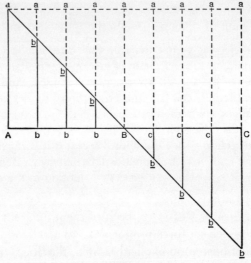

Fig. 22

In his analysis of the wedge, Wallis employed a different set of three measures. Let the resistance, perhaps the tenacity of a piece of wood that is to be divided, be O, and the force applied to the wedge, V. If the thickness of the wedge is to its length as c/a, then the force V will be in equilibrium to the resistance O. 'For since, by Proposition 5, Chapter 2, the forces of motions are proportional [*Motus in ea ratione polleant*] to the product of the motive forces and the forward or reverse displacements in the line of their direction ...,' and since the displacement of the resistance in its direction is to the progress of the force in its direction as c/a; then if $V/O = c/a$, 'the motions will be equal in force [*aequipollebunt*].' Now let a hammer of weight P move with speed C, 'and therefore the *momentum* or force of the hammer be PC; and let

$PC = V = \frac{c}{a}O$: the force of the hammer directly applied to the wedge will equal the obstacle in force [*aequipollebit*]; and therefore when it is increased it will move.'[110] Although he did not designate the product of a force and its displacement explicitly by the word '*vis*', he clearly identified it as a measure of dynamic action. Proposition 5 of Chapter II, which he referred to, had employed the same vocabulary.

> In comparison to each other, the descents of heavy bodies are powerful [*pollent*] in proportion to the products of the weights and their vertical descents.... And universally the forward or reverse displacements of any motive forces whatever are powerful [*pollent*] in proportion to the products of the forces and their forward or reverse displacements measured in the line of direction of the forces.[111]

In his treatment of the windlass, he referred to the same proposition as 'the universal principle from which the forces of all machines are to be calculated.'[112]

By employing different rubrics, Wallis may appear to have distinguished the quantity we call 'work' from 'force', but at most he took a mincing half-step toward removing it from the confused welter of measures attached to force. It was present, after all, as one factor in the equation of equilibrium which he derived in Chapter I as the foundation of his dynamics: $VT = PL$. What was the measure of force? Wallis employed all four of the standard measures, and he employed them interchangeably and without distinction.

Special interest attaches to Wallis' analysis of impact, both because he submitted a résumé of it to the Royal Society at the time when Huygens and Wren presented their papers, and because of the central role of impact in seventeenth-century dynamics. Like the rest of his mechanics, Wallis' approach to impact was straightforwardly dynamic. It rested squarely on the equation derived in Chapter I, $V = PC$ (or $F = mv$), which must have looked forward to the problem of impact from its inception.

> If a heavy body in motion is considered as perfectly hard, and if it directly strikes a firm hindrance or obstacle that is also perfectly hard; and if a force equal in power to the body so moved is less than the force of the obstacle resisting the motion, or even equal to it, the motion will be stopped. But if it is greater, the obstacle will be surmounted and the motion continued, but retarded or diminished in that ratio which the

resistance of the obstacle demands, which a calculation will establish. That is, if the force required to overcome the obstacle is subtracted from the *momentum* (which is composed of the weight and the speed) and the remainder, whatever it may be, is understood to be divided by the weight, the degree of speed remaining will be found.[113]

The force of the obstacle to resist appeared vague in its generalised exposition; it became clear enough when Wallis proceeded to apply the analysis to the impact of two bodies, A and B, both understood to be perfectly hard. Let the second body be at rest before it is struck by A. Because it is indifferent to motion, B must be set in motion by A. How fast will it move? Obviously it cannot move less swiftly than the final velocity of A since A continues behind it. It is equally obvious that it cannot move more swiftly than the final velocity of A, for both bodies are perfectly hard by assumption, and once A has put B into a motion as swift as its own, no force to make them separate is present. Perfect hardness functioned differently in Wallis' treatment of impact than in Huygens'. Whereas the Dutch scientist, in his concern to eliminate dynamic factors, saw in perfect hardness the exclusion of forces in impact, Wallis looked on hardness in its relation to the dynamics of the act of impact itself. He distinguished perfectly hard bodies from soft bodies on the one hand and elastic bodies on the other. In the case of elastic bodies, impact generates forces that cause them to separate again after impact. In the case of soft bodies, no separation occurs, but part of the original force of the moving bodies is absorbed in deforming the soft bodies.[114] Although the latter point contains an insight of potential significance, Wallis had no conceptual tools that could move it from the realm of qualitative assertion to that of quantitative analysis. The loss of force in deformation, moreover, assumes a different measure of force than the one Wallis was using, which in our mechanics is conserved in impact whether bodies are hard or not. Meanwhile, for Wallis, the conservation of force (PC or mv) was a function of perfect hardness. To find the common velocity of the two bodies after impact, he divided the initial force by their combined weight. A similar analysis, too obvious to require restatement here, determined the velocity when the second body is in motion, either in the same or in the opposite direction.[115]

In a passage of considerable interest, Wallis went on to calculate the strength of the blow itself in terms of the change of *momentum*. 'The magnitude of the blow is equal in force [*aequipollet*] to twice the *momentum* lost by the stronger of the bodies colliding head on ...' Consider the stronger body as the percussant and the other as the body struck.

> The body struck receives as much *momentum* as the percussant loses (that is, either in resisting or sustaining the force in the case of a rooted obstacle that cannot give way or of a body moved with an equal contrary impetus; or in receiving a new impetus in the case of a body initially at rest or moving in the same direction; or finally partly the one and partly the other, in the case of a weaker contrary motion; all of which will be shown individually in their places); since both are effects of the blow, the blow equals the two in force, that is, it equals twice the *momentum* lost by the stronger.[116]

Conceived in such terms, the force of the blow bears a family resemblance to the concept of force that Newton defined for dynamics. The force of a blow is to be measured by the total gain and loss of *momentum*, which is equivalent, or nearly equivalent, to the change of motion, Δmv, one of the measures of force found elsewhere in Wallis' *Mechanica*. As such, the force of a blow in his definition is similar to our concept of impulse, the integral $\int F dt$.

In important respects, however, it differs, and in its difference it reveals something of the prevailing approach to dynamics. The force of a blow is measured by twice the loss of *momentum* of the stronger body. As Wallis realised, each of the two bodies strikes equally hard, and the total force of the blow is the sum of the two partial blows, which are most conveniently measured by the loss of *momentum* of the stronger. That is, Wallis was not measuring the force with which A is struck by B, as measured by its change of motion, or B by A, measured in the same way, but the force of the blow, an abstract quantity composed of two equal factors. Moreover, and most importantly, he continued to think of force, not as an external action exerted on a body to change its state of motion, but as the force that a body exerts. In the impact of two bodies, each exerts a force. Each body receives a blow as well, but Wallis concentrated his attention on the exertion of force. The force of the blow is the sum of the two forces exerted. He chose the stronger body's loss of *momentum* merely as the most convenient measure of the half-blow.

Force is exerted only to the extent it is resisted. Should the obstacle not wholly resist the blow, but give way somewhat in doing so, then the half-blow is measured by the 'resistive force' of the obstacle, which equals the other body's loss of *momentum*. On the other hand, if the obstacle remains unmoved, the force of the half-blow measures itself by the *momentum* of the striking body.

> That is, if the weight of the moving body A is mP and its speed rC, and hence its *momentum* or impelling force $mrPC$, a resistance equal to it is in the obstacle (since whatever sustains the total striking force unmoved equals that alone); and since both are part of the blow, the blow equals the two taken together in force, that is, twice the *momentum* of the striking body, or $2mrPC$.[117]

In this case as he defined it, a hard body striking an immovable obstacle, the total change of motion equals only half the force of the blow. Remember that in such passages, according to the formula $V = PC$, *momentum* represents literally the force of the body; it is not a purely conventional term that represents the force necessary to generate an equal velocity in an equal body. For all its resemblance to the Newtonian concept of force, Wallis' force of a blow rather expressed the old idea of the force of percussion. It was the sum of two exerted forces viewed from the bodies exerting them, and not an external action measured by the change of motion it effects.

In the paper he submitted to the Royal Society in November 1668, Wallis confined himself to the impact of perfectly hard bodies, and arrived at conclusions as indicated above which are similar to those we derive for perfectly inelastic impact. In *Mechanica*, he treated elastic impact as well, one of the more penetrating analyses it received before Leibniz. Most writers, he asserted, imagine 'I know not what force' [*Vim nescio quam*] in any body in motion whereby it rebounds in a new direction when its progress is obstructed. Wallis refused to accept such rebound since it appeared to posit a new motion without a new cause. Once a body in motion is brought to rest, only the exertion of a 'positive force' can generate a new motion, and in impact this force can only be the elastic force of one or both of the bodies.[118] 'I call that by which a body deformed by force strives to restore its original figure Elastic Force.'[119] Suppose that a body strikes an elastic but immovable obstacle. As we have seen, for a body of weight mP and speed rC, the force of the blow is $2mrPC$; an elastic force of exactly this amount has been generated when the body comes to rest. Since the elastic force must exert itself in both directions, only half of it is expended in setting the body back in motion. Hence the body rebounds with its original *momentum mrPC*.[120] A similar analysis applies to two bodies approaching each other in opposite directions. In the case of two equal bodies, one of which is at rest, Wallis showed that the one at rest, B, will be set in motion with the velocity of A, and A will be brought to rest. By the analysis of perfectly hard bodies, the two move together first with half the velocity of A. Now the elastic force generated by the impact exerts itself. Acting on A, it destroys the second half of its initial motion, while in B it generates a second increment of motion equal to the first in size.[121] With the concept of elastic force, Wallis returned to the idea of an external action measurable in terms of the motion it generates. It was symptomatic of the confusion surrounding his concept of force that he did not see elastic force as analogous to gravity in his dynamic analysis of free fall. Rather, he could only see it as a total quantity equal to the total increment of motion it generates. Even in those parts of his dynamics that look most familiar to twentieth-century eyes, the ambiguities of the concept of force, derivative almost entirely from the image of the lever, continued to lurk.

★

At much the same time when Borelli in Italy and Wallis in England were writing on mechanics, Edme Mariotte established himself as the leading student of the science in France. One of the original members of the *Académie royale des sciences*, Mariotte demonstrated various properties of the motion of heavy bodies before that group in its early years, and by 1668 he had systematised his conclusions in a *Traité du mouvement des pendules*.[122] In 1673 he published a *Traité de la percussion ou choc des corps*, and some time thereafter he composed a *Traité du mouvement des eaux et des autres corps fluides*, which was only published posthumously. Although Mariotte never undertook a systematic treatment of the whole of mechanics, he did touch on most of the issues central to the science during his period.

Mariotte played a unique, not to say invidious role in the mechanics of his age. His treatise on pendulums opened with a bizarre letter of introduction addressed to Huygens. Mariotte related how he had shown Huygens some demonstrations on the motion of pendulums and heavy bodies only to be informed that Galileo had already demonstrated the same things. Thereupon he had read Galileo and found that their thought agreed so well that Huygens might well have thought he was borrowing – although of course such was not the case.[123] A few years later, when Mariotte published his *Treatise on Percussion*, Huygens complained anew of the same crime – possibly with greater asperity on this occasion since he recognised a source rather closer to home.

It is impossible to read the two works without the feeling that Mariotte's protestations of innocence – the pretence that he had not read Galileo, for example – were more than a little disingenuous, but also that Mariotte understood himself to be doing something different from either of his predecessors. In the letter to Huygens that introduced the *Treatise on Pendulums*, he stated that he gave (or believed he gave) 'the true cause of the acceleration of the motion, whereas Galileo contents himself with assuming acceleration and defining it.' Thus the order of inquiry was reversed, and what Galileo stated about descent on inclined planes as a premise, he demonstrated as a conclusion from dynamic causes.[124] Although he did not venture similar explicit comparisons with Huygens, who was still around to reply after all, Mariotte made it sufficiently evident that he considered mathematical kinematics an interesting exercise at best, whereas dynamics, in its explicit consideration of causes, was the real substance of mechanics. Thus the dynamic assumptions, which Galileo first and Huygens even more sought to avoid because of their vagueness, were placed at the very focus of the inquiry by Mariotte, who paraded his lack of concern for rigour as a positive virtue. Few if any readers from the twentieth century are apt to feel that history has been mistaken in its neglect of Mariotte in favour of Galileo and Huygens. As with Borelli and Wallis, however, his effort to state a dynamics from which accepted kinematics might be derived is in itself revealing.

Although Mariotte did not escape the ambiguities inherent in the intuitive concept of force he employed, one particular idea of force dominated his mechanics with considerable consistency. Typically, it found its clearest expression in his work on percussion. If a body moves more swiftly than another that is equal to it in size, or if it moves with a velocity equal to that of a smaller body, he said, it has 'a greater power [*puissance*] of motion, or a greater quantity of motion.' He went on to call the quantity of motion of a given body 'the force [*force*] of that weight moving effectively with that speed,' and more simply the 'power [*puissance*] of motion.'[125] Obviously related to the 'force of a body's motion' so pervasive in seventeenth-century mechanics,

Mariotte's basic dynamic concept derived instead from more distant sources. Force was not a simple consequence of motion; it was equally a cause. What he called 'force' was nothing other than a late manifestation of the medieval concept of impetus.

Dynamics was not the only respect in which Mariotte distinguished his work on impact from Huygens'. He also looked upon himself as an experimental scientist and believed that his rules of impact derived their validity, not from the internal logical structure of his treatise, but from the experimental evidence he brought to their support. The evidence was considerable, especially the experiments with a pendular instrument that he devised. To support his central dynamic concept, he called upon a more extensive range of evidence. If you throw a ball of lead 'with a great force,' it sinks further into soft ground than it does if you throw it 'with a mediocre force . . .' If you throw two iron balls against something to break it, the heavier of the two (equal velocities being understood) will have a greater effect. A stick carried by a stream is more easily stopped than a beam. A rolling ball of wood is stopped more easily than an iron one of the same size and speed.[126] Implied experiments or appeals to common experience, the examples suggest the sources that called impact immediately to mind as the model of dynamic action. Compared to these intuitive perceptions of force, the model of free fall was abstract and remote.

In the treatise, Mariotte first applied his dynamics to inelastic bodies. Certain cases were perfectly obvious as direct extensions of the concept of force. If a body in motion strikes another at rest, or another moving more slowly in the same direction, the composite body formed by the inelastic impact moves with the total force. More distinctive features of Mariotte's view of mechanics appeared with cases in which the bodies move in opposite directions. He saw such cases in terms of the equilibrium of opposing forces. When the two forces are equal, the impact is identical to static equilibrium and the bodies come to rest. When they are not equal, a partial equilibrium as it were between the smaller force and an equal quantity of the larger results, and the composite body moves with the remaining quantity of the larger force.[127]

Equilibrium played a greater role in his analysis of elastic impact. When two bodies meet, they are both deformed; if they are inelastic, nothing acts either to restore their shapes or to cause them to separate. If in fact they do rebound, only their elasticity can be the cause. Mariotte insisted on examining the dynamics of elasticity itself. As with so much of his mechanics, the examination was long on intuitive insight and short on demonstrative rigour. He imagined a stretched and elastic string the middle point of which, E, is pulled to one side and released so that the string vibrates (see Fig. 23). Call the farthest displacement of a vibration of the string C. As the string returns from C to E, it resumes at each point the velocity it lost as it ascended. Hence

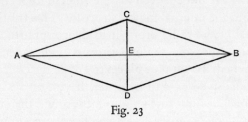

Fig. 23

the string arrives at E with exactly the velocity it had when it passed E moving in the other direction. Now imagine that a body moving perpendicularly to the string hits it and displaces it to C. In restoring itself, the string will restore its initial speed to the body.[128] The example assumes perfect elasticity, of course. Mariotte recognised that perfectly elastic bodies are abstractions quite as much as perfectly inelastic ones, but he confined his discussion of elastic impact to them. The conclusions at which he arrived were identical to Huygens' treatment of perfectly hard bodies.

Let two inflated balls, equal in size, internal pressure, and velocity, strike each other. In addition to Galileo's kinematics and Huygens' analysis of impact, Mariotte also laid claim to Boyle's Law, and the inflated ball represented his image of an elastic body. Two such balls will rebound with their original velocities,

> for ... their simple motion must be entirely destroyed, and it would not be restored if they did not have any elasticity. But since each of the balls is struck with the same force, and since they do not give way to each other, their impact will have the same effect as if each of them had encountered an inflexible and immovable body; and consequently they will push in and flatten each other equally. But according to the preceding proposition, in resuming their original figures by elasticity, at the moment of their complete restoration they will resume the same speed they had before the impact.

Now let the bodies be unequal in size with velocities reciprocally as their weights. Both will again lose all their motion, and each will be put in a state of elastic tension like that it would acquire in hitting an immovable obstacle;

> and establishing then a sort of equilibrium between themselves in acquiring speeds reciprocal to their weights, each will rebound with its original speed.[129]

In an analysis at once redolent of Huygens' basic insight and explicit in its statement of Wren's implicit dynamics, Mariotte then proceeded to show that every elastic impact is identical to that of bodies moving with speeds inversely as their weights. Whatever the means by which they are pressed together, two elastic bodies acquire velocities inversely as their weights when they are released. In impact, the relative velocity of the bodies determines the impression and the 'force of elasticity . . .,' and the force of elasticity in turn produces anew a relative velocity equal to that which generated it. Hence the bodies separate with the original relative velocity, which they divide between themselves inversely as their weights.[130] Impacts that appear to diverge from the pattern above can be reduced to it by adding or subtracting a velocity the two bodies share in common. By referring impacts to the common centre of gravity, one can solve all the problems of elastic impact.

As an example, Mariotte applied the analysis to a problem recognisably Huygensian. Suppose a body A that has a velocity of 100 000 and weighs 100 000 ounces strikes a body B of one ounce which is at rest. It will give B a velocity of 200 000, a result which seems paradoxical. The apparent marvel, Mariotte urged, proceeds from the two rules of nature –

> that the mutual impression of two bodies on each other is always the same when the relative speed with which they strike each other directly is the same; and that when their impact puts them in a state of elastic tension, they divide their relative speed in a ratio reciprocal to their weights . . .[131]

If the case of two bodies radically different in size is reminiscent of Huygens, the entire treatment of elastic impact is pervaded with his insight on the common centre of gravity. Once again, however, Mariotte wanted to demonstrate from his dynamics principles that the Dutch scientist had taken as axioms. Thus Proposition IV of Part II drew on the analysis of elastic impact to demonstrate that the impact of two bodies does not disturb the state of uniform motion or rest of their common centre of gravity.[132] Indeed, Mariotte tried to derive the principle of inertia itself from similar considerations. When an inelastic body strikes another at rest, the two move together with a quantity of motion equal to that of the body initially in motion. 'It follows from this Proposition,' he continued, 'that the motion of a body that does not encounter any resistance does not decay...'[133] Whereas Huygens had seen maximum assurance in kinematic axioms drawn from the principle of inertia, Mariotte chose to rely instead on intuitive perceptions of force.

I have already suggested that Mariotte's treatment of impact explicitly employed the concept of force that was implicit in Wren's analysis. The parallel expressed itself more clearly in Mariotte's discussion of the balance. Bodies are in equilibrium on a balance when their weights are reciprocally as their distances from the fulcrum. In the case of the balance, the phrase 'quantity of weight' designates the force of a weight to move with a velocity that is proportional to the length of the arm. In percussion, 'quantity of motion' designates the force of the weight moving with a corresponding velocity. A weight of six pounds two feet removed from the fulcrum has the same force of weight as four pounds three feet removed; a weight of six pounds with two degrees of velocity has the same force of motion as four pounds with three degrees of velocity.[134] Mariotte introduced this discussion in an *Avertissement* following his assertion of the principle that two bodies with equal quantities of motion stop each other in inelastic impact. Significantly, he saw the law of the lever, not as the justification of the equilibrium of opposed forces, but as its consequence. In his treatise on fluids, he proposed the same relation in the context of two general rules of mechanics – the first that a body resists only that motion which removes it further from the centre of the earth, and the second that two inelastic bodies with equal quantities of motion 'establish equilibrium' in impact.

> From the latter, the principle of Mechanics, which has been badly demonstrated [*mal prouvé*] by Archimedes, by Galileo, and by many authors, is easily proved; to wit, that when weights on a balance are reciprocally as their distances from the fulcrum, they are in equilibrium. For let there be a balance BAC such that A is the centre of motion, AC is four times as great as AB, and the weight B is four times as great as the weight C. I say that one of these weights will not lift the other. For let the weight B raise the other if that is possible. Now in descending it can only move with some speed in the arc BD if it makes the weight C move four times as fast in the arc CE, since the radius AC is four times the

radius *AB*; and hence the quantities of motion of the two bodies would be equal, and one quantity of motion would have overcome another equal to it, which is impossible since by the second rule they must be in equilibrium. For the same reason, the weight *C* will not be able to descend; but if it is removed a little further from *A*, it will descend, for then it will give the other weight a quantity of motion less than the quantity it will assume, and consequently it will overcome it.[135]

Central to Mariotte's perception of mechanics was a concept of force measured by quantity of motion, and most of the problems to which he addressed himself seemed soluble to the extent that an equilibrium or a disequilibrium of such forces could be established.

In this respect, Mariotte's sallies into the domain of fluid dynamics are revealing. Considering a fluid as a collection of discrete particles, he applied his analysis of impact to fluid mechanics, thereby establishing a pattern that Newton later exploited in some detail, not altogether to the benefit of fluid mechanics. To Mariotte, a fluid was like a solid in that all its particles move with the common velocity of the mass of fluid; it was unlike a solid because the parts of the fluid act separately in impact. In typical fashion, he proposed a simple experiment to demonstrate the point. Taking a tube eight or ten feet long, he filled it with water and held it above one end of a balance. He propped the other end of the balance so that it could not descend, and on it placed a weight only one quarter or one fifth as heavy as the water in the tube. When he removed his thumb from the bottom of the tube, the balance did not stir until the final portions of water, those which had fallen the full length of the tube, raised the weight by their impact. A quantity of water that falls from two or three feet, he added, makes less impression to raise a weight than a ball of wax that weighs only half as much falling from the same height. The wax acts as a single unit; the water does not.[136] It follows that equal jets of the same fluid are able to sustain weights proportional to the squares of their velocities. In a jet twice as swift as another, twice as many particles strike an obstacle in a given time, and therefore 'a jet moving twice as fast as another must make twice as much effort solely because of the number of particles which strike; and because they move twice as fast, they make twice as much effort by their motion . . .' The force of a jet is therefore proportional to the square of its velocity.[137]

At first glance, Mariotte's line of inquiry appears potentially fruitful. As he pursued it, however, it took a disconcerting turn totally unexpected by the twentieth-century reader. Impact was comprehensible to him as a species of equilibrium, and similarly his approach to fluid dynamics looked, not for a measure of force in terms of the quantity of the fluid's motion that is destroyed, but rather for an equilibrium of forces like that he saw in impact. His rules for fluid jets, a revealing phrase stated, allow one to calculate 'the equilibrium they make between themselves and with the solid bodies they strike . . .'[138] Thus, in examining water wheels, he employed his conclusions on inelastic impact. He noted that such wheels appear to move only half as fast as the stream that turns them, 'which amounts to the same thing as when a weight in motion strikes another of the same weight at rest and attaches itself to it . . .' Hence he concluded that the friction of the axle and the friction of the grinding together with the weight of the wheel itself 'amounts roughly to the resistance of a weight equal to that of the water which strikes . . .'[139]

Mariotte's basic problem in fluid mechanics was the force of a jet, and his basic device to analyse it was a container with a hole in the bottom from which the fluid could flow. The container might be filled with air, in which case a weighted piston at the top drove it out ('with violence' as he put it), or with water which was driven out by its own weight. Under the hole he placed one end of a balance with equal arms. His problem was to find the weight on the other end that established equilibrium with the jet. In the case of air forced out by a weight P on the piston, when the weight G on the balance is to P as the area of the hole N is to the area of the piston, the balance is in equilibrium.

> For if by a bellows with an aperture equal to the hole N, you push air against the opening N with a force equal to that of the air that the weight P forces out, the two forces will be in equilibrium, and the weight P will not descend because no air will emerge from the opening of the hole N; and hence, the air pressed by the bellows filling that opening will support its part of the weight P as the other parts of the base BC sustain the rest of the weight; and the part that the pressed air will sustain will be to the entire weight P as the ratio of the opening N to the entire size of the base BC. Hence conversely the air emerging from that opening after the bellows will have been removed will establish equilibrium by its impact with a weight that will be to the weight P as the opening N is to the base BC.[140]

Mariotte reached his solution by converting the dynamic action of the jet into a problem in static equilibrium.

We must remember that to Mariotte static equilibrium was a dynamic action. He saw the two postulated conditions of the problem with the fluid jet as identical. The two streams of air holding each other in equilibrium recall two bodies of equal force bringing each other to rest. The equilibrium of the emerging jet with the weight on the balance reminds us that the balance itself was comprehensible only in terms of dynamic equilibrium. If the descending air is in motion against the balance, Mariotte was equally convinced that the weight G on the other end is in implicit motion as well. How is this possible? Because a body that begins to fall starts to move with a finite velocity. Between rest and motion there is no middle ground. In order to be in motion at all a body must move with a determinate velocity.

To establish the point, Mariotte compared the fall of a body with the descent of another down an inclined plane, and the motion of a small body on the long arm of a balance with that of a large body on the short arm. In both cases, the initial motion of the body moving more swiftly must be larger than the initial motion of the slower – 'from which it follows that the beginning of that of the small weight [i.e., on the long arm of the balance] was not infinitely slow [*de la dernière lenteur*] . . .'[141] Let the cause of fall be the impact of subtle matter, an internal principle, or something else; 'that natural agent, whatever it may be, has a determinate action; hence its effect, that is, the initial speed impressed on the body that falls, will also be determinate . . .'[142]

Mariotte illustrated the meaning of this conclusion by another problem with a jet of water. Imagine a body C suspended on a cord above a vertical jet of water just as the cord is cut.

> Now if the first motion of the body C in falling were infinitely small, in so far as the ratio of the weight C to the weight of the first particles of water that strike it is not infinitely great and the ratio of the speed of the jet of water to that of the weight [C] when it begins to fall is infinite by supposition, if the suspending cord is cut, the weight will not be able to fall . . . for the quantity of motion of the first particles of the jet of water will be greater than that of the weight when it begins to fall, and consequently the weight will rise . . .[143]

Inevitably, Mariotte had tried the experiment. Since the weight he used did in fact fall, it followed that its initial motion was finite. Central to the argument was the implicit disequilibrium of forces, both measured by the product of weight and velocity. Standing on the ambiguous ground that separates statics from dynamics, Wallis had tried to lead mechanics to clarity by incorporating dynamics into statics. In contrast, Mariotte's line of march converted statics into dynamics.

Confident in the illumination of experiment, Mariotte even calculated that the initial velocity of descent, on which static equilibrium depends, is about four lines (one-third of an inch) per second. By experiment we know that a drop of water falls about twelve feet in one second, reaching a velocity of twenty-four feet per second. A jet of water flowing from the bottom of a tank twelve feet high has the same velocity, a velocity great enough to carry it back to the level of the surface. Such a jet is also able by its percussion to sustain a cylinder of water equal to the jet in cross-section and twelve feet high. The crucial passage from dynamics to statics, identical to the argument presented earlier in which a bellows held a jet of air in equilibrium, takes place at this point – the passage from dynamics to statics in our eyes, the statement of a crucial insight in Mariotte's. It follows by simple arithmetic that a jet one inch in diameter with the velocity in question will sustain a weight of 72 ounces. How much of the water is in action at any moment? Mariotte estimated it to be a thickness of two lines, which is $\frac{1}{864}$th of twelve feet. Hence 72 ounces is to this little cylinder two lines high as the velocity of the jet is to the weight's initial velocity of fall. 864 is to 1 as 24 feet per second is to *c*. 4 lines per second. Again, a sheet of glass two lines thick laid on a flat piece of marble is crushed by a weight of 400 pounds. A weight of two pounds two ounces falling seven inches breaks a similar piece. 400 is to $2\frac{1}{8}$ as 830 lines per second (the velocity acquired in a fall from seven inches) is to *c*. 4 lines per second – 'from which you can judge that the initial speed of a weight of 400 pounds, which begins to fall in still air, is such that it would be able to cover four lines in one second if it continued to move uniformly with that initial speed.'[144]

By invoking the relativity of motion – hardly a legitimate procedure for one who equated motion and force – Mariotte applied the analysis above to the question of terminal velocity when a body falls through a resisting medium. The resistance of the medium is identical whether the body moves through it or it moves against the body.

> Hence if the speed of the air upward can establish equilibrium with the first effort that a heavy body makes to descend with its small initial velocity, so that it is suspended without falling, when the body will have acquired the same speed in motionless air which the air that sustained it had, there will still be equilibrium between the resistance

that the air makes to the motion of the body and the same initial effort or virtue to fall which always remains in the body. And consequently the initial effort, which causes acceleration by constantly adding the small initial speed which it has to produce to the acquired speed, will no longer be able to add it, and for that reason the body will continue to fall uniformly with the speed which it will have acquired from the top of its fall to the point of equilibrium.[145]

Aside from the fact that Mariotte considered terminal velocity to be a state that is reached and not merely approached, his conception of mechanics in terms of opposing forces encouraged his insight into the equilibrium that obtains in this case. Unmentioned, perhaps even unperceived, in the solution lay the unfortunate consequence that air must be constantly beginning to rise with a finite determinate velocity of about four lines per second.

Although he spontaneously sought out problems that contained an equilibrium, real or apparent, as the object of his mechanics, Mariotte could not be unaware of the phenomenon of accelerated motion which Galileo had placed at the centre of seventeenth-century mechanics. In a few places he dealt with it explicitly. Inevitably his treatment was dynamic, and from it emerged, as from every other dynamic analysis of uniformly accelerated motion, statements recognisably similar to Newton's second law. In some passages, Mariotte applied analogies from his work on impact. When a ship is set in motion by the wind, imagine a discrete gust every sixtieth of a second. Like a thrust from the oars of a galley, each gust imparts a new increment of velocity, and if the wind blows constantly 'with the same force,' the motion is uniformly accelerated, at least in the beginning.[146] To the uniform acceleration of free fall, which Galileo had merely described, he applied the same analysis.

> Here is how I conceive it. If there is some very light body which strikes a body 100 times heavier, it will give it a 100th part of its speed; and hitting it a second time, it will give it still another 100th; with the result that if the striking body has 101 degrees of speed, the body struck will take one degree of the speed at the first blow, and its quantity of motion will be 100; and being struck a second time with the same speed of 101 degrees by the light body, it will receive a new degree of speed from it which will be two degrees when it is joined to the first; the third blow will add still another to it, and so on continually as was proved in the *Treatise on the Impact of Bodies*. The same thing will happen if some

> weak power pulls a very heavy body to it, pulling with discrete tugs [*par reprises*]. Now, whether bodies be pulled or pushed by a subtle fluid matter, it must happen that if, in the first moment of its effort, it traverses one line with a uniform speed, in the second moment and with the second blow it will traverse two, in the third moment three, etc.[147]

And he went on to show that such a motion is equivalent, for a large number of instants, to Galileo's uniformly accelerated motion.

In fact, the image of a subtle fluid did not correspond to Mariotte's conception of gravity. The pulling 'power' was a better figure in his eyes, and for the most part, he preferred to speak of a body's 'natural power to descend toward the centre of the earth . . .'[148] Obviously the power was equated to a given quantity of motion – 'weight is nothing but a power to descend with a certain speed.'[149] To one who was trying to exploit a concept of impetus, such a view of accelerated fall presented problems, and Mariotte tried to solve them with a distinction between natural and acquired power. His work on pendulums began with the statement of two 'natural principles.' The first stated that bodies begin to fall with a constant finite velocity. The second maintained that if a body 'is carried by any cause whatever for a short distance with a uniform speed, the body will continue its motion in the same direction with the same speed through a space equal to the first even though the cause ceases, as long as it is not impeded by another cause.' If the second clause appears to state the principle of inertia, the first invokes a cause that carries a body with a uniform speed. Indeed he went on to say as much about the continuing motion, after he supported the principle by reference to Descartes, Galileo, and experiment. 'We will call this power by which the body continues its motion acquired power.'[150] There is much in this analysis to remind us of Torricelli. Like the Italian scientist, Mariotte conceived of ultimate indivisible units both of space and of time corresponding to the ultimate units of velocity. In Mariotte's case, however, all was done with infinitely less finesse. Torricelli demonstrated a means of reconciling at least linear accelerations with an impetus mechanics. In Mariotte's case, we stare blankly at a conundrum which presumes baldly to assert that one and the same force of gravity causes both uniform and accelerated motions.

In Mariotte's mechanics, the problem of force was entirely dominated by the model of impact. Even the uniform acceleration of free fall, as I have indicated, was intelligible to him in terms of the force of the falling body – an 'acquired power' composed from discrete increments of 'natural power', the latter also understood as the product of the body's size and its initial velocity of fall. The 'force of a body's motion' always invited an equation with the external force that produces it, and we can find a number of such passages in Mariotte. When equal bodies are projected upward 'by forces that differ [*forces différentes*]' they acquire velocities proportional to the forces.[151] It is easier merely to stop a rolling ball than to reverse its motion 'which proceeds from the fact that beyond the force necessary to stop it, another is required to restore its original speed to it in reverse.'[152] In both cases, the implicit measure of force is Δmv, and the force, an external force that operates to change a body's motion, is understood to be equal to, perhaps in some sort of equilibrium with, the internal force it generates, measured by mv. Of the various ambiguities besetting the concept of force in the seventeenth century, this one involved the least disabling confusion, but it imposed a sterile pattern on dynamics by its incompatibility with the principle of inertia and the relativity of motion.

In one respect of immense importance for the future development of dynamics, Mariotte's mechanics explored significant new ground. In his writings, the concept of mass as it was defined and used by Newton in the *Principia* first began to emerge. On the one hand, Mariotte understood that the quantity of matter in a body, which he tended to refer to misleadingly as its weight, involves two factors – both the size of the body and its density.[153] What was far more important, he associated with the quantity of matter a 'resistance' whereby a body opposes the sudden acquisition of motion. When a body at rest is smaller than another that strikes it, 'it will resist the motion less' than it will if it is equal in size, 'and the heavier the body at rest is, the more it will resist the motion.' Is the resistance of the medium in which the bodies are found the ultimate cause of this observed phenomenon? Mariotte cited the fact that a two pound ball of lead is smaller than a one pound ball of wood, and therefore implicates less of the medium, but it resists motion more. Moreover, the principle of heaviness does not cause the resistance, for a body resists horizontal motion in which the principle of heaviness is not involved. 'But the true cause of this effect is the same which makes the body heavier, that is the greater quantity of its matter.' To illustrate the concept, he employed the analogy of a piece of hot iron quenched now in one pint of water, now in three. The single pint of water becomes hotter because it contains less matter to be heated, and similarly a body of one pound receives a greater velocity from the same blow than one of three pounds. 'It can be remarked,' he continued, 'that a body however light greatly resists the sudden reception of a large speed.' Hang a knife by a thread so that the point is horizontal and hit it with a tin plate – the inevitable experiment, and in this case a compelling one. Instead of being set in motion, the knife penetrates the plate. A bullet pierces a weather vane because it is easier to break it than to set it suddenly in so swift a motion.[154] Mariotte's *Treatise on Impact* contained the first explicit modification of the concept of indifference in seventeenth-century literature. The elaboration of a successful quantitative dynamics demanded a further exploration of the path he opened.

As we have seen, Mariotte was not the only mechanician exploring the meaning of indifference. Indeed we can say that before Leibniz and Newton, promising advances had been made in the understanding of both the technical problems confronting dynamics, the role of mass and the factors in circular motion. Equally, more than one student of the science had illustrated the possibilities inherent in free fall as the model of dynamic action, although to be sure the model of impact continued to dominate dynamic thought to its detriment. As far as the ambiguities besetting the science are concerned, almost no progress had been made toward their resolution. The beguiling clarity of the lever continued to confound statics and dynamics, and the perplexed definition of force was an inherent aspect of the confusion. What the men examined in this chapter demonstrate above all is the incredible capacity to mislead that intuitive ideas of force possessed. What dynamics required above all was the 'touch of cold philosophy', systematic analysis in all its rigour such as Descartes had earlier imposed on the idea of motion. The roles of Leibniz and Newton in dynamics are directly related to their success in applying such analysis.

*

Another problem that rational mechanics continued to face during the second half of the seventeenth century was the inadequacy of traditional mathematical tools for many of the problems the science was attempting to handle. Although several of the men examined in this chapter were mathematicians of note and one of them, John Wallis, played a minor role in the evolution of the calculus, the mathematics they applied to rational mechanics was the same geometry with which Galileo approached it at the beginning of the century. They had surpassed Galileo in their willingness to abstract quantitatively. Whereas Galileo almost always based his diagrams on the line representing the path of motion, a technique had developed whereby any quantity involved in the problem could be represented diagrammatically without the need to picture the trajectory itself. This capacity was a considerable

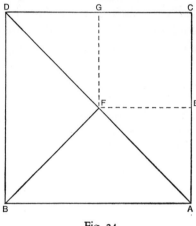

Fig. 24

victory for flexibility of conceptualisation and undoubtedly aided dynamics in its growth. Whatever the diagram, however, its quantities were handled by geometric ratios, and a number of problems obstinately refused to accommodate themselves to such treatment. As long as velocity or acceleration was uniform, Galileo's methods yielded results. Non-uniform accelerations, such as vibratory motions and motions through resisting media, problems that were engaging the attention of mechanicians, were the object of repeated attack and the source of continued bewilderment.

Consider, for example, Borelli's attempt to supplement Galileo's kinematics of vertical motion by introducing the effect of air resistance. Let a body be projected straight up with an impetus AB (see Fig. 24). In time AC it would rise a distance represented by the area $ABDC$ if nothing hindered it. In the ideal case in which there is no resistance, the impetus of gravity increases from nought to CD, and at time C the body ceases to rise. Since gravity has carried it down through a distance equal to $\triangle ACD$, its actual rise to its highest point is only the distance measured by $\triangle ABD$. So far there was nothing here to stifle yawns in the *Accademia del Cimento*, but Borelli now tried to go beyond the received lesson of the master. Let the resistance of the air be such that in time *AE* it reduces the original impetus upward to *FE* (equal of course to the increasing impetus downward of gravity, so that in this case time *E* represents the end of the ascent). On the diagram, $AE = \frac{1}{2} AC$, an assumption for which Borelli offered no justification. It contributed enormously to simplifying his mathematical problems, however; and as we shall see, they were such that we should not begrudge him whatever help he could find. In time *AE*, the decreasing impetus upward carries the body through a distance equal to *ABFE*; gravity causes it to fall *AEF*, and it actually rises a distance *ABF*. As the body now begins to fall, the remaining impetus upward, *FE*, is no longer resisted, and in time *EC* would carry the body up a distance *ECGF*. Meanwhile the increasing impetus of gravity causes it to fall *ECDF*, and the actual fall is represented by $\triangle FDG$. If the air does not diminish the impetus of gravity, $\triangle FDG = \triangle AEF$; since the air does oppose the motion, the real distance of descent is less than $\triangle AEF$ and much less than $\triangle ABF$. Hence the distance of descent is less than the distance of ascent in equal time. Moreover, *DG*, what he chose to call the 'final impetus acquired at the end of the descent,' is less than *AB*, the impetus with which the body began to rise. Whatever the other defects of the analysis, the final conclusion had the unnoticed, but not for that less unhappy effect of leaving the body, at the end of its descent, still in mid-air.[155]

Borelli's analysis of a log bobbing upright in a fluid of twice its specific gravity, an analysis which figured prominently in his celestial dynamics and one of the pioneering endeavours to deal with simple harmonic motion, reveals the same inability to handle non-uniform accelerations with any degree of rigour. Borelli imagined that the log was raised until its lower end touched the surface of the water and was then released (see Fig. 25). In the diagram, *AC* can be taken as the surface of the water, and the verticals *BA*, *MH*, *NI*, . . . as the positions of the log after successive equal periods of time *BM*, *MN*, *NF*, . . . Already Borelli had problems enough in his representation of time. He increased them further by concentrating exclusively on the use of the verticals to represent effective force, thus failing to realise that they must equally represent vertical displacement, and do represent it exactly in so far as they represent effective force. Instead he lit upon the unlikely intervals of the line *AE* as the indicators of displacement. During the

SCIENCE OF MECHANICS IN THE LATE SEVENTEENTH CENTURY

first instant *BM* after the log is released, the excess of its weight over the buoyant force of the water is represented by the trapezoid *BQ*. With this force the log acquires the speed *Fβ* with which it traverses the distance *AQ*. The line *Fλ*, unequally divided, but only intuitively so, embodied the extent of Borelli's success in dealing with the non-uniform motion he recognised to be present. In the second instant, *MN*, the impelling force, *MR*, impresses an increment of speed *βε*,

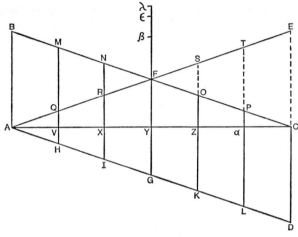

Fig. 25

which is less than *Fβ*, and in the third instant, *NF*, the force *NFR* impresses a last, still smaller increment *ελ*. Alas, the measures of distance traversed in both instants, *QR* and *RF*, are equal to the distance *AQ* of the first instant. Either Borelli failed to notice the anomaly, or he had no idea of what to do with it, and he proceeded regardless. At *F*, the excess of the downward force over the buoyant force disappears, but the impetus acquired carries the log on toward *E*. The force in the fourth instant, *SFO* is equal and opposite to that in the third instant, and the forces in the fifth and sixth to those in the second and first. Hence they operate to subtract increments of speed equal to those the original forces impressed, and the log comes to rest at *E*, whereupon the cycle begins in reverse.[156] Borelli's choice of increments on the line *AE* to express displacement was unfortunate, but in fact the lines *VH, XI, YG*, ... (or their differences, all equal to *VH* on the diagram) are no less linear. Clearly the interpretation of the diagram was not his greatest problem. Non-uniform accelerations – whether they followed a sine function as in this problem, or an exponential function as in the other – were simply beyond the power of his mathematics.

Essentially the same two problems presented themselves to others among the men dealt with in this chapter. As I mentioned before, the Cutlerian lecture that contained Hooke's law also contained an effort to apply it to the analysis of vibratory motions. A specific degree of force or power corresponds to every degree of flexure of a spring. Consequently all those powers, 'added together into one sum, or aggregate,' are proportional to the square of the strain or flexure.

> The spring therefore in returning from any degree of flexure, to which it hath been bent by any power receiveth at every point of the space returned an impulse equal to the power of the Spring in that point of Tension, and in returning the whole it receiveth the whole aggregate of all the forces belonging to the greatest degree of that Tension from which it returned; so a Spring bent two spaces in its return receiveth four degrees of impulse, that is, three in the first space returning, and one in the second; so bent three spaces it receiveth in its whole return nine degrees of impulse... Now the comparative Velocities of any body moved are in subduplicate proportion to the aggregates or sums of the powers by which it is moved, therefore the Velocities of the whole spaces returned are always in the same proportions with those spaces, they being both subduplicate to the powers, and consequently all the times shall be equal.

That is, Hooke was applying his basic dynamic formula – 'that the proportion of the strength or power of moving any Body is always in a duplicate proportion of the Velocity it receives from it' – to the vibrations of a spring; and if we choose to interpret 'velocity of the whole spaces returned' as the final velocity, we can see the expression as mathematically equivalent to the work-energy equation. Unfortunately, Hooke wanted to get time into the equation as well, and to do this he had, in the last sentence, to employ a relation valid only for uniform motion. He was fully aware that the motion is not uniform, however, and he undertook further to calculate both the velocity and the elapsed time at every point along a single oscillation. At any point, the velocity is proportional to the root of the aggregate powers impressed. Let AC represent the total strain and CD the power of the spring at C (see Fig. 26). Then BE must represent the power of the spring at B. The area of the triangle ACD expresses the total aggregate of powers for the half-oscillation CA, and the trapezoid $BCDE$ the powers expended on the motion from C to B. Hence if the total strain is ten units, the velocity at the end of the first unit is proportional to $\sqrt{19}$ ($\sqrt{10^2-9^2}$); at the end of two units it is proportional to $\sqrt{36}$ ($\sqrt{10^2-8^2}$).[157] Let $AC=a$ and $AB=b$. The area of the trapezoid is proportional to a^2-b^2. Now construct a quadrant of a circle,

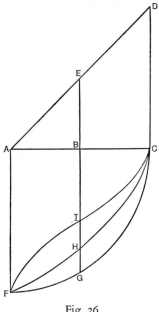

Fig. 26

CGF, on the centre A. Then $(BG)^2 = a^2 - b^2$, and BG is the velocity at B. Compared to Borelli's intuitive perception that the increments of velocity decrease at some unexpressed rate, Hooke's solution embodied a major analytic advance. Time still remained to be expressed, however, and time continued to baffle him. Velocity, he reasoned, is in the same proportion to the root of space as the root of space is to time.[158] Construct now a parabola CHF on the axis AC with its vertex at C. $(BH)^2 \propto CB$. Let CIF be a curve for each point of which $GB/HB = HB/IB$. Then IB represents the time elapsed in the motion from C to B.[159] More clearly than Borelli, Hooke saw that vibratory motions are dynamically equivalent to the oscillations of a pendulum. His attempt to analyse the motion in traditional geometric terms, however, foundered on the complexity of the problem.

Mariotte's analysis of terminal velocity represents a limping half-effort toward a quantitative treatment of resisted motion. From what height and in what time must a body given in size, shape, and composition fall to acquire its terminal velocity? Mariotte confessed that the problem was difficult, but the difficulties he went on to specify were related, not to the mathematical complexities, but to the physical constants of the problem. It is difficult to measure the exact distance a body falls in one second; it is difficult to distinguish the different speeds of bodies that vary in density as they fall in air; it is difficult to measure the different densities of air at divers places. To overcome the difficulties, Mariotte simply assumed that in air a ball of lead six lines in diameter falls fourteen feet in one second and would fall fifteen feet in a vacuum. He also assumed, in direct contradiction to his earlier analysis, that resistance is proportional to velocity. The assumption greatly simplified his mathematical difficulties; as with Borelli, these were sufficiently great that rather than object we should rejoice with him over every comfort. By Galileo's kinematics, in a vacuum the ball would fall 60 feet in two seconds and 135 feet in three. If the resistance were uniform, it would fall 56 feet and 126 feet respectively. The problem, of course, lies in the fact that the resistance increases with the velocity. In the first second, by assumption, it takes away one foot of motion. We might be inclined to proceed again with Galileo's formula and say that in two seconds it will destroy eight units of velocity (the resistance, 2, due to the increased velocity, multiplied by 4, the square of the total time). Since this conclusion assumes that the maximum resistance, reached only at the end of the second second, operates throughout the two seconds, the figure is clearly too large. What we must rather use is the resistance reached in the middle of the second second, $1\frac{1}{2}$, which multiplied by the square of the total time, gives 6. Hence in two seconds, the ball falls 54 feet in the air, and in the second second alone, 40 feet. For a fall of three seconds, we use the median resistance of the period of time after the first second (for two seconds the resistance is $1 + \frac{1}{2} \times 1 = 1\frac{1}{2}$; for three seconds, $1 + \frac{1}{2} \times 2 = 2$; for four seconds, $1 + \frac{1}{2} \times 3 = 2\frac{1}{2}$). In a fall of three seconds, the resistance destroys 18 units of velocity (2×3^2), and the body falls 117 feet; since it falls 54 feet in two seconds, it falls 63 feet in the third. Proceeding in this manner, Mariotte constructed a table through eleven seconds. In the ninth second, the ball falls 138 feet, in the tenth, 140 feet, in the eleventh, 139 feet. Hence acceleration ceases in the middle of the eleventh second and the terminal velocity is 140 or 141 feet per second.

Without manifest qualms, he was prepared to extrapolate to other materials. He had demonstrated to his own satisfaction that terminal velocity is proportional to the square root of density. Wax is one-eleventh as dense as lead. Its terminal velocity therefore is about $3\frac{1}{3}/11 \times 140$ or about 42 feet per second. Assuming that wax falls 12 feet in the first second, he constructed a table like that for lead which showed that it reaches terminal velocity after about three and a half seconds. Cork is only one quarter as dense as wax; therefore its terminal velocity is about 21 feet per second. To construct a table for cork, he employed quarter-seconds as the units of time and found that it reaches its terminal velocity in about six quarter-seconds.[160]

It should be obvious from the examples above that mechanics was trying to deal with a set of problems that defied the relative inflexibility of traditional geometry. In the end, the calculus would offer it an instrument to deal directly and effectively with a vast range of problems beyond those responsive to simple geometry, and the calculus would play a central role in the history of dynamics. It has been my constant argument, however, that the problems of dynamics were primarily conceptual, and the application of the calculus by itself would not remove such problems. It is significant that the conceptual difficulties appear in these examples beside the mathematical ones but not as their consequence. Borelli's analysis of vibratory motion employed two incompatible conceptions of force – force in our sense and force as the integral $\int F dt$. Mariotte's analysis of terminal velocity assumed that it is a state which is reached and not merely approached. The calculus would greatly facilitate the treatment of problems in dynamics and greatly extend the range of the science. The conceptual problems had to be resolved, however, before it could be applied with any useful effect.

★

Beyond the conceptual ambiguities and the problems of mathematical expression stood always the ultimate obstacle that the science of dynamics had to confront and surmount, the metaphysical question of the ontological status of force. Here lay the basic issue of the status of exact mathematical description in the existent order of nature. Assuming a Newtonian stance that rejected the need for hypotheses, John

Wallis refused to discuss the causes of elasticity and heaviness as he proceeded with the quantitative description of their consequences. When the objection was raised to his paper on impact that it gave merely mathematical rules and not physical causes, he replied acidly that 'ye Hypothesis I sent, is indeed of ye *Physical* Laws of Motion, but *Mathematically* demonstrated. For I do not take ye Physical and Mathematical Hypothesis to contradict one another at all. But what is Physically performed, is Mathematically measured. And there is no other way to determine ye Physical Laws of Motion exactly, but by applying ye Mathematical measures & proportions to them.'[161] Wallis' own example sufficiently demonstrates that the mere refusal to bog down in imaginary causal mechanisms did not in itself solve the conceptual problems of dynamics. The objection to his paper, however, testifies to the continuing presence of causal preoccupations that did stand in the way of a successfully quantified science.

The prevailing mechanical philosophy of nature set its face resolutely against any idea of attractions and repulsions between bodies, the idea which ultimately did function as midwife in the birth of dynamics. One senses the presence of mechanistic preoccupations behind the objection to Wallis' mathematical rules of impact. A confirmed mechanist, Borelli endorsed the standard prejudice against any action at a distance. 'No Attraction or Tractive Force exists in Nature,' he asserted. Nothing was more frequent among doctors and philosophers than the phrase 'attractive virtue', and nothing 'more truly absurd' if you examined carefully what they meant by it. For they contended that since they perceived no material agent when a magnet draws iron to itself, or an electric body tiny particles, or a cupping glass humours in the body, there must therefore be an immaterial attractive quality.

> But who will believe that an incorporeal virtue can move and draw some body by a natural force that acts immediately and without a material instrument? For in what way can something that is incorporeal and hence indivisible grasp, seize, constrain, and impell an extended body, since we know by the light of nature that no motion or physical action can be communicated without contact, and we know equally that the corporeal is not really touched by the incorporeal. Hence it is necessary that an attraction be performed through the mediation of some corporeal instrument.[162]

Borelli went on to describe a micromechanism utilising an internal aether that appears potentially quantifiable in its effects. Such devices invariably diverted attention from quantifiable effects to the mechanism itself, however. The discussion above appeared in chapter six of his work on the motion of heavy bodies, and he referred heaviness to such an internal aethereal wind. He was not led thereby to seize on free fall as the model of dynamic action, and it is surely indicative of the predilections fostered by the mechanical philosophy that Borelli adhered constantly to the model of impact through which the 'force' of a corpuscle in motion expressed itself.

Although the mechanical philosophy dominated scientific thought during most of the seventeenth century, it never fully suppressed and eliminated the Hermetic tradition, in which the appeal to sympathies and antipathies between bodies, that is, to attractions and repulsions, was as reasonable as it was absurd within the mechanical tradition. In the summer of 1669, the *Académie royale* arranged a formal discussion of the cause of heaviness in bodies in which three members presented formal papers. Huygens' paper, the original version of his essay on that subject published more than twenty years later, vigorously proclaimed the basic tenets of the mechanical philosophy. Significantly both of the other participants, Bernard Frénicle de Bessy and Giles Personne de Roberval, supported with comparable vigour the proposition that heaviness is caused by the attraction of the earth for terrestrial bodies. The more eloquent of the two was Roberval, a prominent mathematician of Descartes' generation who, in 1669, was an old man near the end of his career. Defining weight as that which causes a body to descend toward a centre, he went on to deny that he meant to attribute any virtue to a centre as such. Rather he argued for 'an attractive quality, which is mutual between all the parts of a total body, to unite together as much as they can.' By striving to unite, they constitute a centre of gravity, and weight is the force by which each individual member of the whole seeks the centre. For the terrestrial system there is a terrestrial gravity, as for the lunar system there is a lunar gravity and for the system of Jupiter a jovial gravity.[163] Attraction, he said elsewhere, 'is a quality that is general in nature.'[164]

Crucial to the understanding of Roberval's conception of gravity, which was virtually identical to that given by Kepler sixty years earlier, is his suggestion of a particular gravity unique to each planet, a mutual attraction whereby the parts of that system form a united whole. That is, gravity is the attraction of like matter for like, and as such it is indistinguishable from the sympathies of Hermetic natural philosophy.

> It is certain that all bodies whatsoever, [Francis Bacon had asserted] though they have no sense, yet they have perception: for when one body is applied to another, there is a kind of election to embrace that which is agreeable, and to exclude or expel that which is ingrate: and whether the body be alterant or altered, evermore a perception precedeth operation; for else all bodies would be alike one to another.[165]

Nearly a quarter of a century before the session in the *Académie*, during the period of his scientific creativity, Roberval himself had revealed the Hermetic sources of his conception in a work entitled *Aristarchi Samii de mundi systemate, partibus, & motibus eiusdem, liber singularis*. An attempt to account for the heliocentric system in wholly naturalistic terms, Roberval's *Aristarchus* sought to ward off theological objections by the transparent pretense that it was an ancient manuscript discovered and published by Roberval. Fundamental to the account was a concept of attraction similar to that he presented in 1669. He imagined first of all a 'mundane matter,' rather like the aethers so dear to seventeenth-century mechanist cosmologies, in which the planets swim as they circle the sun. The parts of this matter are endowed with 'a certain property or certain accident, by the force of which all parts are borne toward each other by a continual endeavour and mutually attract each other so that they cohere tightly, nor can they be separated from each other except by a greater virtue.'[166] Because of the attraction, the universe assumes a spherical shape around the sun, while the sun's heat, by causing the matter near it to rarify, establishes a zone of varying density which determines the distances at which the individual planets circle it.

Accompanied by their own sets of elements, each constituting a unique system, the divers planets float in the sea of mundane matter around the sun. To describe the terrestrial system, Roberval employed virtually the identical phrases he had applied before to the mundane matter – the parts of the system are endowed with 'a certain property or certain accident, ... by the force of which all the parts are obviously collected into one whole, and are borne toward each other and mutually attract each other so that they cohere tightly . . .'[167] The earth, he said, 'binds its elements, from which it cannot be separated, by the fetters of gravity . . .'[168] If earth itself is very dense, water is less so, and air extremely rare, and the three form a total terrestrial body, the average density of which causes it to float at its given distance from the sun. Since the entire thrust of the system was to distinguish matter into several distinct forms, each one of which has its own unique bond of mutual attraction, it is not clear what could constitute the density of the terrestrial system vis-à-vis the sun, by which Roberval sought to fix its distance. For that matter, since the heat of the sun rarefied the centre of the system, greater density vis-à-vis the universal system would carry a planet toward the periphery instead of the centre. Never mind, these were the least of his problems. Suffice it to say that each of the planets also constituted a unique system with a different composite density by which it maintained its given solar distance.

> Therefore as the earth, properly speaking, stands in the middle of its elements, that is, the water and the air, so we understand that Mars, for example, stands in the middle of elements of its nature or Martian elements, Jupiter in the middle of Jovial, Saturn in the middle of Saturnine, the moon in the middle of lunar elements; and so of the rest. Correspondingly, moreover, each one of the planets is so bound to its elements that it cannot be separated from them without violence, but constitutes with them one system, all the parts of which cohere together by their common bond; which bond indeed is a certain quality analogous to terrestrial gravity, that is, which acts in its system in the same way as gravity acts in the terrestrial system.[169]

Obviously the moon constituted a special problem in Roberval's system. Not only does it circle the earth instead of the sun, but apparently it lies within the terrestrial system since Roberval set the diameter of the system at more than a hundred times the earth's diameter. The moon is as close to the earth as it is, he said, because the two systems have nearly the same densities. Moreover, 'they have some affinity [*aliquam affinitatem*] in terms of certain qualities by which they seek to unite with each other . . .' Hence the two systems coalesce. But they do not mix with each other, as water does with water and oil with oil, so as to form one body, 'but so that the system of the moon immerses itself entirely in the terrestrial system but does not mix, as a ball of wax immerses in water but does not mix.'[170] Consequently the moon swims about the earth at a distance set by its density, but the lunar system retains its unique identity. The examples of oil, water, and wax were among the classic phenomena of sympathy and antipathy. Roberval's system rested directly on the conception, Hermetic but not uniquely Hermetic, of the attraction of like for like. Its denial of the uniform matter of nature, which the mechanical philosophy asserted as a basic tenet, expressed the radical nominalism, the unique specificity of natural agents, which the Hermetic tradition embodied.

When *Aristarchus* appeared in 1644, Descartes was unable sufficiently to express his scorn for its obscurantism, and when Roberval expounded similar ideas before the *Académie* twenty years later, Huygens' rejection of them in the name of mechanical orthodoxy was no less indignant. Certainly, if we examine *Aristarchus* closely, we find that the attraction of like for like went hand in hand with other concepts perhaps even more calculated to convince mechanical philosophers of Roberval's occultist tendencies. With no apparent reluctance, he embraced the idea that the earth and the planets are animate, that the earth is endowed with a soul 'which is able to perceive what things are helpful to its system and what are harmful.'[171] Such a faculty is common to animals and plants, and it is found equally in stones and bodies commonly thought to be inanimate.[172] Once he had publicly clasped the occult to his bosom, he went on to deny the very possibility of rational science. *Aristarchus* was concerned essentially with the motions of celestial bodies, and Roberval proclaimed that their spontaneous actions could never be reduced to science. The rotation of the earth is irregular and therefore the measure of time uncertain. The equinoxial points are unstable and hence astronomical observations unsure. Let no one then delude himself that a perfected astronomy is possible, for there are many irregularities 'the hidden causes of which are so abstruse that uncovering them or even understanding them far exceeds human capacity.'[173] Small wonder that mechanical philosophers rejected ideas that embodied the very denial of the ends they pursued. Nevertheless, it is possible to imagine a concept of attraction, especially gravitational attraction, shorn of its more extreme Hermetic accoutrements and quantified. Although such a concept did not appear in Roberval, its possibility constituted one route toward a quantitative dynamics, a route effectively closed to mechanical philosophers of strict persuasion.

In the writings of Robert Hooke, ideas essentially similar to Roberval's did in fact take further steps in the direction of quantification. The pamphlet on capillary action which inaugurated Hooke's career announced a principle of congruity and incongruity that asserted a role in his natural philosophy throughout his life. The ascent of water in narrow glass tubes derives from the lesser pressure of the air inside the tubes, and the decreased pressure in turn arises from 'a much greater inconformity or incongruity (call it what you please) of Air to Glass, and some other Bodies, than there is of Water to the same.' Hooke defined conformity or congruity to be a 'property of a fluid Body, whereby any part of it is readily united or intermingled with any other part, either of itself or of any other Homogeneal or Similar, fluid, or firm and solid body: And unconformity or incongruity to be a property of a fluid, by which it is kept off and hindred from uniting or mingling with any heterogeneous or dissimilar, fluid or solid Body.'[174] To support the existence of such a principle, Hooke cited a number of phenomena much like those Roberval had pointed to. As many as eight or nine different fluids can be made to float on one another in layers without mixing. Indeed incongruous fluids cannot be made to mix; when they are shaken together, they remain separated in drops. Water forms into spherical drops when it is in air, and air into spherical bubbles when it is in water. The principle is not confined to fluids. Water stands on greased surfaces but sinks into wood; in contrast, mercury stands on wood and sinks into metals. Hooke concluded the pamphlet by inquiring whether the principle of congruity might not be found to play a general role in the operations of nature.[175]

Mechanical philosopher that he was, Hooke suggested that congruity and incongruity might be manifestations of harmonious and inharmonious vibrations of the particles of matter. Continually, however, the principle tended to push his thought down paths that mechanical philosophers tended to shun. In the *Micrographia*, he called congruity 'a kind of attraction,' and he even used the despised terms, 'sympathy' and 'antipathy.'[176]

In his *Attempt to Prove the Motion of the Earth*, Hooke applied the idea, as Roberval had done, to the concept of gravity. The system of the world that he was preparing, he said, was different from any yet known, and answered to the rules of mechanics.

> This depends upon three Suppositions. First, That all Coelestial Bodies whatsoever, have an attraction or gravitating power towards their own Centers, whereby they attract not only their own parts, and keep them from flying from them, as we may observe the earth to do, but that they do also attract all the other Coelestial Bodies that are within the sphere of their activity; and consequently that not only the Sun and Moon have an influence upon the body and motion of the Earth, and the Earth upon them, but that ☿ also ♀, ♂, ♄, and ♃ by their attractive powers, have a considerable influence upon its motion as

in the same manner the corresponding attractive power of the Earth hath a considerable influence upon every one of their motions also.[177]

As an apparent statement of the principle of universal gravitation, the passage has attracted considerable attention. When we examine it closely in the light of the concept of congruity, however, we cannot avoid wondering whether Hooke did in fact have universal gravitation in mind. Certainly he had taken a giant stride in that direction by vastly extending the range through which the forces of attraction operate and by imagining that the gravities of individual planets can attract other planets. When he emphasised that planets attract 'their own parts toward their own Centers' and exercise a mere 'considerable influence' on other planets, however, he suggested that the lingering influence of the concept of particular gravities stood in the way of a complete notion of universal gravitation.

More important than the partial step toward rendering the concept of gravitation universal was Hooke's realisation of how it would fit in the science of dynamics. The second supposition, already quoted above, stated the principle of rectilinear inertia and asserted that planets following curvilinear orbits must be deflected continually from their rectilinear paths by an attraction toward the centre. The third supposition assumed that the attraction acts more powerfully on bodies closer to the centre, although he confessed his inability to discover the function that relates force to distance.[178] It was the idea of attraction that allowed Hooke to reconceptualise the mechanics of orbital motion and to recognise curvilinear motion as continuously accelerated. Moreover, if his own quantification of attraction bore no immediate fruit, it suggested the possibilities inherent in the step.

As far as the idea of attraction itself was concerned, Hooke never freed it entirely of the particularistic tendencies associated with the principle of congruity. In *Cometa*, published four years after the *Attempt to Prove the Motion of the Earth*, he proposed again that a mutual attraction between the sun and the planets holds them in their orbits, an attraction which he compared explicitly to that between a magnet and iron. If a magnet attracts iron, however, it has no effect whatever on 'a bar of Tin, Lead, Glass, Wood, &c.,' and on other bodies it has 'a clean contrary effect, that is, of protrusion, thrusting off, or driving away, as we find one Pole of the Magnet doth the end of a Needle touched on the opposite part.' He went on to propose that an internal agitation in Comets 'may confound the gravitating principle...' Jumbled and confused, the parts of the Comet 'become of other natures than they were before, and so the body may cease to maintain its place in the Universe, where first it was placed.'[179] Hence the non-periodic aspect of comets – unlike planets which are held in closed orbits, they are repelled and driven out of the system.

Never a systematic thinker, Hooke left his notion of gravity enveloped in considerable ambiguity. His *Discourse of the Nature of Comets* (1682) defined gravity in terms suitable only to particular gravities such as Roberval had supported.

> By *Gravity* then I understand such a Power, as causes Bodies of a similar or homogeneous nature to be moved one towards the other, till they are united; or such a Power as always impels or drives, attracts or impresses Motion into them, that tends that way, or makes them unite.[180]

Other passages in the same work moderated the particularistic tone in a way that suggested a scale of congruity which determines the extent of one body's action on another. A brief sketch proposed a general system of nature in which light and gravitation, expansion as it were and contraction, are the principle manifestations of the laws that govern the cosmos.

> These two Powers seem to constitute the Souls of the greater Bodies of the World, *viz.* the Sun and Stars, and the Planets, both such as move about the Sun, and such as move about any other Central Body: And both these are to be found in every such Body in the World; but in some more, in some less; in some one is predominant, in others the other; but no one without some Degree of both: For as there is none without the Principle of *Gravitation,* so there is none without some degree of *Light.*[181]

Hooke also proposed that bodies vary in their receptivity to gravity. Bodies composed of particles that are greatest in bulk and closest in texture are most subject to its action, and quantity of matter alone does not determine the gravity of a body.[182] A paper labelled 'Of Gravity' and containing only a few suggestive headings expressed the ultimate dominance of the concept of congruity in his idea of gravity: 'Similar Bodies join together more easily.'[183] Much as the mechanical philosophy opposed the idea of attractions, the particularistic aspect of the concept of congruity which denied the ultimate identity of natural bodies, was perhaps even more fundamentally incompatible with the basic principles of seventeenth-century science.

Through his influence on Newton, who imbibed at once the concept of congruity and the examples by which it was illustrated, Hooke played a significant part in the development of the concept of universal gravitation. More important than the idea of attraction as such, was his indication of the role it might fill in a quantified dynamics. Hooke's own quantification of attraction contained ineradicable defects. He attempted to apply his definition of force, a definition similar to our concept of work, to the intensity of gravitational attraction at discrete places. Thus his derivation of the inverse square relation rested on a bastardised substitution of this definition of force into Kepler's abortive law of velocities.[184] As a contribution to rational mechanics, Hooke's effort was of minimal value. As a program, however, it was of crucial importance. In the famous correspondence with Newton in 1679-80, he proposed the problem basic to the *Principia*, the demonstration of the path of a planet held in a closed orbit by a central attraction that varies in intensity with distance. In its refusal to be confined by the mechanical philosophy's unbending rejection of attractions, Hooke's proposal opened a new path to the science of dynamics. It is not too much to say that the dynamics we know was found at the end of the road he indicated.

NOTES

1. Ignace Gaston Pardies, *Discours du mouvement local*, 3rd ed. (La Haye, 1691), separately paginated section of Pardies' *Oeuvres de mathematiques*, pp. 9-24.
2. Ibid., p. 27.
3. Ignace Gaston Pardies, *La statique ou la science des forces mouvantes*, 3rd ed. (La Haye, 1691), section of Pardies' *Oeuvres de mathematiques* paginated consecutively with *Mouvement local*, pp. 174-6.
4. Pardies, *Mouvement local*, p. 52.
5. The passage continues: 'Et certainement puisque une boule en frappant contre une autre boule qui lui est égale, peut la mouvoir, & en la mouvant lui donner toute sa vîtesse, comme tout le monde en convient; si nous venons à considerer cette seconde boule jointe à une troisième qui n'ajoûte aucune nouvelle résistance; n'est-il pas visible que la même force qui suffisoit pour mouvoir cette seconde boule quand elle étoit seule, suffira aussi pour la mouvoir avec la même vîtesse quand elle est jointe à cette triosième, qui n'apporte aucune nouvelle difficulté?' It is true that we require a great deal more effort to move a large stone than a small one, but this fact arises from the resistance of the weight. If

the large stone had the same weight as the small one, we could move them both equally well. Pardies, *La statique*, pp. 173–4.
6. Huygens, *Oeuvres*, **16**, 207. Cf. Chapter IV, fn 72.
7. Pardies, *Mouvement local*, pp. 53–8.
8. Ibid., pp. 28–9.
9. Ibid., pp. 30–2.
10. Ibid., p. 41.
11. For example, Pardies spoke frequently of the '*force*' of structures to resist various stresses (such as the '*force*' of the wind), and in the same context he discussed how '*forts*' they are (*La statique*, pp. 118, 206, 229, 235–6). The meaning was scarcely different when, in the context of the simple machines, he said that two bodies that prevent each other's motion have equal '*forces*', and if they are not equal the one that is '*plus fort*' prevails (ibid., p. 129). A lever, the fulcrum of which is at one end, exerts more '*force*' the nearer the load is to the fulcrum (ibid., p. 159). Sometimes this usage sounds deceptively precise. For example, a discussion of the motion of a ship stated that if it moves a hundred times more easily in the line of its keel than side ways, a hundred times more '*force*' will be necessary to push it to the side than forward. The same passage examined the '*force*' of the wind for divers settings of the sail (ibid., pp. 241–4). On close examination, the precision largely evaporates – in both cases '*force*' is hardly distinguishable from a vague sense of strength, and in the latter case the '*force*' of the wind of which he spoke is the effective component of the strength. Nevertheless, in similar instances one sees a small movement from the intuitive meaning toward the ultimate precise technical one. A similar movement toward a different meaning is found in his assertion about the simple machines, that the same amount of '*force*' necessary to raise a hundred pounds one foot is needed to raise one pound a hundred feet (ibid., pp. 180–1). It is interesting also to note that '*force*' still carried the connotation of violence. A discussion of motion through a fluid spoke both of the '*force*' with which the fluid presses on the body and of the body being carried '*avec . . . violence . . .*' (*Mouvement local*, pp. 54–6.)
12. Pardies, *La statique*, p. 107.
13. Ibid., p. 117.
14. Ibid., pp. 129–31.
15. Ibid., pp. 176–7.
16. Ibid., pp. 180–1.
17. Ibid., pp. 130–1.
18. Claude-François Milliet de Chales, *Cursus seu mundus mathematicus*, 3 vols. (Lugduni, 1674); *Tractatus sextus mechanice*, **I**, 395–432; *Tractatus VII*; *Statica seu de gravitate terrae*, **I**, 433–570.
19. Cf. the following account of acceleration, in which the word '*potentia*' does not, to be sure, appear. In the account, de Chales referred to the '*Velocitas aut impetus aequabiliter in temporibus aequalibus acquisitus . . .*' Rather than speak of velocity, he went on to add, he preferred to speak of impetus, '*quasi de causa motus, & velocitatis . . .*' From the acceleration of heavy bodies, then, one concludes that they are acquiring impetus '*qui sit huius accelerationis causa.*' But here lies the difficulty. The gravity of a heavy body remains the same and continues to pro-

duce the same effect. Therefore, in order to explain accelerated motion, it is necessary to fall back on something '*quod ipso motu acquiratur*'. Impetus itself endures as long as it is not resisted or destroyed by a contrary impetus. A heavy body that is restrained from falling continually produces impetus that is continually destroyed, with the result that none accumulates. '*At vero dum cadit, quia illi non resistitur omnino, impetus semper crescit. Nihilque, aut parum illius deperit; si ergo impetus sit causa velocitatis, ita ut eodem modo augeatur quo augetur velocitas, dico impetus . . .*,' etc., etc. In this passage, gravity rather than *potentia* is the cause of impetus and impetus the cause of the motion (*Cursus*, **1**, 464–5).

20. I shall not give an extended survey of de Chales' usage of terms. Let me offer two examples, however. In discussing the parabolic trajectory of a cannon ball, he spoke of the uniform horizontal motion caused by the '*impetu a pulvere pyrio recepto*,' which is continually deflected '*vi gravitatis . . .*' (ibid., **1**, 488). In his consideration of pendulums, he noted that a longer pendulum moves more swiftly than a shorter one, even though its period is longer. There are some who explain this by the longer pendulum's greater '*momentum*', which derives from the greater distance of its bob from the support. De Chales rejected this explanation because every difference in '*momentum*' does not produce a difference in velocity. Two bodies of different weights, for example, fall equally though they differ in '*momenta*' (ibid., **1**, 518–19).

21. Ibid., **1**, 402. He went on to expand on his meaning. One pound that descends two feet can overcome the resistance of one pound to rise one foot. 'Therefore, when two unequal weights are equally removed from the fulcrum, the larger raises the smaller because there are more parts of motion down than of motion up; but equally when equal weights are attached to a balance so that one is more distant [from the fulcrum], there will be more parts of motion down in one than parts of motion up in the other; therefore the weight which is farther from the fulcrum will raise the other one that is equal to it in size.' (Ibid., **1**, 402.)

22. Ibid., **1**, 404–5.

23. Cf. two passages in his *Statica* in which impetus was seen to vary in proportion to the square root of height, although the heights in question were not those of free fall. In regard to pendulums, he noted that their periods vary as the square root of their lengths. He went on to ask about their velocities. The velocity of a pendulum is an ambiguous conception, of course. His diagram envisages arcs equal in angular displacement, so that the lengths of arc for three different pendulums are proportional to the lengths of the pendulums. Since the path of the pendulum nine times as long is nine times as great, while it takes three times as long to traverse it, de Chales concluded that its 'velocity' is three times as great. If balls are placed so that the bobs of the pendulums strike them, the '*impetus in iis productus*,' which equals the '*impetus seu velocitas*' of the pendulums, will be proportional to the square root of the length (ibid., **1**, 470). Equally, when water flows from a hole in the bottom of a container, its velocity or '*impetus*' is proportional to the square root of the depth of water in the container above the hole (ibid., **1**, 472).

24. De Chales' calculations shed some light on the level of his mathematical sophistication. He arrived at the velocity of 144 feet per second by using a value of 3.0

for π. To calculate the time necessary to attain that velocity in free fall, he began with a pendulum three feet long that vibrated in one second. In the time of half a vibration, a body fell a bit more than four feet, which by Galileo's proportion yielded a fall of 17 feet in one second. Rather than proceed from this figure to the acceleration and from the acceleration to the time necessary to generate a velocity of 144 feet per second, he laboriously calculated the distances of fall in each successive second – 51 in the second, 85 in the third, 119 in the fourth, 153 in the fifth. The distance travelled in the fifth second indicates a velocity of at least 144 feet per second. The sum of distances in the five seconds is 425 feet (ibid., **1**, 431).

25. Ibid., **1**, 430–2.
26. Ibid., **1**, 395.
27. Ibid., **1**, 406.
28. *Philosophical Transactions*, No. 43 (11 January 1668/9), 867–8.
29. Thomas Birch, *The History of the Royal Society of London*, 4 vols. (London, 1756–7), **2**, 338–9.
30. R. T. Gunther, *Early Science in Oxford*, 14 vols. (Oxford, 1923–45), **8**, 186–7, reprinted by permission of A. E. Gunther, F.G.S., Museum of the History of Science, Oxford. Nevertheless, as I have indicated, Hooke was able to contradict himself on his own principle. Sometimes he set force directly proportional to velocity. He read a paper on the 'force of falling bodies' to the Royal Society in 1663; in it, he asserted 'that a body moved with twice the celerity acquires twice the strength and is able to move a body as big again.' (Birch, *History*, **1**, 195–6.) In *De potentia restitutiva* (1678), he contended that motion and body 'always counterballance each other in all the effects, appearances, and operations of Nature, and therefore it is not impossible but that they may be one and the same; for a little body with great motion is equivalent to a great body with little motion as to all its sensible effects in Nature.' Hence in vibrating particles, motion is in reciprocal proportion to the size of a body. (Gunther, *Oxford*, **8**, 339, 342.) In a lecture examining the motion of ships, delivered to be sure in 1690 when he had passed his prime, he employed the same formula in a way that indicates how powerfully the image of the balance imposed itself on mechanical discussions in the seventeenth century. He began by noting that the specific gravity of air is to that of water as 1 to 800 or 900. If equal quantities of both fluids, moving with equal velocities, flow against the same body, they will communicate motion in the same ratio. Therefore if air is moved with a velocity 28·3 ($\sqrt{800}$) times as great as water, 28·3 times as much air will strike the body in a given time, 'then 28·3 × 28·3 will produce an equality of Motion with the eight hundred Gravitating parts in the Water mov'd with one degree of Velocity.' Let *ab* be a sail set perpendicular to the wind and let the wind move a distance *da* in a given time, so that a total volume of wind *abcd* moves against the sail. Moreover, let *abon* represent a prism of water on the same base *ab*, such that *na*, the length of the prism of water, equals $\frac{1}{30}da$, the length of the prism of air. (For this problem, he had switched to the figure of 900 for the ratio of specific gravities.) He assumed that the water moves (against the sail!) in a direction opposite to the motion of the air. It followed that the two hold the sail in equili-

brium. The situation is the same if the water moves against the sail or the sail against the water. Therefore, if the air is one degree swifter, it will overcome the water and drive the sail before it. (*The Posthumous Works of Robert Hooke, M.D. S.R.S. Geom. Prof. Gresh. &c.*, pub. Richard Waller, (London, 1705), pp. 565–6.) Thus the problem reduced itself to the equilibrium of a balance, an equilibrium upset by the superiority of one side which allows it to overpower the other and to move it. Even Hooke, who had seen the relevance of Galileo's kinematics for dynamics, slipped comfortably back into the old pattern.

31. Birch, *History*, **2**, 126.
32. Gunther, *Oxford*, **8**, 184.
33. Birch, *History*, **2**, 91–2.
34. Gunther, *Oxford*, **8**, 27–8.
35. Hooke to Newton, 24 November 1679; *The Correspondence of Isaac Newton*, ed. H. W. Turnbull, 4 vols. continuing, (Cambridge, 1959 continuing), **2**, 297, reprinted by permission of the Cambridge University Press.
36. Hooke to Newton, 6 January 1679/80; ibid., **2**, 309.
37. Gunther, *Oxford*, **8**, 349–50.
38. Giovanni Alfonso Borelli, *De motu animalium*, 2 vols. (Roma, 1680–1), **1**, 56.
39. Ibid., **1**, Dedication.
40. Ibid., **1**, 2.
41. Giovanni Alfonso Borelli, *Theoricae mediceorum planetarum ex causis physicis deductae*, (Firenze, 1666), p. 29.
42. Ibid., p. 75.
43. Ibid., p. 49.
44. Giovanni Alfonso Borelli, *De vi percussionis liber*, (Bononiae, 1667), p. 173.
45. Giovanni Alfonso Borelli, *De motionibus naturalibus, a gravitate pendentibus*, (Lugduni Batavorum, 1686), pp. 270, 272.
46. See Appendix E.
47. Borelli, *De vi percussionis*, p. 46. Note that DF at the beginning of the problem is the velocity that A loses. The common velocity after impact would be EF.
48. Borelli, *Mediceorum*, p. 58. Cf. *De vi percussionis*, pp. 62–3, where he says the same thing and even uses Descartes' argument that motion does not fade away of itself any more than a triangular body ceases to be so unless an external agent changes its shape.
49. Borelli, *De motu animalium*, **2**, 130.
50. Ibid., **1**, 277–80. Cf. ibid., **1**, 297–9, where the same concept of an occult jump is used in the analysis of the flight of birds.
51. Ibid., **1**, 142. Cf. ibid., **1**, 125–6: When the balance AB of equal arms AC and BC is in equilibrium with weights R and S, the two weights remain at rest; '*& talis quies non dependet ab inertia, sed ab exercitio actuali potentiarum integrarum R, & S, quatenus pondus R tanta vi comprimit librae radium CA, quanta est energia, qua pondus S nititur flectere deorsum radium CB* ...'
52. Borelli, *Mediceorum*, p. 56. A full exposition of Borelli's celestial dynamics is found in Alexandre Koyré, *La révolution astronomique. Copernic, Kepler, Borelli*, (Paris, 1961).
53. Borelli, *Mediceorum*, pp. 61–3.

54. Ibid., pp. 76, 47. Ch. Serrus, 'La mécanique de J.-A. Borelli et la notion d'attraction,' *Revue d'histoire des sciences*, **1** (1947), 9–25, discusses the relation of Borelli's natural instinct to the concept of attraction.
55. Borelli, *Mediceorum*, p. 77.
56. Giovanni Alfonso Borelli, *Responsio ad considerationes quasdam et animadversiones R.P.F. Stephani de Angelis . . . de vi percussionis*. Appended to Borelli's *De vi percussionis*, editio prima Belgica, (Lugduni Batavorum, 1686), p. 247.
57. Borelli, *Mediceorum*, p. 49.
58. Ibid., pp. 63–5.
59. Ibid., p. 65.
60. Ibid., pp. 70–3.
61. Ibid., p. 77.
62. Borelli, *De vi percussionis*, Proemium.
63. Ibid., p. 199.
64. Ibid., pp. 251–2.
65. Ibid., pp. 288–9.
66. Ibid., p. 90.
67. Ibid., p. 68.
68. Ibid., p. 69.
69. Ibid., pp. 295–6. Cf. the discussion of specific gravity in *De motionibus naturalibus*, pp. 114–15:
    Definitio I. Et primo noto, quod corpus sive similare, & homogeneum, sive heterogeneum, tunc vocatur existimaturque rarius specie, quam aliud, quando sumptis aequalibus molibus eorumden illud minorem copiam materialis substantiae corporeae, & sensibilis comprehendit in eodem spatio, quam istud, quod profecto concipi potest, si intelligatur minor copia materiei sensibilis in majori spatio corporis rarioris, extensa per interpositionem inanium spatiolorum.
    Definitio II. Si vero moles aequales, sive inaequales non considerentur, & raritas in una earum contenta major fuerit raritate alterius, tunc dicetur illa raritas absolute major reliqua, sive excessus raritatis extensive in majori mole multiplicetur, sive intensive in minori mole augeatur.
    Suppositio VII. Praeterea suppono ex Aristotele, raritatem alicujus corporis multiplicari & augeri in infinitum posse, prout substantialis moles corporea, quae in eodem spatio continebatur, successive imminuitur, & post diminutionem extenditur expanditurque ut repleat idipsum spatium, quod prius a non imminuto corpore occupabatur.
70. Borelli, *De ve percussionis*, p. 31.
71. Ibid., p. 43.
72. Ibid., pp. 49–50.
73. Ibid., p. 49.
74. Ibid., pp. 71–3.
75. Ibid., p. 78.
76. Ibid., p. 120.
77. Ibid., pp. 122–3.
78. Ibid., p. 126.
79. Ibid., p. 62.

80. Ibid., pp. 127–31.
81. Ibid., pp. 146–9.
82. Ibid., pp. 141–4.
83. Ibid., p. 142.
84. Ibid., pp. 163–4.
85. Ibid., pp. 168, 180.
86. Ibid., pp. 183–4.
87. Ibid., p. 183.
88. Ibid., p. 185.
89. Ibid., pp. 246–7.
90. Ibid., p. 244.
91. Borelli, *De motionibus naturalibus*, p. 282.
92. The paper submitted to the Royal Society was published in the *Philosophical Transactions*, No. 43 (11 January 1668/9), 864–6.
93. John Wallis, *Mechanica: sive, de motu, tractatus geometricus*, (London, 1670–1), pp. 2–3.
94. Ibid., p. 18.
95. Ibid., pp. 3–4.
96. Ibid., pp. 4–5. Cf. the first two propositions of Chapter II in which he asserted that heavy bodies act in proportion to their weights and descend if not restrained. Both propositions explicitly generalised to '*vires motrices quaelibet*'.
97. Ibid., p. 612.
98. Ibid., pp. 646–7.
99. Ibid., p. 651.
100. Ibid., p. 21.
101. Ibid., p. 26.
102. The lever is principally of use, he said, 'in raising loads . . . [*ad onera in altum levanda* . . .]' (ibid., p. 575). In discussing the windlass, he spoke of related machines that work according to the same ratios – 'as the perimeter of the axle, to which the weight to be moved [*Pondus movendum*] is applied, is to the perimeter of the outer circle to which the motive force is applied . . . so, vice versa, is the force maintaining equilibrium to the weight to be moved [*Vis aequipollens, ad Pondus Movendum*].' (Ibid., p. 610.) In the latter case, the juxtaposition of the phrases, '*vis aequipollens*' and '*pondus movendum*' summarises the paradox.
103. Ibid., pp. 61–2.
104. Ibid., p. 770.
105. Ibid., p. 31.
106. Ibid., p. 30.
107. Ibid., p. 63.
108. Ibid., pp. 645–6.
109. Ibid., p. 650.
110. Ibid., p. 684.
111. Ibid., p. 37.
112. Ibid., p. 611.
113. Ibid., p. 660.

114. Ibid., pp. 661–2.
115. Ibid., pp. 662–5.
116. Ibid., p. 666.
117. Ibid., p. 667.
118. Ibid., p. 690. Cf. Wallis to Oldenburg, 3 December 1668: 'That all rebounding comes from Springyness, is my opinion . . .' (*The Correspondence of Henry Oldenburg*, ed. A. Rupert and Marie Boas Hall, 6 vols. continuing, (Madison, The University of Wisconsin Press, 1965, reprinted by permission of the Regents of the University of Wisconsin), **5**, 218.)
119. Wallis, *Mechanica*, p. 686.
120. Ibid., p. 689.
121. Ibid., pp. 697–8.
122. The treatise was not published until Mariotte's *Oeuvres* appeared in 1717. The *Treatise on Pendulums* was introduced in the *Oeuvres* by a letter dated 1 February 1668, which appears adequately to date its completion.
123. Edme Mariotte, *Oeuvres*, (Le Haye, 1740), p. 558. I am citing all of Mariotte's works from this edition. The *Traité du mouvement des pendules* is found in pp. 557–66; the *Traité de la percussion*, pp. 1–116; the *Traité du mouvement des eaux*, pp. 321–482.
124. Ibid., p. 558.
125. Ibid., pp. 11–14.
126. Ibid., p. 11.
127. Ibid., pp. 19–21.
128. Ibid., p. 23.
129. Ibid., p. 29.
130. Ibid., pp. 30–44.
131. Ibid., pp. 50–1.
132. Ibid., p. 62.
133. Ibid., p. 17.
134. Ibid., p. 14.
135. Ibid., pp. 356–7. Mariotte proceeded to a third rule which stated that when two weights are constrained to move together, only the vertical components of motion are relevant in calculating the quantities of motion. He summarised the rule in a 'Universal Principle of Mechanics': 'Whenever two weights or other powers are so arranged that one cannot move without making the other move, if the distance that one of the weights must traverse in its proper and natural direction is to the distance that the other must traverse in its proper and natural direction during the same time reciprocally as the second weight is to the first, equilibrium will be established between the two weights; but if one of the weights is greater in proportion to the other, it will compel it to move.' (Ibid. p. 360.) In the continuing discussion, he applied the same concept of equilibrium to demonstrate why fluids stand at the same level in separate legs of a connected vessel. First he analysed a U tube of constant cross-section. Cylinders of fluid of the same height in the two legs are equal in weight and must move with the same velocity; hence it is easy to prove by the principle of equilibrium of forces that one **cannot** move down, forcing the other up. Now imagine that one leg is four times

greater in diameter than the other. When two cylinders of equal height are taken, one weighs sixteen times as much as the other, but if they were to move, the smaller would have to move sixteen times as fast. 'Hence their speeds would have been reciprocal to their weights, and they would have had an equal quantity of motion; which is impossible; for by the universal principle these cylinders of water must establish equilibrium, and the one cannot make the other move since they are so arranged that they move with equal quantities of motion in the same direction.' (Ibid., p. 367.)

136. Ibid., pp. 394–5.
137. Ibid., p. 74.
138. Ibid., p. 392.
139. Ibid., p. 403.
140. Ibid., p. 67. The passage is cited from Mariotte's treatise on impact. Virtually the identical analysis appeared also in his work on the motion of fluids. (Ibid., pp. 395–6.)
141. Ibid., p. 77.
142. Ibid., p. 77.
143. Ibid., p. 78.
144. Ibid., pp. 79–80.
145. Ibid., pp. 99–100.
146. Ibid., p. 76.
147. Ibid., p. 393.
148. Ibid., p. 560.
149. Ibid., p. 564.
150. Ibid., p. 560.
151. Ibid., p. 4.
152. Ibid., p. 28.
153. Ibid., pp. 11, 371.
154. Ibid., pp. 12–13.
155. Borelli, *De vi percussionis*, pp. 257–8.
156. Borelli, *Mediceorum*, pp. 72–3.
157. As I have indicated, this result is exactly equivalent to the work-energy equation: $\frac{1}{2}mv^2 = \int F dx$. For simplicity, I set the origin at A. (For Hooke, if we can speak of an origin in his formulation, it is at C.) $F = -kx$. $\frac{1}{2}mv^2 = -\frac{1}{2}kx^2 + K = -\frac{1}{2}kx^2 + \frac{1}{2}kx_{max}^2$.
158. Apparently the foundation of this assertion lay in Galileo's two conclusions – $v^2 = 2as$, $s = \frac{1}{2}at^2$ (in algebraic form). In so far as I understand Hooke's procedure, he ignored the fact that the constants in the two expressions are different and combined them to eliminate acceleration. $v/\sqrt{s} = \sqrt{s}/t$ – the relation he expressed. Hence in a problem dealing with non-uniformly accelerated motion, he arrived at a ratio equivalent to the formula for uniform motion, $s = vt$.
159. Gunther, *Oxford*, 8, 349–50. I know of no place in which Hooke addressed himself to motion through a resisting medium. In the *Posthumous Works*, he tried to analyse the dynamics of a sailboat, but he approached it in terms of two prisms of fluid, air and water, in equilibrium according to their quantities of motion. (Cf. above, fn. 30.) As he stated it, the problem was concerned with the steady

160. Mariotte, *Oeuvres*, pp. 106–12. Mariotte did not try to analyse vibratory motion. He did write a treatise on pendulums, however, and some of the problems there are distantly analogous. Following Galileo, he 'demonstrated' that the circle is the curve of swiftest descent. Unlike Galileo, he was bold enough to give an actual figure. If the time to descend the chord of a quadrant is 100 000, he calculated the time along two and three chords in the quadrant and concluded by extrapolation that the time along the quadrant of the circle (and hence by implication along any arc less than a quadrant) is about 93 000, and therefore that the ratio of time along the circle to time on the cord is as 13/14 'à peu près' (ibid., p. 565). The final phrase almost epitomised Mariotte's mechanics.

In addressing himself to the role of resistance in the motion of pendulums, he followed Galileo again in imagining a hole through the centre of the earth. If a body were dropped in such a hole, it would accelerate to terminal velocity and then fall uniformly until it reached the centre. Beyond the centre it would move a space equal to that it traversed at the beginning of fall until it reached terminal velocity (ibid., p. 566). Neither discussion suggests much comprehension of the complexities of the issue.

161. Wallis to Oldenburg, 5 December 1668; Oldenburg, *Correspondence*, **5**, 221. Cf. Wallis to Oldenburg, 31 December 1668: Wallis insisted that he had not undertaken to determine the causes of springiness and gravity. Both do in fact exist, and whatever their causes, he had given an account of their effects. The one is the principle of the motion of restitution, the other of the tendency downward. Although Descartes had tried to assign a cause for both, Wallis had never seen an hypothesis that satisfied him, and hence he had not tried to construct one himself (ibid., **5**, 287–8). A scholium in the *Mechanica* said substantially the same thing about gravity. He refused to enter into the question of its cause or even to consider whether it acts according to a regular law. Others had constructed hypotheses to explain gravity; he intended to avoid the question altogether. Experiments had demonstrated that gravity either is constant, as he assumed, or very nearly so. Accepting that evidence, he had proceeded to construct a mathematical theory of its consequences. (*Mechanica*, p. 650.)

162. Borelli, *De motionibus naturalibus*, pp. 166–7. Cf. *De motu animalium*: Why does air enter the lungs when the thoracic cavity is enlarged? It is not attracted by the lungs, '*cum nulla vis attractiva detur in natura.*' (**2**, 166.) In *De vi percussionis*, he examined magnetic action. Whereas it is vulgarly attributed to an attraction, those '*qui magis physice philosophantur & non acquiescunt nominibus non perceptis aut nil significantibus*' trace it to some effluvium. Borelli himself went on to suggest that when a piece of iron is placed within a magnet's sphere of activity, the action '*ab effluvio halituum magnetis*' causes a fermentation within the iron whereby all the particles of a certain type line up and exert their endeavour in the same direction (pp. 185–7). In the same work, he also described the mechanism, utilising an aethereal medium within the pores of bodies, that he understood to be the cause of elasticity (pp. 235–44).

163. Roberval's mémoire on the cause of weight, read before the *Académie* on 7

August 1669, is printed in Huygens, *Oeuvres*, **19**, 628–30. Both Frénicle's and Huygens' papers are published in the same section of the *Oeuvres*.

164. Léon Auger, *Un savant méconnu: Giles Personne de Roberval (1602–1675)*, (Paris, 1962), p. 79. Auger does not make it clear from what source he is quoting this phrase, but I take him to indicate the manuscript treatise *De mechanica* (B.N. Man. lat. n. acq. 2 341).
165. Francis Bacon, *The Works of Francis Bacon*, ed. James Spedding, Robert Leslie Ellis, and Douglas Denon Heath, 15 vols. (Boston, 1870–82), **5**, 63.
166. Giles Personne de Roberval, *Aristarchi Samii de mundi systemate, partibus, & motibus eiusdem, liber singularis*, 2nd ed., p. 2. *Aristarchus* appears as a separately paginated section in Marin Mersenne, *Novarum obervationum physico-mathematicarum tomus III*, (Paris, 1647).
167. Ibid., p. 4.
168. Ibid., pp. 5–6.
169. Ibid., p. 7.
170. Ibid., pp. 7–8.
171. Ibid., p. 59. Cf. his explanation of the tides. He suggested that the earth is '*animata*' and that the tides are the effect of its respiration or of some similar motion related to its life. Because of the closely related qualities of the two systems, the moon serves to excite the motion by its presence. The animation of the earth also explains the sundry exhalations and vapours, hot and cold, wet and dry, that the earth spontaneously emits '*vel ad exteriores suas partes calefaciendas. refrigerandasve, aut exsiccandas, aut humectandas; prout sibi conducere nativo sens,' deprehendit: vel certe tanquam excrementa sibi inutilia, atque fortassis nocentia .. u* (Ibid., p. 33.)
172. Ibid., p. 49.
173. Ibid., pp. 61–2.
174. Robert Hooke, *An Attempt for the Explication of the Phaenomena, Observable in an Experiment Published by the Honourable Robert Boyle, Esq.; in the XXXV. Experiment of his Epistolical Discourse Touching the Aire*, (London, 1661). A facsimile reproduction is published in Gunther, *Oxford*, **10**, 1–50; the passage quoted here appears on pp. 7–8 of the Gunther reproduction. Virtually the entire pamphlet, with some additional material, was included in the *Micrographia*, (London, 1665), pp. 11–31.
175. Gunther, *Oxford*, **10**, 41.
176. Hooke, *Micrographia*, pp. 15, 16.
177. Gunther, *Oxford*, **8**, 27–8.
178. Ibid., p. 28.
179. Ibid., pp. 228–9.
180. Hooke, *Posthumous Works*, p. 176.
181. Ibid., p. 175.
182. Ibid., p. 182.
183. Ibid., p. 191.
184. See above in this chapter, fn. 36.

Chapter Six

Leibnizian Dynamics

DURING the last two decades of the seventeenth century, two men untangled the set of conceptual knots that bound the nascent dynamics and created a successfully quantified science. Gottfried Wilhelm Leibniz and Isaac Newton did not start building on cleared and levelled ground. The tradition of rational mechanics during the century had laid the foundations on which their structures were raised. Both men were conscious of the tradition; both drew upon it. Nevertheless, their work cannot be summarised merely as the inevitable product of the century's labour. As I have argued, the conceptual problems to be resolved were neither obvious nor easy, and not the least part of the reputation of either man rests on the creative insight that he applied to them. It is significant in this respect that both Leibniz and Newton created their own sciences of dynamics, sciences that employed different concepts of force and presented different images of dynamic action. If mechanics since that time has learned to reconcile the two and embody the insights of both, the late seventeenth century found it less than clear that such would be the case. At least to some extent, the *vis viva* controversy of the eighteenth century was the offshoot of their divergent visions. Meanwhile, the very fact that two such different sciences of dynamics were proposed suggests how far from obvious and inevitable the solutions to the problems were.

Leibniz was the younger of the two men, born some four years after Newton. A polymath of incredible versatility and a cosmopolite acquainted through the entire intellectual community of Europe, he was a prominent diplomat and public servant even while he built his enduring reputation as a philosopher and mathematician. Beside his contributions in those two fields, his work in mechanics may seem to shrink in importance, but I have already suggested that he stood beside Newton as one of the creators of modern dynamics. Moreover, his dynamics cannot be isolated from his philosophy, which was in many ways a generalised statement of its principles, or from his infinitesimal calculus, which furnished the language in which alone his dynamics could find adequate expression. Instead of a general treatise on dynamics, Leibniz produced a considerable number of essays, most of them quite brief, in which he expounded his conclusions. The most important of the essays, as far as dynamics is concerned, stemmed from the decade between 1685 and 1695.

The very word 'dynamics' (in French, '*dynamique*') was of Leibniz's coinage, from a Greek root of course, and he employed it in the titles of two of his most important works on the science – *Essay de dynamique* (1692) and *Specimen dynamicum* (1695). His interest in the science of mechanics dated from more than twenty years earlier when he had composed the *Hypothesis physica nova* (1671), a work that sketched faint outlines of his mature position among a tangle of opinions he was soon to disown. The effective birth date of what we know as Leibnizian dynamics belongs to the year 1686 when Leibniz, too old now to be a prodigy and already beginning to shake the learned world with the publication of his calculus, shook it further with the '*Brevis demonstratio erroris memorabilis Cartesii et aliorum circa legem naturalem*' in the *Acta eruditorum*. To be, like Leibniz, brief, Descartes' memorable error had been the identification of quantity of motion (effectively mv) with force. Leibniz undertook to show that in fact only mv^2 can be the measure of force. The demonstration, as Leibniz took it to be, of Descartes' error was the central event in Leibniz's career as a student of mechanics. From everything he composed on dynamics thereafter, one gains the impression that his free hand was gesticulating wildly as he wrote, calling attention to his achievement; and he did not allow many pages to pass without reminding the reader that it was he, Gottfried Wilhelm Leibniz, who had set Descartes right. Perhaps he protested too much. The quantity mv^2 owed more to Huygens, on whose work in mechanics Leibniz drew most heavily. To say as much is in no way to deny Leibniz's crucial role in the history of dynamics, however. In regard to the quantity mv^2, what was a mere number to Huygens was invested by Leibniz with cosmic significance. The central character in the drama of Leibnizian dynamics, living force or *vis viva*, entered upon its public career with the 'Brief Demonstration.'

Leibniz started from the assumption 'that a body falling from a certain altitude acquires the same force which is necessary to lift it back to its original altitude . . .' The assumption, a basic tenet of seventeenth-century natural philosophy, was not likely to be challenged. He assumed, second, 'that the same force is necessary to raise the body A . . . of 1 pound to the height CD of 4 yards as is necessary to raise the body B of 4 pounds to the height EF of 1 yard.' As he noted, Cartesians and others agreed on this assumption. Once the two assumptions were granted, Leibniz had only to feed the numbers into Galileo's kinematics of free fall and turn the crank. Let body A fall the distance CD and acquire a velocity of two units. Let body B fall the distance EF. According to Galileo, it will acquire a velocity of one unit. If force is equivalent to quantity of motion, the force of B (4×1), applied to A (since A is of 1 pound, its velocity would be 4), will cause it to rise, not four feet, but sixteen, a manifest absurdity.

> There is thus a big difference between motive force and quantity of motion, and the one cannot be calculated by the other, as we undertook to show. It seems from this that *force* is rather to be estimated from the quantity of the effect which it can produce; for example, from the height to which it can elevate a heavy body of a given magnitude and kind but not from the velocity which it can impress upon the body.[1]

Central to the demonstration was the second assumption – that the same force which lifts one pound four yards can lift four pounds one yard. Leibniz was aware that Descartes had insisted on the same measurement of force in his treatise on the simple machines; in effect, he employed Descartes to refute Descartes. The point of the argument extended beyond the clever manipulation of Descartes to his own destruction, however. Like Descartes, Leibniz realised the special significance of the product of weight times vertical height. Here was a quantity – work, we call it – the applicability of which was not confined to the simple machines, one which offered a universally valid measure of the effects that causes produce. Ultimately, the demonstration against Descartes rested on that insight.

> I assume it to be certain . . . that nature never substitutes for forces something unequal to them but that the whole effect is always equal to the full cause.[2]

In replying to Leibniz, Cartesians asserted that doubling the velocity of a body exactly doubles the effect of the moving force it embodies.

Leibniz denied the assertion of course. Doubling the mass of a moving body does double its force. If A and B are equal bodies with equal velocities, obviously the two together have twice the force of one alone. The quantity of matter involved is simply doubled while nothing else is changed. Although Cartesians argued that a doubled velocity is wholly analogous, Leibniz insisted otherwise. In the first case, a perfect doubling of the original body in both quantity of matter and quantity of motion allows a direct comparison of forces. There is no similar homogeneous doubling in the second case, and the two states have to be compared indirectly, by the effects they can produce. Raising weights vertically is a homogeneous effect that allows the comparison to be made. Imagine that a body of four pounds is divided into four equal parts, which are lifted successively a distance of one yard each by a pulley arrangement. It is immaterial if the first weight is lifted a second yard or a second weight of one pound is lifted one yard. The two effects are fully homogeneous. With velocity, we cannot make the same assertions with assurance. Velocity and time, related quantities, can be misleading. A ball rolling up a gently inclined plane ascends as high as it does on a steeply inclined plane, but it takes longer to do so. A vertical ascent that wholly exhausts the force of a moving body is thus the effect by which to measure that force. Body A of four pounds with a velocity of one unit can rise one foot. Body B of one pound with a velocity of two can rise four feet, 'and that effect, in the two cases, will be total, will exhaust the power of its cause, and will hence be equal to the cause which produces it. But in regard to power or force, the two effects are equal to each other: the elevation of four pounds (body A) to the height of one foot and the elevation of one pound (body B) to the height of four feet exhausts the same power. Consequently the causes, that is A of mass 4 and speed 1 and B of mass 1 and speed 2, are also equal in force or power . . .'[3]

What lay behind Descartes' error? Of this Leibniz had no doubt whatever. It was the simple machines which had led him to equate quantity of motion with force. 'Seeing that velocity and mass compensate for each other in the five common machines, a number of mathematicians have estimated the force of motion by the quantity of motion or by the product of the body and its velocity.'[4] The conclusion holds only in the case of statics, however, and it holds there only because the two proposed measures of force (mv and mv^2) chance to coincide in the case of dead forces. When two bodies are in equilibrium on a balance, their possible vertical displacements are inversely proportional to their weights.

> And it happens only in the case of equilibrium or of dead force, that the heights are as the velocities, and that thus the products of the weights by the velocities are as the products of the weights by the heights. This, I say, happens only in the case of *dead force*, or of the infinitely small motion which I am accustomed to call *solicitation*, which takes place when a heavy body tries to commence movement, and has not yet conceived any impetuosity; and this happens precisely when bodies are in equilibrium, and trying to descend, are mutually hindered. But when a heavy body has made some progress in descending freely, and has conceived some impetuosity or *living force*, then the heights to which this body might attain are not proportional to the velocities, but to the squares of the velocities. And it is for this reason that in the case of living force the forces are not as the quantities of motion or as the products of the masses by the velocities.[5]

A significant part of Leibniz's achievement in dynamics lay in his perception of its distinction from statics. By breaking the tyranny of the lever in mechanics – and what is perhaps more, recognising consciously that he was doing so – Leibniz freed the concept of force from the major source of ambiguity that had dogged its path throughout the century. It is not too much to date the birth of dynamics as a science from the moment when its distinction from statics was adequately defined. Leibniz himself served the cause of historical justice by coining the name of the science he effectively created.

*

Some five years after the publication of his 'Brief Demonstration,' Leibniz observed that people were beginning to accept its truth and ceasing to believe in the conservation of the quantity of motion. A consequent inconvenience was arising. Going too far toward the other extreme, those who accepted his demonstration failed to recognise the conservation of anything absolute which might replace the Cartesian quantity. 'But our minds look for this, and it is for this reason that I remark that philosophers who do not enter into the profound discussions of mathematicians have difficulty in abandoning an axiom such as this of the quantity of conserved motion without giving themselves another to which they may hold.'[6] Never fear for the shallow philosophers. Leibniz was not one to abandon them to the perplexity in which the profound mathematicians, such as Leibniz, had plunged them. With the rest of the seventeenth century he shared the vision of a universe which is eternal, a machine which functions forever without the continued intervention of an external mover. That the universe is in fact such was a premise too basic to be called into question. As he said, his mind looked for an absolute quantity to be conserved. Indeed, the gravamen of his objection to Descartes lay in his demonstration that by Descartes' definition of force the self-sustaining character of the universe was placed in jeopardy. The continued operation of the machine demands the conservation of force, and the conservation of force demands that the effect always equal the power of the cause – 'for, if the effect were greater, we should have mechanical perpetual motion, while, if it were less, we should not have perpetual physical motion.'[7]

Perpetual physical motion was equivalent to the self-sustaining universe, the fundamental datum. That force should continually decay and perish at last 'is without doubt contrary to the order of things.'[8] One of the measures of Leibniz's clarifying genius was his ability to distinguish this ultimate metaphysical level from the phenomenal realm of actualised mechanics. If perpetual physical motion cannot be questioned, perpetual mechanical motion is impossible 'because then the force of a machine, which is always diminished a little by friction and must therefore soon come to an end, would restore itself and consequently increase itself without any new impulsion from without.'[9] The recognised reality of friction in every mechanical motion, a fact which in no way challenged the perpetuity of 'physical motion', dictated the particular form of Leibniz's argument and invested it with peculiar cogency. A perpetual mechanical motion demands 'an effect more powerful than the cause . . .' By its own force, body A of four pounds with a velocity of one unit can raise itself to a height of one foot, and the same force can raise body B of one pound four feet. If force is equivalent to quantity of motion, as Descartes contended, body B with a velocity of four units will be able to raise itself by its own force to a height not of four but of sixteen feet.

> By the force of B we would then be able not only to raise A again to the height of one foot from which it would descend to regain its original speed, but also to produce several other effects, which indeed constitutes a perpetual mechanical motion since a disposable excess remains after the original force has been restored.[10]

Since such a result is manifestly absurd, force must be measured by mv^2 instead of mv; and the ultimate reality of physical nature, the law of conservation on which its continuing uniformity rests, is the conservation of living force, *vis viva*.

If in one sense Leibniz corrected Descartes by turning him against himself, we can speak with equal justice of Leibniz employing Galileo to refute Descartes. Galileo's kinematics of free fall, employed as a body of universally accepted conclusions which required no further justification, furnished a necessary step in his argument. Had he not been able to extend his dynamics beyond that single confined point, it would have been a science crippled from birth by its limitations, a science not likely to have promoted its inventor into the ranks of the great contributors to mechanics. It was Leibniz's good fortune that he had yet a third source on which to draw, Christiaan Huygens, his mentor and friend, whose investigation of impact added the dimension needed to make the concept of *vis viva* truly significant.

Leibniz himself added no important quantitative relation to dynamics. His one effort in that direction, a concept of 'moving action' which was at once artificial and sterile, contributed nothing to the further development of the science. His role was rather to bring prior achievements together in new patterns of comprehension. In no way does such a judgment belittle his achievement. Quite the contrary, it suggests the extent to which he epitomised the progress of dynamics through the seventeenth century. It has been my argument that the new conception of motion formulated in the first half of the seventeenth century, primarily by Galileo and Descartes, implied a new dynamics, although it did not embody one. The creation of modern dynamics was a matter, not of experimentation and new discovery, but of drawing consequences from accepted conclusions, of clarifying ambiguities and resolving conceptual tangles, above all of breaking through the rigidity of received intuitive conceptions of dynamic action. For such tasks had the terrifying clarity of Leibniz's mind been called into being. In relation to Huygens' work, he made correlations that had not been fully perceived by their author explicit, and in so doing generalised the concept of *vis viva*.

With Huygens, Leibniz agreed that Descartes' rules of impact were incorrect. He had proposed an alternative approach in his *Hypothesis physica nova* before he met Huygens in Paris in 1672. A variety of shortcomings in the early work led him to abandon it, and his ultimate treatment of impact was identical to Huygens' in its solutions of given problems. His final objections to Descartes' rules, however, rested on grounds that were his own. Whereas Huygens rejected Descartes because his rules contradicted both each other and experience, Leibniz played his role as philosophic critic by exploring the rules' deficiencies in the light of fundamental principles. His examination of the rules by the principle of continuity remains today one of the most devastating critiques in the entire history of science.

Leibniz spoke of continuity as a principle of general order which is absolutely necessary in geometry and applicable as well in physics, since God acted as a perfect geometrician in the creation. With such a principle, the philosopher can criticise a scientific theory as it were from the outside, before he begins to examine it in detail.

> *When the difference between two instances in a given series or that which is presupposed can be diminished until it becomes smaller than any given quantity whatever, the corresponding difference in what is sought or in their results must of necessity also be diminished or become less than any given quantity whatever.* Or, to put it more commonly, *when two instances or data approach each other continuously, so that one at last passes over into the other, it is necessary for their consequences or results (or the unknown) to do so also.* This depends on a more general principle: that, *as the data are ordered, so the unknowns are ordered also.*[11]

Compare Descartes' first two rules with the principle of continuity in mind. Rule one states that when bodies B and C come together with equal speeds and body B is equal to body C, both rebound with their original speeds. Rule two states that when B is greater than C and they come together with equal speeds, C rebounds and B continues with its motion unchanged. By the principle of continuity, the speed of B after impact should vary continuously with its size, so that as B approaches C in size, its speed of rebound should approach its speed when the two are equal. By Descartes' rules, a sudden discontinuity separates all impacts when the bodies are unequal from the one impact when they are equal. Rules four and five, in which body C is initially at rest, can be compared to similar effect. In rule four, when B is less than C, B always rebounds with its initial speed. In rule five, however, when B is greater than C, B continues in the same direction and C moves with it, their speed being such that their quantity of motion equals that of B before impact. Imagine B to approach C in size, both when it is smaller and when it is larger. By the principle of continuity, its speeds after impact should also approach a median figure continuously, but by Descartes' rules they do not.[12] Bringing the phenomena of impact into conformity with the principle of continuity had the net effect of adopting Huygens' insight, that the common centre of gravity of the two bodies in impact remains unchanged in its inertial state. By means of the principle of continuity, Leibniz supplied the insight with a philosophical foundation.

Although Leibniz's rules of impact were identical to Huygens', his approach was wholly different. As befitted one concerned with the conservation of force, he saw impact in dynamic terms and spoke in consequence, not of perfectly hard bodies, but of perfectly elastic ones. He rejected the distinction between quantity and determination of motion, which both Descartes and Huygens had accepted, insisting that quantity and determination of motion maintain each other mutually. A body tends to conserve its determination with all the force and all the quantity of its motion. What it loses in speed it also loses in determination even though its direction remains unaltered, 'for in advancing more slowly in the same direction, the body is also less determined to conserve it.' Moreover, greater opposition is necessary to reflect a body in the direction opposite to its motion than merely to stop it. If body A hits body B at rest, A continues in the same direction when B is smaller; it comes to rest when B is equal in size; it rebounds (though not with its initial speed) when B is larger. Determination then is like motion itself; it cannot be changed, as Descartes and Huygens wished to contend, without a dynamic action.

> And this determination of a body in motion, that is, its effort to advance in the same direction, has itself a certain quantity which it is more easy to diminish than to reduce to zero, that is, to rest, and more easy (that is, it requires less opposition) to stop and reduce to rest than to reverse direction and transform into an opposite motion...[13]

Hence the vectorial nature of momentum also emerged from the dynamic treatment of impact.

In the *Essay de dynamique*, Leibniz expressed the rules of impact in three general equations. (a and b represent the masses of the two bodies; v is the velocity of a before impact, and x its velocity after; y is the velocity of b before impact; and z its velocity after. v is always positive, and velocities that proceed in the opposite direction are negative.)

$$\text{I.} \quad v - y = z - x.$$

The 'Linear equation' expresses 'the conservation of the cause of the impact, or of the relative velocity . . .' In effect, it states the assumption of perfect elasticity, Leibniz's replacement of the perfect hardness he rejected.

2. $\quad av + by = ax + bz.$

The 'Plane equation' expresses 'the conservation of the common or total progress of the two bodies . . .' It is identical to our principle of the conservation of momentum and differs from the Cartesian conservation of quantity of motion in its explicit recognition of the vectorial nature of momentum. Unlike the first equation, the plane equation holds for all cases of impact, perfectly elastic, imperfectly elastic, perfectly inelastic.

3. $\quad avv + byy = axx + bzz.$

The 'Solid equation' expresses 'the conservation of the total absolute force or of the moving action . . .' Since the quantities are squared, differences of sign disappear.

> And it is for that reason that this equation gives something absolute, independent of the relative velocities, or of the progressions from a certain side. The question here concerns only the establishment of masses and velocities, without troubling ourselves in what direction the velocities tend. And this it is which satisfies at the same time the rigour of mathematicians and the aspiration of philosophers, experiments and arguments drawn from different principles.[14]

With the concept of 'moving action', Leibniz attempted to place the conservation of *vis viva* in perfectly elastic impact on an independent foundation. He learned of the role of mv^2 in impact from Huygens, and Huygens had derived it from Galileo's kinematics of free fall applied to the common centre of gravity. Leibniz wanted to derive the result from a principle valid specifically for uniform motion, and to that end he developed the concept of 'moving action' [*action motrice*]. Whatever changes take place among bodies in impact, he asserted, there must be the same quantity of moving action in the same interval of time among those bodies taken alone. To measure moving action, we must first measure the 'formal effect' of motion, which consists in what is changed, that is, the transfer of mass through space. When a mass of two moves three feet, for example, the formal effect is equal to that when a mass of three moves two feet. Formal effect must be distinguished from violent effect.

> For violent effect consumes the force and is exercised upon something without; but formal effect consists in the body in motion, taken in itself, and does not consume the force, and indeed it rather conserves it since the same translation of the same mass must always be continued, if nothing from without prevents it; it is for this reason that the absolute forces are as the violent effects that consume them, but nowise as the formal effects.

'Formal effect', then, concerns itself with uniform motion. 'Moving action', in turn, is measured simply by the velocity with which a formal effect is produced. When one hundred pounds moves one mile in one hour, the moving action is double to that when it accomplishes the same motion in two hours. From this point it was, as Leibniz said, a simple matter to demonstrate the conservation of moving action. 'I have already proved elsewhere that the same force is conserved, and ... at the bottom the exercise of force or the force drawn into the time is action, the abstract nature of force consisting only in that. Thus since the same force is conserved, and since action is the product of force by time, the same action will be conserved in equal times.'

$$\text{Action} = Ft = mv^2 t.$$

Under conditions of uniform motion, $v = s/t$. Hence,

$$\text{Moving action} = msv$$

as Leibniz's analysis had already concluded.

By means of the first two equations for impact, Leibniz could show that moving action is conserved in perfectly elastic impact. Once the conservation of moving action was established, the merest step carried him on to the conservation of *vis viva*. Since velocity is proportional to space traversed in unit time when a body moves uniformly, moving action can be expressed by either ms^2 or mv^2.

> Consequently, it is proved that the moving action is conserved, without speaking of other proofs by which I have shown elsewhere that the forces are conserved, and that the forces are as the products of the masses by the squares of the velocities, while the actions are as the products of the forces by the times, so that if we did not otherwise know this measure and conservation of force, we might learn it here....[15]

It is impossible to take the concept of moving action and its dependent derivation of the conservation of *vis viva* in perfectly elastic impact seriously. Patently concocted to reach a conclusion already known, the demonstration manipulated formulae without conviction as it covertly slipped the conclusion into the definition of 'action'.[16] That unhappy gambit excepted, however, Leibniz's treatment of impact is a masterpiece of scientific precision. While its foundation rested squarely on Huygens, the exposition raised Huygens' conclusions to a higher level of generality and succinctness. What was more important for Leibniz's purposes, it translated Huygens' kinematics of perfectly hard bodies into a dynamics of perfectly elastic bodies, in which the conservation of relative velocity, momentum, and *vis viva* all rested directly on the recognition of the dynamic action of impact itself.

In many ways, perfectly elastic impact presented the conservation of *vis viva* more satisfactorily than the dynamics of vertical motion did. In vertical ascent, the violent effect consumes the force; and though the force can always be retrieved in free fall, *vis viva* is not in fact conserved throughout the cycle of ascent and descent. In elastic impact, on the other hand, the force of the motion transposes itself into elastic force. Elasticity itself was not an arbitrary assumption on Leibniz's part. The principle of continuity demanded it. In their elasticity, macroscopic bodies must be continuous with their microscopic parts. Whereas Huygens asserted that the composition of particles concludes in ultimate parts which are perfectly hard, the principle of continuity led Leibniz to deny the very possibility of ultimate parts. Elastic bodies are composed of particles still smaller, and so on in an infinite series from which the discontinuities both of hardness and of ultimate particles themselves are banished.[17] Equally, the principle of continuity excluded the instantaneous changes of velocity demanded by the impact of perfectly hard bodies. In the necessity of elasticity lay the foundation of the dynamics of impact and the conservation of force.

> We must thus recognize that if bodies A and B collide and come from A_1 and B_1 to the place of collision A_2B_2, they are there gradually compressed like two inflated balls, and approach each other more and more as the pressure is continuously increased; but that the motion is weakened by this very fact and the force of the conatus carried over into the elasticity of the bodies, so that they then come entirely to rest. Then, as the elasticity of the bodies restores itself, they rebound from each other in a retrograde motion beginning from rest and increasing continuously,

at last regaining the same velocity with which they had approached each other, but in the opposite direction . . .[18]

In elastic impact, the *vis viva* of the composite bodies, what Leibniz also called their 'directive force', converts itself into the *vis viva* of the parts, their 'relative force', from which the *vis viva* of the composite body generates itself anew. Although the concept of *vis viva* derived initially from Galileo's kinematics of vertical motion, its conservation appeared with particular clarity in the analysis of elastic impact.

Of course, all bodies are not perfectly elastic. Indeed no bodies are perfectly elastic, as Leibniz knew very well. A good part of the brilliance of his formulation of impact lay in the facility with which he accommodated the discrepancy from the ideal. Bodies are imperfectly elastic to the extent that their parts are not united tightly enough to transfer their motion to the whole. 'Whence it comes that in the impact of such bodies a part of the force is absorbed by the small parts that compose the mass, without this force being given to the whole; and this must always happen when the pressed mass does not recover perfectly . . .'[19] The reader in the twentieth century invariably says 'energy' when he sees the phrase *vis viva*, and especially when he sees the formula mv^2. In nothing does Leibniz's concept approach energy more closely than in his analysis of imperfectly elastic impact. Some force is 'absorbed by the parts'; the first and third equations cease to hold 'since that which remains after the impact has become less than what it was before the impact, by reason of a part of the force being turned elsewhere.'[20] Perhaps it is a needed corrective to note that Leibniz did not realise that the 'force' diverted to the parts can be measured as heat. This is no more than to say that he too was unable to leap a century beyond his time. In no way does it dull the lustre of his insight, which provided the cornerstone of the ultimate principle of the conservation of energy.

> But this loss of the total force, or this failure of the third equation, does not detract from the inviolable truth of the law of the conservation of the same force in the world. For that which is absorbed by the minute parts is not absolutely lost for the universe, although it is lost for the total force of the concurrent bodies.[21]

*

In comparing *vis viva* to energy, I am indicating what is of course obvious, that 'force' (or 'living force') in Leibniz's conception has nothing to do with 'force' in the Newtonian conception we are accustomed to use. Like virtually the entire seventeenth century, Leibniz associated force and dynamic action with a moving body. It was not entirely by chance that he thought to define the 'formal effect' of a uniform motion and the 'moving action' of a force under such conditions. 'Force' is something that a body has, associated with its motion; and because it has force, it can act (or appear to act) on another in impact, the sole means Leibniz acknowledged whereby one body can alter (or appear to alter) the state of another. That is, Leibniz's conception of force expressed the model of impact. Although he insisted that the measure of force in that model is mv^2 instead of mv, in no way did he challenge the model itself. At the same time, the model of free fall also played a central role in his dynamics. Had he not known how to exploit free fall, his dynamics would have ended where countless others had, on the rubbish heap of unsuccessful attempts to construct a dynamics on the intuitive perception of the force in a moving body. The unique achievement of Leibniz lay in his utilisation of free fall to dissolve the ambiguity surrounding the concept of force and to correct the measure of force conceived on the model of impact.

The unique achievement of his dynamics was also its limitation. Despite the manifest opportunities that the calculus opened to him, and despite his tentative efforts verbally to generalise, his dynamics, considered as an aspect of rational mechanics, never freed itself from the case from which its quantitative elements were initially derived, the kinematics of free fall. With the phenomena of impact, he strove to reach a broader realm of generality, but I have argued that this attempt to derive the conservation of *vis viva* in elastic impact from independent principles was a delusion and that the asserted conservation rested, as it had with Huygens, on the potential vertical ascent of the common centre of gravity. Leibniz liked to imply the generality of his dynamics by citing other mechanical actions that were comprehensible in its terms. Vertical ascent, which renders force measurable by wholly consuming it, is only one among many violent effects that consume the forces causing them. For example, he mentioned the production of tension in a spring, the impulsion of a body to motion, and the retardation of a body in motion as analogous violent effects.[22] In all such passages, however, Leibniz merely assumed the analogy and the ultimate necessity of mv^2 as the measure of force. In no case did he attempt to derive it in general terms. *Vis viva* played the central conceptual role in Leibniz's dynamics, but he never effectively based the quantitative measure of *vis viva* on any other foundation than the kinematics of free fall.

With the vantage of hindsight, we can see that Leibniz's failure to generalise the dynamics of *vis viva* stood in intimate relation with his failure sufficiently to explore and develop his correlative concept of dead force. Within the domain of rational mechanics, the concept of kinetic energy, as we now call it, attains its full utility when it is yoked with concepts of work and potential energy. In no sense do Leibniz's brief and frequently obscure references to dead force adequately fill the same function. In isolated passages directed toward specific problems, 'dead force' appears identical to 'force' as we are accustomed to use it, and one can find no obstacles inherent in the concept that forbade its development. The peculiar conformation of Leibniz's dynamics and metaphysics, however, did in fact encourage him not to carry the development through.

The relation of dead force, or as he also called it 'solicitation' and '*conatus*', to living force repeated the relation of statics to dynamics.

> I call the infinitely small efforts or *conatus* by which the body is so to speak solicited or invited to motion solicitations; such is, for example, the action of weight or centrifugal tendency, of which an infinity is required to compose an ordinary motion.[23]

Dead force is especially visible in static situations, where no impetus has been generated and the first tendencies to motion can appear. There is, then, something of virtuality about dead force, as there is supremely something of actuality about living force, when the summation of repeated solicitations has produced a finite velocity. Leibniz had no doubt that Descartes' error in regard to the measure of force stemmed from his mistaking conditions of equilibrium for dynamics. In the case of dead forces, the infinitesimal descents that correspond to the first efforts to move are proportional to the elements of velocity, as they cease to be after a finite velocity is generated. It is a general rule

> *that forces are proportional to the product of the weights and of the heights to which these weights can raise themselves by means of their forces.* And it is relevant to consider that equilibrium consists in a simple effort (conatus) before motion, and this is what I call *dead force*, which has the same ratio in respect to *living force* (which is in the motion itself) as a point to a line. Now at the beginning of descent, when the motion is infinitely small,

> the velocities or rather the elements of the velocities are as the descents, whereas after integration, when force has become living, the descents are as the squares of the velocities.[24]

The language of the passage above reminds us that from its inception Leibniz's dynamics found its expression through the infinitesimal calculus. By itself alone, the calculus raised his dynamics to a new plane of sophistication and rigour where, at least potentially, it acquired the capacity to handle a whole range of problems simply beyond the competence of a mechanics confined to traditional geometry. In the language of the calculus, dead force is the element or differential of motion, and the infinite repetition of the *conatus*, that is, the integration of dead force in respect to time, gives rise to motion. If we allow the concept of mass, which never presented a major problem in Leibniz's dynamics, and was certainly clear after the publication of Newton's *Principia*, we can translate passages such as that cited above into the basic dynamic relations:

$$F = m\frac{dv}{dt},$$

$$mv = \int F dt.$$

Leibniz's clarity in the resolution of ambiguities was one of his major contributions to dynamics. By means of differential analysis, he was able (when he chose) to cut away the confusion surrounding the concept of force (dead force in his terminology), to dissolve the perplexity of statics and dynamics, and to define force as we define it today.

> And the impetus of living force is related to bare solicitation as the infinite to the finite, or as lines to their elements in my differential calculus.... Meanwhile as the law of equilibrium is found always to apply to differentials or increments, so also in the integrals by the admirable art of nature the same *vis viva* is found to be conserved according to the law of equivalence [of cause and effect] ... Consequently, in the case of a heavy body which receives an equal and infinitely small increment of velocity in each instant of its fall, dead force and living force can be calculated at the same time: to wit, velocity increases uniformly as time, but absolute force as distance or the square of time, that is, as the effect. Hence according to the analogy of geometry or of our analysis, solicitations are as dx, velocities as x and forces as xx or $\int x dx$.[25]

At first glance, the passage appears to propose contradictory definitions. The initial sentence seems to state that dead force is the differential of *vis viva*, and *vis viva* the summation or integral of dead force. In contrast, the final sentence poses a repeated integration in which velocity (or better motion) is an intermediate quantity, at once the integral of dead force and the differential of living force. Perhaps Leibniz was seeking to maintain that distinction when in the opening sentence he referred, not simply to *vis viva*, but to the 'impetus' of *vis viva*.[26] In this interpretation, impetus (mv) expresses what Leibniz frequently referred to as the impetuosity of *vis viva*, its live capacity to act, but is related to it mathematically as its differential.

Unfortunately, the neat twofold integration whereby dead force passes into living force via the intermediary quantity of impetus was frequently short-circuited into a single integration. Living force, 'appearing as a consequence of an infinity of degrees of dead forces, is to them as a surface is to a line. Dead forces, such as weight, elasticity, and centrifugal tendency, do not consist in a finite velocity but only in an infinitely small velocity which I call solicitation and which is only the embryo of living force which the continuation of the solicitations brings to birth.'[27] Again in the *Specimen Dynamicum*, he discussed the distinction of solicitation and impetus, which arises from the repetition of solicitations, in terms of the circular and centrifugal impulses of a ball in a tube rotating in a horizontal plane about one end.

> Hence force is also of two kinds: the one elementary, which I also call *dead* force, because motion does not yet exist in it but only a solicitation to motion, such as that of the ball in the tube or a stone in a sling even while it is still held by the string; the other is ordinary force combined with actual motion, which I call *living* force. An example of dead force is centrifugal force, and likewise the force of gravity or centripetal force; also the force with which a stretched elastic body begins to restore itself. But in impact, whether this arises from a heavy body which has been falling for some time, or from a bow which has been restoring itself for some time, or from some similar cause, the force is living and arises from an infinite number of continuous impressions of dead force.[28]

In one sense it is fruitless to torture individual passages for the exact implication of their words. Leibniz's capacity to carry through elementary integrations can hardly be questioned, and whenever he attended carefully to the terms of his definitions of dead force and living force, he saw clearly enough that a single integration of dead force

$\left(m\dfrac{dv}{dt}\right)$ over time cannot yield living force (mv^2). The crucial point rather is the summation through an interval of time in every integration. For dynamic problems in which force (that is, dead force) is a function of time, Leibniz's approach to dynamics yielded results on which we have no improvement to offer. Consider his reply to the objection that force (as he would have said, living force) must be equivalent to quantity of motion since two bodies with equal quantities of motion – say A of mass 2 and velocity 3 and B of mass 3 and velocity 2 – stop each other in impact.

> For although A is weaker than B absolutely, A being able to raise a pound only 12 feet, if B can raise a pound 18 feet; nevertheless in impact they can stop each other, the reason of which is that bodies are hindered only according to the laws of dead force or of statics. For being elastic as we suppose, they act on each other in impact only by dead forces or according to equilibrium; that is to say, by infinitesimal changes, because in pressing, resisting, and continually weakening each other more and more until they come to rest, at each moment they destroy one another only by infinitely small motion or by dead force, equal on both sides; now the quantity of dead force is measured according to the laws of equilibrium by the quantity of motion, infinitely small in truth, but the continual repetition of which exhausts at last the whole quantity of motion of the two bodies, which being supposed equal in both bodies, each quantity of motion is exhausted in the same time, and consequently the two bodies are reduced to rest in the same time by the pressures of their elasticities, which, restoring themselves afterwards, reproduce the motion. It is in this continual diminution of the quantity of motion according to equilibrium in the impact of the two elastic bodies that the cause of this paradox consists, that two forces that are absolute unequal but have equal quantities of motion must stop each other, because this happens in an action that is relative where the contest takes place only according to infinitely small quantities of motion continually repeated.[29]

It is one of the advantages of analytic procedure that we can reduce Leibniz's verbosity on this occasion to

$$\int F dt = \Delta mv.$$

The central quantity in Leibniz's dynamics, however, derived from a case in which (dead) force is a function of distance. What was readily accessible to his mathematics was obstructed by philosophical considerations, and Leibniz never suggested that the generation of *vis viva* in free fall be seen as the integration of dead force over distance. No concept in his dynamics corresponds closely to the ideas of work or potential energy. His closest approach, the notion of a violent effect that consumes *vis viva*, had different connotations. Indeed the whole tenor of Leibniz's philosophy set itself against the possibility of a functional relation of force and distance. He saw force as a body's property, a momentary state in the law of its development – the summation of the past which it embodies, the source of the future with which it is great. Dead force as such appeared to Leibniz less as a crucial term in rational mechanics than as a single moment in the series that defines the internal law of a body, a law which speaks more eloquently through living force. The very adjectives 'living' and 'dead' express the emphasis of Leibniz's thought. He did not trouble himself extensively to investigate the full implications of dead force as a concept within dynamics. Although the mathematical expression of dead force $\left(m\frac{dv}{dt}\right)$ is identical to the definition of force we are accustomed to employ, in conception the two differ radically. Force as we understand it is the logical correlate of the principle of inertia, an external action that alters the state of a body unable to initiate such a change of itself. In contrast, Leibniz rejected the very possibility that anything external can influence the autonomous law that governs a substance. The word *'conatus'* which he employed as a synonym for dead force bore a freight of significance. Dead force is a body's 'effort' or 'endeavour' to move, just as its living force expresses itself in the impetuosity of motion achieved. Conceived in such terms, force was uniquely a function of time. The law that governs a substance unrolls its course through time rather than space. Whereas the summation over time of a body's endeavours repeats mathematically the law of its being, a similar summation over space lacks all meaning. When Jean Bernoulli demurred at the measurement of *vis viva* by the height to which it can raise a body, objecting that the effect is an accident which depends on the law of gravity and the motions of the aether that cause it, an accident that cannot measure absolute force since God could have made conditions otherwise, he invited Leibniz in effect to state the equivalent of the work-energy equation, that the increment of *vis viva* is equal to the integral of any dead force over distance. Leibniz's reply is in that all the more revealing – he evaded the presumed difficulty by reminding Bernoulli that he had demonstrated the conservation of *vis viva* elsewhere from the *a priori* principle of moving action.[30]

As Leibniz insisted on the metaphysical reality of force, he insisted equally on mechanical causation on the phenomenal level. The idea of action at a distance, so intimately related to the concept of force as a function of distance, was no less anathema to him than to Huygens. As a consequence, the motion of free fall could present itself to him as the summation of differential elements of velocity but not as the summation of differential elements of work. Ironically, Leibniz never tired of pointing out that the particular utility of vertical rise as a violent effect that measures force is its independence both of velocity and of time. Whatever the path, an equal vertical ascent consumes the same quantity of force. Nevertheless, vertical motion did not appear to him as something analogous to our concept of work, as an integral $\int F ds$. Perhaps the absence of the coefficient $\frac{1}{2}$ which we find so puzzling in his expression of *vis viva* is indicative. Surely the fact that even in the *Essay on Dynamics*, which he composed after the *Principia*'s publication, he could say that the force of a body is measured by the product 'of the mass or the weight' and the height to which the body can rise is significant.[31] The central quantity in his dynamics depended directly on the fortunate accident of Galileo's kinematics of free fall, which in its recognition that a body can rise to the height from which it fell embodied a concept of conservation that could not fail to attract Leibniz. He never generalised the dynamics of free fall in a way that allowed him to apply it quantitatively to the other dynamic actions he referred to as analogous.

In its ultimate form, Leibniz's dynamics proposed general principles based primarily on analyses of two major phenomena, free fall and elastic impact. The relation of the analyses to each other could scarcely have been more anomalous. As far as his central concept, the conservation of *vis viva*, was concerned, elastic impact offered the best illustration. In elastic impact, the forces of the bodies in motion (their *vires vivae*) translate themselves into elastic forces from which the original forces of motion are constituted anew. Elastic force in turn was understood as the manifestation of aethereal particles in motion. Here indeed one could talk intelligibly of the conservation of *vis viva*. Unfortunately, Leibniz could only talk about it. His discussions of elasticity were uniformly non-quantitative. At no place did he attempt the sort of analysis of a vibrating spring that Hooke undertook; it is worth remarking that the analysis would have confronted him with a functional relation of (dead) force to distance such as he avoided in dealing with free fall. His derivation of the conservation of *vis viva* in elastic impact from the concept of moving action was a sham. The conclusion rested in fact where it had in Huygens' work, on the potential ascent of the common centre of gravity.

In contrast to elastic impact, free fall provided the terms of a quantitative dynamics inextricably set in a context that displayed the conservation of force ambiguously at best. Although his argument against Descartes rested on the impossibility of an effect exceeding its cause, he never indicated what it is in the elevation of a heavy body that is analogous to the stress of an elastic body. The central quantitative factor in his dynamics, mv^2, emerged, not from a generalised analysis, which would have involved a more thorough exploitation of the concept of dead force, but from a direct transposition of Galileo's kinematics. The generalised analysis of elastic impact in turn yielded only a qualitative and verbal dynamics.

*

If the fortunate accident of Galileo's kinematics furnished Leibniz with a quantitative measure of force, no similar piece of luck aided his treatment of circular motion. Indeed the particular formulation of his dynamics virtually foreclosed any fruitful insight into this critical problem of seventeenth-century dynamics. To pose the issue in our terminology, uniform circular motion involves the continuous action of force without a consequent increase in kinetic energy. The concept of work is essential to the resolution of the apparent enigma. In uniform circular motion, no work is performed; there is no integration of force over distance. As we have seen, in Leibniz's dynamics dead force integrates through time alone. If, as we believe, circular motion is uniformly accelerated, then by his standards it must involve an increase in *vis viva*, a consequence he knew to be false. His distinction of violent and formal effects repeated the classic distinction of vertical and horizontal motion in a universe organised in spheres around centres, exactly as Galileo had stated it at the beginning of the century. Hence circular motion, that motion in which the constancy of velocity indicates the constancy of *vis viva*, presented itself to Leibniz, not quite as inertial motion, for he understood inertial motion to be rectilinear, but as similar to inertial motion, a state of equilibrium analogous to static equilibrium.

In this context, as passages quoted above reveal, Leibniz frequently cited centrifugal force as an example of dead force. Imagine a ball in a tube rotating in a horizontal plane about one end. As the tube rotates,

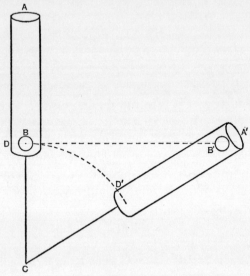

Fig. 27

the ball experiences a centrifugal force; and it begins to move radially with a *conatus* that is infinitely small at the beginning in comparison to the impetus of rotation by which the tube moves from D to D' (see Fig. 27). 'But if the centrifugal impulsion proceeding from the rotation is continued for some time, there must arise in the ball, from its own progression, a certain complete centrifugal impetus $D'B'$ comparable to the impetus of rotation DD'.'[32] If the example was dubious, it was not for that less revealing of the continuing sway of the Cartesian tradition. Like Descartes, in this case Leibniz treated a tangential motion as the resultant of a circular motion (in which the ball does not in fact participate) and a centrifugal tendency. In order for circular motion to take place, of course, the centrifugal tendency must be checked.

> Since every moving body which describes a curved path tends to recede along the tangent, we may call this an outward conatus, as in the motion of a sling for which an equal force is required which constrains the moving body, lest it wander away.[33]

Leibniz liked to trace the analysis of circular or vortical motion beyond Descartes to Kepler. Be that as it may, vortices played a role in his natural philosophy quite as central as their role in Descartes'. He was prepared to speak of 'the law of nature . . . that rotating bodies tend to recede from their centers along the tangent . . . ,'[34] and with Descartes he traced the cause of gravity to the centrifugal force of an invisible aether. He went beyond Descartes in deriving solidity from the same cause. Every body is a microvortex of aethereal particles striving to fly off along the tangent. Their centrifugal pressure creates a counter-pressure in the matter surrounding them, which crowds them back.[35] Hence the elasticity of bodies, an essential factor in Leibniz's universe, depends on the centrifugal tendency of their parts, and physical reality consists of vortices within vortices within vortices in infinite series. As a mechanist, Leibniz believed that it is 'unworthy of the admirable workmanship of God, to assign special intelligences to the stars [i.e., planets] directing their courses, as if God lacked the means of doing this by the laws of bodies . . .' Since, in addition, the crystalline spheres have been rejected, and 'sympathies and magnetic and other abstruse qualities of the same kind are either unintelligible, or where intelligible are thought to be the visible effects of a corporeal impression; I think that there is nothing else left except to say that the cause of the motions of celestial bodies arises from the motions of the aether, or, speaking astronomically, from the deferent orbs, which are fluids.'[36] Throughout his *Tentamen de motuum coelestium causis* (*Essay on the Causes of the Motions of the Heavenly Bodies*), he insisted that the attraction of gravity, as he allowed himself to call it, which influences the motions of the planets, must be caused by the aether. Something must constrain the planets' *conatus* to recede, but there is nothing contiguous except the aether, and no *conatus* is constrained except by something contiguous.[37] It is difficult to ignore the connection in Leibniz's writings between the priority ascribed to centrifugal force and a mechanistic natural philosophy. When every centripetal impulse in nature was derived from the centrifugal tendency of a subtle matter, it was impossible to think of circular motion as a rectilinear motion continually diverted by a centripetal force. Since his dynamics was not able to accommodate a view of circular motion as uniformly accelerated, it was just as well for the internal consistency of his thought that he viewed such motion as he did.

In the *Tentamen*, composed shortly after the publication of Newton's

Principia, Leibniz applied his dynamics, which he had hitherto confined to terrestrial phenomena, to the resolution of the celestial motions. As I have mentioned, he considered the existence of a circulating aether essential and beyond question. Drawing on Kepler's corrected law of velocity, by which the component of a planet's velocity that is perpendicular to the radius varies inversely as the radius, Leibniz posited an aether moving in what he called harmonic circulation. Obviously, in harmonic circulation, the linear velocities of aethereal shells are inversely proportional to their radii. Such an aether is incompatible with the solar system as it is constituted, and Leibniz recognised as much, although he pretended that it involved no serious difficulties for his system. In fact, his *Tentamen* confined itself to the consideration of the elliptical orbit of a single planet. To extend his solution to the entire solar system, one would have to postulate a distinct zone for each planet within which an harmonic relation holds, though it cannot hold for the solar vortex considered as a whole. Meanwhile, within its harmonically circulating medium, a planet follows two motions, on the one hand the circulation of the aether, on the other a radial (or 'paracentric,' as Leibniz called it) motion whereby it approaches and recedes from the sun. The planet's velocity of circulation varies as a result of its paracentric motion, of course, since the planet obeys the motion of the aether in which it finds itself. Leibniz easily demonstrated that the law of areas must follow as a consequence – scarcely an exciting result sixty years after Kepler's *Epitome*. To Huygens he extolled the 'excellent privilege' of harmonic circulation, setting it firmly near the heart of the best of all possible worlds:

> that it alone is able to conserve itself in a medium that circulates in the same way and to harmonize the motion of the solid and the ambiant fluid permanently. And this is the physical rationale of this circulation that I promised to give sometime, bodies having been determined by it to accord better with each other. For harmonic circulation alone has the property that a body circulating in that manner keeps precisely the force of its direction or preceding impression exactly as if it were moved in the void by its impetuosity alone joined with its weight. And the same body is also moved in the aether as though it swam in it tranquilly without any impetuosity of its own or any residue from preceding impressions and did nothing but absolutely obey the aether around it as far as circulation is concerned (paracentric motion excepted) . . .[38]

Leibniz deluded himself in contending that his circulating medium carried a planet exactly as it would move in a vacuum under the influence of gravity – a manifest allusion to Newton's *Principia*. Failing to comprehend the difference in their views of circular motion, he separated the action of gravity from the harmonic circulation of the aether and ascribed to the latter the primary function that Newton assigned to gravity. In Leibniz's system, gravity has nothing to do with diverting a planet from its rectilinear path; gravity operates solely to vary its distance from the sun. He assigned the function of constraining the planet to follow a closed orbit to the circulating aether and omitted that action from the analysis. Hence his celestial mechanics emerges as a revision of Kepler's, as Leibniz himself suggested, or perhaps more recently of Borelli's. He brought a more sophisticated mathematics to the task, but the conceptual framework remained unaltered.

The central problem of the *Tentamen*, to which Leibniz primarily applied his analytic gifts, was the paracentric motion. Whereas Borelli had traced the radial motion to a fluctuating centrifugal tendency that is alternately greater than and less than the constant tendency of the planet toward the sun, Leibniz, faced with Newton's demonstration that gravity obeys an inverse square law, found a more complicated relation between two factors that vary. You may be sure that the ultimate equation for the centripetal acceleration at which he arrived contained a factor r^2 in the denominator. Under conditions of harmonic circulation, centrifugal force varies inversely as the cube of the radius. One of the keys to Leibniz's analysis lies in his insistence that centrifugal *conatus* is relative solely to the component of circulation and not to the curvature of the orbital path. Since velocity is inversely as distance in harmonic circulation, Huygens' discovery that centrifugal acceleration is proportional to v^2/r led directly to the inverse cube relation. Gravitational attraction decreases as the square of the distance, centrifugal endeavour as the cube – hence the possibility of the necessary alternating motion with gravity prevailing at the greater distances and drawing the planet toward the sun whereupon the centrifugal tendency prevails and drives it away.

The specific form in which the central equation of Leibniz's analysis emerged is nevertheless perplexing. In his words, it states that 'the element of paracentric impetus (that is, the increase or decrease of the velocity of descent toward the center, or of ascent from the center) is the difference between ... the paracentric solicitation (that is, the impression made by gravity ... or a similar cause) and twice the centrifugal conatus (which arises from the harmonic circulation itself.)'[39] The consequence follows, as Leibniz explicitly recognised, that in the simplest figure, the circle, centripetal force is twice centrifugal force. We want to say that uniform circular motion is therefore uniformly accelerated. Leibniz saw it quite otherwise. The idea of a paracentric impetus or velocity (and consequently a *vis viva*) generated by the actions of a dead force continued over a period of time was acceptable to his dynamics. The idea of an accelerated motion without the generation of *vis viva* was unacceptable. He simply ignored the apparent import of his own equation and preserved the equilibrium of circular motion by equating the effective centrifugal force with twice the *conatus*. Thus, in the case of an elliptical orbit, he said that after perihelion the velocity of recession increases 'until that new impression to recede, or double the centrifugal conatus, again becomes equal to gravity or the impression to approach...'[40] As it turned out, moreover, this particular formulation also provided a neat distinction of the conics. When the centrifugal tendency is half the attraction of gravity, a circular orbit is traced. When the centrifugal force at perihelion is greater than half but less than the full force of gravity, an ellipse results. When it is equal to gravity at perihelion, the curve is a parabola, and when it is greater, an hyperbola.[41] Uniform circular motion, then, embodies an enduring equilibrium of opposite forces; a varying disequilibrium, based on the different rates at which the two decrease with distance, produces the other conic curves.

Consider the elliptical path, the one of crucial importance in celestial dynamics. Leibniz demonstrated that the element (or differential) of the angle between the major axis and a planet's radius vector is proportional to $1/r^2$, the ratio by which the paracentric acceleration caused by gravity also varies. Since the sums of proportionals are proportional, the angle of circulation, as he called it, is proportional to the 'impetus which the planet conceives from the continuous attraction of the sun while it is moving ...' Let us start from aphelion (see Fig. 28). At that point, A, the attraction of gravity is greater than twice the centrifugal tendency. 'Hence the planet descends toward the sun along the path $AMEW\Omega$, and the impetus to descend increases continuously as in the acceleration of heavy bodies falling [*in gravibus acceleratis – sic*!], as long as the new solicitation of gravity remains stronger than twice the new centrifugal *conatus*; for until then the impression to approach is

LEIBNIZIAN DYNAMICS

greater than the impression to recede and therefore the velocity of approaching increases absolutely until the planet reaches the point when the two new contrary impressions equal each other . . .' This point is at *W* on the line *XW* through the lower focus perpendicular to the major axis. From *W* to Ω, the planet continues to approach the sun, but its paracentric velocity now decreases, 'until the centrifugal

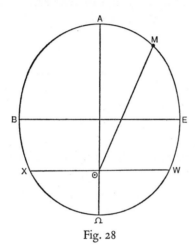

Fig. 28

impressions from the initial point *A* to the point in question, collected into one sum, exactly consume the impressions of gravity, also summed from the beginning to that point, or when the total impetus to recede (conceived from the individual centrifugal impressions collected together) finally equals the total impetus to approach (conceived from the continuously repeated impressions of gravity), at which point the approach comes to an end . . .' This point, of course, is perihelion.

During the second half of the orbit, exactly the reverse process occurs. The velocity of receding increases as long as twice the centrifugal *conatus* is greater than the attraction of gravity. Recession itself continues as long as 'the total impetus to recede, or the sum of all the impressions to recede acquired from Ω until the point in question, is greater than the total impetus to approach impressed anew from Ω to that point.' During the second part of the recession, the attraction of gravity exceeds twice the centrifugal *conatus*, and the sum of the impressions increases until it equals the sum of centrifugal impressions at *A*, 'where they are mutually destroyed' and the recession stops. 'And thus when all the earlier impressions have been consumed by the compensation of equal contraries, the situation returns to its original state, and the entire process perpetually repeats itself anew.'[42] In the ellipse, the equality of the integrals of centripetal and centrifugal impressions through each half orbit, reproduce the equilibrium of the two in circular motion.

Nothing could illustrate the basic animation of Leibniz's dynamics more clearly. Although he presented both of the forces deemed to be present as functions of distance and followed their interplay along the orbit, he was still able to think of them only in terms of the equation,

$$v = \int a dt,$$

that governed the paracentric velocity. Such an approach was not capable of shedding new light on the dynamics of circular motion.

Leibniz's impotence to deal effectively with curvilinear motion indicates the full extent of the riddle it posed to any dynamics that looked to impact for its model of force. In some sense, the force of a body's motion always expressed the sum of the forces that had generated the motion. In Leibniz's terms. *vis viva*, the integral of the dead forces that have acted, embodies the past states the body has known. The nub of the issue with circular motion, in an inertial mechanics, lies in the circumstance that the continuous action of force in diverting a body from its rectilinear course adds nothing to the force of its motion. Half a century before Leibniz, Torricelli had shown that the model of impact, even when expressed by the concept of impetus, could yield a satisfactory quantitative treatment of uniformly accelerated motion in a straight line. Had Torricelli attempted to deal with circular motion, his dynamics of *momento* would surely have foundered on the same reefs that brought Leibniz's *vis viva* to grief. By itself, the dynamics of *vis viva* was incapable of solving all the problems that confronted rational mechanics at the end of the seventeenth century. Their solution required that the concept of force drawn from the model of impact be supplemented by another which saw force as an external action altering a body's inertial state.

*

Behind the individual problems in Leibniz's dynamics ever looms the imposing presence of his metaphysics. This is scarcely the place to enter upon a detailed examination of the sophisticated and perplexed issues of Leibniz's philosophy, nor would I propose myself as a commentator uniquely qualified to do so. Nevertheless the topic cannot be entirely avoided. The science of mechanics in the seventeenth century stood in intimate relation to the prevailing mechanical philosophy, at once peculiarly concerned with the basic causal processes that the mechanical philosophy admitted in nature, and peculiarly constrained in its development by the categories of being that the mechanical philosophy imposed. In Leibniz, the relation entered upon a new stage. On the one hand, like every other creative scientist of the late seventeenth century, he experienced the liberating influence of the mechanical philosophy, an influence in his case which was both profound and permanent. On the other hand, Leibnizian dynamics set its foundation squarely on the repudiation of the mechanical philosophy's ultimate categories of physical thought.

On the surface, on the phenomenal level, all appears to be the same; but under the surface, at the level of metaphysics, everything has changed. The universe still presents itself overtly as an infinite machine, but behind the appearance the principle of universal harmony introduces an articulated, organic whole. All motion intends a purpose; the vectorial nature of velocity on the phenomenal level expresses its pursuit of a goal on the metaphysical plane. The motion of any point whatever in the world, he stated in a passage that neatly captures the dual point of view, 'takes place in a line strictly determined in nature, which this point has taken once for all and which nothing can ever make it abandon.'[43] In a similar way, his statements of the principle of inertia contain supplementary clauses which remind us that under metaphysical scrutiny the appearance of inertia dissolves into self-activity.

> Thus a thing not only remains in the state in which it is, insofar as it depends on itself, but also continues to change when it is in a state of change, always following one and the same law. But in my opinion it is the nature of created substance to change continually following a certain order which leads it spontaneously (if I may be allowed to use this word) through all the states which it encounters, in such a way that he who sees all things sees all its past and future states in its present.[44]

More fundamental than the question of motion was the question of substance. As the passage cited above suggests, substance in Leibniz's view had little to do with the inert matter basic to the mechanical philosophy of nature. 'I hold that an active principle which is superior to material concepts and, so to speak, vital exists everywhere in bodies

...'[45] The very heart of his objection to the prevailing philosophy of nature had to do with its denial, as Leibniz believed, of activity and force to created beings. In no sense does the ascription of force and activity to creatures compromise the unique position of God, for they owe their powers to God the creator. Quite the contrary, the ontologically impotent matter of the mechanical philosophy, a pretended substance unable to initiate any activity of its own, is an affront to the dignity of the creator. To be and to act are synonymous. If God creates anything, his creatures must be substances able to act, substances worthy of his omnipotence. 'Otherwise I find . . . that God would produce nothing and that there would be no substances beyond his own – a view which would lead us back into all the absurdities of Spinoza's God. It also seems to me that Spinoza's error comes entirely from his having pushed too far the consequences of the doctrine which denies force and action to creatures.'[46] The *vis viva* with which rational mechanics concerns itself, becomes then something utterly different from the mere quantity that Huygens calculated. A derivative of the primitive force which is the ultimate reality of existence, *vis viva* in its conservation expresses on the phenomenal level the eternal conservation of the force and vitality of individual substances. To the mechanical philosophy of nature, matter is inert and passive. In Leibniz's philosophy, matter is composed of foci of activity. '*Substance* is a being capable of action,' he asserted. '. . . As a result, the whole of nature is full of life.'[47]

As is so frequently the case with Leibniz, elastic impact furnishes the classic expression of the continuity of his dynamics with his metaphysics. In the collision of bodies, he stated, 'each suffers only from its own elasticity, caused by the motion which is already within it.'[48] To understand the statement, which initially appears deliberately idiosyncratic, we must start where all of Leibniz's mechanics started, with Huygens. The ultimate insight on which Huygens' treatment of impact rested stemmed directly from the relativity of motion. From the point of view of the common centre of gravity, a point of view we can always assume because of the relativity of motion, two perfectly hard bodies rebound with their original motions intact. Leibniz appropriated the conclusion but transformed its significance. Huygens' perfectly hard bodies became his perfectly elastic ones, and the kinematic solution converted itself into a dynamic one. The conservation of the sum of the products mv^2 in all frames of reference became, of course, the conservation of *vis viva*. The insight that each body rebounds with its initial speed when the impact is viewed from the common centre of gravity became the prime mechanical illustration of the basic proposition that each body conserves its own force always. It follows from the phenomena of impact, Leibniz stated, that

> *every passion of a body is spontaneous or arises from an internal force, though upon an external occasion.* But I mean by this the passion proper to it, which arises from percussion, or which remains the same whatever hypothesis may be chosen [*i.e.*, whatever inertial frame of reference may be assumed] or to whatever body we may ascribe rest or motion. For since the percussion is the same regardless of what body the true motion may belong to, it follows that the effect of percussion will be equally distributed between both, and thus that *both act equally in the collision*, so that half of the effect comes from the action of one, the other half from the action of the other. And since half of the effect or passion is also in one and half in the other, it suffices to derive the passion which is in one from the action which is in it, so that we need no influence of one upon the other, even though the action of one provides an occasion for the other to produce a change within itself. Certainly when A and B collide, the resistance of the bodies combined with elasticity causes them to be compressed through the percussion, and the compression is equal in both, whatever may be the hypothesis about their original motion. Experiments show this, too, if we let two inflated balls collide, whether both are in motion or one is at rest, and even if the one at rest is suspended from a string so that it can swing back with ease, for if the velocity of approach or relative velocity is always the same, the compression or elastic tension will be the same and will be equal in both. Then the balls A and B will restore themselves by the force of the active elasticity compressed within them, repel each other, and burst apart as if driven by a bow, each being driven back from the other with equal force, and thus receding, not by the force of the other, but by its own force. But what is true of inflated balls must be understood of every body insofar as it suffers in percussion. Repercussion and repulsion, namely, arise from elasticity within the body itself, or from the motion of an ethereal fluid matter which permeates it, and so from an internal force existing within it.[49]

Thus the derivative force of a perfectly elastic body repeats on the phenomenal plane the autonomous unrolling of the primitive force of the individual monads which constitute ontological reality.

Leibniz expressed the relation of dynamics and metaphysics through the concepts of derived and primitive forces. Primitive force refers to the ultimate active principle of a monad. It endures unchanged forever. Derivative force, on the other hand, is the momentary manifestation of the primitive force as it is constrained and limited by the presence of other monads with which it must fit harmoniously into the organised whole that is the universe. The essence of the monadology lies in the assertion that the limitation and constraint of primitive force, like an elastic ball in impact, does not follow in any way from the action of one monad on another. Rather it is the result of the internal law of the monad unfolding itself in complete autonomy according to a determinate sequence, as the very concept of a substance demands.

> *The complete or perfect concept of an individual substance involves all its predicates, past, present, and future.* For certainly it is already true now that a future predicate will be a predicate in the future, and so it is contained in the concept of the thing.... It can be said that, speaking with metaphysical rigor, *no created substance exerts a metaphysical action or influence upon another.* For to say nothing of the fact that it cannot be explained how anything can pass over from one thing into the substance of another, it has already been shown that all the future states of each thing follow from its own concept. What we call causes are in metaphysical rigor only concomitant requisites.[50]

By the principle of pre-established harmony, the concomitant requisites, or the occasions as he called them elsewhere, themselves the predicates of equally autonomous substances, appear spontaneously at the right moment to give rise to the appearance of causes. The science of mechanics deals only with derivative forces, the instantaneous predicates of the primitive forces that are the concepts of the autonomous substances in the universe. By the principle of pre-established harmony, the sum of the derivative forces, the *vires vivae* of bodies, remains constant. The ultimate foundation of the conservation of *vis viva*, however, rests in the eternal persistence of the primitive force that is the law of each monad.

As far as the explanations of natural phenomena were concerned, Leibniz was a convinced mechanist. The Scholastics had been wrong in attempting to account for the properties of bodies by substantial forms when they needed rather to examine their manner of operation. 'This is as if one were content to say that a clock has a time-indicating property proceeding from its form, without inquiring wherein this property consists.'[51] With mechanical philosophers, he rejected any appeal to action at a distance as occult; and when Newton's *Principia* appeared, no Cartesian outbid the scorn with which Leibniz greeted the notion of gravitational attraction. For all that, Leibniz played on the mechanical philosophy the same trick it had played on qualitative natural philosophy half a century before. Where Descartes had argued that mechanism is the reality behind qualitative appearance, Leibniz now contended that mechanism as well is mere appearance, a phenomenon the ultimate reality of which has nothing to do with mechanism. Extension itself dissolved away into the multiplied contemporaneous resistance of *materia prima*. 'Extension, motion, and bodies themselves, insofar as they consist in extension and motion alone, are not substances but true phenomena, like rainbows and parahelia.'[52]

If extension, motion, and bodies themselves are only phenomena, what then is absolute and real? There is nothing real in motion 'except that momentaneous state which must consist of a force striving toward change. Whatever there is in corporeal nature besides the object of geometry, or extension, must be reduced to this force.'[53] It scarcely needs to be repeated that 'force' in Leibniz's philosophy bore a meaning different from that we give it in mechanics today. Rather than the cause and measure of changes of motion, it was more akin to our concept of energy, a relation still more clear in Leibniz's references to the monad as '*a real and animated point, . . .* an atom of substance which must include a certain active form to make a complete being,' and similar phrases.[54] The full appreciation of his dynamics had to wait for the middle of the nineteenth century, when the assimilation of the Newtonian concept of force allowed its relation to Leibnizian force to emerge in full clarity, and Leibnizian dynamics could contribute its due measure to the principle of the conservation of energy. Meanwhile, in the seventeenth century, it demonstrated immediately that the elevation of dynamics to the status of a precise and quantitative science entailed profound modifications of the prevailing mechanical philosophy.

Despite his professed mechanistic convictions, in Leibniz's philosophy the Pythagorean tradition dominated the Democritean completely. He devoted almost no attention to the favourite pastime of mechanical philosophers, the imagination of micromechanisms. Although he agreed that such must exist, he merely acknowledged their necessity while he treated the phenomena they allegedly cause in mathematical terms. His use of Galileo's kinematics of free fall to refute Descartes' measure of the causal particle's force symbolised the relation of the two traditions in Leibniz's thought. Where so much of seventeenth-century dynamics bogged down in the swamp of causal mechanisms into which it rushed headlong, Leibniz announced the pre-established harmony of efficient and final causality and neatly skirted the entire issue. And the idea of the monad, defined in terms of the mathematical model of infinite series, encouraged the exploration of precise quantitative relations, by tracing them back to the source to which the Pythagorean tradition was always wont to turn, the geometrising God.

*

For the science of dynamics, Leibniz's redefinition of the nature of substance had a further consequence. His conception of substance projected a corresponding conception of matter. If the monadology has not dominated modern metaphysics, Leibniz's conception of matter has survived in the idea of mass employed by the science of mechanics.

The two dominant elements in the prevailing conception of matter were extension and indifference to motion. In Leibniz's opinion, the two ideas demanded each other. If matter were equivalent to extension, it would lack entirely the property of substantial activity that he found essential to it and contrary to the idea of indifference. Neither 'motion (or action) nor resistance (or passive force) derive from extension...'[55] As far as action was concerned, he had shown to his own satisfaction that being and the capacity to act are one and the same. For the later conception of mass, his idea of passive force was to be more important. Passive force, or resistance, denied one of the central tenets of seventeenth-century mechanics, that bodies are indifferent to their states of motion or rest.

Imagine that a body in motion overtakes another moving more slowly in the same direction and strikes it. If bodies are indifferent to motion, the swifter ought to carry the slower along with it, suffering no diminution whatever in its own motion. Imagine now that the second body is at rest. Again the moving body ought to set the body it strikes in motion without suffering any loss of motion itself, and the two should move together with its initial speed. Moreover, the relative sizes of the two bodies should not affect the outcome of the impact in any way if bodies are wholly inert and indifferent to motion. Hence 'the largest body at rest will be carried away by a colliding body, no matter how small, without any retardation of its motion, since such a notion of matter involves no resistance to motion but rather indifference to it. Thus it would be no more difficult to move a large body than a small one, and hence there would be action without reaction, and no estimation of power would be possible, since anything could be accomplished by anything.'[56] In short, if matter is indifferent to motion, a quantitative science of dynamics is impossible.

In fact, Leibniz had explored the concept of indifference to its limit before he rejected it. In this youthful *Theory of Abstract Motion* (a component part of the *Hypothesis physica nova*), he had attempted to derive the laws of impact from such assumptions. A concept of *conatus* provided the basis of the essay, and it defined *conatus* solely in terms of instantaneous velocity without regard to the size of the body in motion. A *conatus* can be opposed only by another *conatus*. A body at rest or a body moving in the same direction offers no resistance because of matter's indifference to motion. Hence the *Theory of Abstract Motion* proposed solutions to the problems of impact in terms of a set of rules for subtracting and compounding *conatus*.[57] Leibniz's early notion of *conatus* was reminiscent in many ways of Hobbes' similar concept, although the German philosopher pursued the logical consequences of indifference more relentlessly. The *Theory* concluded in a peculiarly dynamical kinematics, in which the dynamic connotations of *conatus*, foreshadowing the later concept of force, coloured the purely kinematic rules at which the treatise arrived.

Leibniz soon realised, moreover, that the conclusions were absurd. They clashed egregiously with all experience. Manifestly, more effort is required to set a large body in motion than a small one. Moreover, the perception that a resistance of some sort is associated with matter meshed harmoniously with all the tendencies that thrust Leibniz's thought toward a dynamic concept of substance. In his mature philosophy, force reveals itself under two guises, one active and the other passive. Active force we have seen. As primitive force, it is the entelechy or law of the individual monad; as derivative, it manifests itself in the dead force, impetus, and living force with which the science of mechanics concerns itself. Leibniz called primitive passive force *materia prima*. Extension itself is an appearance derived from the spatial replication of *materia prima*. Impenetrability as well, which is not, Leibniz maintained, a consequence of extension itself, is a derivative manifestation of passive force. Above all, derivative passive force constitutes the resistance of matter to the imposition of motion, the effort of matter to remain in its current state. In a word, derivative passive force is mass.

Matter, Leibniz said, 'resists motion by virtue of a natural inertia, as Kepler suitably called it, with the result that it is not, as is commonly believed, indifferent to motion or rest, but requires more force to be moved in proportion to its quantity.'[58] The phrase 'natural inertia' is revealing. Although Leibniz's concept is not identical to Kepler's, it is closer to it than to the principle of inertia as we use it today. Significantly, he never used the word 'inertia' to refer to the uniform motion of a body. He did not see such motion as a consequence of the inertness of matter; quite the contrary, it is a product of the activity essential to substance, a formal effect that does not consume its cause, but in no sense an effect that is uncaused.

Equally, natural inertia, in Leibniz's use of the phrase, referred to the essential activity of matter. It fits comfortably into the central pattern of his dynamics, the mutual interplay, on the level of phenomena of course, of active and passive forces. Since there is a natural inertia in bodies, 'everything that is acted upon must act reciprocally, and everything that acts must suffer some reaction, and consequently a body at rest cannot be set in motion by another without changing something of the direction and speed of the agent.'[59] In establishing the equality of action and reaction, it supports the conservation of force in the universe. Active force, which alone can generate motion, insures that the quantity of force will not diminish. Passive force, which can only resist, insures that the quantity of force will not increase. The principle of continuity also required the natural inertia of matter. The resistance of bodies to changes in their states of motion or rest imposes a gradualness on the changes, in contrast to the instantaneous alterations that Descartes and Huygens envisioned. Especially, since there is a resistance associated with matter, effects are proportional to causes, and a quantitative science of dynamics is possible.

The last point is of some importance. Together with the problem of circular motion, the role of mass in dynamic action was one of the critical problems the solution of which was basic to the satisfactory establishment of the science of dynamics. In the late seventeenth century, dynamics was discovering that it could not progress without a conception of matter that included resistance. Mariotte came to exactly the same conclusion as Leibniz and for the same reasons, if we confine ourselves to considerations internal to rational mechanics. As he composed the *Principia*, Newton found himself constrained to assert a similar resistance. No one analysed the problem with greater clarity than Leibniz, and no one contributed more to the understanding of it. His legacy to dynamics in this respect was not without its dimension of irony, however. In Leibniz's case, the passive force of matter was one aspect of a philosophy that defined substance in terms of activity. In contrast, the science of mechanics since the seventeenth century has based itself on a principle of inertia that implies the utter inactivity of matter. It has concealed the resistance of matter in the concept of mass, where it survives today as one of the strangest anomalies in the entire structure of modern science.

NOTES

1. 'Brief Demonstration'; Gottfried Wilhelm Leibniz, *Philosophical Papers and Letters*, tr. and ed. Leroy E. Loemker, 2 vols. (Chicago, 1956), **1**, 456–7, reprinted by permission of D. Reidel Publishing Company.
2. '*Specimen Dynamicum*'; ibid., **2**, 726.
3. '*Remarques sur Descartes*'; Gottfried Wilhelm Leibniz, *Opuscules philosophiques choisis*, tr. Paul Schrecker, (Paris, 1954), p. 45, cited by permission of Librairie A. Hatier. The English translation is mine.
4. 'Brief Demonstration'; *Papers and Letters*, **1**, 455. Cf. p. 457 in the same work. Also the 'Essay on Dynamics'; Gottfried Wilhelm Leibniz, *New Essays Concerning Human Understanding*, tr. Alfred G. Langley, (New York, 1896), p. 659. In a letter to Arnauld, he acknowledged the passage in Descartes' essay on the simple machines in which he had insisted on the primary importance of vertical displacement; had Descartes only remembered it when he wrote his *Principles*, he might have avoided the errors about the laws of nature into which he fell. 'But he happened to eliminate the consideration of velocity exactly where he ought to have kept it, and to have retained it in the cases where it led to errors. For in regard to the forces that I call dead (as when a body makes its first effort to descend without having yet acquired any impetuosity by the continuation of the motion) and when two bodies are in equilibrium (for then the first efforts that they make against each other are always dead), it happens that the velocities are as the distances; but when the absolute force of bodies that have some impetuosity is considered (which must be done to establish the laws of motion), the measurement must be made by the cause or the effect, that is, by the height to which it can rise in virtue of this speed, or by the height from which it must fall to acquire this speed.' Leibniz to Arnauld, 28 November/8 December 1686; Gottfried Wilhelm Leibniz, *Philosophischen Schriften*, ed. C. I. Gerhardt, 7 vols. (Berlin, 1875–90), **2**, 80.
5. 'Essay on Dynamics'; *New Essays*, pp. 659–60. Although I have chosen to cite the 'Essay on Dynamics' from the one published English translation, I have taken the

liberty to modify the translation wherever I have seen fit. Cf. a similar passage in 'Supplement to the Brief Demonstration'; *Papers and Letters*, **1**, 460.
6. 'Essay on Dynamics'; *New Essays*, pp. 657–8.
7. 'On the Elements of Natural Science'; *Papers and Letters*, **1**, 430.
8. 'Essay on Dynamics'; *New Essays*, p. 661.
9. 'Discourse on Metaphysics'; *Papers and Letters*, **1**, 482.
10. '*Remarques sur Descartes*'; *Opuscules*, pp. 45–6.
11. 'On a General Principle Useful in Explaining the Laws of Nature'; *Papers and Letters*, **1**, 539. This particular passage, though concerned with the laws of impact, was addressed directly to Malebranche instead of Descartes. In his '*Remarques sur Descartes,*' Leibniz noted that the principle of continuity implies that motion, as it continually decreases, fades at last into rest, and that rest is therefore a special case of motion (*Opuscules*, p. 52). Thus the principle of continuity led to one of the basic corollaries of the concept of inertia.
12. '*Remarques sur Descartes*'; ibid., pp. 53–60.
13. '*Remarques sur Descartes*'; ibid., pp. 49–50.
14. 'Essay on Dynamics'; *New Essays*, pp. 666–8.
15. 'Essay on Dynamics'; ibid., pp. 661–6.
16. Cf. a similar judgment in Martial Guéroult, *Dynamique et metaphysique leibniziennes*, (Paris, 1934), pp. 131–2.
17. 'Essay on Dynamics'; *New Essays*, pp. 668–9. Cf. Leibniz to Huygens, 10/20 March 1693; Gottfried Wilhelm Leibniz, *Der Briefwechsel mit Mathimatikern*, ed. C. I. Gerhardt, (Berlin, 1899), pp. 713–14.
18. '*Specimen Dynamicum*'; *Papers and Letters*, **2**, 730–1.
19. 'Essay on Dynamics'; *New Essays*, p. 669.
20. 'Essay on Dynamics'; ibid., p. 670.
21. 'Essay on Dynamics'; ibid., p. 670.
22. 'Supplement to the Brief Demonstration'; *Papers and Letters*, **1**, 462.
23. 'Deux problemes'; Gottfried Wilhelm Leibniz, *Mathematische Schriften*, ed. C. I. Gerhardt, 7 vols. (Berlin, 1849–63), **6**, 234.
24. '*Essay de dynamique*'; another version of the 'Essay' published originally in the nineteenth century and recently republished by Pierre Costabel, *Leibniz et la dynamique*, (Paris, Hermann, 1960), p. 104.
25. Leibniz to de Volder, c. January 1699; *Philosophischen Schriften*, **2**, 154–6.
26. This interpretation is proposed by Guéroult, *Dynamique et metaphysique*, pp. 40–3. Cf. Costabel, *Leibniz*, pp. 92–3.
27. Leibniz to l'Hôpital, 4/14 December 1696; *Mathematische Schriften*, **2**, 320.
28. '*Specimen Dynamicum*'; *Papers and Letters*, **2**, 717.
29. 'Essay on Dynamics'; *New Essays*, p. 660.
30. Bernoulli to Leibniz, 8/18 June 1695; *Mathematische Schriften*, **3**, 189. Leibniz to Bernoulli, 24 June 1695; ibid., **3**, 193. Guéroult suggests that Leibniz developed the concept of motive action in order to have a derivation of mv^2 that employed the dimension of time instead of vertical displacement (*Dynamique et metaphysique*, p. 119).
31. 'Essay on Dynamics'; *New Essays*, p. 659.
32. '*Specimen Dynamicum*'; *Papers and Letters*, **2**, 716.

33. 'An Essay on the Causes of the Motions of the Heavenly Bodies,' tr. Edward J. Collins; in I. Bernard Cohen, *Readings in the Physical Sciences*, (Cambridge, Mass., 1966), p. 172. The entire 'Essay' is not translated in the Cohen volume. I cite from it in all cases in which it includes the passages I need.
34. 'Essay on the Heavenly Bodies'; ibid., pp. 166.
35. '*Specimen Dynamicum*'; *Papers and Letters*, **2**, 735.
36. 'Essay on the Heavenly Bodies'; Cohen, *Readings*, p. 167.
37. 'Essay on the Heavenly Bodies'; ibid., pp. 168, 177.
38. Leibniz to Huygens, late 1690; *Briefwechsel*, p. 608.
39. 'Essay on the Heavenly Bodies'; Cohen, *Readings*, p. 174.
40. '*Tentamen*'; *Mathematische Schriften*, **6**, 159.
41. '*Tentamen*'; ibid., **6**, 161.
42. '*Tentamen*'; ibid., **6**, 158–60.
43. 'Reply to the Thoughts on the System of Pre-established Harmony Contained in the Second Edition of Mr. Bayle's Critical Dictionary'; *Papers and Letters*, **2**, 938.
44. 'Clarification of the Difficulties which Mr. Bayle Has Found in the New System'; ibid., **2**, 800.
45. '*Specimen Dynamicum*'; ibid., **2**, 722.
46. 'Reply to Thoughts in Mr. Bayle's Critical Dictionary'; ibid., **2**, 949.
47. 'The Principles of Nature and of Grace'; ibid., **2**, 1033–4.
48. 'New System of the Nature and the Communication of Substances'; ibid., **2**, 750.
49. '*Specimen Dynamicum*'; ibid., **2**, 733–4. Cf. a passage in 'First Truths', which dates from the early 1680's and preceded the initial assault on Descartes. To support the assertion that nothing can exert a metaphysical influence on another, he cited elastic impact in which the bodies recede from each other 'by force of their own elasticity and not by any alien force . . .' The presence of another body is necessary as a concomitant requisite to set the elasticity to work (ibid., **1**, 415).
50. 'First Truths'; ibid., **1**, 414–15. Cf. a passage from the 'Discourse on Metaphysics' – one from the almost unlimited number of similar passages that Leibniz's works afford. Anyone who understands perfectly the concept of a subject will know all of the predicates contained in it. 'This being premised, we can say it is the nature of an individual substance or complete being to have a concept so complete that it is sufficient to make us understand and deduce from it all the predicates of the subject to which the concept is attributed.' Knowing the individual concept of Alexander the Great, God would know everything that could be affirmed of him – for example, that he would conquer Darius. 'Thus, when we well consider the connection of things, it can be said that there are at all times in the soul of Alexander traces of all that has happened to him and marks of all that will happen to him and even traces of all that happens in the universe, though it belongs only to God to know them all.' (Ibid., **1**, 472.)
51. 'Discourse on Metaphysics'; ibid., **1**, 473.
52. 'First Truths'; ibid., **1**, 417.
53. '*Specimen Dynamicum*'; ibid., **2**, 712.
54. From a revision of the 'New System'; ibid., **2**, 1185.
55. '*Remarques sur Descartes*'; *Opuscules*, p. 33.
56. '*Specimen Dynamicum*'; *Papers and Letters*, **2**, 720.

57. 'Abstract Theory of Motion'; ibid., **1**, 221–2. Cf. Leibniz's later discussion of his early theory in '*Specimen Dynamicum*'; ibid., **2**, 719–20.
58. '*De la nature en elle-meme*'; Oeuvres de Leibniz, ed. A. Jacques, new ed. 2 vols. (Paris, 1842), **1**, 462.
59. A letter of June 1691; ibid., **1**, 449.

Chapter Seven

Newton and the Concept of Force

WITH the concept of *vis viva*, Leibniz not only raised the science of mechanics to a new plane of sophistication, but he also expressed the conviction that the ontology of nature proclaimed by the mechanical philosophy required a basic revision. Not matter in motion, but force is the fundamental category of being. Nevertheless, *vis viva* was not as far removed from the mechanical philosophy as Leibniz perhaps thought. What it raised to metaphysical status, and gave a new quantitative statement, was a direct descendant of the concept of force prevalent throughout the seventeenth century, the concept known, among other names, as the force of a body's motion. If the seventeenth century tended to think of force as a property of a body associated with its motion, the new idea of motion, embodied in the principle of inertia, implied quite a different definition of force, an external action on a body producing a change of motion. Contrary to the prevalent usage in the seventeenth century, the science of mechanics has come to reserve the word 'force' for this second concept, largely because of the achievement of Isaac Newton. Lines of distinction are seldom clearly drawn. If Newton, who became Leibniz's bitterest antagonist, adopted a different idea of force, his mechanics, like Leibniz's, built from a metaphysical platform that revised the ontology of the mechanical philosophy. As in the case of Leibniz, Newton's dynamics was interwoven with his natural philosophy to the extent that the one cannot be understood in isolation from the other.

It is fitting, then, that both Newton's philosophy of nature and his mechanics should trace their history to a common source. While he was still an undergraduate at Cambridge, Newton was introduced to the writings of the leading figures of the scientific revolution, and the notes that the reading occasioned marked the first steps in his scientific career. A packrat instinct which led him to save virtually every paper to which he touched pen and ink, including such things as a schoolboy notebook and sheets of raw calculations, makes it possible to follow the development of Newton's scientific thought to a degree of detail possible with few if any other figures of comparable stature. From every period of his life papers survive in which he speculated on the nature of physical reality, drawing on the philosophic tradition of the seventeenth century, but moulding the tradition to fit his own distinctive view. Within the framework set by his developing philosophy of nature, he attacked the conceptual difficulties internal to the science of mechanics. Unlike his philosophic speculations, which were sustained in an unbroken sequence from his undergraduate years until his death, providing as it were the warp of his intellectual life, on which the woof of specific inquiries was woven, Newton's active work in mechanics confined itself largely to two distinct periods. Both strands of thought, technical mechanics and philosophy of nature, lead back to the same source, a set of notes begun during his undergraduate years, the origin of the career in science which spanned the following sixty years and achieved results no hyperbole can exaggerate.

It was the fashion for undergraduates in Cambridge in the seventeenth century to keep notebooks devoted to various subjects. Young Isaac Newton started to fill a number of them. One contained his notes on theological reading, and another his notes on philosophy. Later, he would devote similar notebooks to mathematics and to chemistry, and he jotted down occasional reflections, and accounts of his expenses, in other notebooks, one of which, at least, dated back to schoolboy days in Grantham. In addition, he had a huge volume that his stepfather, the Reverend Barnabas Smith, had apparently once set out to fill as a 'commonplace book'. Then, as now, such projects were conceived in order that they might be abandoned, and Newton inherited an unlimited expanse of blank paper, which he renamed the *Waste Book*, and which he too barely began to fill. In it he entered significant passages on mechanics and mathematics on pages originally intended by the Reverend Mr Smith for more spiritually edifying topics beginning appropriately with 'Deus', and continuing with 'Abstinentia', 'Abnegatio nostri', and so on, a catalogue of piety exhaustive and exhausting in its proportions. For my present purposes, the philosophy notebook is of greatest interest. On the flyleaf, Newton entered his name and the date 1661, the year of his matriculation, and in one end of the book he began a set of notes labelled '*Ex Aristotelis Stagiritae Peripateticorum principis Organo.*' The notes themselves are in Greek and give the appearance of having been as much exercises in the language as lessons in philosophy. From the other end of the notebook, the undergraduate began another set of similar notes, likewise labelled in Latin '*Ex Aristelis [sic] Stagiritae Peripateticorum principis Ethice.*' Other notes follow these, including a set from G. J. Vossius' *Partitionum oratorarium libri*. The traditional curriculum of the universities of Europe, from the time of their foundation in the thirteenth century, had been built on the cornerstones of logic, ethics, and rhetoric. We know from other sources that the traditional curriculum remained largely intact at Cambridge past the middle of the seventeenth century, and Newton's notebook testifies that his undergraduate career was directed along well established lines.[1] Traditionally, the cornerstones supported the edifice of Aristotelian philosophy, and Newton's tutor introduced him to standard texts which were widely used at the time. In addition to Vossius' rhetoric, notes from the ethics of Eustachius of St Paul, the axioms of Daniel Stahl, and the physics of Johannes Magirus appear with the notes from Aristotle.[2] All of this was ordinary fare. During the seventeenth century, thousands of undergraduates, not only in Cambridge but in a score of universities across Europe, must have taken similar notes from similar books. The university curriculum in 1661 was thoroughly reactionary, devoted to a philosophic tradition that had been rejected by the vanguard of European thought. The notes that Newton dutifully recorded looked to the past instead of the future, and one will search them in vain to find a hint of their author's subsequent career.

What was unique about the notebook was the passage in its centre. Entering material from each end, Newton had left approximately a hundred pages in the middle blank, when a page of notes on Cartesian metaphysics abruptly announced a change in tone. A following page carried the title '*Questiones quaedam Philosophiae*,' and under the forty-five headings which he set at the heads of pages, he entered notes from a wholly different body of reading. A motto was later added above the general title: '*Amicus Plato amicus Aristoteles magis amica veritas.*' The motto was copied verbatim from Walter Charleton's *Physiologia Epicuro-Gassendo-Charletoniana*, a hybrid précis-translation of Gassendi's atomistic philosophy, and the contents of the opening *quaestio*, 'Off y^e

first mater,' derived from the same source. Not only had Newton read Charleton's aureate rendition of Gassendi, and perhaps Gassendi also,[3] but he had consumed Descartes, and digested him as well, though not without some consequent flatulence. He had made the acquaintance of the English authors of the age who were attempting to formulate systematic philosophies of nature – Thomas Hobbes, Kenelm Digby, and Henry More. Newton had read those works of Robert Boyle which had appeared by this time, just as he was to read Boyle's further books as they appeared during the following quarter of a century. Joseph Glanvill's *Vanity of Dogmatizing* deposited a modest freight of notes. Among foreign authors beyond Descartes and Gassendi, references from Galileo's *Dialogue* at least (in the Salusbury translation) found their way into the *Questiones*. The *Veritas* who had seduced Newton's affections from Plato and Aristotle was none other than the brazen lady, *philosophia mechanica*.

The *Questiones quaedam philosophiae* stand like a monument marking the point at which Newton made his personal revolution against the established, albeit crumbling, intellectual order, the revolution that every one of the authors who contributed to the *Questiones* had himself had to make. Nothing suggests that the educational process established in Cambridge and in Trinity College introduced him to the new body of reading. The formal curriculum contributed the notes at either end of the notebook, the legacy of the peripatetic influence which had fostered the European universities in the thirteenth century and continued to dominate them. The *Questiones* were precisely Newton's rejection of that influence. At least one source informs us that Descartes enjoyed a considerable vogue among Cambridge undergraduates in the 1660's, and the prominence of Descartes throughout the *Questiones* makes it easy to believe that Newton began with him.[4] Henry More, a prominent figure in the university at the time and an influence throughout the *Questiones*, may also have provided Newton's entrance into the new world of thought. Once he entered, there would have been no problem in following the path laid out in the books themselves toward the ever widening horizon of reading to which the *Questiones* testify. We cannot even date the *Questiones* with assurance. Changes in ink, and above all changes in hand, argue persuasively that entries were made over a period of time. A personal observation of a comet, which was not among the earliest entries, was dated in December 1664, establishing a time before which the *Questiones* were begun. Nothing establishes a prior limit with assurance. The early entries were all recorded in a peculiar hand that was transitional between the one Newton used when he matriculated in 1661 and the tiny perpendicular hand that he adopted toward the end of 1664. Perhaps early 1664 is the most reasonable conjecture. Whatever the exact date, well before the end of his undergraduate career, Newton discovered the basic writings of the mechanical philosophy of nature and of the new quantitative science which it supported. He was converted forthwith. The *Questiones* contain, not only his introduction to the new science of motion, but also the first instalment of the sustained speculations on natural philosophy which set the framework of his work in mechanics.

The first two headings of the *Questiones*, and the material entered under them, provide an interpretative key to the whole. 'Off ye first mater,' 'Of Attomes' – together they examine the ultimate nature of the material from which the physical world is composed. As to first matter, Newton offered four possibilities: 'Whither it be mathematicall points: or Mathematicall points & parts: or a simple entity before division indistinct: or individualls i.e. Attomes.'[5] In fact, the four alternatives were taken almost verbatim from Charleton's version of Gassendi,[6] and the discussion of them that followed was drawn from Charleton and from Henry More. What shall we call the two entries? To a considerable degree, they are notes from Newton's reading, but they are more than notes. To a considerable degree, they are essays, but they depend too heavily on their sources properly to qualify as essays. In their ambiguous form, they characterise the *Questiones*. I have traced the majority of the entries to the works mentioned above, and I am confident that most of the remaining entries derived from works I have not succeeded in finding. But already Newton was at work shaping the heritage he received into something that would be his own. The two 'essays' on matter reveal an earnest young man setting out to philosophise. Naturally and properly, he is dependent to a large extent on what books tell him, but he proposes not passively to accept what they tell him. He will weigh. He will choose. He will accept what seems good. And starting with those pieces, he will put together his own philosophy of nature. As the first two 'essays' also tell us, Newton confined his attention to alternative mechanical systems and confronted initially a fundamental choice between the Cartesian philosophy and atomism. Following Charleton, he rejected out of hand the first two of the four possibilities he listed, and was left to decide between the Cartesian plenum on the one hand and atoms and the void on the other. He decided for atoms, not Gassendi's atoms to be sure, but Henry More's *minima naturalia*, a strange notion of bodies at once indivisible and small as a thing can be, which attempted to combine the better properties of an atom with the advantages of a point. Later he crossed the passage out, although the view of matter that he ultimately adopted was reminiscent of it. Meanwhile he had decided for atoms and the void. Already he had set his course down a particular road from which he would never turn.

The debate between Cartesianism and atomism was one characteristic of the *Questiones*. It was not yet an acrid debate. If later in his career Newton was to assume an increasingly vitriolic tone with Descartes, such was not the case in 1664. The name *Questiones* adequately represents the content. His tone was not that of adamant answer, but rather of genuine questioning. Nevertheless, he directed some exquisitely pointed questions to Descartes. Consider the passage under the heading, 'Of ye Celestiall matter & orbes.'

> Whither Cartes his first element can turne about ye vortex & yet drive ye matter of it continually from the ☉ [sun] to produce light & spend most of its motion in filling up ye chinkes betwix ye Globuli. whither ye least globuli can continue always next y^e ☉ & yet come always from it to cause light & whither when y^e ☉ is obscured ye motion of ye first Element must cease (& so whither by his hypothesis y^e ☉ can be obscured) & whither upon ye ceasing of ye first elements motion ye Vortex must move slower. Whither some of ye first Element comeing (as he confesseth y^t hee might find out a way to turne ye Globuli about theire one axes to grate ye 3^d El into wrathes like screws or cockle shells) immediately from ye poles & other vortexes into all ye parts of o^r vortex would not impel ye Globuli so as to cause a light from the poles of those places from whence they come.[7]

Whatever the accuracy of the views he ascribed to Descartes, Newton did not pose the questions in a way to suggest that he expected a satisfactory answer. The same spirit displayed itself in other passages. Under the heading, 'Of Water & Salt,' he asked if fresh water could consist of long flexible particles and salt of rigid ones, and in this case he detailed six reasons why they could not.[8] The question of water led on to the question of tides and to Descartes' explanation of them. Newton proposed to test the theory by observing whether the pressure of the moon, which Descartes contended depresses the seas, also affects a barometer.[9] In fact, the proposed experiment derived from Christopher Wren via the writings of Boyle, but to Wren's experiment Newton added his own – the pressure of the moon ought to move the earth from the centre of its vortex with the consequence that a monthly parallax of Mars should be observable.[10] He rejected the Cartesian explanation of light. It cannot be pressure, for then we should see as well at night as in the day; the sun could not be eclipsed by the moon; a man running at night should see light; day and night should be reversed since the greatest pressure of the vortex on the earth must be on the side away from the sun. In the cases of water and of tides, and even in the case of the vortex, Newton was rejecting Descartes more than accepting rival atomistic theories. In the case of light, however, atomism tended to promote a corpuscular view – certainly Gassendi adopted one – and Newton's further proposed experiment was suggestive of the conception of light he would accept: 'Whither ye rays of light may not move a body as wind doth a mill saile.'[11]

The extent of the implicit criticism of Descartes cannot obscure the extent of the implicit agreement, however. With Descartes, and with Gassendi, Newton's *Questiones* embraced a mechanical conception of nature as the only possible system, as the only feasible basis for scientific discussion. The pointed questions addressed to Descartes, and the equally pointed if less numerous questions addressed to Gassendi[12] confined themselves to details and were only intelligible within the general context of the mechanical philosophy. Hence Newton's scientific career began from a position that embodied accepted solutions to a number of issues that were to be crucial to his mechanics and to his concept of force. Attractions and repulsions are only apparent phenomena. Invisible mechanisms must exist whereby particles in motion cause what appear to be attractions and repulsions. Take gravity – *gravitas*, heaviness – the tendency of bodies at the surface of the earth to descend – a phenomenon destined to play a major role in Newton's thought.

Of Gravity & Levity

The matter causing gravity must pass through all ye pores of a body. it must ascend againe. i for else ye bowels of ye earth must have had large cavitys & inanitys to conteine it in, 2) or else ye matter must swell it. 3 ye matter y^t hath so forcibly borne down ye earth & all other bodys to ye center (unles you will have it growne to as gross a consistance as ye Earth is, & hardly y^n) cannot if added to gether be of a bulke so little as ye Earth, for it must descend exceeding fast & swift as appeares by ye falling of bodys, & exceeding weighty pressure to ye Earth. It must ascend in another forme y^n it descendeth or else it would have a like force to beare bodys up y^t it hath to press y^m downe & so there would bee no gravity It must ascend in a grosser consistence y^n it descends 1 because it may be slower & not strike boddys wth so greate a force to impell y^m upward 2 y^t it may onely force ye outside of a body & not sinke into every pore & y^n its densness will little availe it because it will yeild from ye superfecies of a body wth ease to run in an easier channell as though it never strove against y^m. if it should ascend thinner it can have onely this advantage y^t it would not hit bodys wth so weighty a force but y^n it would hit more pts of ye body & would have more pts to hit w^th & hit wth a smarter force: & so cause ascension wth more force y^n ye others could do descension. Wee know no body that not sinke into ye pores of body's finer y^n aire & it will sink into most if it be forcibly crouded in. ye stream descending will lay some hould on ye streame ascending & so press it closer & make it denser & therefore twill rise ye slower. ye streame descending will grow thicker as it comes nigher to ye earth but will not loose its swiftnesse untill it find much opposition as it hath helpe from ye following flood behind it. but when ye streames meete on all sides in ye midst of ye Earth they must needs be coarcted into a narrow roome & closely press together & find very much opposition one from another so as either to turne back ye same way y^t they came or croud through one anothers streames wth much difficulty, & pressure & so be compacted & ye descending streame will keepe y^m so by continually pressing y^m to ye Earth till they arise to ye place from whence they came, & there they will attaine theire former liberty.[13]

In a mechanical universe, the tendency of bodies called heavy to descend could only be ascribed to the impulsion, however delivered, of some invisible matter. Descartes had traced the heaviness of tangible bodies to the greater centrifugal tendency of the subtle matter in the vortex around the earth. Gassendi had imagined grappling hooks on lines of magnetic force. Newton's descending aethereal shower, an idea possibly borrowed from Kenelm Digby, expressed equally well the basic proposition of mechanical philosophies, that all the phenomena of nature are caused by particles of matter in motion, acting on each other by contact alone.

One of the intriguing facets of the *Questiones* is their reflection of Newton's fascination with perpetual motion, a dream for which the mechanical philosophy appeared to offer a rationale. That endless flux of aethereal matter causing gravity, for instance – if one could only tap it, the perpetual motion of matter in a mechanical universe could drive a tangible perpetual motion machine. 'Whither ye rays of gravity may bee stopped by reflecting or refracting y^m,' Newton asked himself, 'if so a perpetuall motion may bee made one of these two ways.' Two drawings, without further discussion, embodied the two possibilities. In one, a gravitational shield reflected the 'rays of gravity.' A wheel, hung freely on a horizontal axis, was placed so that half of it extended beyond the shield; since the exposed half would always be heavier, the wheel would turn without further input of power (see Fig. 29). The second device proposed to yoke the rays to a perpetual

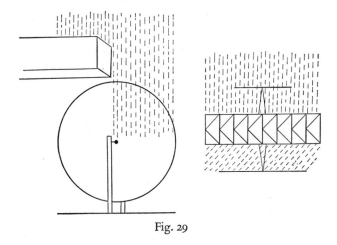

Fig. 29

motion machine by 'refracting' them, to use Newton's word. A wheel with vanes was mounted on a vertical axis, a horizontal windmill as it were, turned by the descending blast of gravitational rays.[14] Under the subject of 'Attraction Magneticall,' assumed to be caused by analogous streams of subtle matter, he imagined another set of similar perpetual motion devices.

Magnetic attraction, indeed, was a crucial phenomenon to proponents of mechanical philosophies of nature – the very epitome of the occult sympathies and antipathies with which Renaissance naturalists had populated the universe. Unless mechanical philosophers could explain it away, the assertion that nature consists solely of matter in motion was, without further ado, shown to be false. To that purpose, Descartes had elaborated an intricate fable of screw-shaped particles.

It is revealing of Newton's early philosophic commitments that he also elaborated a fable, and one no less intricate than Descartes. It is not found in the *Questiones* but in a separate paper which, to judge from its hand, was composed around 1667. The problem of magnetism, which made it at once the epitome of what had been meant by sympathy and antipathy and also a thorny phenomenon to explain away in mechanical terms, was its bipolar nature. To explain the two poles, Descartes had imagined some particles with right hand threads and others with left hand threads. Newton opted for two distinct streams of magnetic matter, one of which in passing through the earth acquires a certain 'odour' that renders it more 'sociable' to iron and loadstones, the other a different 'odour' whereby it is less 'sociable' to iron and more 'sociable' to the aether filling its pores. Hence the two streams can pass by each other in the magnet following different paths. When a loadstone with passages prepared by the terrestrial magnetic stream is taken from the earth, it diverts the streams it meets in the air as they return around the earth, breaking the streams into many small ones as they pass through the passages in the stone; these small streams can be easily deflected when they leave the stone. Perhaps the pressure of the aether, which is denser in free space than in the pores of bodies, diverts the fragmented streams; perhaps the streams excite an electric spirit by their friction, and the spirit in turn, as it shrinks again back into the iron, deflects the streams. Deflected and turned back they are by some means, and the resulting circulation of the magnetic streams around the magnet causes its apparent attractions.[15] The concept of sociability, which Newton appropriated from Hooke and which made its first appearance in his papers at this point, was to play a long and central role in his speculations. Meanwhile, the interest of the paper on magnetism lies in its testimony to Newton's philosophical allegiance. Whether Descartes' screw-shaped particles departed more wildly from empirical fact than Newton's odiferous streams is an open question, but a question beside the point. The mechanical philosophy rested on the proposition that ultimate reality differs from the phenomenal world, and its conviction that ultimate reality consists solely of matter in motion demanded that phenomena such as magnetic attraction and gravity be explained away by some imagined mechanism.

Although the *Questiones quaedam Philosophiae* were directed, as their name implies, to natural philosophy and not to mechanics, other entries under 'Gravity' indicate Newton's familiarity with the basic propositions of the new science of mechanics as found in Galileo's analysis of free fall.

> According to Galilaeus a iron ball of 100^l fflorentine (y^t is 78^{li} at London of Averdupois weight) descends an 100 braces fflorentine or cubits (or 49.01 Ells, perhaps 66^{yds}) in 5″ of an hower....
>
> The gravity of bodys is as their solidity, because all bodys descend equall spaces in equall times consideration being had to the Resistance of ye aire &c.

The central problem of the science of mechanics in the seventeenth century was to reconcile Galileo's conclusions – both the equal fall of all heavy bodies and their uniform acceleration – with the demand for causal mechanisms. Another entry on the same page suggests that Newton would have to confront the same issue. 'In ye descention of a body,' he noted, 'There is to be considered ye force wch it receives every moment from its gravity (wch must be least in a swiftest body) & ye opposition it receives from ye aire (wch increaseth in proportion to its swiftnesse).'[16]

One of the fascinating aspects of the *Questiones* is the extent to which they foreshadow the major features of Newton's career in science – as though already as an undergraduate he had perceived the issues that would command his continuing attention. The basic idea on which his work in optics was built – that phenomena of colours arise, not from the modification of light, but from the separation of heterogeneous rays disposed to excite the different sensations of colour – occurred to him first as he entered material under the heading 'Of Colours.'[17] Significant as this passage was for his conception of light, other entries, which called attention to a number of phenomena that occupied a central position in his speculations of natural philosophy, are more germane to my present concerns. Optics always had a two fold bearing in Newton's mind; it was concerned at once with the nature of light and with the structure of matter. Under the heading 'Of Perspicuity & Opacity,' the *Questiones* began to explore the second issue by asking why a wet bladder is opaque when both a dry bladder and water are transparent, and why water is clearer than vapours. He offered the opinion that transparency is effected in glass, crystal, and water differently than in air, aether, bladders, and paper.[18] Another question related to the structure of bodies was their cohesion, and Newton examined various theories that had been proposed to account for it.

'Whither ye conjunction of bodys be from rest.' He rejected the Cartesian explanation on the grounds that a pile of sand should then become a united whole. 'Whither it be from ye close crouding of all ye matter in ye world.' This solution he affirmed. We know how air pressure can hold two polished pieces of marble together. From Boyle's experiments, we also know that air pressure is quantitatively limited. 'but ye pressure of all ye matter twixt ☉ & us made by reason of its indevor from ☉ being farr greater (& it may be some other power by wch matter is kept close together &c),' when particles once touch so that no other matter is between them, such a pressure would be sufficient to hold them together.[19] The Cartesian reference to centrifugal force aside, the passage looks forward to the two explanations of cohesion, the one mechanical, the other not, which Newton later proposed at different times. A note on relative densities – that gold is 19 times as heavy as water and mercury 14 times as heavy, whereas water is 400 (or 'perhaps 2000') times as heavy as air[20] – introduced another set of data that played a central role in shaping Newton's conception of matter.

In addition to the cohesion of bodies, capillary action, or 'ffiltration' as he called it in the *Questiones*, also remained a central question in Newton's speculations.

> Whither filtration be thus caused. The aire being a stubborne body if it be next little pores into wch it can enter it will be pressed into y^m (unles theye be filled by something else) yet it will have some reluctancy out wards like a peice of bended whale bone crouded into a hole wth its middle pte forwards, if y^n water whose (pts are loose & pliable) have opportunity to enter y^t hole ye aire will draw it in by strivei[ng] it selfe to get out. The aire too being continually shaken & moved in its smallest parts by vaporous particles every where tossed up & down in it as appeares by its heate, it must needs strive to get out of all such cavity wch doe hinder its agitation: & this may be the cheife reason sponges draw up water. But in paper ropes hempe theds fiddle-strings betwixt whose particles there is noe aire or but a little & it so pend up y^t it can scarce get out the cause may be this. y^t ye parts of those bodys are crushed closer together y^n there nature will well permit & as it were bended like ye laths of crosbows so y^t they have some reluctancy against y^t position & striv to get liberty wch they cannot fully doe unless some oth^r bodys come betweene y^m as aire or water but where aire cannot enter water will (as appeares in y^t it will get through a bladder wch aire cannot doe &c) wherefore when opportunity offers it selfe by striveing to get assunder they draw in ye pts of water betwixt y^m.[21]

Other problems appeared more briefly. He noted the phenomenon of surface tension ('why ye superficies of water is lesse divisible y^n tis within in so much y^t what will swim in its surface will sinke in it') and the expansion of gases ('What is ye utmost naturall dilatacon of ye aire may be known by Torricellius his Experiment').[22] Although Newton had only scratched the surface of chemistry at the time when he composed the *Questiones*, he had come upon two phenomena which continued to occupy a prominent place among a constantly expanding set that he considered similar – mercury sinks into metals but into nothing else, and oil mixes with most bodies whereas water does not.[23]

For any philosophy of nature there are bound to be critical problems, problems that appear to challenge the philosophy, problems the solution of which are necessary to its viability. Already in the *Questiones*, Newton had singled out a set of phenomena crucial to a mechanical philosophy in their apparent challenge to its assertion that all the phenomena of nature are produced by particles of matter in motion. The cohesion of bodies, capillary action, surface tension, the expansion of gases, and chemical reactions displaying elective affinities and generating heat – this set of phenomena remained at the centre of Newton's attention through a lifetime of inquiry and speculation. Of the crucial phenomena, only chemical reactions generating heat had not asserted their significance when he composed the *Questiones*. He regarded the tendency of oil to mix with other bodies more readily than with water as a phenomenon identical to chemical affinity. Beyond the crucial phenomena, he had already laid hold on the materials from which he would build first one philosophy of nature, and then, when the same materials had seemed awry in the structure, a second one on a different foundation. Each in its turn exercised an influence on his mechanics.

*

As far as Newton's early philosophy of nature was concerned, the *Questiones* indicate its basic character with a clarity that cannot be misunderstood. Newton was a mechanical philosopher. In so far as the *Questiones* transcend the level of notes, and strive to assume the dignity of connected discourse, they propose another mechanical system, drawing on the work of Descartes, Gassendi, Hobbes, and Boyle, but differing from them in sundry details. For a period of roughly fifteen years following the initial *Questiones*, Newton elaborated the system they implied. Central to that system was a subtle aether, pervading the whole of nature and proffering invisible mechanisms for the production of the crucial phenomena. Some such subtle medium was a *sine qua non* of every mechanical philosophy – a form of matter, unobservable by definition, from which the imagination could construct fictitious mechanisms to generate intractable phenomena. Newton did not invent the concept of a subtle aether. It pervaded mechanical philosophies as freely as, in their view, it pervaded the universe. Numerous passages in the *Questiones*, such as those cited above on gravitation and cohesion, assumed its presence. In his fully elaborated system of the 1670's, the aether functioned, among other things, to explain the mechanics of optical phenomena. Some entries under 'Of Reflection undulation & refraction' indicate that he also perceived this possibility from the beginning.

> Whither ye backsid of a clear glas reflect light in vacuo
> Since there is refraction in vacuo as in ye aire it follows y^t ye same subtile matter in ye aire & in vacuo causeth refraction
> Try Whither Glasse hath ye same refraction in Mr Boyles Receiver, ye aire being drawn out, wch it hath in ye open aire.

What has the word 'undulation', used in the heading, to do with this, given that the *Questiones* already treat light as corpuscular? The next entry suggests the connection.

> How long a pendulum will undulate in Mr Boyles Receiver? &c.[24]

If light reflects from the back side of a piece of glass in a vacuum, reflection cannot be caused by light impinging on solid matter as prior mechanical philosophers had held. Such reflection then 'demonstrates' the presence of an aether in 'Mr Boyles Receiver', and equality of refraction in air and in a vacuum 'demonstrates' the same conclusion. Newton believed that the motion of a pendulum reveals the presence of an aether even more clearly, and the word 'undulation' referred, not to light, but to the pendulum. A similar proposal appears under the heading 'Of Motion.'

> How much longer will a pendulum move in ye Receiver then in ye free aire. Hence may be conjecttured w^t bodys there bee in the receiver to hinder ye motion of the pendulum.[25]

The use of the pendulum to demonstrate the existence (or the non-existence) of the aether – here was another enduring element in Newton's speculations.

A subtle material medium was not the only device by which to explain otherwise inexplicable phenomena, and in a treatise undertaken some time not long after his undergraduate career, Newton began almost inadvertently to explore another alternative.[26] *De gravitatione et aequipondio fluidorum* addressed itself primarily to the issue the *Questiones* had not finally resolved – as a mechanical philosopher, was Newton a Cartesian or an atomist? The answer now given was clear beyond any doubt – he was an atomist. Perhaps that is a misleading way to state the issue. From the terms he used – the charge of atheism hurled grimly at Descartes, for example – it is manifest that Newton was carrying on more than a discussion of natural philosophy. In the *Questiones*, the influence of Henry More was particularly strong in those passages intended to refute the possibility of a material order autonomous and independent of spiritual control. Perhaps the influence of More stood behind Newton's further development into violent anti-Cartesianism.[27] Whatever the animating influence, the result, as far as his philosophy of nature was concerned, was the definitive triumph of atomism.

In *De gravitatione*, Newton engaged Descartes first on the question of relative motion. Four definitions with a Gassendist flavour set the stage. 'Place is a part of space which something fills evenly.' 'Body is that which fills place.' 'Rest is remaining in the same place.' 'Motion is change of place.'[28] As Newton went on to state, the definitions explicitly distinguished body from space. Using them as a springboard, he leaped forthwith into the fray with Descartes. Not only was Descartes' conception of motion 'confused and incongruous with reason,' but it had 'absurd consequences' as well. It was filled with internal contradictions, which Newton proceeded to detail at some length. Underlying much of the discussion at this point was the specific definition of motion that Descartes had devised in response to Galileo's condemnation by the Inquisition. Even though Newton bore no responsibility for the situation, it is not greatly illuminating to watch him, one step further removed from the Inquisition in time, place, and church, scoring points at the expense of Descartes' dilemma. But the issue cannot be reduced entirely to this level, for whatever the details of Descartes' definition, the necessities of his system dictated that he consider motion as relative, and its relativity was the ultimate nub of Newton's objection. As he insisted, it follows from Descartes' conception 'that no one motion can be said to be true, absolute and proper in preference to others, but that all, whether with respect to contiguous bodies or remote ones, are equally philosophical – than which nothing more absurd can be imagined.'[29]

To support this assertion, Newton introduced a series of considerations associated in one way or another with force. The 'true, absolute, and proper' motion of the earth is that one alone whereby it endeavours to recede from the sun. Not only in circular motion but in all motion, force is the absolute factor whereby motion can be identified. 'Force is the causal principle of motion and rest,' he stated in definition five.[30] The weakness of Descartes' position lay in the possibility that motion, according to his definition of it, could be generated without the exertion of force or that force could be exerted without the generation of motion. Thus God might stop the vortex in which the earth moves without stopping the earth; by Descartes' definition, the earth would begin to move although no force had been applied to it. Or God might exert an immense force to turn all the heavens about the earth, although the heavens would continue to be at rest according to Descartes' view of motion. 'And thus physical and absolute motion is to be defined from other considerations than translation, such translation being designated as merely external.'[31] Descartes' effort to define motion in purely relative terms had left him with no means to determine whether a body is truly in motion or not. How can a body be said to move when the bodies from whose neighbourhood it is transported are not, and cannot, be seen to be at rest?

> Lastly, that the absurdity of this position may be disclosed in full measure, I say that thence it follows that a moving body has no determinate velocity and no definite line in which it moves. And, what is worse, that the velocity of a body moving without resistance cannot be said to be uniform, nor the line said to be straight in which its motion is accomplished. On the contrary, there cannot be motion since there can be no motion without a certain velocity and determination.[32]

Suppose we ask where Jupiter was a year ago. The particles of fluid matter around it have moved. The fixed stars do not furnish a frame of reference since they too are afloat in an aethereal sea. Indeed no bodies in the world stay at rest in relation to each other.

> And so, reasoning as in the question of Jupiter's position a year ago, it is clear that if one follows Cartesian doctrine, not even God himself could define the past position of any moving body accurately and

geometrically now that a fresh state of things prevails, since in fact, due to the changed positions of the bodies, the place does not exist in nature any longer.[33]

To avoid the Scylla of relativity, Newton embraced the Charybdis of absolute space. Every motive impelled him in this direction. As an atomist, he was concerned to assert the existence of voids and thus of extension distinct from matter. As a Gassendist who had accepted the degree of scepticism which held that knowledge of the ultimate essences of things is foreclosed to man, he was concerned to deny the intelligibility of matter inherent in the Cartesian equation of it with geometric extension. And for personal reasons, apparently in the end abhorrence of the insecurity implied by the relativity of motion and place, he demanded an absolute frame of reference as some sort of anchor in the infinite universe which he too accepted.[34] In Gassendi, he found a concept of space answering all these needs.[35] Extension is neither substance nor accident. It is not a substance because it does not exist independently and cannot act on other substances. It is not an accident because it can exist without a subject. God did not create space; for if that were the case, God would not have been anywhere before He created it. Space, extension, is an 'emanent effect of God, or a disposition of all being . . .'[36] As the emanent effect of an eternal God, space also endures eternally.

> Space is a disposition of being *qua* being. No being exists or can exist which is not related to space in some way. God is everywhere, created minds are somewhere, and body is in the space that it occupies; and whatever is neither everywhere nor anywhere does not exist. And hence it follows that space is an effect arising from the first existence of being, because when any being is postulated, space is postulated.[37]

Which is equivalent to saying that absolute space is constituted by the infinite presence of God and absolute time by His infinite duration.

If matter is not equivalent to extension, what is it?

> Of this, however, the explanation must be more uncertain, for it does not exist necessarily but by divine will, because it is hardly given to us to know the limits of the divine power, that is to say, whether matter could be created in one way only, or whether there are several ways by which beings similar to bodies could be produced.[38]

Here spoke the voice of Gassendist scepticism denying what Descartes had demanded, the transparency of physical nature to reason. Finite man is forever denied knowledge of the intimate essences of things, knowledge open solely to the infinite understanding of their Creator. Matter in the form posited here, Newton asserted, requires only 'extension and an act of the divine will . . .'[39] As we are conscious of our power to move our bodies, so we cannot deny the same power to God. Let us suppose that he prevents bodies from entering a given space. In what way could we distinguish such a space from body? Being impenetrable, it would be tangible; reflecting light, visible. If struck, it would resonate because the neighbouring air would be moved. Further suppose that there are several such spaces, and that they move according to definite laws – in no way would they differ from particles of matter. 'And so if all this world were constituted of this kind of being, it would seem hardly any different.'[40]

In accordance with this view, Newton defined matter as 'determined quantities of extension which omnipresent God endows with certain conditions.' The conditions are three in number. The quantities are mobile. They are impenetrable, and when they meet each other they reflect according to certain laws. They can excite sensations in a mind and can in turn be moved by a mind.[41] Do the particles he described correspond in fact to material reality? Newton refused to answer, of course. Gassendist sceptic to the core, he merely repeated that God could have created still other beings that display all the actions and phenomena of bodies. Hence he was unable to say what the nature of bodies is, 'but I rather describe a certain kind of being similar in every way to bodies, and whose creation we cannot deny to be within the power of God, so that we can hardly say that it is not body.'[42]

In the discussion of matter, Newton laid special emphasis on the analogy of the human mind to God.

> Thus I have deduced a description of this corporeal nature from our faculty of moving our bodies, so that all the difficulties of the conception may at length be reduced to that; and further, so that God may appear (to our innermost consciousness) to have created the world solely by the act of will just as we move our bodies by an act of will alone; and, besides, so that I might show that the analogy between the Divine faculties and our own is greater than has formerly been perceived by Philosophers.[43]

Because of this analogy, man is said to be created in the image of God; the power of his mind to move his body is a faint incarnation of that power of God to move matter, which, in Newton's eyes, was equivalent to His power to create matter. The crux of the charge of atheism surreptitiously aimed at Descartes lay in the assertion that Cartesian matter was independent of God whereas matter as Newton defined it could not be conceived without supposing the existence of God. Extension is not independent of God; it is 'eminently contained in God . . .'[44] Equally the whole material universe, which is composed of determined quantities of extension endowed by God with certain properties, is eminently contained in God, for the infinite extension of absolute space is constituted by the omnipresence of God.

One of the significant consequences of this doctrine for natural philosophy lay in the possibility that God might replace the aether Newton had already learned to employ. The role he attributed to God was exactly the role he attributed to the aether – an invisible medium that moves particles of matter in the manners they are observed to move. The influence of Henry More and the Cambridge Platonists, with their anxiety over the autonomy of the material order which they saw implicit in the mechanical philosophy, can perhaps be glimpsed behind Newton's move. At one point, he referred explicitly to the possibility of a 'soul of the world,' a concept he rejected as an unnecessary intermediary between the universe and God.[45] The immaterial, or rather the divine medium offered the possibility of escape from the inherent limitations of material media that made the reconciliation of aetherial mechanisms with mathematical kinematics so difficult. Compare the *Questiones'* discussion of acceleration in free fall, cited above, with the possibilities implicit in the assertion that God moves the determined quantities of extension 'hither and thither according to certain laws . . .'[46] In *De gravitatione* Newton did not see the presence of God as an immaterial substitute for a material aether.[47] Nevertheless he had begun to entertain an idea which could lead in that direction, and however much he continued to embroider aetherial mechanisms, he remembered the idea as he remembered every idea.

*

De gravitatione et aequipondio fluidorum was apparently conceived as a treatise on fluid mechanics. What Newton completed was primarily the introduction, the discussion of space, motion, and body directed against Descartes. Before he stopped, however, he added a number of further definitions and two propositions. In the development of Newton's mechanics, *De gravitatione* occupies a special position, not only for the inherent interest of its contents, but also as the last product apparently of his first period of inquiry in mechanics. Following *De gravitatione*, he undertook no sustained work in mechanics for a period of about fifteen years, until he picked up his thread of thought again in 1684. To understand the import of *De gravitatione* for his mechanics it is necessary to look initially at his earlier papers.

Newton's first recorded interest in the science of mechanics is found, of course, in the *Questiones*. I have already cited a number of items, having to do mostly with heaviness and the descent of heavy bodies, that bear on mechanics. An entry under the heading 'Of Aer' considered briefly the angle to the wind at which the sail of a windmill should be set, and in so doing showed some familiarity with the principles of composition (or resolution) of motions.[48] By far the most important passage dealing with mechanics is found in the extended essay 'Of violent Motion.' On the evidence of the hand in which it was written, it belongs with the earliest entries in the *Questiones*. Newton considered three alternatives for the continuation of violent, that is, projectile motion after the projectile has separated from the projector, the classic problem to which the new conception of motion addressed itself. At some length, he rejected the theory that air can act as the mover, dealing not with the Aristotelian theory but with its vulgarised version, antiperistasis. Second, he considered whether the motion can be sustained by 'a force imprest.' The discussion is curious in that he took impressed force to mean the continued action of the original motor at a distance, and he 'refuted' the theory by demonstrating the impossibility of such a force being communicated by either material or immaterial means. Hence he concluded for the third alternative, that a projectile is moved by its 'naturall gravity.' The concept of gravity as an internal motive principle derived from the atomist tradition, a fact made even more clear by Newton's discussion. Instead of extending the dynamic considerations that he had applied to the first two cases, he examined solely the question whether motion in a vacuum is possible; that is, he argued that change of place can be defined in a void, a passage which looked forward to the idea of absolute space. It is not without significance meanwhile that in the essay 'Of violent Motion,' Newton's effective introduction to the science of mechanics, uniform motion was treated, as it was in the atomist tradition, as the product of a force internal to the moving body.[49]

On 20 January 1665, Newton undertook an extensive examination of questions of motion and impact, which he recorded in the *Waste Book* as a set of definitions, axioms, and propositions entitled 'Of Reflections.' Quite a different idea of motion animated these notes. Whereas the atomist tradition stood behind the essay 'Of violent Motion,' the form in which Newton cast 'Of Reflections' bears unmistakable marks of the influence of Descartes. After a couple of false starts, he put down eleven definitions of such things as motion and quantity of motion, followed by a series of axioms.

> 1 If a quantity once move it will never rest unlesse hindered by some externall caus.
> 2 A quantity will always move on in ye same streight line (not changing ye determination nor celerity of its motion) unlesse some externall cause divert it.[50]

Further on, a revision cast the two axioms into one.

> Ax: 100 Every thing doth naturally persevere in y^t state in wch it is unlesse it bee interrupted by some externall cause, hence axiome 1st, & 2^d, & γ, A body once moved will always keepe ye same celerity, quantity & determination of its motion.[51]

Perhaps the verb 'persevere', with its ambiguous connotations at once active and passive, echoes the internal force he called 'gravity' earlier. Beyond that, only an explicit reference to the state of rest would have been required to convert Axiom 100 into an affirmation of inertia indistinguishable from that in the *Principia*'s first law of motion.

As the title, 'Of Reflections,' indicates, Newton's purpose was to solve the problem of impact with which Descartes had grappled unsuccessfully. In the two false starts, he had tried to come directly to grips with it. Now, with a foundation of definitions laid, he attempted to build his treatment of impact on the principle of inertia, stated in terms of 'perseverance'.

> If 2 equall bodys (*bcpq* & *r*) meete one another wth equall celerity (unlesse they could pass through one y^e other by penetration of dimensions) they must mutually (since y^e one hath noe advantage over y^e other they must) equally hinder y^e one y^e others perseverance in its state. likewise if y^e body *aocb* be = & equivelox w^{th} *r*, they meeting would equally hinder or oppose y^e one y^e others progression or perseverance in their states therefore y^e power of y^e body *aopq* [*aocb* + *bcpq*] (when tis equi-

velox wth r) is double to y^e power y^t r hath to persever in its state. y^t is y^t efficacy force or power of ye caus wch can reduce *aopq* to rest must bee double to ye power & efficacy of ye cause wch can reduce r to rest, or ye power wch can move ye one must be double to ye power wch can move ye other soe y^t they be made equivelox.

Hence in equivelox bodys ye powers of persevering in their states are proportionall to their quantitys.[52]

Like most students of the question in the seventeenth century, Newton approached impact by seizing on exactly that aspect of the Cartesian approach which Huygens sought constantly to avoid. Newton's treatment, from its inception, was frankly dynamic. Already he had defined quantity of motion as the product of a quantity (that is, the quantity of matter in a body) and its velocity. Basic to his dynamics of impact was to be a unit of power or force defined in terms of quantity of motion. As he said in an earlier version of the passage above, 'the motion of one quantity to another is as their powers to persever in that state.'[53]

A concept of force measured in terms of quantity of motion was hardly a novelty to seventeenth-century mechanics. As the 'force of a body's motion', it had been a staple of seventeenth-century discussions, those of Descartes among others. When Newton seized on quantity of motion as his unit of measure, he was only indicating another dimension of the debt to Descartes which the entire passage reveals. The unique position of the *Waste Book* in the history of dynamics derives from its recognition that a dynamics built on the principle of inertia demands a concept of force different from the prevailing one. He realised that the 'force of a body's motion' can be seen from another perspective. In impact, the force of one body's motion functions in relation to the second body as the 'external cause' mentioned in axioms 1 and 2 as the sole means that can alter its state of motion or rest. Newton made the perspective of the second body his primary one. Thus he was the first man fully to comprehend the implication of inertia for dynamics, that the prime necessity of an operative dynamics was a conceptual unit to measure the 'external cause' of changes of motion. 'Of Reflections' set out to convert the available idea of force to that use. After the initial set of axioms defining inertial states, he added two more that began the definition of force.

> 3 There is exactly required so much & noe more force to reduce a body to rest as there was to put it upon motion: et e contra.

> 4 Soe much force as is required to destroy any quantity of motion in a body soe much is required to generate it; & soe much as is required to generate it soe much is alsoe required to destroy it.[54]

It followed then that when equal forces move unequal bodies, their velocities are inversely as their quantities, and when unequal forces move equal bodies, their velocities are proportional to the forces.

> 104 Hence it appears how & why amongst bodys moved some require a more potent or efficacious cause others a less to hinder or helpe their velocity. And ye power of this cause is usually called force.[55]

As an historical statement, the last sentence will scarcely bear examination. In the sentence itself, Newton found himself compelled to use the word 'power' as a synonym for force, and in the preceding axiom he had referred to 'ye power or efficacy vigor strength or virtue of ye cause' which generates motion.[56] Nevertheless the sentence marked a turning point in the terminology of mechanics. Primarily because Newton seized it from the welter of available terms, 'force' became the accepted word to designate the cause of changes of motion.

The concept was far more important than the word chosen to name it, of course. In 'Of Reflections,' Newton began to develop an abstract conception of force, separated from its cause and treated solely as a quantitative concept able to enter into a quantitative mechanics. In a context set by the introductory assertion of the principle of interia and by the problem of impact, there was apparently only one quantity that could be the measure of force – change in motion. It is obvious, of course, that change of motion is not identical to rate of change of motion. Newton was seeking to quantify the phenomena of impact; the total change of motion was the quantity that presented itself. Thus he measured force, not by ma or $\frac{d}{dt}mv$, but by Δmv. A generation earlier, Descartes had established the ultimate foundation of the new mechanics by setting all motion, as motion, on the same plane. To this principle Newton now added a second, that all changes of motion are equivalent and measurable on a linear scale. "Tis knowne by the light of nature,' he asserted, 'y^t equall forces shall effect an equall change in equall body.'[57] As much force as is necessary to generate a given motion, so much exactly is needed to destroy it. 'For in loosing or to [sic] getting ye same quanty of motion a body suffers ye same quantity of mutaion in its state, & in ye same body equall forces will effect a equall change.'[58] In a body already in motion, an equal force will generate another increment of motion equal to the first, 'since tis noe greater change for (*a*) to acquire another part of motion now it hath one y^n for it to acquire y^t one when it had none . . .'[59] He summed up the definition of force in axiom 23:

> If y^e body *bace* acquire y^e motion q by y^e force d & y^e body f y^e motion p by y^e force g. yn $d:q::g:p$.[60]

One of the by-products of the new definition of force was an initial step toward the definition of mass. In his original set of definitions, Newton employed the word 'body' – for example, one 'body' has so much more motion than another as the sum of the spaces through which each of its parts moves is to the sum of the spaces through which each of the parts of the other moves in the same time.[61] He then went back over the definitions, striking out the word 'body' where it appeared in order to substitute 'quantity,' a step in the direction of finding a more precise concept to indicate the object on which force is exerted. In the context of the mechanical philosophy, the word 'quantity' used for 'body' involved assumptions about matter that left a permanent mark on mechanics. Matter was qualitatively neutral and homogeneous, differentiated solely by quantity. It followed then – and Newton assumed it without pausing – that matter is inertially homogeneous as well. The linear scale for changes of motion requires as its complement a second linear scale whereby an increase in the quantity of matter implies a proportional increase in the force required to generate a given velocity.

The concept of force that dominated dynamics in the seventeenth century before Newton was a legacy from impetus mechanics and the common sense perceptions which it formalised. It was a legacy ultimately beyond reconciliation with the principle of inertia that was basic to the new science of motion. Seen with the vantage of hindsight, the idea of inertia appears to have demanded a concept of force measuring not motion but change of motion such as Newton proposed in the *Waste Book*. Did his formulation of such a concept signify his rejection of the point of view recorded in the essay 'Of violent Motion,' and his definitive endorsement of the principle of inertia which, in its statement as the first law of motion, was to function as the cornerstone of his mature mechanics? The *Waste Book* suggests that rather than seeing the

NEWTON AND THE CONCEPT OF FORCE

one concept of force as the denial of the other, Newton sought to reconcile them in a unitary dynamics.

> The force wch ye body (*a*) hath to preserve it selfe in its state shall bee equall to ye force w^{ch} [pu]t it into y^t state; not greater for there can be nothing in ye effect wch was not in ye cause nor lesse for since ye cause only looseth its force onely by communicateing it to its effect there is no reason why its should not be in ye effect w^n tis lost in ye cause.[62]

Pursued to its ultimate implication, the passage seems to dissolve kinematics in a universal science of dynamics, in which the force of a body at any moment, expressed in its motion, is the sum of the forces that have acted upon it.

Would it be possible to reconcile such a dynamics with the principle of inertia on which the entire passage attempted to base itself? Newton's statement of inertia in the *Waste Book* derived immediately from Descartes; and for all his talk of the 'force of a body's motion', Descartes had drawn from the principle of inertia the conclusion of the relativity of motion, a conclusion impossible to reconcile with the universal dynamics toward which Newton was groping. Moreover, the uniformly linear relation of motion and force that he adopted, whereby no less force is required to add a second increment of motion to a moving body than to start it from rest and give it a first increment, implied the dynamic identity of uniform motion and rest. Apparently Newton perceived the incompatibility as well. At least he was uneasy enough with the passage above to strike it out, and to replace it with another in which he seemed deliberately to abandon the idea that the force of a body's motion could be a meaningful absolute quantity in an inertial mechanics.

> 112 A Body is saide to have more or lesse motion as it is moved wth more or lesse force, y^t is as there is more or lesse force required to generate or destroy its whole motion.[63]

Although he continued to equate quantity of motion with force in the *Waste Book*, axiom 112 suggests that when he said a body 'is moved wth more or lesse force,' we should understand that he meant to relate force, not to the motion itself, but to its generation or destruction, that is, to change of motion. The concept of the 'force of a body's motion' began to lose its connotation of a constant force sustaining a uniform motion and to express something nearly equivalent to our 'momentum'.

Its quantity has meaning only in a given inertial frame of reference, and it expresses the force necessary to generate or destroy that quantity of motion in an inertial frame of reference, but not a force that sustains the motion. Axiom 118, which derived a general formula relating force, mass, and velocity, repeated this interpretation of force. Given a body p moved by a force q and a body r moved by a force s, what is the ratio of their velocities, v and w? Newton demonstrated that

$$\frac{v}{w} = \frac{qr}{ps} \quad \text{or} \quad \left(\frac{v_p}{v_r} = \frac{f_p m_r}{f_r m_p}\right)$$

to wit, that 'y^e motion of p is to y^e motion of r as y^e force of p to y^e force of r.' In the following sentence, however, he insisted again that he understood changes of motion to be in question.

> And by y^e same reason if y^e motion of p & r bee hindered by y^e force q & s, y^e motion lost in p is to ye motion lost in r, as q is to s. or if ye motion of p be increased by ye force q, but ye motion of r hindered by ye force s; then as q, to s::so is the increase of motion in p, to ye decrease of it in r.[64]

The purpose of the entire exercise was the treatment of impact, and exactly the treatment of impact served to emphasise a concept of force as the measure of changes of motion instead of the measure of motion itself. Examining what he called 'ye mutual force in reflected bodys,' Newton asserted that 'soe much as p presseth upon r so much r presseth on p. And therefore they must both suffer an equall mutation in their motion.'[65] Except for the special case in which p and r have equal motions, the conclusion that they press each other equally had seemed obviously false to the great majority of those who before 1665 had attempted to analyse the force of percussion. The *Waste Book* suggests that Newton came to deny the intuitively obvious by recognising that for every impact there is a frame of reference, that of the common centre of gravity, in which the two bodies do have equal forces. That is, his treatment of impact depended on his accepting a relativity of motion in terms of which the idea of an absolute force of a body's motion is meaningless. A series of propositions established that two bodies have equal motions in relation to their common centre of gravity, and that the centre of gravity of two bodies in uniform motion is also in uniform motion (or at rest), both when the two bodies are in the same plane and when they are in different planes. He was now ready to demonstrate that their centre of gravity also continues to move uniformly when the two bodies meet in impact and rebound (see Fig. 32).

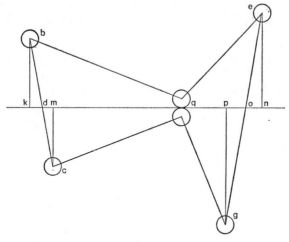

Fig. 32

> For y^e motion of b towards d y^e center of their motion is equall to y^e motion of c towards d ... therefore y^e bodys b & c have equall motion towards y^e points k & m y^t is towards y^e line kp [the line of motion of their center of gravity before impact]. And at their reflection so much as (c) presseth (b) from ye line kp; so much (b) presseth (c) from it ... Therefore e & g [b and c at some time after impact] have equall motions from ye point o. w^{ch} ... must therefore be ye center of motion of ye bodys b & c when they are in ye places g & e, & it is in ye line kp.[66]

In the history of mechanics, 'Of Reflections' occupies a role of special importance. It was here, early in 1665, that Newton formulated the first imperfect version of the concept of force that has functioned as part of the foundation of modern dynamics. The concept of force proposed in 'Of Reflections' was internal to the science of mechanics, a concept that promised to expand the competence of mathematical mechanics by providing a quantitative measure of the causes of changes of motion, but one that asserted no claim to special ontological status. In the *Waste Book*, Newton's concept of force did not challenge the prevailing mechanical philosophy in any way. Quite the contrary, he saw 'force' as a clarification of that philosophy, a means of adding quantitative rigour to the one interaction allowed to take place between bodies. 'Of Reflections' assumed that impact, and impact alone, is the occasion whereby one body acts on another to change its motion. As to the ontological status of force, it had none. 'Force' was merely a quantitative concept useful in treating impact. 'Force,' he asserted, 'is ye pressure or crouding of one body upon another.'[67]

At another point, he indicated again the extent to which his discussion assumed the mechanical philosophy. In the opening definitions, he set out to define rebound (or 'reflection') in terms of the degree of elasticity of the bodies or the medium between them, whichever it is that causes the rebound. To the degree that the 'spring' is more 'vigorous' the velocity of separation approaches that with which they come together. A note in the margin interrupted at that point to deny the whole discussion. 'Noe motion is lost in reflection,' he decided. 'ffor y^n circular motion being made by continuall reflection would decay.'[68] Clearly he was thinking of the orbital motion of planets in the vortex. If their impacts with the matter of the vortex are not perfectly elastic, the motions of the planets could not continue undiminished as observations show them to have done, at least since ancient times. Newton never forgot that insight, although he later turned it to quite a different use.

*

If Newton continued to speak of the force of motion in the *Waste Book*, even while he formulated a concept of force as that which changes motion, his attempt to extend the analysis of impact into the mechanics of circular motion may have been responsible. When a body revolves in a circular path, its constant endeavour to recede from the centre suggested obvious analogies with the force of its motion. Imagine a ball rolling around the inside of a cylindrical surface. At any given moment, the determination of its motion is along the tangent to the circle it traces; and if the cylindrical body should cease to check it, the ball would move uniformly along the tangent and recede obliquely from the centre. If the cylinder opposes itself to the tendency of the ball and keeps it in the circular path, 'that is done by a continued checking or reflection of it from ye tangent line in every point of ye O . . .' But the cylinder cannot check the determination of the ball 'unless they continually presse upon one another.' The same can be said of a body restrained into circular motion by a string.

> Hence it appears y^t all bodys moved Olarly have an endeavour from ye center about wch they move . . .[69]

So far Descartes, on whom Newton was drawing, had gone. Starting from the same position, Huygens had drawn upon the geometry of the circle to arrive at a quantitative statement of the endeavour from the centre. Newton attacked the problem through his analysis of impact.

> The whole force by wch a body . . . indevours from ye center *m* in halfe a revolution is double to the force wch is able to generate or destroy its motion; that is to ye force wth wch it is moved.[70]

When a body has revolved through half a circle, its determination has been exactly reversed. Accepting the implications of Descartes' analysis of circular motion and rejecting his dicta, given elsewhere, that no action is required to change the determination of a body's motion, Newton reasoned that revolution through half a circle is equivalent to a perfectly elastic rebound which requires a force great enough first to stop a body's forward motion and then to generate an equal motion in the opposite direction. When further analysis had informed him otherwise, he added the words 'more y^n' before the phrase 'double to the force . . .'

Whatever the analogy with impact, there were obviously differences as well. The force of a body's circular motion, its endeavour to recede from the centre, does not dissolve away at the wave of a magic inertial wand. Whatever the inertial frame of reference, it remains constant. Here was a powerful motive deriving from the conceptual structure of mechanics to retain the idea of the force of motion. Another difference from impact is the fact that a body in revolution never stops and that the endeavour away from the centre is exerted uniformly over a period of time, unlike the force of an impact which produces a single finite change of motion. In our terms, the 'force' from the centre is dimensionally incommensurable with the 'force' (or impulse) exerted in impact. Newton's reference to the 'whole force' during half a revolution (equivalent dimensionally in our terms to the integral $\int F dt$) constitutes an implicit recognition of the fact that he was dealing with two different quantities.

One of the fascinating questions arising from Newton's analysis of circular motion concerns his conceptualisation of the problem. In an analysis that followed and flowed from a concept of force as that which changes uniform motion, he spontaneously set out to quantify, not the external force which diverts a body from its inertial path, but the endeavour of a body in circular motion to recede from the centre. Implicit in his understanding, of course, was the necessity of the external force, equal and opposed to the centrifugal endeavour. But this is only to say that he conceived of circular motion in terms of an equilibrium of forces, a view prevalent in the seventeenth century, as we have seen, though to us a strange conclusion at which to arrive in an analysis that began with the principle of inertia. Why did Newton, like other early students of circular motion, conceptualise it in this way? Partly, no doubt, the example of Descartes' pioneering effort, the starting point for everyone in the following generation, led them all to think of the problem in his terms, so that a major effort of the imagination was required by Hooke to restate it in a different form. It appears to me that Newton's specific context, the problem of impact, and the whole mechanical philosophy which stood behind it, also helped him to see Descartes' formulation of the problem as natural and proper. Although he saw that the change of direction of the revolving body requires the exertion of an external force, its change of motion is not associated with an equal and opposite change in another body's motion, which is the case in impact. Except for a sharply confined set of examples, such as the ball inside the cylinder, the external force is not even associated with one body. For example, in planetary motion as Newton then conceived of it, the force diverting the planet from its rectilinear path derives from impact with innumerable tiny bodies. In the circumstances, it was far easier for Newton to think of the force away from the centre exerted by the body in motion than to think of the force exerted on it. Thus the man who later coined the word 'centripetal' began with the idea, if not the word, of 'centrifugal' force, and arrived at a measure of a force we regard as fictitious which is quantitatively identical with the formula we still use in the dynamics of circular motion.

The choice of quantity of motion as the measure of force, and its application to circular motion, also reflected the continuing influence of the lever on dynamics. In the very process of formulating a conception of force that would finally distinguish dynamics from static equilibrium, Newton seized on the quantity that analyses of the lever had isolated as the measure of force. In the case of circular motion, he concluded that since the 'whole force' by which a body tends from the centre in one revolution is equal to something more than six times 'ye force by wch y^t body is moved,' then the whole force of one body in one rotation is to that of another as the motion of the one body to that of the other. Since the common centre of gravity of two bodies is the point on the line connecting their centres such that their distances are inversely as their bulks, it follows that when they turn about their

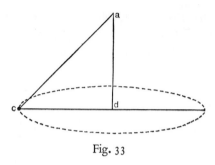

Fig. 33

centre of gravity, they will have equal motions – the law of the lever. Hence their endeavours from the centre will be in equilibrium. Centrifugal tendency suddenly revealed a new dimension in an old proposition.[71]

The tendency to see an equilibrium of opposed forces in circular motion made the distinction of dynamics from statics in its case most difficult. In the *Waste Book*, for example, Newton suggested using the conical pendulum to compare 'ye pressure of ye rays of gravity to ye force by wch a body hath any given motion . . .' The proposal reveals incidentally how Newton failed fully to probe the insight implicit in his phrase, 'total force' in a revolution, and how he applied the word 'force' indiscriminately to dimensionally incommensurable concepts. In a sketch that accompanies the proposal, *ad* represents the vertical height of the cone and *dc* the radius of the circle the bob *c* traces (see Fig. 33). 'And *ad*:*dc* :: force of gravity to ye force of *c* from its center *d*.'[72] What the diagram represents, of course, is the condition of static equilibrium.

At another place in the *Waste Book*, Newton extended the analysis of circular motion and arrived at a quantitative statement of the tendency away from the centre which is identical to his later formula for centripetal force.

FORCE IN NEWTON'S PHYSICS

If ye ball b revolves about ye center n ye force by wch it endeavours from ye center n would beget so much motion in a body as ther is in b in ye time y^t ye body b moves ye length of ye semidiameter bn. [as if b is moved wth one degree of motion through bn in one seacond of an hower y^n its force from ye centre n being continually (like ye force of gravity) impressed upon a body during one second it will generate one degree of motion in y^t body.] Or ye force from n in one revolution is to ye force of ye bodys motion as :: Periph : rad

To demonstrate this proposition, Newton imagined a square to be circumscribed around the circle and the ball to follow a path inside it, rebounding from each side at the point where the circle touches the

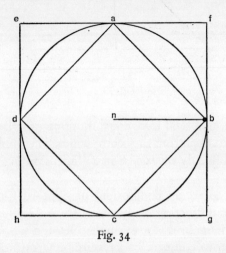

Fig. 34

circumscribing square, so that its path is a second square inscribed in the same circle (see Fig. 34). Taking the component of motion perpendicular to the side, he set down an expression which compared the force of one impact, in which that component is reversed, to the force of the ball's motion. '$2fa:ab::ab:fa::$ force or pression of b upon fg at its reflecting: force of b's motion.' In one complete circuit, there are four reflections, so that the total force in one circuit is to the force of the body's motion as $4ab:fa$, that is, as the length of the path to the radius of the circle. At this point, Newton perceived (although it can easily be demonstrated, he did not do so in the *Waste Book*) that if the number of sides of the inscribed and circumscribed polygons is increased, the ratio of force for one circuit continues to equal the ratio of the length of the path to the radius.

And soe if body were reflected by the sides of an equilaterall circumscribed polygon of an infinite number of sides (i.e. by y^e circle it selfe) y^e force of all y^e reflections are to y^e force of y^e bodys motion as all those sides (i.e. y^e perimeter) to y^e radius.[73]

That is, for one complete revolution $F/mv = 2\pi r/r$, or $F = 2\pi mv$. Force here refers to the total force in one revolution, of course, and if it is divided by the time for one revolution ($2\pi r/v$), the 'force by wch it endeavours from y^e center' at each instant – 'like y^e force of gravity' as Newton explained, so that if it were continually impressed on a body over a period of time it would generate a finite quantity of motion – equals mv^2/r, our formula for centripetal force.[74]

Via the problem of circular motion, Newton's incipient dynamics was brought face to face with another possible model for the conceptualisation of force, uniformly accelerated motion as Galileo's analysis of free fall presented it. When he took impact as the model, Newton had defined force as Δmv. Nevertheless, the acceleration of a falling body is obviously a change of motion quite as much as that which a body suffers in impact, an identification made all the easier by the conviction that the multiple impacts of invisible bodies cause gravity. Without evident hesitation, Newton identified gravity (and the *conatus* from the centre) as 'force', recognising the dimensional incompatibility only in so far as he spoke of 'total force.' What then was the measure of force, ma or Δmv? The ambiguity in the definition of force which appeared in his dynamics at this point continued to plague it to the end.

*

At about the same time that he undertook the examination of impact in the *Waste Book*, Newton also addressed himself to the problem Galileo had attempted to answer in the *Dialogue*. If the earth rotates daily on its axis, as the Copernican system asserts, why does the centrifugal tendency not project bodies off the surface of the earth?[75] Newton's first approach to the endeavour from the centre had been to set the whole force in half a revolution equal to twice the force of the body's motion since whatever checks and turns a body through half a revolution has the effect of destroying its original motion and generating one opposed to it. On the separate paper in question here, what Herival has named the Vellum Manuscript, he used a quarter of a revolution instead of a half, and imagined that a point on the equator is uniformly decelerated during a quarter of a revolution, coming completely to rest while it travels a distance equal to one quarter the equatorial circumference. From Galileo's assertion that a heavy body falls 100 braces in five seconds, he then calculated how long it would take gravity to generate an equal motion. He reasoned that the two forces, gravity and centrifugal endeavour, are inversely as the times in which they could generate equal motions. On his first try, Newton made two errors – he confused degrees with hours and multiplied $90 \times 60 \times 60$ to get the number of seconds in a quarter rotation, and he calculated the time for a body to fall a distance equal to a full quarter of the equatorial circumference instead of half that distance (since in uniformly accelerated motion from rest to velocity v, $s = \frac{1}{2}vt$). The calculation led to the conclusion that 'y^e force from gravity is 159·5 times greater y^n y^e force from y^e Earth's motion at y^e Equator.' On a second try he corrected both errors and arrived at the figure 15 – gravity is fifteen times as great.[76]

The deficiencies of the approach are obvious enough. Newton recognised them also, and the new attack on circular motion that he entered on page one of the *Waste Book* probably came at this time. The same page contains references to the conical pendulum, which Newton also utilised in the Vellum MS. to correct his figure for the acceleration of gravity. As far as the endeavour from the centre was concerned, when he returned to the problem on the Vellum MS., he knew that the crucial distance was not a quarter rotation but a rotation of one radian. The centrifugal force of a revolving body is such that an equal force, applied to a body of equal mass during the time that the body revolves through one radian, would generate an equal linear velocity in the other body and move it from rest through half the length of a radian, that is, through half the length of the radius of revolution. Applying his new analysis to Galileo's problem and Galileo's figures, he calculated 'y^t ye force of ye Earth from its center is to ye force of Gravity as one to 144 or there abouts.'[77]

At this point, Newton decided to check Galileo's figure for the acceleration of gravity. Perhaps the insight of the *Waste Book*, that with the conical pendulum and the formula for centrifugal force he could calculate g with a precision not to be reached by a direct measure of free fall, dictated the attempt. Whatever prompted him, the MS. contains the diagram of a conical pendulum with a string 81 inches long. In some way not disclosed, he convinced himself that he had measured a number of swings with the thread inclined at 45°. A series of ratios led to the figure of 1 512 'ticks' in an hour. In roughly three-eighths of a second, the bob of the pendulum travels the length of the radius of its circle, and in the same time a body falling from rest would travel half as far, or about 50 inches in half a second and therefore 200 inches in one second, an excellent approximation to our value. Since the figure from Galileo that he had been using was roughly two and a half yards in a second, he chose to round his own off at five yards. He then returned to the earlier calculation and doubled his result.[78]

Throughout the calculation, Newton employed Galileo's result which set distance travelled from rest under uniform acceleration proportional to the square of time. Hence his references to the 'force of gravity' have the implicit effect of translating Galileo's kinematics into dynamics. Here, in a context removed from impact and its finite changes of motion, Newton first employed the relation that is commonly called his second law of motion, that acceleration of a given body is proportional to force. The problem, of course, was to reconcile this use of force with the other two he employed, force as the measure of motion, and force as the measure of changes of motion.

In another separate paper, somewhat later than the Vellum MS. but by the evidence of its hand dating from the years immediately following his undergraduate career, Newton attacked the question of circular motion again and began to apply his quantitative analysis to the revolutions of heavenly bodies. He derived the formula for centrifugal *conatus* in a more economical and more elegant style. A body A revolves uniformly around the centre C in the circle $ADEA$ (see Fig. 36). Because its motion is uniform, lengths of arc can be taken as the measure of time. When A moves through AD, '(which I set very small),' its *conatus* from the centre would carry it through the distance DB away from the circumference in the same period of time, since it would diverge that distance from the circle if it moved freely along the tangent without hindrance. Again Newton applied a dynamic interpretation to Galileo's kinematics. This endeavour, 'provided it were to act in a straight line in the manner of gravity, would impel bodies through distances which are as the squares of the times . . .' Therefore, in order to know how far a body would be impelled in the time of one revolution $ADEA$, he calculated the length of a line which would be to BD as $(ADEA)^2$ (the circumference squared) is to $(AD)^2$. A simple geometric proportion established that the line is equal to $(ADEA)^2/DE$, or $2\pi^2 R$.

As in the Vellum MS., he calculated how much greater 'the force of gravity' is than the endeavour from the centre of the earth, arriving this time at the conclusion that it is 350 times as great. In this paper, he went a step further. He compared the 'endeavour of the Moon to re-

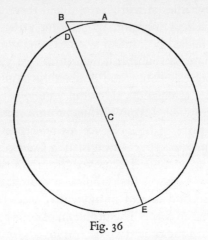
Fig. 36

cede from the centre of the Earth' with the force of gravity on the surface of the earth and found that gravity is more than 4000 times as great.

> Finally since in the primary planets the cubes of their distances from the sun are reciprocally as the squares of the numbers of revolutions in a given time: the endeavours of receding from the Sun will be reciprocally as the squares of the distances from the Sun.[79]

The basis of the inverse square relation, which became the foundation of the law of universal gravitation, was the substitution of the formula for centripetal force into Kepler's third law. When Hooke later claimed that Newton had stolen the inverse square relation from him, Newton rejected the charge indignantly, asserting that he had known it earlier. The paper in question here could well have been the basis of his assertion. It appears to be the paper that Gregory described in 1694. Written before 1669, Gregory wrote, the paper contained 'all the foundations of his philosophy . . .: namely the gravity of the Moon to the Earth, and of the planets to the Sun: and in fact all this even then is subjected to calculation.'[80] If indeed this is the paper both men referred to, in what sense could Newton and Gregory have been right? Certainly there is no mention of universal gravitation in the paper.

Nevertheless, the paper explicitly considered that the endeavour from the centre arises because a constraint holds the body in a circular path, and it assumed that the endeavour from the centre is equal and opposite to the constraining force. When he compared the endeavours of the several planets from the sun, Newton explicitly considered them to be subject to common dynamic factors, and if he did not apply the word 'gravity' to the planets, he did compare the endeavour of the moon from the earth to gravity. In the *Principia*, the correlation of the moon's motion with the acceleration of gravity provided the crucial link between the cosmos and the force at the surface of the earth whereby the name of the latter was generalised and the conception of 'universal gravitation' justified. In the paper under discussion, of course, the comparison did not reveal an exact inverse square relation. Using a lunar distance of sixty earth radii, Newton found gravity to be, not 3600 times the endeavour of the moon, but more than 4000. We know why the discrepancy arose. Because the figure he used for the size of the earth was too small, gravity at the surface of the earth was too large in comparison to the endeavour of the moon. The discrepancy does correspond, however, both with Newton's later assertion that he found it to 'answer pretty nearly' and with the testimony of three independent sources that Newton long believed there were other factors at work on the moon's orbit.[81] Even in the case of the moon, however, there is no suggestion in the paper that the earth attracts the moon. What we must remember is the explicit context. The endeavour of the moon from the earth arises because something constrains the moon to circle the earth, and the endeavour is equal in force to the constraint. If Newton did not say that the earth attracts the moon, neither did he believe, at the time he wrote the paper, that gravity on the earth is caused by an attraction. It is the result of a mechanical action – what, in the *Waste Book*, he called the pressure of the rays of gravity. It is difficult to imagine why Newton would have undertaken the comparison at all had he not been convinced that common dynamic factors operate, both to hold the planets and the moon in their orbits, and to cause heavy bodies to descend to the earth. The paper confirmed the conviction by revealing the existence of a simple quantitative relation among the motions in question, exact in the case of the planets, only approximate in the crucial case of the moon.

The mechanical philosophy made it impossible to consider that common dynamic factor to be an attraction. Both the power of precedent, stemming from Descartes' analysis of circular motion, and factors internal to Newton's dynamics made it easier to think of circular motion in terms of the moving body's endeavour to recede from the centre. The impossibility of attractions in a mechanical universe encouraged the same approach. But the exact quantitative relation he had discovered added only another to the growing list of exact quantitative relations that strained the invention of those condemned by their philosophy of nature to imagine causal mechanisms.

*

In another paper, which dates from the later half of the 1660's (and which went through more than one draft before it arrived at a more or less completed form), Newton brought together his various efforts in mechanics under the title, 'The Lawes of Motion.'[82] Again the context helps toward understanding the paper. The laws of motion that Newton attempted to formulate culminated in a fully generalised treatment of impact, the one form of interaction allowed between bodies in a mechanical universe. Whereas the treatment of impact in the *Waste Book* had confined itself primarily to the translational motions of bodies, Newton now expanded the generality of the treatment by allowing the bodies rotational motions as well. Recognising that the total motion of a body is composed both of its 'progressive' motion and of its 'circular', he attempted to supply an exact definition of the latter which would reduce it to the same quantitative terms that measure the former. Imagine that a body rotates on an axis through its centre of gravity and strikes another body equal in mass so placed that the other body is set in motion while the rotating one comes to rest. Since the real quantity of circular motion 'is more or lesse accordingly as the body hath more or lesse power and force to persevere in that motion,' he decided that the circular motion is equal to the product of the velocity of a point on its 'Equator of circulation' and its mass (or 'bulke').[83] To this definition, he added the principle of the conservation of angular momentum, here enunciated for the first time in the history of mechanics.

> Every body keepes the same reall quantity of circular motion and velocity so long as tis not opposed by other bodys.[84]

As far as progressive motion was concerned, he stated again the principle of inertia which he had adopted in the *Waste Book*.

With the ground work laid, Newton now turned to the problem of impact. Let any two bodies A and a, each with any progressive and any rotational motion, strike each other.[85] Assuming that the velocity of separation of the points of contact equals the velocity of approach, that is, assuming a perfectly elastic impact, he set down an expression, Q, equal to twice the velocity by which the points of contact approach each other. Q then is equal to the total change of velocity of the points of contact. Another expression, P, is the sum of the reciprocals of the masses and the reciprocals of the quantities Newton employed in substitution, as it were, for what we call moments of inertia. P, he said, represented the sum of the 'smallnesse of resistance' to each of the changes of motion possible.[86] Q/P was then analogous to the total change of motion, and this he parcelled out among the four factors in proportion to 'the easinesse (or smallnesse of resistance) with which those velocitys are changed . . .' That is, the total change of progressive motion of A would be equal to Q/AP, and so on. If the difference between his 'real quantity of circular motion' and our 'moment of inertia' is allowed for, his results correspond to a modern solution of the same problem.[87] Although 'The Lawes of Motion' did not employ the concept of force, the critical steps in its development flowed from the dynamic analysis of impact in the *Waste Book*. Like the rest of Newton's work on impact, it contrasted with Huygens' kinematic approach.

Aside from its steps forward – the generality of its treatment of impact and its enunciation of the principle of conservation of angular momentum – 'The Lawes of Motion' is notable also for its step backward from the relativity of motion. In the *Waste Book*, in his treatment of impact, he had seemed ready to abandon the earlier atomistic view of absolute motion and to accept the Cartesian view of inertia with its dynamic equivalence of motion and rest. 'The Lawes of Motion,' in contrast, began with a reaffirmation of absolute motion and space.

> There is an uniform extension space or expansion continued every way with out bounds: in which all bodys are each in severall parts of it: which parts of space possessed and adequately filled by them are their places. And their passing out of one place or part of space into another, through all the intermediate space is their motion.[88]

How far can attention to the problems of circular motion have dictated Newton's apparent reversal of opinion? In the *Waste Book*, examination of the problem of simple impact had led inexorably toward the relativity of motion, until the analysis of circular motion suggested complexities less readily amenable to treatment in terms of the concept of force he developed there. Is the force of a body to recede from the centre relative to a frame of reference as the force of its motion is? In 'The Lawes of Motion,' Newton seemed to assert that it is not. Not only does every body preserve the same quantity of circular motion, but it keeps the same axis of rotation as well

> if the endeavour from the axis which the two opposite quarters twixt the Equator and every meridian of motion have, bee exactly counterpoised by the opposite endeavours of the 2 side quarters, and then also its axis doth always keepe parallel to it selfe. But if the said endeavours from the axis bee not exactly counterpoised by such opposite endeavours: then for want of such counterpoise the prevalent parts shall by little and little get further from the axis and draw nearer and nearer to such a counterpoise, but shall never be exactly counterpoised. And as the axis is continually moved in the body, so it continually moves in the space too with some kind or other of spirall motion, always drawing nearer and nearer to a center or parallelisme with it selfe, but never attaining to it.[89]

Not only did he find it necessary to posit an absolute frame of reference to correspond to the absolute centrifugal endeavour, but in discussing the motion of the axis of an unbalanced body, he treated the endeavour as a force able to affect the motion of a body. Here was a step in the direction of accepting again an internal force of motion.

Although neither the 'Lawes of Motion' nor *De gravitatione* carry a date, the development of thought suggests that *De gravitatione* came later. Swept along by the violence of its anti-Cartesianism, Newton renounced completely his flirtation with the Cartesian conception of motion to which the *Waste Book* bears witness. Physical and absolute motion, he asserted against Cartesian relativity, 'is to be defined from other considerations than translation, such translation being designated as merely external.'[90] If translation is merely external, what can be essential? What except force?

> Force is the causal principle of motion and rest. And it is either an external one that generates and destroys or otherwise changes impressed motion in some body; or it is an internal principle by which existing motion or rest is conserved in a body, and by which any being endeavours to continue in its state and opposes resistance.[91]

It is true that the only examples he gave of forces that distinguish true from apparent motions were centrifugal forces. Nevertheless, when he said that 'physical and absolute' motion is to be defined by something other than translation, Newton had more than circular motion in mind. The purpose of his argument, be it remembered, was to establish the existence, not of absolute space, but of absolute motion – absolute motion in refutation of relative motion, of which one could assert with assurance neither its velocity nor its determination. In circular motion, he found his readiest example, but he clearly intended it as no more than an example. Among his further definitions, that of inertia as 'force within a body, lest its state should be easily changed by an external exciting force' could be reconciled with the relativity of motion and looked forward to Newton's mature mechanics. The definition of impetus, however, reasserted the concept of an internal force associated with uniform motion. 'Impetus is force in so far as it is impressed on a thing.'[92] He also defined gravity and conatus in terms of force. The basic message of *De gravitatione* is the displacement of relativistic kinematics by absolutistic dynamics in which force rather than translation expresses ultimate reality.

With *De gravitatione*, Newton's early work in mechanics reached its terminus. During the following fifteen years, he put mechanics aside until the composition of *De motu* in 1684 announced a new period of intense activity from which the *Principia* emerged. When Newton picked up the thread again, he found the tangles of the 1660's still there. Above all, two major ambiguities associated with the concept of force remained. Is force the measure of motion or the measure of change of motion? If it is the latter, is its paradigm case impact or free fall, is it measured by Δmv or by ma? These questions remained to plague the composition of the *Principia*.

★

Meanwhile, before ever he dreamed of the *Principia*, Newton's philosophy of nature underwent further development and a fundamental change which also played a role in the evolution of his dynamics. By the early years of the 1670's, the inchoate philosophy of nature perceptible in his early papers had taken on the tangible form we know as the 'Hypothesis of Light.' Although its major features appeared in a manuscript that dates from 1672,[93] the paper called 'An Hypothesis

Explaining the Properties of Light' received its final form late in 1675 when Newton sent it to the Royal Society together with his observations of optical phenomena in thin films. Ostensibly, he intended the 'Hypothesis' as an explanation of the observations recorded in the other paper, and optical phenomena occupied more of its space than any others. Nevertheless, in order to explain the optical phenomena, he found it necessary first to sketch in the outlines of a complete system of nature.

It was a mechanical system, conceived in the orthodox spirit of the predecessors who had shaped Newton's natural philosophy. Central to it was an aethereal medium, much like air but far rarer, subtler, and more strongly elastic. As I have argued, and as the 'Hypothesis' demonstrates anew, such an aether was the *sine qua non* of a successful mechanical philosophy of nature. By means of it, Newton explained electrical attraction, gravitation, the cohesion of bodies, elasticity, sensory perception, animal motion, heat, and, of course, optical phenomena. In proof of the aether's existence, he cited the fact that a pendulum in the empty receiver of an air pump stops swinging nearly as quickly as it does in the open air.

> Perhaps the whole frame of Nature [he suggested] may be nothing but various Contextures of some certaine aethereall Spirits or vapours condens'd as it were by praecipitation, much after the manner that vapours are condensed into water or exhalations into grosser Substances, though not so easily condensible; and after condensation wrought into various formes, at first by the immediate hand of the Creator, and ever since by the power of Nature, wch by vertue of the command Increase & Multiply, became a complete Imitator of the copies sett her by the Protoplast. Thus perhaps may all things be originated from aether.[94]

For example, he explained electrical attraction in terms of aethereal currents set up when friction vaporises the aether condensed in electric bodies – a passage which exercised some influence on the future history of electricity. The descent of aether (or of some still more subtle aethereal spirit diffused through the aether) toward the earth, caused by its condensation in fermentations and fires ('the vast body of the Earth ... may be every where to the very center in perpetuall working...[95]) bears down bodies and causes the phenomena of gravity. The sun also imbibes the aether as fuel to conserve its substance and heat, and the movement thus set up may retain the planets in their orbits. In 1686, Newton referred Halley to this passage in response to Hooke's charge of plagiary. Meanwhile, in his cosmic circulation, Newton had reproduced another basic feature of seventeenth-century mechanical philosophies, the uniformity of matter. The aether was not a substance distinct from matter in Newton's view, but a subtler form of the homogeneous matter from which all bodies in the universe are formed.

As far as optical phenomena are concerned, the crucial property of the aether is its variation in density. Aether stands rarer in the pores of bodies. When corpuscles of light pass through the aether, the varying zones of density alter the direction in which the corpuscles move, and all the phenomena of reflection, refraction, and inflection (or diffraction) derive from the varying density of the aether and its operation on the motion of corpuscles of light. From Newton's point of view, such an explanatory mechanism had considerable advantage over existing ones. It could account for the uniform regularity of reflection and refraction. Whereas previous mechanical systems had thought of light striking the constituent particles of bodies, Newton contended that reflections and refractions could not be uniform if this were the case since the particles of bodies are not arranged in a uniform pattern. The aether filling the pores, like water in sand, constitutes a uniform surface, however, and causes uniform changes of direction. Moreover, his mechanism could explain the total internal reflection of light, even in the apparently difficult case of a piece of glass in a vacuum. No previous mechanism had been able to touch this phenomenon. Above all, it could explain the periodic phenomena of thin films which Newton had been the first to examine in detail. Every thin film, such as a film of air confined between two pieces of glass, is also a thin film of aether, a film distinguished from the ambient aether by its specific density. Although his explanations of refraction and reflection assumed a zone at the edges of bodies in which the density of aether changes gradually, he now assumed a distinct surface with the properties that surfaces of water possess. When a corpuscle of light strikes such a surface, it sets up vibrations like those that a rock causes when it falls into water. Newton insisted that these vibrations are not light itself, but that they influence the motion of light. In twentieth-century terminology, they are longitudinal vibrations – pulses of density and rarity – and they travel faster than light. Thus when the corpuscle arrives at the second surface of the film, it finds a surface either rare enough to let it pass or dense enough to reflect it, and since a periodic wave controls the density of the surface, the optical phenomena of thin films are periodic.[96] The explanation of optical phenomena in the 'Hypothesis of Light' was one of the classic exercises in mechanical philosophising in the seventeenth century.

In the course of the 'Hypothesis,' Newton brought up virtually all of the crucial phenomena around which his speculations on nature revolved. The cohesion of bodies, the expansion (or pressure) of gases, surface tension, capillary action, elective affinities – they all appeared either as phenomena to be explained by the aether or as examples illustrative of principles that functioned in other explanations. Of the set of crucial phenomena, only chemical reactions generating heat were not included. Newton's serious and detailed study of chemistry did not begin long before 1675, and the chemical phenomena that seized his attention and dominated his speculations from the late 1670's had not yet assumed such importance. In a letter of 1676 to Oldenburg, which ought to be read in conjunction with the 'Hypothesis,' he discussed Boyle's famous mercury that grew hot when it was mixed with gold.[97] Here, for the first time, he attempted to treat such chemical reactions in terms of the same principles that governed his entire system of nature.

Despite his apparent success in reducing all the difficult phenomena to mechanical principles, Newton's 'Hypothesis of Light' reveals a number of internal tensions that suggest why his speculations did not harden permanently in this form. When he came to light, he argued that it consists of a multitude of extremely small corpuscles that spring from shining bodies and are 'continually urged forward by a Principle of motion, wch in the beginning accelerates them till the resistance of the Aethereall Medium equal the force of that principle . . .' Many optical phenomena do seem to require such a principle of motion. When light, understood as a stream of particles, is refracted away from the normal as it passes into a rarer medium, its velocity decreases as Descartes had shown. Decrements of velocity offered no problems to mechanical philosophers. They could imagine that the corpuscles meet a greater resistance; and as long as they did not examine the details too closely, they might consider the problem solved. Refraction is a reversible process, however; when the beam enters the denser medium and is refracted toward the normal, the corpuscles of which the beam is composed must be accelerated if the refraction is to conform to the sine law. From what source could the new motion come? In his derivation of the sine law, Descartes had assumed a new blow to increase the velocity; he had conveniently neglected to explain what in nature could correspond to the blow. Newton's 'principle of motion' answered to those phenomena in optics, like refraction toward the normal, which seemed to demand a source of activity in the particles of light. As a good mechanical philosopher, however, he insisted that such a principle could be conceived in mechanical terms. 'Some would readily grant this may be a Spiritual one; yet a mechanical one might be showne, did not I think it better to passe it by.'[98]

The principle of motion in the corpuscles of light was not the only active principle in Newton's mechanical system. The 'aethereall Spirit' diffused through the main body of the aether had similar characteristics, and Newton compared it to the 'vital aereall Spirit requisite for the conservation of flame and vitall motions' diffused through the air.[99] The phrase recalls John Mayow's nitro-aerial particles; indeed the entire 'Hypothesis of Light' bears strong resemblances to Mayow's theory. Mayow had taken the active principles of Helmontian philosophy and restated them in the vocabulary of mechanical philosophy. His nitro-aerial particles were the source of all activity in nature – of animal life, animal motion, vegetable life, elasticity, and much else. Without facing the question of what an active principle could be in a mechanical universe, he had merely tricked it out in a particulate costume and deluded himself that the issue was solved. Perhaps there was less self-delusion in Newton's 'Hypothesis,' but ultimately it expressed a similar determination that active principles could be translated into particles of matter. He stated as much in regard to the principle of motion in light, and when he mentioned the 'vital aereall Spirit' he hastened to add parenthetically '(I mean not ye imaginary volatile saltpeter).'[100] Meanwhile, the 'Hypothesis' dealt with a similar range of 'active' phenomena – electric attraction, gravity, fermentation, elasticity.

In one sense, Newton was facing the generic problem of mechanical philosophies. Whereas the tradition of Renaissance Naturalism had spoken of active principles, the mechanical philosophy insisted that the reality of nature does not have to be identical to its appearance. Both traditions had perforce to deal with the same phenomena, however, and Newton as a mechanical philosopher was acutely aware of the apparent presence of active principles in nature. Indeed, he was aware of them to the point that some commentators have been willing to describe the 'Hypothesis' as an alchemical cosmology.

He was aware also of the specificity of certain phenomena. Water and oil penetrate wood and stone whereas quicksilver does not. Quicksilver penetrates metals whereas water and oil do not. Water and acid spirits penetrate salts; oil and spirit of wine do not. Oil and spirit of wine penetrate sulphur, and acid spirits do not. A wide range of similar, or apparently similar phenomena had long been known; Renaissance naturalism had dealt with them under the rubrics of sympathy and antipathy. When Galileo had confronted the tendency of water in air to take the form of round drops, he had referred it to the antagonism of water and air, but he had deliberately shunned the word 'antipathy' and had spoken instead of a 'certain incompatibility' between them.[101] Hooke had extended Galileo's suggestion in order to explain capillary action. Positing general principles of 'congruity' and 'incongruity' whereby similar bodies readily unite and dissimilar ones do not, he had argued that air is more incongruous to glass than water is, with the result that the pressure of the atmosphere decreases in narrow pipes.[102] Newton picked up the concept from Hooke, renamed it 'sociability,' and like Hooke employed it to explain both capillary action and the chemical phenomena cited above.

Although the mechanical tradition offered various devices, such as the shapes of particles and pores, that could be drawn upon in such cases, Newton explicitly rejected them. Liquids and spirits, he asserted, 'are disposed to pervade or not pervade things on other accounts then their Subtility . . .' Some fluids such as oil and water 'though their pores are in freedome enough to mix with one another,' nevertheless do not mix. Newton called it 'some secret principle of unsociableness . . .'[103] He returned to it in his letter to Oldenburg of 26 April 1676. Written in response to Boyle's communication in the *Philosophical Transactions* that he possessed a mercury that became hot when it was mixed with gold [104] the letter contained Newton's first discussion of a topic that would bulk ever larger in his future speculations – chemical reactions that generate heat. In the letter, he explained the heat by a combination of ordinary mechanical philosophy and the principle of sociability. It seemed to him 'yt ye metalline particles wth wch yt ☿ [mercury] is impregnated may be grosser yn ye particles of ye ☿ & be disposed to mix more readily with ye ☉ [gold] upon some other account then their subtility, & then in so mixing, their grossness may enable them to give ye parts of ye gold ye greater shock, & so put ym into a brisker motion then smaller particles could do.' He compared

Boyle's impregnated mercury with a 'corrosive liquor' such as *aqua fortis*. In the 'Hypothesis,' he had suggested that unsociable bodies can frequently be made sociable by the mediation of a third body; thus lead, which does not mix with copper when they are melted together, mixes with it readily when a little tin or antimony is added. In a similar way, the saline palticles in corrosive liquors, like the metallic particles in Boyle's mercury, 'may be of a middle nature between ye liquor wch they impregnate & ye bodies they dissolve & so enter those bodies more freely & by their grossness shake ye dissolved particles more strongly then a subtiler agent would do.'[105]

Neither in the 'Hypothesis' nor in the letter to Oldenburg did Newton attempt to reduce the 'secret principle' of sociability to mechanical terms, although he employed it in mechanical contexts where it aided the power of such mundane factors as size. Redolent of hermetic tradition, it refused to be made sociable to the mechanical philosophy and stood out starkly against its background. In the case of the principles of motion or of activity mentioned in the 'Hypothesis,' Newton asserted their mechanical nature, although he did not venture to interpret how that might be. With their immediate Helmontian forebears, they too suggested the lingering presence in his thought of a tradition alien to the mechanical. His intensive study of alchemical literature during the latter years of the 1670's may well have intensified these influences.[106]

*

There is some reason to think that Newton's response to Boyle's communication in the *Philosophical Transactions* was conceived primarily as an indirect effort to confirm the acquaintance of a man whom he had met the previous spring and whose every work he had read and digested. Whether or not he intended it so, the result was such. In November of 1676, Boyle sent Newton a book, probably *Experiments, Notes, &c. about the Mechanical Origin or Production of Divers Particular Qualities*, his most recent publication.[107] Apparently the book became the subject of discussion between the two men; when Newton wrote to Boyle on 28 February 1678/9, the well known letter on the aether, he began with a reference to such a discussion.

> Honoured Sr
> I have so long deferred to send you my thoughts about ye Physicall qualities we spake of, that did I not esteem my self obliged by promise

> I think I should be ashamed to send them at all. The truth is my notions about things of this kind are so indigested yt I am not well satisfied my self in them, & what I am not satisfied in I can scarce esteem fit to be communicated to others, especially in natural Philosophy where there is no end of fansying.[108]

Certainly the content of the letter resembles the content of Boyle's *Mechanical Origin of Qualities* in the topics it considered, though not always in the solutions it proffered. The second sentence quoted above is a typical example of Newton's pretence of unconcern by which he sought to forestall criticism. In fact, the letter marks a further stage in his uninterrupted speculation on the nature of physical reality, devoted in this case almost exclusively to the essentially chemical phenomena of cohesion, dissolution, and vaporisation.

As in 1675, he began by positing the existence of an aether, strongly elastic, varying in density, and far more subtle than air. Although it pervades the pores of bodies, it stands rarer in them than in free space, and this property he employed to explain such phenomena as capillary action and the coherence of two polished marbles in the exhausted receiver of an air pump. In passing, Newton explained refraction and diffraction as he had in 1675; and at the end of the letter, he offered to explain gravity (i.e., heaviness at the surface of the earth) in terms of differential aethereal densities and differential sizes of aethereal particles. The latter explanation diverged from that in the 'Hypothesis' but shared with the former its mechanical orthodoxy. Newton's main interest in the letter did not focus on such matters, however. He was concerned primarily with the dissolution of bodies in acids, the generation of aerial substances, and, as a necessary preliminary to these, the cohesion of bodies.

The greater rarity of aether in the pores of bodies does not terminate exactly at their surface; it extends beyond them through a narrow zone of increasing density. When two bodies are brought close to each other, the aether between them must be rarified before they can touch, and the rarefaction of the aether requires the application of force. In fact, it is difficult to bring two bodies into immediate contact. His investigation of the colours in thin films had shown Newton that considerable pressure is necessary to bring two pieces of glass into contact, even when one of them is convex. Because of the aether, bodies have an 'endeavour to recede from one another . . .' Hence flies can walk on water without wetting their feet, a pile of dust does not cohere even when pressed, and vapours and exhalations expand when pressure is removed from them. If enough force is applied to overcome the reluctance of the aether to be rarefied, so that the bodies do in fact touch, then the pressure of the denser aether around them holds them together. There is a critical distance at which the endeavour to recede, due to the reluctance of the aether to be rarefied, is balanced by the pressure of the surrounding aether; inside the critical distance, bodies show an 'endeavour to accede . . .' and force is necessary to part two bodies in contact with each other. Separated beyond the critical distance, however, they spring apart. By employing such principles, Newton explained the dissolution of bodies. Suppose a soluble body is put in water. The particles of water enter the pores of the body and by entering equalise the pressure of the aether on all sides of a particle, whereupon any movement can shake it loose.[109]

So far so good – but all bodies are not soluble in water. Water cannot enter the pores of metal, for example, to dissolve it.

> Not yt water consists of too gross parts for this purpose, but because it is unsociable to metal. For there is a certain secret principle in nature by wch liquors are sociable to some things & unsociable to others.[110]

To confirm the existence of the principle, Newton cited the same list of phenomena he had cited three years earlier. Water sinks into wood and not into metals; quicksilver sinks into metals and not into wood. *Aqua fortis* dissolves silver but not gold; *aqua regia* dissolves gold but not silver. It is interesting to note in passing that virtually every one of the chemical reactions (and the phenomena he considered identical) cited by Newton, both in the letter to Boyle and in later writings such as Query 31, are found in Boyle's works, and Newton's knowledge of them undoubtedly derived from that source. What he brought to chemistry was not a new body of information drawn from his own experimentation, but a new approach to information generally available.[111] When a liquid is unsociable to a body, he continued, it can frequently be made sociable by the mediation of a third substance. Thus saline spirits in water enable it to dissolve metals. By their sociableness, such particles can enter the pores of the metal, carrying water with them, and dissolution is effected in the manner described above. If a new substance such as salt of tartar, which is more sociable than the metal to the saline particles, is added to the solution, it seizes the saline particles which mediated between the metal and the water, and the metal precipitates.

Departing from the explanatory model of his earlier letter to Oldenburg, Newton now employed his 'endeavour to recede' to account for the heat generated in the acid solution of metals.

> In ye solution of metals, when a particle is loosing from ye body, so soon as it gets to that distance from it where ye principle of receding ... begins to overcome ye principle of acceding ... the receding of ye particle will be thereby accelerated, so yt ye particle shall as were wth violence leap from ye body, & putting ye liquor into a brisk agitation, beget & promote yt heat we often find to be caused in solutions of Metals.[112]

When the violence is great enough to hurl the particle out of the solution, it generates an aerial substance.[113] In the case of some substances, such as water, the particles of which are small, heat alone is sufficient to shake particles free. We call its aerial form vapour. Because of their size, the endeavours of vaporous particles to recede are weak, and vapours quickly condense again into water. More permanent aerial substances are made of larger and denser particles, which require in turn a greater force than mere heat to separate them initially. Their source is fermentations.

> Which has made me sometimes think that ye true permanent Air may be of a metallic original: the particles of no substance being more dense then those of metals. This I think is also favoured by experience for I remember I once read in ye P[hilosophical] Transactions how M. Hugens at Paris found that ye air made by dissolving salt of Tartar would in two or three days time condense & fall down again, but ye air made by dissolving a metal continued without condensing or relenting in ye least. If you consider then how by the continual fermentations made in ye bowels of ye earth there are aereal substances raised out of all kinds of bodies, all wch together make ye Atmosphere & that of all these ye metallic are ye most permanent, you will not perhaps think it absurd that ye most permanent part of ye Atmosphere, wch is ye true air, should be constituted of these: especially since they are ye heaviest of all other & so must subside to ye lowest parts of ye Atmosphere & float upon ye surface of ye earth, & buoy up ye lighter exhalation & vapours to float in greatest plenty above them. Thus I say it ought to be wth ye metallic exhalations raised in ye bowels of ye earth by ye action of acid menstruums, & thus it is wth ye true permanent air....

> The air also is ye most gross unactive part of ye Atmosphere affording living things no nourishment if deprived of ye more tender exhalations & spirits yt flote in it: & what more unactive & remote from nourishment then metallick bodies.[114]

In a paper recently published for the first time, which was intimately associated with the letter to Boyle, Newton's speculations advanced another pace.[115] *De aere et aether* appears to have been an effort to express the central principles of the letter to Boyle in the more systematic form of a treatise. With the announcement that he would begin a treatise on the nature of things by considering heavenly bodies and among them the ones most accessible to the senses, to wit the air, in order that he might proceed under the guidance of his senses,[116] he proceeded to list the principal characteristics of air. The most remarkable of its properties is its capacity for great rarefaction and condensation. Newton suggested that there are several causes for the rarefaction of air. For example, air seeks to avoid other bodies; it is rarer in proximity to them. It stands rarer in the pores of bodies as the various capillary phenomena (which, he added, do not occur in a vacuum) demonstrate. As air seeks to avoid bodies, so bodies tend to fly from each other. It is difficult to bring two lenses into contact with each other. Dust does not cohere when it is pressed together. Insects walk on the surface of water without wetting their feet. Divers opinions have been offered concerning this repulsion. The intervening medium may give way with difficulty. God may have created an incorporeal nature that repels bodies; bodies may have a hard nucleus surrounded by a sphere of tenuous matter which does not readily admit other bodies. 'About these matters I do not dispute at all. But as it is equally true that air avoids bodies, and bodies repel each other mutually, I seem to gather rightly from this that air is composed of the particles of bodies torn away from contact, and repelling each other with a certain large force.'[117] With this in mind, we can understand the phenomena that air presents. Air is condensed or rarefied according to the pressure on it. Halve the pressure, and the volume doubles. Diminish the pressure to one hundredth or even to one thousandth, and the volume increases a hundred or a thousand times. Such an expansion hardly seems possible, Newton remarked, if the particles of air are in mutual contact;

> but if by some principle acting at a distance [the particles] tend to recede mutually from each other, reason persuades us that when the distance between their centers is doubled the force of recession will be halved,

when trebled the force is reduced to a third and so on, and thus by an easy computation it is discovered that the expansion of the air is reciprocal to the compressive force.[118]

The generation of air is explained by the same principles. Whenever the particles of a body are separated, by heat, by friction, or by fermentation, they spring apart. Various chemical reactions in which heat is generated illustrate the point, and the explosion of gun powder best of all – 'almost all the substance of the mixt is changed by vehement agitation into an aerial form, the huge force of this powder arising from its sudden expansion, as is the nature of air.'[119] Moreover, aerial substances differ according to the bodies from which they are generated. Vapours, which are the lightest and least permanent, arise from liquids. Exhalations, which are of a middle nature, come from vegetable matter. True permanent air, which is also the heaviest, is generated from metals through subterranean corrosions. As the indestructible nature of metals demands, true permanent air serves to sustain neither life nor fire.[120]

De aere et aethere consists of two chapters. Chapter one, *On Air*, went further than anything Newton had heretofore written in its assertion of forces between particles. Whereas the letter to Boyle had spoken of an 'endeavour to recede,' he now asserted flatly that bodies repel each other at a distance. Nevertheless the meaning of the assertion is ambiguous, for he named chapter two '*De aethere*,' and in the chapter he apparently intended to explain repulsions by means of aethereal mechanisms. Already in chapter one, he had in passing referred the cohesion of bodies to the pressure of the aether. Chapter two now set out to describe the aether more fully. As terrestrial bodies can be broken into particles and converted into air, so particles of air can be further broken and converted into a more subtle air called aether. That such an aether exists, subtle enough to penetrate the pores of bodies, is shown by Boyle's experiment in which a metal gained weight when heated in a closed flask. 'And that in a glass empty of air a pendulum preserves its oscillatory motion not much longer than in the open air, although that motion ought not to cease unless, when the air is exhausted, there remains in the glass something much more subtle which damps the motion of the bob.' [sic][121] Magnetic and electric effluvia demonstrate its existence as well. And at this point, not yet at the bottom of the page and in the middle of a sentence, the treatise stopped.

Who can say why Newton stopped? Perhaps he was called away to dinner, accompanied friends to a tavern for the evening, and lost the impulse to continue before he saw the paper again. It is also possible – and I propose it as a possible interpretation – that the tensions developing in Newton's philosophy of nature now reached the breaking point. I have been arguing that these tensions were present – first the active principles and the secret principle of sociability and unsociability, now forces of repulsion between particles, though perhaps such forces were not distinct from either the active principles or the principle of unsociability. I have argued as well that the tensions expressed themselves in his approach to a set of crucial phenomena – the cohesion of bodies, surface tension, capillary action, the expansion of gases, certain chemical reactions – the keys, in Newton's mind, to unlock the secret operations of nature. In the 'Hypothesis of Light,' they all found their place in the mechanical theory of the aether. In the letter to Boyle, they were related to the endeavours to accede and recede. They appeared again in *De aere et aethere*, expressions now of forces of repulsion (and implicitly attraction), which were apparently to be explained in turn, as the 'endeavours' of the letter to Boyle, by a subtle aether. The next important speculation on nature that has survived, dating from about 1686 and the composition of the *Principia*, revolved around the same set of phenomena, all of which now unmistakably revealed attractions and repulsions between particles as the ultimate processes of nature. Did the abrupt termination of *De aere et aethere* mark the point at which Newton's philosophy of nature underwent its decisive change in direction? Certainly we can say that what he had written so far offered food for thought should he have paused to ruminate. Compared to the inherent complications of aethereal mechanisms, the concept of repulsion offered considerable economy. As the discussion of Boyle's Law demonstrated, it disclosed an avenue leading toward quantitative precision which was otherwise obscured by a cloud of aethereal dust. Above all, *De aere et aethere* seemed to have embarked on an infinite regress. If the aether were, as he asserted, an elastic fluid like air, and if the elasticity of air were due to the mutual repulsion of its particles, what would he gain by referring the repulsion of aerial particles to the aether? What made aethereal particles repel each other?

In 1685 or '86, in writing the *Principia*, Newton described an experiment he had performed some years before. Although he did not indicate the exact date, he did say he had lost the paper on which he had recorded it and was forced to recite it from memory. It is not unreasonable to speculate that the experiment was connected with the reversal of opinion that I have tentatively placed about 1679.

> Lastly, since it is the most commonly held opinion of philosophers in this age that there is a certain aethereal medium extremely rare and subtile, which freely pervades the pores of all bodies; and from such a medium, so pervading the pores of bodies, some resistance must needs arise; in order to try whether the resistance, which we experience in bodies in motion, be made upon their outward surfaces only, or whether their internal parts meet with any considerable resistance upon their surfaces, I thought of the following experiment.[122]

He set up a pendulum eleven feet long with an empty wooden box as its bob, taking considerable pains to minimise friction at the point of suspension. He then pulled it aside about six feet and carefully marked the points to which it returned on the first three swings. Having weighed the box together with half the string, and having added to their sum the calculated weight of the air inside, he filled the box with metal and found that it weighed seventy-eight times as much. Since the string stretched, he adjusted it to the original length, and pulling the pendulum aside to the same point as before, he counted the swings until it returned successively to the three marks. It took seventy-seven swings to return to the first, seventy-seven more to the second, and as many again to the third. Since the box filled had seventy-eight times as much inertia, he concluded that its resistance full was to its resistance empty in the ratio of 78/77. By calculation, he concluded that the resistance on the external surface of the empty box was more than 5 000 times as great as the resistance on its internal parts.

> This reasoning [he concluded] depends upon the supposition that the greater resistance of the full box arises from the action of some subtle fluid upon the included metal. But I believe the cause is quite another. For the periods of the oscillations of the full box are less than the periods of the oscillations of the empty box, and therefore the resistance on the external surface of the full box is greater than that of the empty box in proportion to its velocity and the length of the spaces described in oscillating. Hence, since it is so, the resistance on the internal parts of the box will be either nil or wholly insensible.[123]

What the pendulum had given the pendulum had taken away. Newton's aethereal hypothesis had rested on the existence of an aether able to act with great effect on the innermost parts of bodies, and the experiment of the pendulum in a vacuum had been a major piece of evidence confirming the existence of an aether with such capabilities. Apparently the resistance it offered was much greater than that of the air since the removal of air scarcely affected the time required for the oscillations to die out. Now an experiment considerably more refined pointed to exactly the opposite conclusion. The interpretation most favourable to the aethereal hypothesis put the resistance 5 000 times less than that of the air. Such an aether would have been useless for Newton's purposes, and in fact he interpreted the experiment to prove that there is no aether at all. The conclusion possibly reached as he wrote *De aere et aethere* was confirmed. Bodies act on each other at a distance, attracting and repelling, as he put it, 'with a certain large force.'

*

The significance of the redirection of his philosophy of nature undertaken by Newton about 1679 can scarcely be overstated. What he proposed was an addition to the ontology of nature. Where the orthodox mechanical philosophy of seventeenth-century science insisted that physical reality consists solely of material particles in motion, characterised by size, shape, and solidity alone, Newton now added forces of attraction and repulsion, considered as properties of such particles, to the catalogue of nature's ontology. As I shall argue later, the ultimate ontological status of forces in Newton's conception of nature is a complex and involved question. In his published writings, he chose to refer to his true opinion, as I understand it, only obliquely, and we must consult unpublished manuscripts to seize his meaning. As far as the works he published are concerned, and above all, as far as the conceptual tools he employed in scientific discussion are concerned, he treated forces as entities that really exist. In orthodox mechanical philosophy, the moving particle was the ultimate term of explanation. In confronting magnetic and static electric phenomena, for example, the kinetic mechanical philosopher refused to employ terms such as 'attraction', which were held to be occult, and turned in preference to the imaginary construction of invisible mechanisms whereby particles of matter produce the motions observed. Newton's philosophy of nature did not cease to be mechanical. Physical reality still consisted of material particles in motion, but the ultimate term of explanation was now the force of attraction or repulsion that altered a particle's state of motion. Newton's conception of nature has been called a dynamic mechanical philosophy. In the *Principia*, he referred to magnetic and electric phenomena, which he had earlier traced to effluvial mechanisms, as examples readily observable of the forces he posited throughout nature. Whereas the moving particle had been the term that itself required no explanation, active principles, forces, now played that role in a radically reoriented philosophy of nature.

> Have not the small Particles of Bodies certain Powers, Virtues, or Forces, by which they act at a distance, not only upon the Rays of Light for reflecting, refracting, and inflecting them, but also upon one another for producing a great Part of the Phaenomena of Nature? For it's well known, that Bodies act one upon another by the Attractions of Gravity, Magnetism, and Electricity; and these Instances show the Tenor and Course of Nature, and make it not improbable but that there may be more attractive Powers than these. For Nature is very consonant and conformable to her self.[124]

Published initially in 1706 with the first Latin edition of the *Opticks*, the opening sentences of Query 31 (*Quaestio* 23 of the 1706 *Optice*) state the central proposition of the dynamic mechanical philosophy. Although the Query was composed more than twenty years after the major reorientation of Newton's philosophy of nature, it embodied the classic statement of the outlook that remained essentially unaltered during the rest of his life. I have argued that the change of outlook occurred about 1679, not long after the letter to Boyle and the associated paper *De aere*. Whether or not it occurred in 1679, it had occurred by 1686–7 when, as he completed the *Principia*, Newton composed his next extensive speculations on the nature of physical reality. Intended originally as a formal conclusion to the work, then as a preface, and finally suppressed entirely, the papers ranged over a set of phenomena less extensive but otherwise similar to those on which Query 31 later built.[125] For us, their significance lies in the fact that they expounded a philosophy of nature based on forces between particles – forces neat, undiluted by aethereal fluids. In the 1690's, Newton wrote a brief paper that later appeared in Harris' *Lexicon Technicum* under the title of *De natura acidorum*.[126] During the 1690's, he also wrote and rewrote the papers that finally emerged as the queries attached to the *Opticks*; and after the English and Latin editions of the *Opticks* (1704 and 1706), he both completed the further set of eight Queries (numbers 17–24), which he added to the second English edition of 1717, and composed the General Scholium, which he appended to the second edition of the *Principia* in 1713. In all, the papers amounted to a body of manuscripts more considerable by far than the speculative writings before 1679 but less varied in their thrust. Once and for all, the decisive turn had been made. Newton's speculations had reached their destined goal. More fully than any other document, Query 31 embodies his ultimate philosophy of nature.

Newton did not easily bring himself to publish his most private thoughts. As mentioned above, he drafted extensive discussions of his philosophy of nature to include in the *Principia*, and then suppressed them. As he put the *Opticks* together from his earlier papers on light, he projected a fourth book in which he would move from the role of attractions and repulsions in optical phenomena to the affirmation of the general presence of such forces in nature. Again he had second thoughts; and in the first edition of the *Opticks*, he confined himself to the briefest assertions of the interactions of light and matter in the sixteen Queries the edition contained. Finally two years later with the Latin edition, he screwed up his courage and carried the publication through. Meanwhile, the projected Book IV of the first edition contained an important statement which further illuminates his view. Newton intended Book IV to begin with a series of propositions about light – twenty in one version, eighteen in another.[127] Following the propositions, a series of 'Hypotheses' generalised their meaning. Hypothesis 2 is the most revealing of Newton's understanding of the significance of the philosophy he proposed.

> Hypoth 2 As all the great motions in the world depend upon a certain kind of force (wch in this earth we call gravity) whereby great bodies attract one another at great distances: so all the little motions in ye world depend upon certain kinds of forces whereby minute bodies attract or dispell one another at little distances.
>
> How the great bodies of ye Earth Sun Moon & Planets gravitate towards one another what are ye laws & quantities of their gravitating forces at all distance from them & how all ye motions of those bodies are regulated by their gravities I shewed in my Mathematical Principles of Philosophy to the satisfaction of my readers. And if Nature be most simple & fully consonant to her self she observes the same method in regulating the motions of smaller bodies wch she doth in regulating those of the greater. This principle of nature being very remote from the conceptions of Philosophers I forbore to describe it in that Book leas[t it]

should be accounted an extravagant freak & so prejudice my Readers against all those things wch were ye main designe of the Book: & yet I hinted [at them] both in the Preface & in ye book it self where I speak of the [refraction] of light & of ye elastick power of ye Air: but [now] the design of y^t book being secured by the approbation of Mathematicians, [I have] not scrupled to propose this Principle in plane words. The truth of this Hypothesis I assert not because I cannot prove it, but I think it very probable because a great part of the phaenomena of nature do easily flow from it wch seem otherwise inexplicable: such as are chymical solutions precipitations philtrations, detonizations, volatizations, fixations, rarefactions, condensations, unions, separations, fermentations: the cohesion texture firmness fluidity & porosity of bodies, the rarity & elasticity of air, the reflexion & refraction of light, the rarity of air in glass pipes & ascention of water therein, the permiscibility of some bodies & impermiscibility of others, the conception & lastingness of heat, the emission & extinction of light, the generation & destruction of air, the nature of fire & flame, ye springnesse or elasticity of hard bodies.[128]

Compared to his earlier speculations, one of the distinctive features of his later writings, including Query 31, is the vastly expanded set of chemical phenomena referred to. More than anything else, the conclusion that forces are the ultimate causal agents in nature emerged from Newton's ruminations on chemistry.[129] In Query 31, after the initial paragraph asserting the general existence of forces, he turned at once to the evidence offered by chemical phenomena. Salt of tartar runs *per deliquium* (that is, it deliquesces) because of the attraction its particles exert on particles of vapour in the air, and although water alone distils with a gentle heat, the attraction of salt of tartar is such that water distils from it only with the application of great heat. Common salt and saltpetre, in contrast to salt of tartar, do not run *per deliquium* because their particles do not attract vapour. When compound spirit of nitre is poured on a heavy animal or vegetable oil, the two liquids become so hot as to emit flame. Does not 'this very great and sudden Heat,' Newton asked,

argue that the two Liquors mix with violence, and that their Parts in mixing run towards one another with an accelerated Motion, and clash with the greatest Force? And is it not for the same reason that well rectified Spirit of Wine poured on the same compound Spirit flashes; and that the Pulvis fulminans, composed of Sulphur, Nitre, and Salt of Tartar, goes off with a more sudden and violent Explosion than Gunpowder, the acid Spirits of the Sulphur and Nitre rushing towards one another, and towards the Salt of Tartar, with so great a violence, as by the shock to turn the whole at once into Vapour and Flame?'[130]

Even ordinary sulphur, powdered and mixed with an equal weight of iron filings and enough water to make a paste, becomes too hot to be touched in five or six hours. Newton was convinced that such experiments offer the key to understanding earthquakes and thunderstorms. Sulphurous steams in the bowels of the earth ferment with minerals, and if confined, explode causing earthquakes. If they escape into the atmosphere and ferment there with acid vapours which abound in the air, they take fire and cause lightning, thunder, and fiery meteors. Newton regarded heat as the motion of the particles of which bodies are composed; when heat is generated from the mixture of two cold substances, new motion has appeared.

> Now the above-mention'd Motions [he said in regard to earthquakes and storms] are so great and violent as to shew that in Fermentations the Particles of Bodies which almost rest are put into new Motions by a very potent Principle, which acts upon them only when they approach one another, and causes them to meet and clash with great violence, and grow hot with the motion, and dash one another into pieces, and vanish into Air, and Vapour, and Flame.[131]

In addition to reactions that generate heat, those that reveal elective affinities grasped Newton's attention. Salt of tartar precipitates metals from acid solutions. 'Does not this argue that the acid Particles are attracted more strongly by the Salt of Tartar than by the Metal, and by the stronger Attraction go from the Metal to the Salt of Tartar?' Iron, copper, tin, or lead, added to a solution of mercury, precipitate the mercury and go into solution themselves. Copper precipitates silver. Iron precipitates copper. In each case, the metal that goes into solution attracts the acid more strongly than the one it precipitates. For the same reason, iron requires more *aqua fortis* to dissolve it than copper, and copper more than the other metals; and of all metals iron rusts most easily, and after iron, copper.[132] *Aqua fortis* dissolves silver but not gold; *aqua regia* dissolves gold but not silver. Is this not due to the fact that the acid in each case, *aqua fortis* in regard to gold and in regard to silver *aqua regia*, is subtle enough to penetrate the metal 'but wants the attractive Force to give it Entrance.' When spirit of salt precipitates silver from *aqua fortis*, it not only attracts the acid and mixes with it, but it may repel the silver as well.

And is it not for want of an attractive virtue between the Parts of Water and Oil, of Quicksilver and Antimony, of Lead and Iron, that these Substances do not mix; and by a weak Attraction, that Quick-silver and Copper mix difficultly; and from a strong one, that Quick-silver and Tin, Antimony and Iron, Water and Salts, mix readily? And in general, is it not from the same Principle that Heat congregates homogeneal Bodies, and separates heterogeneal ones?[133]

As already intimated, attractive forces were not alone in Newton's world. Similar evidence established the existence of repulsive forces as well. When salt is dissolved in water, it does not sink to the bottom even though it is heavier than water, but rather diffuses evenly through the whole solution. This implied to Newton that 'the Parts of the Salt ... recede from one another, and endeavour to expand themselves, and get as far asunder as the quantity of Water in which they float, will allow.'[134] The fact that saline solutions evaporate to leave crystals of regular shapes confirms the belief that the particles of salt in solutions are held apart at regular distances. The immense expansion of which air is capable argues for the existence of similar repulsions between its particles.

Newton was beginning now to move beyond the range of chemical phenomena into the other crucial phenomena around which his speculations on nature had always revolved. In 1675, they had all been associated in one way or another with the aether. All were now explained by forces of attraction and repulsion between particles of matter. A work entitled Opticks invites comparison with a paper named 'An Hypothesis of Light,' and in fact the contrasting treatments of optical phenomena reflect the redirection of Newton's philosophy. What density gradients accomplished in the 'Hypothesis,' attractions performed in the Opticks. 'Do not Bodies act upon Light at a distance, and by their action bend its Rays ...?' Query 1 asserted under the guise of a rhetorical question.[135] As it turned out, optics done with forces had nearly all the advantages of optics done with an aether. The integrated force of multiple particles presented a uniform surface to cause uniform refractions and reflections just as the aether before had appeared to smooth out the rough surface of random particles. Attractions could even explain the reflection of light from the backside of a piece of glass. One thing they could not account for was the periodicity of the phenomena of thin films. Vibrations in the aether had been the foundation of Newton's explanation of periodic phenomena, and he could imagine no way to make forces vibrate in a similar way. Hence periodicity, one of his notable contributions to optics and originally the focus of his interest in thin films, underwent a considerable devaluation in the *Opticks*; and under the implausible and enigmatic term, 'Fits of easy transmission and reflection,' it was presented as an unexplained empirical fact.[136]

The explanation of Newton's other crucial phenomena underwent an analogous change. The cohesion of bodies obviously arises from the attractions of particles for each other. To account for cohesion, Newton argued, some philosophers invent hooked atoms, which only beg the question. Others contend that bodies are glued together by the quiescence of their particles relative to each other – that is, he added with a sly thrust which could not have been lost on his audience, by an occult quality.

> I had rather infer from their Cohesion, that their Particles attract one another by some Force, which in immediate Contact is exceeding strong, at small distances performs the chymical Operations above-mention'd, and reaches not far from the Particles with any sensible Effect.[137]

Analogous to the internal cohesion of bodies, and cited in earlier papers by Newton, were such facts as the cohesion of two polished pieces of marble even in a vacuum and the cohesion of mercury well purged of air to the top of a tube as much as seventy inches long if it were carefully and gently erected.[138] He now accounted for both phenomena by attractions, and he even explained how an aether that penetrates bodies could not possibly account for them.[139]

The smallest particles, which cohere by the strongest attractions, compose larger particles with weak attraction, and these in turn form still larger particles of still weaker virtue. The attractive forces in question here reach only a small distance from the particles in any case, in Newton's opinion. 'And as in Algebra, where affirmative Quantities vanish and cease, there negative ones begin; so in Mechanicks, where Attraction ceases, there a repulsive Virtue ought to succeed.'[140] The reflection of light, together with the expansion of air and the diffusion of salts in solution cited above, were evidence for such repulsions. So also were capillary phenomena. The repulsion between particles of air and glass causes air to be rarer in narrow glass pipes with the result that water in them rises to a higher level.[141] As Newton liked to observe, the fact that capillary phenomena do not take place in the exhausted receiver of an air pump confirms this explanation. Sometime after the first appearance of the Query in the Latin edition of 1706, Newton tried the experiment (or had Hauksbee try it), an experiment the outcome of which he had confidently affirmed on the authority of Hooke for more than forty years.[142] When it turned out that capillary action does in fact occur in a vacuum, he silently altered the passage in subsequent editions to an affirmation of attractive forces between liquids and glass. The difficulty of bringing two convex lenses into contact with each other further demonstrates the existence of repulsions, as does the ability of flies to walk on water without wetting their feet and the impossibility of making powder cohere merely by pressing it together.

The conviction that forces between particles do in fact exist led Newton radically to alter his conception of matter as well. Perhaps he did not alter his conception of matter as such, in that he continued to assert the almost universally held opinion that all bodies are composed of one and the same matter, matter which is qualitatively neutral and differentiated solely by its division into particles of given sizes and shapes. To the standard attributes of such particles, he added, of course, the forces discussed above. In the beginning, God created matter in 'solid, massy, hard, impenetrable, moveable Particles' of sizes and shapes befitting His ends.[143] Atomist that he was, Newton maintained that the particles endure through all time unchanged. What makes atoms immutable? He replied, in effect, that bodies divide where they have interstices and pores. By definition, atoms have none and are therefore incapable of division.[144] Just to be sure, he added that no 'ordinary Power' is able to divide what God Himself made one in the creation.[145]

Although he started with the same matter, the bodies that he composed from it in his imagination departed wildly from the accepted model. He rejected first the idea that bodies are composed of 'irregular particles casually laid together like stones in a heap . . .'[146] Quite the contrary, the matter of bodies is highly textured – not a pile of stones, but a crystal lattice. The same God who formed animals, formed matter as well, and He textured it according to the purposes for which He intended it. 'Bodies,' Newton asserted further, 'are far more rare than is commonly believed.'[147] Gold, the densest substance we know, cannot be completely solid since it absorbs mercury and reveals itself to be translucent when it is beaten into thin foils. Assume for the moment that gold is, in volume, half solid matter and half voids. Gold is nineteen times as dense as water. Thirty-seven parts out of thirty-eight in a volume of water must then be empty space, yet water cannot be compressed. In fact water must be rarer by far than the comparison with gold suggests. Consider its transparency. Whatever light may be, if a ray passing through a medium were to strike a solid particle, it would be turned aside from its path, and we can hardly imagine the circumstances by which it could be turned again exactly into the original one. Water, however, is not only transparent, it is transparent from every angle. 'How bodies may be sufficiently porous to give free passage *every way* in right lines to the rays of light is hard to conceive but not impossible.'[148] We must have recourse, he concluded, to 'some wonderful and very skillfully contrived texture of particles' whereby bodies are composed 'on the pattern of nets [*more retium*].'[149]

> Suppose a body were composed of smaller parts lying together with as much pores between them, & that these parts were composed of smaller parts lying together with as much pores between them & that these parts were composed of parts still smaller with as much pores between them & so on for as many compositions as you please untill you come at solid parts void of all pores. And if in this progression there were ten compositions, the body would have above a thousand times more pores then parts; if twenty compositions, it would have above a thousand thousand times more pores then parts; if thirty compositions it would have above a thousand thousand thousand times more pores then parts & so on perpetually.[150]

As he remarked, whoever is able to explain how water can be as porous merely as the comparison of its density to that of gold demands, and yet be incapable of compression, can doubtless explain by the same hypothesis how gold can be as porous as he pleases. In contrast to our own kinetic view of matter, Newton's model was static – diaphanous threads tenuously strung in a three dimensional net. With the exception of that point, however, his conception of matter embodied a startling premonition of future conclusions. In contrast to Descartes' plenum, Newton's universe was a vast expanse of empty space seasoned with the subtlest suggestion of solid matter. Radically exhausted beyond any performance of Mr Boyle's machine, the immensity of space was now populated solely by limited sections of the slenderest threads. Those threads in turn were composed, not of woven fibres, but of point-like particles – reminiscent of Henry More's *parva naturalia* with which Newton had dallied in his undergraduate notebook – laid out in rows and held together, as they could only be held together, by forces of attraction. Populating the universe with forces was the condition for depopulating it of matter.

*

Both in the *Opticks* and elsewhere, Newton expressed some anxiety about the reception of an idea the rejection of which had been one of the basic premises of seventeenth-century natural philosophy. He denied vigorously that forces were equivalent to occult qualities. Quite the contrary, they were 'general Laws of Nature, by which the Things themselves are form'd; their Truth appearing to us by Phaenomena, though their Causes be not yet discover'd.'[151] He recurred frequently to the question of causes – what he called attractions might be caused in many ways, such as the impulses of an intervening medium. Meanwhile the existence of attractions and repulsions, whatever their cause, appeared to him as a valid and demonstrated conclusion.[152] He liked to think of them as general principles, applicable to the whole of natural philosophy, providing a foundation on which it could rest securely. The trouble with the mechanical philosophy was its tendency to fragment into multiple *ad hoc* resolutions of phenomena.

> Could all the phaenomena of nature be deduced from only thre or four general suppositions there might be great reason to allow those suppositions to be true: but if for explaining every new Phaenomenon you make a new Hypothesis if you suppose y^t ye particles of Air are of such a figure size & frame, those of water of such another, those of Vinegre of such another, those of sea salt of such another, those of nitre of such another, those of Vitriol of such another, those of Quicksilver of such another, those of flame of such another, those of Magnetick effluvia of such another, If you suppose that light consists in such a motion pression or force & that its various colours are made of such & such variations of the motion & so of other things: your Philosophy will be nothing else then a systeme of Hypotheses. And what certainty can there be in Philosophy wch consists in as many Hypotheses as there are Phaenomena to be explained.[153]

His aim, in contrast, was to uncover what he called 'general principles' or 'general properties of things'[154] which could explain all phenomena without recourse to similar *ad hoc* assumptions.

> To tell us that every Species of Things is endow'd with an occult specifick Quality by which it acts and produces manifest Effects, is to tell us nothing: But to derive two or three general Principles of Motion from Phaenomena, and afterwards to tell us how the Properties and Actions of all corporeal Things follow from those manifest Principles, would be a very great step in Philosophy though the Causes of those Principles were not yet discover'd: And therefore I scruple not to propose the principles of Motion [forces between particles] above-mention'd, they being of very general Extent, and leave their Causes to be found out.[155]

One cannot avoid the conclusion that Newton had deluded himself when he spoke of two or three general principles of nature. His forces between particles offered an invitation to *ad hoc* hypotheses at least as broad as that of ordinary mechanical philosophy. After all, Boyle had liked to talk of the catholic principles of matter and motion before he set about imagining his multiplicity of particles, and before Boyle, Descartes had conceived of the mechanical philosophy as a set of general explanatory principles. In a similar way, Newton could speak of attractions and repulsions in a general way, but the possible specific attractions and repulsions were legion. And the very intellectual current that had carried him to the point of asserting their existence constantly drove him on toward the fragmentation of his general principles into myriad *ad hoc* ones. As both Query 31 and a host of manuscripts suggest, one of the major bodies of evidence that led him toward the conviction that forces exist was composed of chemical reactions displaying elective affinities. In earlier papers, he had referred similar phenomena to 'a certain secret principle of sociability,' a name that recalls the origin of specific attractions in Renaissance Naturalism.[156] The sympathies and antipathies of Renaissance Naturalism, the occult qualities treated with such disdain throughout the seventeenth century, had attempted to embody and express the specificity so obviously attached to sundry natural agents. The loadstone had presented itself as the very incarnation of the sympathies and antipathies thought to populate nature. It attracts iron but not copper. Unlike poles attract each other but like poles repel. Part of the mechanical philosophy's reaction against Renaissance Naturalism was the repudiation of specificity and the assertion of generality. One common matter, differentiated solely by the shape and size of particles, forms the substance of all bodies. Hence Boyle could assert with confidence that any substance can be made from any other. Despite Newton's confident assertion that he had established general principles, his attractions and repulsions threatened constantly to fragment nature into a multiplicity of specific agents. Whereas the mechanical philosophy affirmed one universal matter, Newton's speculations sometimes spoke of 'homogeneal Bodies' and bodies 'of the same nature.'[157]

As a seventeenth-century philosopher of nature, Newton did not propose lightly to accept this conclusion toward which his speculations seemed to tend. In the '*Conclusio*,' which he intended for the first edition of the *Principia*, he affirmed that 'the matter of all things is one and the same, which is transmuted into countless forms by the operations of nature . . .'[158] Although he did not finally include the '*Conclusio*' in the book, edition one did contain an Hypothesis III which asserted the same thing.

> Every body can be transformed into a body of any other kind whatever and endued successively with all the intermediate grades of qualities.[159]

A manuscript version went on to say that nearly all philosophers affirm as much. All Cartesians, peripatetics, and others, teach 'that all things arise from imposing divers forms and textures on a certain common matter and that all things are resolved again into the same matter by the privation of forms and textures.'[160] At times, Newton seemed to suggest that the variation of forces demanded by his speculations could result from the composition of larger and larger corpuscles out of ultimate uniform particles. The pretended reconciliation with the prevailing mechanical conception of matter was a delusion on his part, however. From homogeneous particles it was impossible to compose corpuscles endowed with the specific attractions and repulsions to which his speculations pointed.[161] In light of the reaction of orthodox mechanical philosophers to the idea of gravitational attraction, he was undoubtedly wise to conceal his full philosophy of nature when he published the *Principia*. If he was not reviving sympathies and antipathies, he was introducing their immediate offspring, and mechanical philosophers would not have failed to perceive the resemblance.

Moreover, his habit of referring to forces as 'active principles' suggests a greater influence of Renaissance Naturalism than he would readily have acknowledged. Attractions are not evenly distributed among substances in his opinion; on the contrary, they are associated especially with acids. Water has only a small dissolving force because it contains so little acid. 'For whatever doth strongly attract, and is strongly attracted, may be call'd an Acid.'[162] Hence the vigour with which acids dissolve. In various substances, such as sulphur, an acid combined with earthy material lies hidden and suppressed, and by its attraction it initiates fermentations and putrefactions. Sulphurous bodies also act upon light more strongly than others.

> Prop. 1 The refracting power of bodies in vacuo is proportional to their specific gravities.
> Note that sulphureous bodies caeteris paribus are most strongly refractive & therefore tis probable y^t y^e refracting power lies in y^e sulfur & is proportional not to y^e specific weight or density of y^e whole body but to that of y^e sulphur alone. ffor $\frac{\triangle}{+}$s do most easily conceive y^e motions of heat & flame from light & y^e action between light & bodies is mutual.[163]

Newton also entertained the thought that corpuscles of light themselves are the ultimate source of active powers.

> Do not bodies & light mutually change into one another. And may not bodies receive their most active powers from the particles of light wch enter their composition? . . . Now since light is the most active of all bodies known to us, & enters the composition of all natural bodies, why may it not be the chief principle of activity in them?

Attractions are always stronger in proportion to bulk in smaller bodies. By considering the velocity of light and the sharpness of its changes of direction, we can calculate that the attractive force of its corpuscles in proportion to their bulk is more than ten hundred thousand thousand million [10^{15}] times greater than the gravity of bodies at the surface of the earth. 'And so great a force in the rays cannot but have a very great effect upon the particles of matter wth which they are compounded, for causing them to attract one another.'[164]

As in the case of his early 'Hypothesis of Light,' Newton's concept of active principles inevitably recalls John Mayow's nitro-aerial particles. Mayow had mechanised Helmontian active principles by casting them into corpuscular form, but his nitro-aerial particles had continued to cause all the 'active' phenomena that had been ascribed to them within the tradition of Renaissance Naturalism. Newton's forces frequently appeared to do the same. Well before he composed the Queries, he had perfected his early mechanics in the *Principia*. There he had developed his early idea of internal and external forces into the concepts of *vis inertiae*, by which bodies resist changes in their state of motion or rest, and impressed forces, which operate from without to generate new motions in bodies. *Vis inertiae* is a passive principle, he stated in Query 31; by it alone there could never have been any motion in the world. 'Some other Principle was necessary for putting Bodies into Motion; and now they are in Motion, some other Principle is necessary for conserving the Motion.' By a brief reference to the composition of motions, he demonstrated quickly that the quantity of motion in the universe, in the Cartesian sense of the concept, cannot be constant. Because of friction, the tenacity of fluids, and imperfect elasticity in bodies, indeed, 'Motion is much more apt to be lost than got, and is always upon the Decay.' The statement must surely stand as the strangest assertion in the vast corpus of papers left by the man who formulated the principle of inertia in its enduring form. It is only comprehensible in the context of an argument that affirms the existence of active principles, which would constantly increase the quantity of motion if motion as such did not constantly decay.

To confirm the assertion that motion constantly decays, Newton cited the impact of imperfectly elastic bodies, that is, of all bodies that really exist. Although he believed that heat is the motion of the particles of bodies, he did not suggest at this point that the motion (or energy) lost could show up as heat. Drawing on his analysis in the *Principia*, he further cited vortical motion. If three round vessels are filled, one with water, one with oil, and one with molten pitch, and all three are stirred into vortical motion, the pitch will come to rest quickly, and the oil less quickly. Although the water will keep its motion longest, it too will lose it in only a short time. Only a vortex of matter free of all tenacity and internal friction would never stop, but such matter does not exist.

> Seeing therefore the variety of Motion which we find in the World is always decreasing, there is a necessity of conserving and recruiting it by active Principles, such as are the cause of Gravity, by which Planets and Comets keep their Motions in their Orbs, and Bodies acquire great Motion in falling; and the cause of Fermentation, by which the Heart and Blood of Animals are kept in perpetual Motion and Heat; the inward Parts of the Earth are constantly warm'd, and in some places grow very hot; Bodies burn and shine, Mountains take fire, Caverns of the Earth are blown up, and the Sun continues violently hot and lucid, and warms all things by his Light. For we meet with very little Motion

in the World, besides what is owing to these active Principles. And if it were not for these Principles, the Bodies of the Earth, Planets, Comets, Sun, and all things in them, would grow cold and freeze, and become inactive Masses; and all Putrefaction, Generation, Vegetation and Life would cease, and the Planets and Comets would not remain in their Orbs.[165]

Throughout the seventeenth century, mechanistic modes of expression disguised the survival of animistic modes of thought from earlier philosophies of nature. Such was especially true in regard to those phenomena – above all, phenomena of life – to which the relatively crude mechanisms of seventeenth-century thought were most inadequate. Animistic philosophies had not been wholly arbitrary constructions; they had expressed common perceptions of apparent sources of spontaneous activity in the physical world. For the most part, the mechanical philosophy of the seventeenth century had no adequate explanation of the same apparent sources of activity, and it covertly introduced its own active principles in the guise of special particles or fluids. Mayow's nitro-aerial particles are only one example of a common procedure. In part, Newton's forces were merely a different device to achieve the same end. In another sense, of course, his forces, as attractions and repulsions, were concepts uniquely adapted to a mechanical philosophy of nature. A mechanical philosophy containing forces differed from the orthodoxy of seventeenth-century science, to be sure, but it still discussed natural phenomena in terms of the motions of particles. As the passages above indicate, however, his forces embodied the same ambiguity as Mayow's particles. As 'active principles', they equally concealed a depth of animistic connotation which could never be translated, in the concepts available to Newton, into pure mechanism.

★

In the later years of his life, Newton's approach to forces in their mechanical function underwent one further development – a change more apparent than real. The basic reorientation of his natural philosophy, which took place about 1679, involved not only the acceptance of forces between particles but also the abandonment of a subtle aether as the locus of invisible mechanisms to account for apparent attractions and repulsions. During the following thirty years, the climactic period of Newton's life when, at the height of his powers, he produced the *Principia* and the *Opticks*, he rejected the existence of such an aether. There is no document from these years of which I know in which he argued unambiguously for the existence of an aether.[166] In the second decade of the eighteenth century, however, an aether reappeared in his papers on natural philosophy, condensing as it were from the still more subtle vapours of his speculations. He referred to it apparently at the end of the General Scholium added to the second edition of the *Principia* in 1713, and in 1717 the second English edition of the *Opticks* contained eight new Queries that discussed it at length. In the preface (or Advertisement) of that publication, he stated that 'to shew that I do not take Gravity for an essential Property of Bodies, I have added one Question concerning its Cause . . .'[167]

In fact, this final development in Newton's philosophy of nature proceeded in two distinct stages; the 'certain most subtle spirit which pervades and lies hid in all gross bodies . . .'[168] that he referred to in the General Scholium was not identical to the aether of the eight new Queries four years later. The first fluid was specifically electrical. In his 'Hypothesis of Light,' Newton had cited some crude static electric phenomena as evidence for an aether. During his years with the Royal Society in London, the experiments of Francis Hauksbee rekindled his memory and reawakened dormant ideas. Static electric phenomena powerfully suggested the action of material effluvia to every early experimenter. Even William Gilbert, who was certainly not a mechanist, saw electric attraction in such terms, distinguishing its material causation from the immaterial causation of magnetic action. When Newton began to draft the passages that ultimately appeared as the General Scholium, he adopted a vaguely similar distinction between magnetic and electric attraction on the one hand and gravity on the other.[169] He had not assigned a cause for gravity, he said, because he had not been able to demonstrate one from phenomena. Certainly it cannot arise from the centrifugal tendencies of a vortex. Mechanical causes in general act on the surfaces of bodies and are proportional to their surfaces. Gravity, in contrast, is proportional to the mass of bodies, and any cause of gravity must penetrate a body to its every particle. 'From the phenomena it is entirely certain that gravity exists and acts on all bodies in proportion to distances according to the laws described above . . .' It suffices to explain the motions of planets and comets and is a law of nature whether or not we know its cause.

> In the same way that the system of the sun, planets, and comets is moved by the forces of gravity and perseveres in its motions, so also smaller systems of bodies seem to be moved by other forces, especially by electrical force, and their parts to be moved variously among themselves.[170]

In its original versions, the General Scholium maintained a distinction, obscured in the Scholium's final form, between gravity, which explains the macrocosmic motions, and the microcosmic forces between particles. The laws by which the latter act 'are completely different from the laws of gravity.' Gravity remains constant; they increase and decrease (or intend and remit). Gravity always attracts; they sometimes repel. Gravity is proportional to the quantity of matter; only iron has magnetic force; while many bodies have electric force, it is not proportional to their masses. Gravity acts at great distances, magnetic and electric forces only at short ones. Magnetic force decreases roughly in proportion to the cube of the distance and is ineffectual at any great distance. Electric force acts only at short distances unless it is excited by friction.[171] At this time, the expansion of the sphere of electric action by friction particularly caught his eye. The electric spirit 'is capable of contraction & dilatation [sic] expanding it self to great distances from the electric body by friction...'[172] Newton then went on to apply the electric spirit to the entire range of phenomena on which he had initially rested his assertion of forces between particles.

> Whoever therefore will have undertaken to explain the phenomena of nature which depend on the fermentation and vegetation of bodies and on the other motions and actions of the smallest particles among themselves will need especially to turn his mind to the forces and actions of the electric spirit, which (if I am not mistaken) pervades all bodies and to investigate the laws that this spirit observes in its operations. Then magnetic attractions will also need to be considered since particles of iron enter the composition of many bodies. These electric and magnetic forces ought to be examined first in chemical operations and in the coagulation of salts, snow, crystal, fluxes [*fluorum*], and other minerals in regular figures and sometimes in the figures of plants, and afterwards to be applied to the explanation of the other phenomena of nature.[173]

The 'subtle spirit' in gross bodies that the last paragraph of the General Scholium introduced was not a full resuscitation of his earlier aethereal system. Confined to solid bodies, the electric spirit deployed itself to cause a set of attractions and repulsions between particles that differed from gravitational attraction.

As he prepared the second English edition of the *Opticks*, Newton intended originally to include a fuller discussion of this spirit. The Latin edition had contained twenty-three Queries. On at least one occasion, he composed two additional ones devoted to the electric spirit. *Quaestio* 23 of the Latin edition (Query 31 as we now know it) had asserted the existence of forces between particles of matter.

> Quaest. 24. May not the forces by wch the small particles of bodies cohere & act upon one another at small distances for producing the above mentioned phaenomena of nature, be electric? ffor altho electric bodies do not act at a sensible distance unless their virtue be excited by friction, yet that virtue may not be generated by friction but only expanded. ffor the particles of all bodies may abound with an electric spirit wch reaches not to any sensible distance from the particles unless agitated by friction or by some other cause & rarefied by the agitation. And the friction may rarefy the spirit not of all the particles in the electric body but of those only wch are on the outside of it: so that the action of the particles of the body upon one another for cohering & producing the above mentioned phenomena may be vastly greater then that of the whole electric body to attract at a sensible distance by friction. And if there be such an universal electric spirit in bodies certainly it must very much influence the motions & actions of the particles of the bodies amongst one another, so that without considering it, philosophers will never be able to give an account of the Phenomena arising from those motions & actions. And so far as these phaenomena may be performed by the spirit wch causes electric attraction it is unphilosophical to look for any other cause.
>
> Quaest 25 Do not all bodies therefore abound wth a very subtile active potent elastic spirit by wch light is emitted refracted & reflected, electric attractions & fugations are performed, & the small particles of bodies cohaere when contiguous, agitate one another at small distances & regulate almost all their motions amongst themselves.[174]

Before he published the second English edition, however, Newton changed his mind, and instead of proposing the existence merely of an electric spirit in gross bodies that causes their mutual actions on one another, he inserted eight new Queries (numbers 17 through 24) that asserted the existence of an 'Aethereal Medium' pervading all space.[175] To this aether he referred the reflection and refraction of light, the propagation of heat, and the transmission of nervous impulses (both sensory and motor). He also employed it to explain gravitational attraction, though a reference to electric 'Exhalations' and magnetic 'Effluvia' implies that he continued to distinguish the cause of macroscopic attractions from that of microscopic ones.[176] Since he mentioned the cause of gravity in the preface, it has been almost universally accepted that he introduced the aether to provide a mechanical explanation of the forces which had appeared so occult to a generation raised on the mechanical philosophy. Two things need to be noticed before this rationale for the aether is accepted. First, he devoted more space in the eight new queries to optics than to gravity, and one of the things the aether offered him was a vibrating medium to explain the periodic phenomena of Book II. Query 17, the first of the new Queries, applied the aether explicitly to this purpose. Second, if the aether had been introduced primarily to explain gravity, the explanation it offered was one hardly calculated to appease mechanical philosophers who took the trouble to read it with care. The aether was extremely rare and extremely elastic. Since pulses had to travel in it faster than the speed of light, and since the speeds of light and sound had been measured, he calculated that the elasticity of the aether in proportion to its density (the crucial factor in the velocity of pulses through elastic media, according to Newton's analysis) must be more than 490 000 000 000 times the corresponding proportion in air. Such a combination of elasticity and rarity could exist only if the particles of the aether 'endeavour to recede from one another' with great force.[177] In fact, the reappearance of the aether changed nothing in Newton's philosophy of nature. Composed of particles repelling each other, the aether embodied the very problem of action at a distance which it pretended to explain. In no way did it constitute a compromise with the basic premise on which Newton's philosophy of nature had rested for more than thirty-five years.

★

The ultimate nature of that basic premise remains to be fully explored. As far as the scientific level of discourse was concerned, Newton was prepared to speak in terms of attractions and repulsions as real forces residing in bodies, and in so doing he separated himself consciously from the prevailing mechanical philosophy. Nevertheless, he constantly inserted disclaimers stating that he did not mean to determine the causes of the phenomena when he used the words 'attraction' and 'repulsion'.

Several such passages appeared in the *Principia*, and he repeated their intent in Query 31.

> How these Attractions may be perform'd, I do not here consider. What I call Attraction may be perform'd by impulse, or by some other means unknown to me. I use that Word here to signify only in general any Force by which Bodies tend towards one another, whatsoever be the cause.[178]

To Richard Bentley, he expressed himself more bluntly.

> Tis unconceivable that inanimate brute matter should (without ye mediation of something else wch is not material) operate upon & affect other matter wthout mutual contact... That gravity should be innate inherent & essential to matter so yt one body may act upon another at a distance through a vacuum wthout the mediation of anything else by & through wch their action or force may be conveyed from one to another is to me so great an absurdity that I beleive no man who has in philosophical matters any competent faculty of thinking can ever fall into it. Gravity must be caused by an agent acting constantly according to certain laws, but whether this agent be material or immaterial is a question I have left to the consideration of my readers.[179]

In fact, he had not left it to the consideration of his readers. He had filled Book II of the *Principia* with arguments meant to demonstrate, not only the impossibility of Cartesian vortices, but the non-existence of any material aether; and in this respect, Book II only repeated what his private papers affirmed with greater vehemence. If not a material aether, then an immaterial aether – the passage to Bentley contrived to suggest it twice. What could an immaterial aether be? To Newton, it was the infinite omnipotent God, who by His infinity constitutes absolute space and by His omnipotence is actively present throughout it.

In the youthful *De gravitatione*, Newton had tentatively explored the idea that infinite space may be the sensorium of God, and the properties and movements of bodies the direct effects of His will. We know by experience that we can move our own bodies; similarly God, to whom all bodies are immediately present, can move them at His will. Far from being forgotten, the idea became the ultimate foundation of Newton's conception of nature. In Queries 28 and 31 of the *Opticks* and in the General Scholium of the *Principia*, he brought himself only so far as an evasive and obscure presentation of his innermost thoughts;[180] their full exposition can be found only in the more intimate surroundings of his private papers.[181] A draft dating from around 1705 for what we know as Query 31 raised the question of how bodies act on each other at a distance. The ancient atomists attributed gravity to atoms without explaining it except by figures. For example, they represented God and matter by Pan and his pipe, by which they seemed to mean that matter depends on God, not only for its existence but also for its laws of motion. The Cartesians make God the author of matter; there is no reason not to recognise Him as the author of its laws as well. Matter is passive. It remains in its state of rest or of motion and cannot move itself. It receives motion in proportion to the force impressed on it, and resists as much as it is resisted. These are purely passive laws, but experience teaches us that there are active laws as well.

> Life & will are active Principles by wch we move our bodies, & thence arise other laws of motion unknown to us. And since all matter duly formed is attended with signes of life & all things are framed wth perfect art & wisdom & Nature does nothing in vain: if there be an universal life & all space be the sensorium of a thinking being who by immediate presence perceives all things in it . . . the laws of motion arising from life or will may be of universal extent.[182]

In fact, the universal presence of God performed all the functions that Newton ascribed to the aether in the 'Hypothesis of Light.' God was an incorporeal aether who could move bodies without offering resistance to them in turn.[183] When two corpuscles appear to attract each other, that is, when they move toward each other, God rather than a material medium is the cause.

> Dividing the whole of nature between body and void, the Epicureans denied the existence of God, but this is plainly absurd. For two planets separated from each other by a great expanse of void do not mutually attract each other by any force of gravity or act on each other in any way except by the mediation of some active principle that stands between them by means of which force is propagated from one to the other. [According to the opinion of the ancients, this medium was not corporeal since they held that all bodies by their very natures were heavy and that atoms themselves fall through empty space toward the earth by the eternal force of their nature without being pushed by other bodies.] Therefore the ancients who grasped the mystical philosophy more correctly taught that a certain infinite spirit pervades all space, and

contains and vivifies the entire world; and this supreme spirit was their numen; according to the poet cited by the Apostle: In him we live and move and have our being. Hence the omnipresent God is recognized, and by the Jews is called 'place'. To the mystical philosophers, however, Pan was that supreme numen.... By this symbol, the philosophers taught that matter is moved in that infinite spirit and by it is driven, not at random, but harmonically, or according to the harmonic proportions as I have just explained.[184]

From the point of view of Newton's ultimate metaphysics, then, forces were no more real entities in the universe than they were from the point of view of orthodox mechanical philosophy. In the one case, apparent attractions and repulsions were the effects of invisible mechanisms, aetherial effluvia of one sort or another which pushed bodies about and created the appearance of attractions and repulsions. In the other case, they were the effects of an incorporeal medium, the infinite God who, in His sensorium, controls and moves the material world even as we control and move our bodies. As Newton said in the General Scholium, 'a being, however perfect, without dominion, cannot be said to be Lord God.'[185] Dominion his God assuredly had – every movement in the world was the immediate effect of His power.

If the net result of the reorientation in Newton's natural philosophy was the substitution of an immaterial aether for a material one, was the change in fact significant at all? The possibility of abstracting the concept of force from the question of its ultimate causation was open to the orthodox mechanical philosopher quite as much as it was open to Newton. In the one case a material aether moves bodies as though they attract each other; in the other case an immaterial medium does the same. Had anything changed? In my opinion, everything had changed. What the new philosophy of nature made possible was less the concept of force than its precise mathematical formulation. After a century of conflict and antagonism in which the mechanical tradition constantly frustrated the Pythagorean, Newton had found a means to their reconciliation, as Leibniz had found a different, but not altogether different, one. To Descartes, Galileo's kinematics of free fall had been a pointless exercise since the mechanism that causes bodies to fall could not possibly generate a uniformly accelerated motion. The vortical mechanism of the heavens was also incapable of yielding Kepler's laws. The demand for causal mechanisms had constantly thwarted the drive to express the mathematical regularities in nature. Newton's immaterial aether, the omniscient God, was free of exactly that shortcoming. Newton realised as much and insisted on it. 'There exists,' he stated, 'an infinite and omnipresent spirit in which matter is moved according to mathematical laws.'[186] Such a convenient medium delivered him immediately from the preoccupation with causal mechanisms and directed his steps down the royal road toward a quantitative, as opposed to a verbal, dynamics.

If in practice Newton did not advance beyond a qualitative consideration of most of the forces he discussed, all of them were quantifiable in principle. His natural bent, moreover, inclined him toward their quantification whenever possible. By comparing measured velocities and radii of curvatures of inflected paths, he calculated that the attractive force of a corpuscle of light is 10^{15} times as powerful in proportion to its quantity of matter as the gravity of a projectile.[187] Capillary action offered another opportunity for quantification. Using two parallel sheets of glass set at different distances from each other, he could compare the quantity of water raised to the distance between the sheets of glass. This in turn suggested a more refined experiment which he performed with Hauksbee's aid. Taking two sheets of glass about twenty inches long, they placed one sheet horizontal and laid the other on it so that they touched at one end and were separated at the other enough to contain an angle of about 10 or 15 minutes. A drop of orange juice placed between the separated ends touched both plates. The drop began at once to move toward the end where the sheets were in contact – an accelerated motion caused, in Newton's opinion, by the attraction of the glass for the drop of juice. If the end toward which the drop was moving were then raised, the drop could be brought to rest when the inclination became steep enough that the weight of the drop balanced the motivating attraction. 'And by this means you may know the Force by which the Drop is attracted at all distances from the Concourse of the Glasses.' Newton concluded that the attraction varies inversely as the distance between the glasses for a constant area of attracting surface. At very short distances, such an attraction could become immense. For example, at a distance of three-eighths of the ten hundred thousandth part of an inch (the distance had relevance to his study of colours in thin films – a film of water this thick reflects almost no light, and according to Newton's theory particles of such diameter are among the smallest from which macroscopic bodies are composed) the attraction exerted over a circular surface an inch in diameter would suffice to support a column of water two or three furlongs in length – another demonstration of the extent to which atomic forces exceed gravitational attraction.[188] For most of the forces on which Newton speculated, data were not available to allow even this crude degree of calculation. Nevertheless, all of them were quantifiable in principle, and all of them potentially could function within a quantitative dynamics.

As far as the history of science after Newton is concerned, the ultimate significance of his philosophy of nature lies in the new possibilities it opened to the science of dynamics. In the case of gravitational attraction, a force not in the least prominent among those which dominated Newton's attention at the time of his change of view, he was able to demonstrate the incredible power of a quantitative dynamics joined to a quantitative kinematics, and to carry into actuality the programme of quantification potentially present from the beginning in his concept of force. And it was possible in this case because he had abandoned the mechanisms of a material aether in favour of the Divine Medium who moves bodies as though they attract each other according to exact mathematical laws.

NOTES

1. Cf. William T. Costello, *The Scholastic Curriculum at Early Seventeenth-Century Cambridge*, (Cambridge, Mass., 1958).
2. University Library Cambridge, *Add. MS. 3996*, ff. 1–26, 34–81.
3. The word for word correspondence of passages, especially under the early headings, establishes Newton's knowledge of Charleton beyond any question. I have not been able to find similar notes from Gassendi, and it may be that Newton's knowledge of him was entirely mediated through Charleton. I find this conclusion impossible to accept, however. As I shall be arguing, the influence of Gassendi's atomism on Newton was profound and enduring. Gassendi's *Opera* were available, as well as his individual works, and Newton's habits of study appear, in my eyes, to preclude the possibility that he would ignore the very source of one of the dominant influences on his life. Whether he read Gassendi or not, the French atomist left his imprint via Charleton.
4. Roger North, *The Lives of the Right Hon. Francis North, Baron Guilford; the Hon. Sir Dudley North; and the Hon. and Rev. Dr John North. Together with the Autobiography of the Author*, ed. Augustus Jessopp, 3 vols. (London, 1890).
5. University Library Cambridge, *Add. MS. 3996*, f. 88.
6. Walter Charleton, *Physiologia Epicuro-Gassendo-Charltoniana: or a Fabrick of Science Natural upon the Hypothesis of Atoms*, (London, 1654), p. 107.
7. University Library Cambridge, *Add. MS. 3996*, f. 93.
8. Ibid., f. 100v.

9. Ibid., f. 111. Cf. *The Works of the Honourable Robert Boyle*, ed. Thomas Birch, 5 vols. (London, 1744), **I**, 27.
10. University Library Cambridge, *Add. MS. 3996*, f. 112.
11. Ibid., f. 103v. Cf. 'Of Species visible,' f. 104v, a perplexing passage in which Newton considered how the old doctrine of antiperistasis might account for the continuing motion of a 'Globulus of light' through the aether.
12. For example, under the heading 'Of Gravity & Levity' Gassendi's conception of calorific and frigorific particles and his identification of magnetism and gravity (with subtle threads causing both) were the objects of proposed experiments: 'Try whither ye weight of a body may be altered by heate or cold, by dilatation or condensation, beating, poudering, transfering to serverall places or sevearll heights or placing a hot or heavy body over it or under it, or by magnetisme. Whither leade or its dust spread abroade, whither a plate flat ways or edg ways is heaviest....' (Ibid., f. 121v.)
13. Ibid., ff. 97, 121.
14. Ibid., f. 121v.
15. University Library Cambridge, *Add. MS. 3970.3*, ff. 473-4.
16. University Library Cambridge, *Add. MS. 3996*, f. 121v. The last of the three passages cited continues: 'To make an experiment concerning this increase of

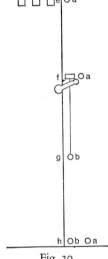

Fig. 30

motion When y^e Globe *a* is falne from *e* to *f* let y^e Globe *b* begin to move at *g* soe y^t both y^e globes fall together at *h*.' (See Fig. 30.) Although at this time Newton correctly drew the consequence that weight is proportional to solidity (or, in his atomic view, mass) from Galileo's conclusion that all bodies fall with the same acceleration, he was apparently not fully convinced of it. Indeed, as we shall see later, he accepted that proportionality finally and conclusively only in 1685. Meanwhile, the heading 'Of Rarity & Density. Rarefaction & Condensa-

FORCE IN NEWTON'S PHYSICS

tion' contains an entry which demonstrates his vacillation at the time of the 'Questiones'. 'two bodys given to find wch is more dense. Upon ye Threds *da* & *ce* hang ye bodys *d* & *e*. & exactly twixt y^m hang ye spring *sbt* by a thred soe y^t it have liberty to move to move [sic] any way. then compress ye spring *bs* to *bt* by ye thred *st*. Then {clipping / cutting} ye thred *st* ye spring shall cast both ye body *d* & *e* from it & they receve alike swiftnes from ye spring if there be ye same quantity of body in both otherwise y^e body *bo* (being fastened to y^e spring) will move towards y^e body wch hath less body in it. wch motion may be observed by compareing y^e motion of y^e point (*o*) to y^e point *p* & other points in y^e resting body

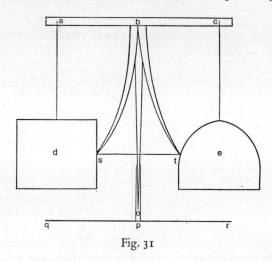

Fig. 31

qr.' (See fig. 31.) (Ibid., f. 94.) Instead of simply weighing the bodies to compare their masses, that is, he devised a mechanism to compare their inertial masses.

17. Ibid., ff. 122–23v. Cf. A. R. Hall, 'Sir Isaac Newton's Note-book, 1661–65,' *Cambridge Historical Journal*, **9** (1948), 239–50, and R. S. Westfall, 'The Development of Newton's Theory of Colour,' *Isis*, **53** (1962), 339–58.
18. University Library Cambridge, *Add. MS. 3996*, f. 94v.
19. Ibid., ff. 90v–91. Under the heading 'Of Softnesse hardnes fflexility Ductility Tractility,' Newton included the further question relative to the cohesion of bodies: 'Whither hard bodys stick together by branchy particles foulded together. Cartes.' (Ibid., f. 95v.)
20. Ibid., f. 121. The figure, '400' is that given by Galileo in the *Discourses* (p. 80) and may be the nearest thing to evidence that Newton had read them. Cf. I. B. Cohen, 'Newton's Attribution of the First Two Laws of Motion to Galileo,' *Atti del Simposio su 'Galileo Galilei nella storia a nella filosofia della scienza,'* Firenze-Pisa, 14–16 September 1964, pp. XXIII–XLII. Since there were other channels through which Galileo's figure could have gotten to Newton, I am not proposing to challenge Cohen's conclusion, that Newton had not read the *Discourses*, on the basis of evidence so slender.

21. University Library Cambridge, *Add. MS. 3996*, f. 103. Both the facts cited here and the theory to explain them derived from Boyle's *New Experiments Physico-Mechanical, Touching the Spring of the Air* (*Works*, **I**, 54.)
22. University Library Cambridge, *Add. MS. 3996*, ff. 100v, 111, 100.
23. Ibid., ff. 111v, 110.
24. Ibid., f. 99.
25. Ibid., f. 117.
26. *De gravitatione et a equipondio fluidorum* (published with an English translation in *Unpublished Scientific Papers of Isaac Newton*, ed. A. R. and Marie Boas Hall, Cambridge University Press, 1962, pp. 89–156) carries no date. On the basis of handwriting alone, everyone who has handled it places it in the late 1660's. Newton's hand went through distinct, identifiable stages, and although one must retain a degree of scepticism in such cases, I think there can be no more than minimal doubt as to the general time of the essay. A reference to a letter of Descartes (p. 147) indicates that it postdated the Clerselier edition of Descartes' correspondence in 1668.
27. Cf. J. E. McGuire, 'Body and Void and Newton's De Mundi Systemate: Some New Sources,' *Archive for History of Exact Sciences*, **3** (1966), 206–48.
28. Hall and Hall, *Unpublished Papers*, p. 122.
29. Ibid., pp. 124–7.
30. Ibid., p. 148.
31. Ibid., p. 128.
32. Ibid., p. 129.
33. Ibid., p. 130.
34. Cf. R. S. Westfall, 'Newton and Absolute Space,' *Archives internationales d'histoire des sciences*, **17** (1964), 121–32.
35. Pierre Gassendi, *Opera omnia*, 6 vols. (Lyons, 1658), **I**, 179–228.
36. Hall and Hall, *Unpublished Papers*, p. 132.
37. Ibid., p. 136.
38. Ibid., p. 138.
39. Ibid., p. 140.
40. Ibid., p. 139.
41. Ibid., p. 140.
42. Ibid., p. 138.
43. Ibid., p. 141.
44. Ibid., p. 143.
45. Ibid., p. 142.
46. Ibid., p. 139.
47. For example, he compared the density of quicksilver to that of the aether, indicating a continued acceptance of it (ibid., p. 147).
48. University Library Cambridge, *Add. MS. 3996*, f. 100.
49. Ibid., ff. 98–98v, 113–14. The essay is published entire in John Herivel, *The Background to Newton's* Principia, (Oxford, The Clarendon Press, 1965), pp. 121–5.
50. *Waste Book* (University Library Cambridge, *Add. MS. 4004*), f. 10v. Herivel has published 'Of Reflections' with the exception of a few passages that Newton cancelled; the present citation is found on p. 141.

51. *Waste Book*, f. 12. *Background*, p. 153.
52. *Waste Book*, f. 12. *Background*, pp. 153–4. The passage is extensively corrected and revised with resulting damage to syntax. Herivel's decision on its final content differs slightly from mine. Newton finally cancelled the entire paragraph.
53. *Waste Book*, f. 10v. *Background*, p. 137.
54. *Waste Book*, f. 10v. *Background*, p. 141.
55. *Waste Book*, f. 12v. *Background*, p. 156.
56. *Waste Book*, f. 12v. *Background*, p. 155.
57. *Waste Book*, f. 12v. *Background*, p. 157.
58. *Waste Book*, f. 13. *Background*, p. 158.
59. *Waste Book*, f. 12v. *Background*, p. 157.
60. *Waste Book*, f. 12. *Background*, p. 150.
61. *Waste Book*, f. 10v. *Background*, p. 141.
62. *Waste Book*, f. 12v. As I mention below, Newton cancelled the passage.
63. *Waste Book*, f. 13. *Background*, p. 157.
64. *Waste Book*, f. 13. *Background*, p. 159.
65. *Waste Book*, f. 13. *Background*, p. 159.
66. *Waste Book*, ff. 14–14v. *Background*, pp. 168–9.
67. *Waste Book*, f. 10v. *Background*, p. 138.
68. *Waste Book*, f. 10v. *Background*, pp. 137–9.
69. *Waste Book*, f. 11v. *Background*, pp. 146–7.
70. *Waste Book*, f. 11v. *Background*, p. 147. The same concept was repeated again in Proposition 24 which came a few paragraphs later (*Waste Book*, f. 12. *Background*, p. 148).
71. *Waste Book*, ff. 11v–12. *Background*, p. 148.
72. *Waste Book*, f. 1. *Background*, p. 131. Since Herivel has not published all of the material on f. 1, the first quotation above is not found in his *Background*.
73. *Waste Book*, f. 1. *Background*, pp. 129–30. Although the passage appears on f. 1, I am prepared to accept Herivel's argument that it must have been entered after the material on ff. 11v–12 (*Background*, p. 88). For the purpose of my argument, however, the relative chronology of the two entries does not matter. My account of the material on f. 1 does not add anything to that of Herivel, who first examined and published it.

Newton's analysis of circular motion is found on a page containing several references to conical pendulums which I find curious. As I have mentioned, he proposed that the conical pendulum could be used to compare 'ye proportion of ye pressure of ye rays of gravity to ye force by wch a body hath any given motion ...' This idea seems intuitive enough, especially when the diagram showing the triangle formed by the string, the radius of the circle traced by the bob, and the vertical from the support to the centre of the circle, is seen. The force of gravity is proportional to the vertical and the force from the centre to the radius of the circle. This way of conceptualising it converts a dynamic problem into a static one, but luckily the result is not incorrect. What are not intuitively obvious are the statements that conical pendulums revolve with the same period as simple pendulums equal in length to the vertical heights of the cones and that all conical pendulums of the same vertical height therefore revolve with equal periods. These assertions are set down without even the suggestion of demonstrations, as though they were statements of generally accepted facts. At the time they were set down, Huygens' treatise on centrifugal force, which arrived at the same conclusions, had not been published.

74. I have gratuitously substituted *m* (mass) in Newton's proportion in place of his vague and undefined 'body'. I have already argued that his alteration of 'body' to 'quantity' was an initial step toward his later definition of mass. Nevertheless, he did not have a clearly formulated conception of mass at this time, and to that extent the expression of his proportion as our formula for centripetal force falsifies the state of his understanding.
75. The figures Newton used for the size of the earth, the distance of the moon, and the acceleration of gravity all come from Salusbury's translation of the *Dialogue*. Indeed they are from the very page on which the *Dialogue* took up the same problem, and they leave no doubt whatever that Newton's treatment of it was inspired by Galileo's statement of the problem, though not by his solution.
76. University Library Cambridge, *Add. MS. 3958.2*, f. 45. The MS. is published in Herivel, *Background*, pp. 183-91.
77. *Add. MS. 3958.2*, f. 45. *Background*, p. 185.
78. *Add. MS. 3958.2*, f. 45. *Background*, pp. 186-8. In a paper on motion in a cycloid which probably dates from the early 1670's, Newton implicitly used the concept of the force of gravity, which he set proportional to the acceleration it generates (see Fig. 35). 'I affirm that the efficacy of gravity or the acceleration of descent,

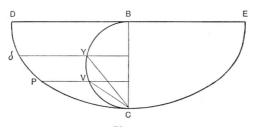

Fig. 35

in the single points D, δ, P, etc. of the descent are proportional to the spaces to be described DC, δC, PC etc. Clearly the obliquity of the descent diminishes the efficacy of gravity so that if two weights [*gravia*] are about to descend to C, the one B directly by the diameter BC, the other Y obliquely by the chord YC: the acceleration of the weight Y will be less on account of the obliquity of the descent in the ratio YC to BC so that both weights will arrive simultaneously at C.' He then demonstrated from the properties of the cycloid that the acceleration at δ is equal to that along YC and that the accelerations at both D and δ are proportional to the displacements DC and δC along the cycloid (*Background*, p. 203).
79. *Ibid.*, pp. 195-7.
80. Gregory's note is quoted in *The Correspondence of Isaac Newton*, ed. H. W. Turnbull and J. F. Scott, 4 vols. continuing. (Cambridge, 1959 continuing), **I**, 301.

81. Newton's own statement (University Library Cambridge, *Add. MS. 3968.41*, f. 85) dates from more than fifty years after the events in question when he was writing one of his innumerable justifications in connection with the calculus controversy. Cf. the memorandum of Abraham DeMoivre written in 1727: After Newton found a discrepancy in his calculation, 'he entertained a notion that with the force of gravity, there might be a mixture of that force which the Moon would have [if] it was carried along in a vortex . . .' (An incomplete copy of the memorandum in the Portsmouth papers contains this passage, *Add. MS. 4007*, f. 707.) Although I stated that there were three independent sources for this opinion, there is no way to demonstrate that the other two were in fact independent of DeMoivre. Henry Pemberton's *View of Sir Isaac Newton's Philosophy* appeared in 1728. In the preface he asserted that because Newton used an incorrect figure for the size of the earth, 'his computation did not answer expectation; whence he concluded, that some other cause must at least join with the action of the power of gravity on the moon.' William Whiston, *Memoirs of the Life and Writings of Mr William Whiston*, (London, 1749) could well have drawn on Pemberton whose account he mentioned. According to him, the failure of the correlation made Newton 'suspect that this Power was partly that of Gravity, and partly that of *Cartesius's* Vortices . . .' (p. 37).
82. University Library Cambridge, *Add. MS. 3958.5*, ff. 85–86v, 81–83v. The earlier version is entitled 'The laws of Reflection.' The second version is published in Herivel, *Background*, pp. 208–15. In his extensive notes, Herivel demonstrates how Newton's rather strange looking formulations can be brought into agreement with modern results.
83. Ibid., pp. 209–10. Herivel shows that Newton's 'radius of circular motion' (which describes the equator of circulation) is equivalent to our radius of gyration (k), and that Newton's real quantity of circular motion $(Mk\omega)$ differs from our angular momentum $(Mk^2\omega)$ by a factor k.
84. Ibid., p. 211.
85. In defining real quantity of circular motion, Newton imagined an impact at the rotating body's radius of circular motion (or radius of gyration). In the supposedly arbitrary impact, it appears to me that he implicitly assumed that both bodies are struck at their radii of circular motion. Not only does this nullify the generality of the treatment, but it imagines a case virtually impossible to realise.
86. Let B be the motion of A perpendicular to the plane of contact, and β that of a (see Fig. 37). Let G be the radius of circular motion of A (γ of a), D the real velocity of A's circular motion (δ of a) and F the distance along the plane of contact from the point of contact to the perpendicular from the centre of gravity (the distance BC in the diagram) and ϕ the corresponding distance for a (F/G then is the sine of the angle between the radius to the point of contact and the perpendicular to the plane of contact).

$$Q = 2B + 2\beta + 2D\frac{F}{G} + 2\delta\frac{\phi}{\gamma}$$

$$P = \frac{1}{A} + \frac{1}{a} + \frac{F}{AG} + \frac{\phi}{a\gamma}$$

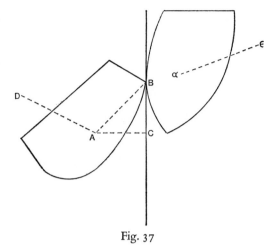

Fig. 37

87. Ibid., pp. 212–13. Herivel solves the problem with modern notation. I find no error in his solution.
88. Ibid., p. 208.
89. Ibid., p. 211.
90. Hall and Hall, *Unpublished Papers*, p. 128.
91. Ibid., p. 148.
92. Ibid., p. 148.
93. University Library Cambridge, *Add. MS. 3970.3*, ff. 519–28. Cf. R. S. Westfall, 'Newton's Reply to Hooke and the Theory of Colours,' *Isis*, **54** (1963), 82–96.
94. *Correspondence*, **1**, 364.
95. Ibid., **1**, 365–6.
96. A set of eleven propositions in an unpublished manuscript embodies a succinct statement of Newton's aethereal hypothesis as it related, among other things, to the phenomena of thin films. It posits a medium by means of which the surfaces of bodies act upon light and light acts upon bodies, but one which does not resist the motion of light through uniform bodies. Light excites vibrations in the medium, shorter vibrations by the more refrangible rays and longer ones by the less refrangible, and on the difference in vibrations depends the difference in reflection and inflection. As elsewhere, he also referred sensation and animal motion to vibrations in the same medium confined within the nerves (University Library Cambridge, *Add. MS. 3970.3*, f. 374v).
97. Newton to Oldenburg, 26 April 1676; *Correspondence*, **2**, 1–2.
98. Ibid., **1**, 370.
99. Ibid., **1**, 365. In his notes from Boyle's *Origin of Forms and Qualities*, Newton recorded an experiment in which Boyle's *menstruum peracutum* dissolved gold and precipitated a white powder that turned out to be silver. Moreover, some chemists had turned most of an ounce of gold into silver by means of a certain sort of *aqua regia*. 'Perhaps there may be some nobler & subtler matter in gold (its anima or tinctura) wch when united to ye particles of silver make them have

407

all ye Phaenomena of gold & wch let goe ye silver when they meete wth some body wth wch they more easily unite.' (University Library Cambridge, *Add. MS. 3975*, p. 78.)
100. *Correspondence*, **1**, 365.
101. Galileo, *Discourses*, pp. 70–1.
102. Hooke, *Micrographia*, pp. 11 f. In his notes on the *Micrographia*, Newton specifically noticed this concept (University Library Cambridge, *Add. MS. 3958.1*, f. 1). His notes on Hooke's *Micrographia* have been published by Hall and Hall, *Unpublished Papers*, pp. 400–13.
103. *Correspondence*, **1**, 368.
104. 'Of the Incalescence of *Quicksilver* with *Gold*, generously imparted by B. R.,' *Philosophical Transactions*, **10** (1675), 515–33. It is found in Boyle, *Works*, **3**, 558–64.
105. *Correspondence*, **2**, 1.
106. The extensive body of alchemical MSS. in Newton's hand are generally undated. Judged by the handwriting alone, they were not all the products of one period of his life, but they do seem to have clustered more closely in the late 1670's than in any other period.
107. Newton to Oldenburg, 14 November 1676; ibid., **2**, 182. The work, published in 1675 and the last book published by Boyle before the gift mentioned in Newton's letter to Oldenburg, is found in Boyle's *Works*, **3**, 565–652.
108. *Correspondence*, **2**, 288.
109. Ibid., **2**, 290–1.
110. Ibid., **2**, 292.
111. Cf. Boyle's orthodox mechanical discussion of similar phenomena in the *Mechanical Origin of Qualities*, the work to which Newton's letter was a distant commentary. Most chemists, Boyle said, pretend that solutions are performed by a certain sympathy between the menstruum and the body dissolved. 'And it is not to be denied, that, in divers instances, there is, as it were, a consanguinity between the menstruum and the body to be dissolved; as when sulphur is dissolved by oils, whether expressed or distilled: but yet, as the opinion is generally proposed, I cannot acquiesce in it, partly, because there are divers solutions and other phaenomena, where it will not take place; and partly, because, even in those instances, wherein it is thought most applicable, the effect seems to depend upon mechanical principles.' What affinity can there be between the oil of vitriol and the wide variety of substances, animal, vegetable, and mineral, that it dissolves, some of them indeed substances held to be antipathetic to each other? So also substances considered antipathetic can dissolve the same third substance. Such consanguinity as there is, then, must be ascribed, not to any substantial form and not to the salt, sulphur, or mercury of the bodies involved, 'but to the congruity between the pores and figures of the menstruum, and the body dissolved by it, and to some other mechanical affections of them.' (*Works*, **3**, 628.)
112. *Correspondence*, **2**, 293. Cf. Boyle's discussion in the *Mechanical Origin of Qualities*: 'It is known, that many learned men, besides several chemical writers, ascribe the incalescences, that are met with in the dissolution of metals, to a conflict arising from a certain antipathy or hostility, which they suppose between the

conflicting bodies, and particularly between the acid salt of the one, and the alcalizate salt, whether fixed or volatile, of the other. But since this doctrine supposes a hatred between inanimate bodies, in which it is hard to conceive, how there can be any true passions, and does not intelligibly declare, by what means their supposed hostility produces heat; it is not likely, that, for these and some other reasons, inquisitive naturalists will easily acquiesce in it. And on the other side it may be considered, whether it be not more probable, that heats, suddenly produced in mixtures, proceed either from a very quick and copious diffusion of the parts of one body through those of another, whereby both are confusedly tumbled and put into a calorific motion; or from this, that the parts of the dissolved body come to be every way, in great numbers, violently scattered; or from the fierce and confused shocks or jostlings of the corpuscles of the conflicting bodies, or masses, which may be supposed to have the motions of their parts differingly modified according to their respective natures: or from this, that, by the plentiful ingress of the corpuscles of the one into the almost commensurate parts of the other, the motion of some etherial matter, that was wont before swiftly to permeate the distinct bodies, comes to be checked and disturbed, and forced to either brandish or whirl about the parts in a confused manner, till it have settled itself a free passage through the new mixture . . .' (*Works*, **3**, 581.)

113. Cf. Boyle's discussion of volatility and fixity: To be volatile, bodies must have a number of properties: their corpuscles must be small, not too heavy to be raised into the air, smooth and conveniently shaped for motion, and only loosely attached in the body. Hence to volatilise a body one can reduce it to minute particles, rub and grind its particles (for example, in fermentations and putrefactions) so as to remove entangling extensions, associate the particles to be raised with others more volatile (what he called the grand instrument of volatilisation), and use fire or some other heat. The properties that make for fixity are the grossness or bulk of corpuscles, their weight or solidity, and their irregular shape. To make a body fixed, then, one can make the particles bigger by means of fire or by adding to them, expel the volatile corpuscles, cool the body so as to preserve the relative rest of the particles among themselves, or add a substance that will entangle the particles (ibid., **3**, 610–24).

114. *Correspondence*, **2**, 294.

115. The Halls, who published *De aere et aethere* together with a translation (*Unpublished Papers*, pp. 214–28), date it to about 1674 (p. 187). It must have postdated a book by Boyle that appeared in 1673 since it referred to the book, and they feel that it definitely preceded the 'Hypothesis' since the paper speaks of forces of attraction and repulsion which the 'Hypothesis' explained by the aether. The thrust of my argument on Newton holds that he started with aethereal mechanisms – the commonplace orthodoxy of the day – and that he ultimately dispensed with them and replaced them by forces. Whereas the development of thought as they see it places the paper before the 'Hypothesis' (which had in fact been drafted in 1672 – a considerable inconvenience given the necessity that *De aere* be post 1673), the development as I see it places *De aere* after the letter to Boyle. I recognise the danger of making one's own interpretation a procrustean

bed to which the pieces of evidence are fitted by brute force. No point can be served by defending my interpretation in this note. It is spread through the text, and I can only leave it to the judgment of the reader. Although the placement of *De aere et aethere* after the letter to Boyle appears to me to strengthen my argument, I do not think it is so crucial that the whole stands or falls with the validity of the dating. In favour of a time near the letter to Boyle, though not necessarily later than the letter, is the considerable identity in content. This will appear in my discussion of the paper.

116. The first sentence, which the Halls did not include entirely because of damage to the MS., appears to me to read as follows: 'De rerum natura scripturus, a caelestibus exordiar, et in ijs a maxime sensibilibus aere scilicet & naturis aeris ut sensu doce [*sic* – the same phrase written originally elsewhere in the sentence and crossed out has "duce"] progrediar.' (University Library Cambridge, *Add. MS. 3970.12*, F. 652.)
117. Hall and Hall, *Unpublished Papers*, p. 223.
118. Ibid., pp. 223–4.
119. Ibid., p. 226.
120. Ibid., pp. 226–7.
121. Ibid., pp. 227–8.
122. *Principia*, p. 325. I have altered the translation of the first clause to conform to the Latin of the first edition: 'Denique cum receptissima Philosophorum aetatis hujus opinio sit . . .' (first ed., p. 352).
123. *Principia*, first ed., p. 353. In edition two and in subsequent editions, Newton omitted the final three sentences possibly because he began to admit the existence of an extremely rare aether at about the time of the second edition. Newton apparently referred to this experiment in a scholium that he inserted in the third version of *De motu* after Problem 5. He had treated the motion of bodies in non-resisting media so far, he said, in order to determine the motions of the heavenly bodies through the aether. For the resistance of the aether is either null or extremely small. Bodies resist in proportion to their densities, water much less than quicksilver, air much less than water. When riding at a gallop, horsemen feel considerable resistance from the air, but sailors in a ship, shielded from the wind, feel no resistance at all from the aether. 'If air were to pass freely among the parts of bodies and were to act as it does not only on the external surface of the whole but also on the surfaces of the individual parts, its resistance would be far greater. The aether passes among the parts with complete freedom but does not offer sensible resistance.' (Herivel, *Background*, p. 298 and p. 302.) He went on to note that even the tails of comets suffer no resistance. Although he used the word 'aether', the obvious conclusion of the passage is that a material aether cannot exist. William Whiston interpreted the result of this experiment in the same way. 'Since pendulous Bodies receive no sensible Resistance in their Internal Parts; and since both Planets and Comets move prodigiously swift, with the utmost Freedom, and without any sensible Resistance through the Aethereal Regions, 'tis certain there is no *Subtile Matter* pervading the Universe, as some have supposed.' (*Astronomical Principles of Religion, Natural and Reveal'd*, (London, 1725), p. 82.)

124. *Opticks*, (New York, 1952), pp. 375-6, reprinted through permission of Dover Publications Inc., New York.
125. They are published together with English translations in Hall and Hall, *Unpublished Papers*, pp. 320-47, 302-8.
126. A reprint of the original publication, which included an English translation, is readily available in *Isaac Newton's Papers & Letters on Natural Philosophy*, ed. I. Bernard Cohen, (Cambridge, Mass., 1958), pp. 255-8.
127. University Library Cambridge, *Add. MS. 3970.3*, ff. 335-6, 337-8.
128. Ibid., ff. 338-8v. The paper is damaged; the words in brackets are my reconstruction of the text.
129. Newton left behind extensive notes on his own chemical experimentation in which one can perceive some of the speculations in Query 31 (University Library Cambridge, *Add. MS. 3973*, ff. 1-49; *Add. MS. 3975*, pp. 101-58). The dates of the notes run from 1678 to 1696. Apparently the notes in 3973 are cruder than the others; he copied some of them over verbatim into 3975, suggesting that he meditated on their meaning at some length. Whatever the role of his own experimentation in his speculations, we must remember that virtually every reaction Newton cited was found in Boyle and that Newton's notes on Boyle record the extent of his debt. We must also recall that his deepest reading in alchemical literature took place between 1675 and 1685 if we can trust the evidence of his handwriting to date it. Although I am not satisfied that I have even begun to penetrate the thrust of his experimentation, now and then a note seems to express alchemical intentions too clearly to be missed. For example, the following two: 'May 10 1681 intellexi Luciferam ♀ et eandem filiam ♄^{ni}, & unam columb^{rum} [?]. May 14 intellexi —♃. May 15, intellexi *Sunt enim quaedam* ☿^{ij} *sublimationes* &c ut & columbam alteram: nempe Sublimatum quod solum foeculentum est, a corporibus suis ascendit album, relinquitur foex nigra in fundo, quae per solutionem abluitur, rursusq; sublimatur ☿^{ius} a [ ]datis corporibus donec foex in fundo non amplius restet. Nonne hoc sublimatum depuratissimum sit —♓?' (*Add. MS. 3975*, p. 121.) Newton later crossed this note out. 'ffriday May 23 [c. 1684] Jovem super aquilam volare feci.' (Ibid., p. 149.)
130. *Opticks*, pp. 378-9.
131. Ibid., p. 380.
132. Ibid., pp. 380-1.
133. Ibid., p. 383.
134. Ibid., p. 387.
135. Ibid., p. 339.
136. When he began seriously to work at composing the *Opticks* about 1690, Newton's first impulse was to transfer the vibrations by which he explained periodic phenomena from the aether, which he no longer accepted, to the refracting medium itself. In the published *Opticks*, this suggestion was inserted so tentatively as to be hardly noticeable (pp. 280-1). In MSS. from about 1690, it was proposed much more firmly. In what I take to be the earliest of the MSS., Proposition 12 asserted that light excites vibrations in bodies, as the phenomena of thin films demonstrate. He cited enough evidence to show that the phenomena are periodic; '& the reason I cannot yet conceive to be any other then that

every ray of light in passing through ye first surface of ye plate stirs up a reciprocal motion wch being propagated through the plate to ye second surface doth alternately increase & diminish the reflecting power of y^t surface, so y^t if ye ray arrive at ye second surface when its reflecting power is encreased by the first impuls or vibration of the motion, it is reflected; but if the plate be a little thicker so that ye reflecting power of ye second surface be diminished before the ray arrives at it, the ray is transmitted . . .' (University Library Cambridge, *Add. MS. 3970.3*, f. 348.) In another paper, which I take to be later, he elaborated this theory in somewhat greater detail.

Prop 12 The motion excited in pellucid bodies by the impulses of the rays of light is a vibrating one & the vibrations are propagated every way in concentric circles from the points of incidence through the bodies.

Prop 13 The like vibrations are excited by ye inflected rays of light in their passage by ye sharp edges of dense bodies: & these vibrations being oblique to the rays do agitate them obliquely so as to cause them to bend forwards & backwards wth an undulating motion like that of an Eele.

Prop 14 As the oblique vibrations excited by inflected rays do agitate the rays sidways wth a reciprocal motion so the perpendicular vibrations excited by the refracted rays do agitate the rays directly wth a reciprocal motion so as to accelerate & retard them alternatively & thereby cause them to be alternately refracted & reflected by thin plates of transparent substances for making those many rings of colours described in ye Observations of the third book. (Ibid., f. 335v.)

For a fuller discussion of Newton's 'Fits' cf. R. S. Westfall, 'Uneasily Fitful Reflections on Fits of Easy Transmission,' *The Texas Quarterly*, **10** (1967), 86–102.

137. *Opticks*, pp. 388–9.
138. Christiaan Huygens had originally discovered this phenomenon in tubes filled with water. Hooke had achieved the same effect with mercury, and Newton learned it from Hooke.
139. *Optice* (the first Latin edition, 1706), p. 337. This passage was eliminated from the translated version of the Query that appeared in the second English edition of 1717.
140. *Opticks*, p. 395.
141. *Optice*, p. 340.
142. Newton's early explanation of capillary action was the central argument of Hooke's initial publication, *An Attempt for the Explication of the Phaenomena, Observable in an Experiment Published by the Honourable Robert Boyle, Esq.*; Newton found it in the *Micrographia*. (Cf. Hall and Hall, *Unpublished Papers*, p. 400.) Perhaps we should regard capillary action as Hooke's revenge.
143. *Opticks*, p. 400.
144. Ibid., pp. 389–90. Cf. a Scholium to Prop. 7, Book III of the *Principia*, composed in the early 1690's but never inserted in a published version of the work. He asserted 'atomos ipsos ob soliditatem et plenum partium contactum ac densitatem summam nec dividi posse nec alteri nec ulla ratione comminui nec augmentum unquam sumere sed immutabilia rerum semina in aeternam manere et inde

fieri ut rerum species perpetuo conserventur.' (University Library Cambridge, *Add. MS. 3965.6*, f. 270.)
145. *Opticks*, p. 400.
146. University Library Cambridge, *Add. MS. 3970.3*, f. 234v.
147. University Library Cambridge, *Add. MS. 3965.6*, f. 266v. Cf. J. E. McGuire, 'Body and Void and Newton's De Mundi Systemate: Some New Sources,' *Archive for History of Exact Sciences*, **3** (1966), pp. 206–48; and Arnold Thackray, '"Matter in a Nut-Shell": Newton's *Opticks* and Eighteenth-Century Chemistry,' *Ambix*, **15** (1968), 29–53.
148. University Library Cambridge, *Add. MS. 3970.3*, f. 296. My italics.
149. University Library Cambridge, *Add. MS. 3965.6*, f. 266v.
150. University Library Cambridge, *Add. MS. 3970.3*, f. 234. In the last clause, Newton originally wrote a fourth 'thousand', a symbolic slip of the pen which suggests where his imagination was tending. In a less startling form, the content of this passage appeared among the addenda to the Latin edition of the *Opticks* in 1706 and then in the body of the second and subsequent editions at the end of Prop. VIII, Part III, Book II, pp. 268–9.
151. Ibid., p. 401.
152. In the '*Conclusio*' of 1687, he had expressed himself with a lesser degree of certainty: 'I have briefly set these matters out, not in order to make a rash assertion that there are attractive and repulsive forces in bodies, but so that I can give an opportunity to imagine further experiments by which it can be ascertained more certainly whether they exist or not.' (Hall and Hall, *Unpublished Papers*, p. 340.)
153. University Library Cambridge, *Add. MS. 3970.3*, f. 479.
154. Ibid., f. 480v.
155. *Opticks*, pp. 401–2.
156. In a delightfully seasoned passage, Galileo had indicated the tenuous distinction between the concept Newton later used and that of sympathies and antipathies. Salviati is discussing how water on leaves forms in drops. It cannot be due to any internal tenacity in the water. Such a property would reveal itself more strongly when water is surrounded by wine, in which it is less heavy than it is in air, but in fact a drop collapses when wine is poured around it on the surface of the leaf. Rather the formation of the drop is due to the pressure of the air with which water has some incompatibility which he does not understand. At this point, Simplicio (speaking, of course, only the words that Galileo chose to put in his mouth) breaks in. Salviati makes him laugh by his efforts to avoid the word 'antipathy'. To the interjection, Salviati replies ironically: 'All right, if it please Simplicio, let this word antipathy be the solution of our difficulty.' (*Discourses*, p. 71.) Clearly Galileo felt there was a world of difference between an incompatibility he did not understand and an antipathy, just as Newton would have argued the same for his secret principle of sociability. For my part, the concepts appear to have had more in common than either man would have admitted. Unsociability (or incompatibility) responded to the same specificities that the concept of antipathy had tried to express. The *ad hoc* mechanisms by which the mechanical philosophy had attempted to explain the same observables had always constituted one of its most glaring deficiencies.

157. 'The Parts of all homogeneal hard Bodies which fully touch one another, stick together very strongly.' (*Opticks*, p. 388.) In one of his MSS. connected with the second English edition of the *Opticks*, he speculated on the necessary dissolution of bodies to their ultimate particles by means of fermentation and putrefaction before the nutritive process can utilise the matter present. 'And when the nourishment is thus prepared by dissolution & subtiliation the particles of the body to be nourished draw to themselves out of the nourishment the particles of the same density & nature wth themselves. ffor particles of one & the same nature draw one another more strongly then particles of different natures do. . . . And when many particles of the same kind are drawn together out of ye nourishment they will be apt to coalesce in such textures as the particles wch drew them did before because they are of the same nature as we see in the particles of salts wch if they be of the same kind always crystallise in the same figures.' (University Library Cambridge, *Add. MS. 3970.3*, f. 235v.) *Quaestio* 22 of the *Optice* (corresponding to Query 30 of the later English editions) concluded with a page, which was not carried over into the English version, introducing the discussion of forces in the following *Quaestio*. On that page, he spoke of bodies 'quae sunt ejusdem generis & virtutis . . .' (*Optice*, p. 320).
158. Hall and Hall, *Unpublished Papers*, p. 341.
159. *Principia*, ed. 1, p. 402. For discussions of this hypothesis, see J. E. McGuire, 'Transmutation and Immutability: Newton's Doctrine of Physical Qualities,' *Ambix*, **14** (1967), pp. 69-95; I. Bernard Cohen 'Hypotheses in Newton's Philosophy,' *Physis*, **8** (1966), pp. 163-84; and Alexandre Koyré, 'Newton's "Regulae Philosophandi",' *Newtonian Studies*, (London, 1965), pp. 261-72.
160. 'Ex materia quadam communi formas & texturas varias induente res omnes oriri et in eandem per privationem formarum et texturarum resolvi, docent omnes . . .' (University Library Cambridge, *Add. MS. 4005.15*, f. 81v.) In the article cited in fn. 159, I. Bernard Cohen argues that Newton inserted Hypothesis III as an *ad hominem* argument in which he did not himself believe and that in passages such as this he was pursuing the argument against Cartesians and Aristotelians. I cannot accept his argument, and believe that Hypothesis III expressed a conviction found elsewhere. For example, *De natura acidorum* affirmed that if gold could be made to ferment and putrefy, thus breaking down its constituent corpuscles into ultimate particles, 'it might be turn'd into any other Body whatsoever' (Cohen, *Papers & Letters*, p. 258). In a later MS. associated with the second English edition of the *Opticks*, Newton wrote out three successive versions of a similar paragraph on gold. The last of the three, itself still much revised and altered, was as follows: 'Si aurum fermentescere posset [& per putrefactionem in particulas minimas resolve, idem, ad instar substantiarum vegetablium & animalium putrescentium, formam suam amitteret, in fimum abiret vegetabilibus nutriendis aptum & subinde per generationem] in aliud quodvis corpus transformari posset. [Et similis est ratio Gemmarum & mineralium omnium.]' The brackets are Newton's and probably indicate his intention to omit those passages. The paragraph has a number of other words and phrases crossed out that I have not attempted to indicate (University Library Cambridge, *Add. MS. 3970.3*, f. 240v).

161. In another context Newton saw as much. Consider a MS. from the early 1690's which proposed a revision in the corollary to Prop. 6, Book III of the *Principia*. In edition one, the corollary had affirmed that differences in specific gravity require the existence of a vacuum; he now proposed to add the following: 'Valet haec demonstratio contra eos qui Hypothesin vel tertiam vel quartam admittunt. [Hypothesis 3 in this revision affirmed that qualities which admit neither intension nor remission are the qualities of all bodies universally, and Hypothesis 4 repeated that all bodies can be changed into any other.] Siquis Hypothesibus hisce repudiatis ad Hypothesin tertiam recurrat nempe materiam aliquam non gravem dari per quam gravitas materiae sensibilis explicetur; necesse est ut duo statuat particularum solidarum genera quae in se mutuo transmutari nequeunt: alterum crassiorum quae graves sint pro quantate materiae et ex quibus materia omnis gravis, totusq; adeo mundus sensibilis confletur, & alterum tenuiorum quae sint causa gravitatis crassiorum sed ipsae non sint graves ne gravitas earum per tertium genus explicanda sit & ea hujus per quartum et sic deinceps in infinitum. Hae autem debent esse longe tenuiores ne per actionem suam crassiores discutiant & ab invicem dissipent: qua ratione corpora omnia ex crassioribus composita cito dissolverentur. Et cum actio tenuiorum in crassiores proportionalis fuerit crassiorum superficiebus, gravitas autem ab actione illa oriatur et proportionalis sit materiae ex qua crassiores constant: necesse et ut superficies crassiorum proportionales sint earum contentis solidis, et propterea ut particulae illae omnes sint aequaliter crassae utq; nec frangi possint nec alteri vel ratione quacunq; comminui, ne proportio superficierum ad contenta solida et inde proportio gravitatis ad quantitatem materiae mutetur. Igitur particulas crassiores in tenuiores mutari non posse et propterea duo esse particularum genera quae in se mutuo transire nequeunt omnino statuendum est.' (University Library Cambridge, Add. MS. 3965.6, f. 267.)
162. *De natura acidorum*; Cohen, *Papers & Letters*, p. 258.
163. From a MS. connected with the first edition of the *Opticks*; University Library Cambridge, *Add. MS. 3970.3*, f. 337.
164. From a MS. connected with the Latin edition: ibid., f. 292.
165. *Opticks*, pp. 397–400. The dichotomy of passive and active played a prominent role in the Clarke-Leibniz correspondence. Since Newton has been demonstrated to have participated in the composition of Clarke's replies to Leibniz, passages in those letters help to illuminate his thought. Cf. Clarke's fifth reply: 'But indeed, all mere mechanical communications of motion, are not properly action, but mere passiveness, both in the bodies that impel, and that are impelled. Action, is the beginning of a motion where there was none before, from a principle of life or activity: and if God or man, or any living or active power, ever influences any thing in the material world; and every thing be not mere absolute mechanism; there must be a continual increase and decrease of the whole quantity of motion in the universe.' *The Leibniz-Clarke Correspondence*, ed. H. G. Alexander, (New York, 1956), p. 110, reprinted by permission of the Philosophical Library. Cf. also pp. 45, 97, and 112.
166. Perhaps the best example of an ambiguous passage is the discussion of the tails of comets in the *Principia*, in which he made use of what he called 'aethereal air

[*aura aetherea*].' In the English translation, this appears as 'ether,' but in the context, which insists that the tail is generated in the vicinity of the sun, it can only refer to a presumed solar atmosphere (*Principia*, pp. 528-9; First ed., p. 505).
167. *Opticks*, cxxiii.
168. *Principia*, p. 547. Cf. Henry Guerlac, 'Francis Hauksbee: expérimentateur au profit de Newton,' *Archives internationales d'histoire des sciences*, **16** (1963), 113-28; and 'Sir Isaac and the Ingenious Mr. Hauksbee,' *Mélanges Alexandre Koyré. L'aventure de la science*, introduced by I. Bernard Cohen and René Taton, (Paris, 1964), **1**, 228-53.
169. University Library Cambridge, *Add. MS. 3965.12*, ff. 350-65.
170. Ibid., f. 357v.
171. Ibid., f. 350v.
172. From a MS. that contains a draft of the additional Queries added to the second English edition of the *Opticks*; University Library Cambridge, *Add. MS. 3970.3*, f. 241v. Cf. from a similar MS. what he labelled there Query 23.

> Qu. 23. Is not electrical attraction & repulse performed by an exhalation wch is raised out of the electrick body by friction & expanded to great distances & variously agitated like a turbulent wind. & wch carrys light bodies along with it. & agitates them in various manners according to it own motions, making them go sometimes towards the electric body, sometimes from it & sometimes move with various other motions? And when this spirit looses its turbulent motions & begins to be recondensed & by condensation to return into the electric body doth it not carry light bodies along with it towards the electrick body & cause them to stick to it without further motion till they drop off? And is not this exhalation much more subtile then common Air or Vapour? For electric bodies attract straws & such light substances through a plate of glass interposed, tho not so vigorously. And may there not be other Exhalations & subtile invisible Mediums which may have considerable effects in the Phaenomena of Nature?' (Ibid., f. 293v.)

173. University Library Cambridge, *Add. MS. 3965.12*, f. 351. Cf. a list of twelve propositions on sheets that contain drafts and redrafts of the General Scholium's paragraph on God.

Prop. 1. Perparvas corporum particulas vel contiguas vel ad parvas ab invicem distantias se mutuo attrahere. Exper 1. Vitrorum parallelorum. 2 Inclinatorum. 3 fistularum. 4 Spongiarum. 5 Olei malorum citriorum.

Prop. 2. vel Schol. Attractionem esse electrici generis.

Prop. 3. Attractionem particularum ad minimas distantias esse longe fortissimam (Per exper 5) & ad cohaesionem corporum sufficere.

Prop. 4. Attractionem sine frictione ad parvas tantum distantias extendi ad majores distantias particulas se invicem fugere. Per exper 5. Exper. 6. De solutione metallorum

Prop. 5. Spiritum electricum esse medium maxime subtilem & corpora solida facillime permeare. Exper. 7. Vitrum permeat

Prop. 6 Spiritum electricum esse medium maxime actuosum et lucem emittere Exper 8.

NEWTON AND THE CONCEPT OF FORCE

Prop. 7   Spiritum Electricum a luce agitari idq; motu vibratrorio, & in hoc motu calorem consistere. Exper 9. Corporum in luce solis.

Prop. 8.   Lucem incidendo in fundum oculi vibrationes excitare quae per solida nervi optici capillamenta in cerebrum propagatae visionem excitant.

Schol.   Omnen sensationem omemq; motum animalem mediante spiritu electrico peragi.

Prop. 9.   Vibrationes spiritus electrici ipsa luce celeriores esse.

Prop. 10.   Lucem a spiritu electrico emitti refringi reflecti et inflecti.

Prop. 11.   Corpora homogenea per attractionem electricam congregari heterogenea segregari.

Prop. 12   Nutritionem per attractionem electricam peragi. (Ibid., ff. 361v–362v.)

Cf. also, a sheet headed '*De Motu et sensatione Animalium*,' which is concerned with speculations like those for the second English edition of the *Opticks* and apparently dates from about 1710–15.

1   Attractionem electricam per spiritum quendam fieri qui corporibus universis inest, et aquam vitrum Crystallum aliaq; corpora solida libere permeat libere pervabit. [*sic*] Nam corpora electrica fortiter attrita aurum foliatum per interpositam aquae vel vitri substantiam trahunt.

2.   Spiritum hunc electricum dilatari et contrahi et propterea elasticum esse eundemq; in nervis animalium latentem esse medium quo objecta sentimus et ictu oculi membra movemus. [Nam Spiritus quos vocant animales ob densitatem tarde moventur.] & vibrationes per eundem quam celerrime propagari.

Two further paragraphs briefly discuss the capillamenta containing the electric spirit that carry sensations and impulses for movement. (University Library Cambridge, *Add. MS. 3970.3*, f. 236.)

Another MS., associated, as the last one with the second English edition of the *Opticks* and also from around 1710–15, is entitled '*De vita & morte vegetabili.*'

1   Corpora omnia vim habent electricam & vim illam in superficiebus particularum fortissimam esse sed non longe extendi nisi frictione vel alia aliqua actione cieatur.

2   Corpora vi electrica plerumq; trahi quadoq; vero dispelli per experimenta constat; et particulas aeris & vaporum sese dispellere. Particulas etiam olei dispellere particulas aquae.

3   Particulas corporum per vim electricam diversimode coalescere & cohaerere. Et particulas minores fortius agere & artius cohaerere.

Particulas menstrui quae vi electrica particulas linguae fortissime agitant sensationem acidi ciere.

Menstruum acidum corpora densa dissolvere per particulas suas **acidas**, vi attractrice in interstitia partium ultimae compositionis inventes & partem unam quamq; circumeuntes ut cortex nucleum vel atmosphaera terram. partes vero acido circumdatas corpus suum linquere et in menstruo fluitare, atq; acido ambiente linguam pungere excitandi sensationem salis. Nam acidum a nucleo incluso attractum retentum & impeditum minus agit in linguam quam prius.

417

The paper goes on, as other somewhat similar passages do, to apply these concepts to corruption and putrefaction, by which bodies are separated into their ultimate particles, and to nutrition and generation, in which the ultimate particles are joined together by attractions into new substances. It ends with the following generalisation: 'Per fermentum itaque corpora mortua dissolvuntur, viva nutriuntur & crescunt. Mortua propter debilem partium attractionem vincuntur a menstruo pervadente: viva propter fortem partium attractionem non dissolvuntur sed menstruum vincunt.' (Ibid., ff. 237-237v.)

174. Ibid., f. 235. At the top of the sheet is the heading, '- conjungi queant ut cohaerescant. p. 340. lin. 27.' The words referred to, in the Latin edition of the *Opticks*, come in Quaestio 23 at the end of the discussion asserting the existence of forces. As the Query was, and is, published, the next paragraph turns to broader cosmological issues in which Newton asserted that the universe would run down if there were no 'active principles.' Thus when he drafted these two queries, he intended his argument to move from the assertion of forces between particles to the consideration of their electrical nature.

175. *Opticks*, 349. In MSS. concerned with additions and corrections to the third edition of the *Principia*, Newton contemplated adding three definitions, of phenomena, body, and void, after the *Regulae philosophandi* at the beginning of Book III. The definition of body made his conception of an aether clear. He defined body in terms of its tangibility and its resistance to bodies touching it whereby it can be perceived. Thus effluvia, which have resistance proportional to their density, are bodies. Mathematical solids are not. A quintessence different from the four elements and not perceived by any sense is not a phenomenon and therefore is not body. 'Materia subtilis in qua Planetae innatent et corpora sine resistentia moveantur, non est phaenomenon. Et quae phaenomena non sunt nec ullis sensibus obnoxia, ea in Philosophia experimentali locum non habent.' (University Library Cambridge, *Add. MS. 3965.13*, f. 422.) And in contrast, Newton's aether was in principle subject to perception and was therefore material. He also contemplated altering Corollary 2, Prop. 6, Book III. In the second edition, this corollary had argued against a mechanical aether, not heavy itself, that would cause gravity. Perhaps because he had meanwhile acknowledged the existence of an aether (though not one with the properties specifically discussed in the corollary), he preferred at one point to change the word 'aether' to the word 'air' (ibid., f. 505). In fact, the change was not inserted into the third edition, nor were the definitions. Cf. Henry Guerlac, 'Newton's Optical Aether,' *Notes and Records of the Royal Society of London*, **22** (1967), 45-57.

176. *Opticks*, p. 353.
177. Ibid., p. 352.
178. Ibid., p. 376.
179. Newton to Bentley, 25 February 1692/3; *Correspondence*, **3**, 253-4.
180. Newton even recalled the Latin *Optice* to modify an assertion in *Quaestio* 20 (Query 28) that space is the sensorium of God. As it was originally published, the passage stated: 'Annon Spatium Universum, Sensorium est Entis Incorporei, Viventis, & Intelligentis; quod res Ipsas cernat & complectatur intimas, totasq; penitus & in se praesentes perspiciat...' In the revision, he stated instead: 'Annon

ex phaenomeniis constat, esse Entem Incorporeum, Viventem, Intelligentem, Omnipraesentem, qui in Spatio infinito, tanquam Sensorio suo, res Ipsas intime cernat, penitusq; perspiciat, totasq; intra se praesens praesentes complectatur . . .' Needless to say, the revision has survived in the English translation, which states that 'there is a Being incorporeal, living, intelligent, omnipresent, who in infinite Space, *as it were* in his Sensory, sees the things themselves intimately, and thoroughly perceives them, and comprehends them wholly by their immediate presence to himself.' (*Opticks*, p. 370; my italics.) A few copies escaped without the revision. Cf. Alexandre Koyré and I. Bernard Cohen, 'The Case of the Missing *Tanquam*,' *Isis*, **52** (1961), pp. 555–66.

181. With the published writings, these papers stress two aspects of the idea as it was originally presented in *De gravitatione* that are tangential enough to my present purpose as not to demand extensive discussion here. First, he continually emphasised our ability to move our bodies as an analogy to the power of God over the physical universe. Cf. a MS. from the early 1690's: 'Hypoth 5. The essential properties of bodies are not yet fully known to us. Explain this ... by ye metaphysical power of bodies to cause sensation, imagination & memory & mutually to be moved by o^r thoughts.' (University Library Cambridge, *Add. MS. 3970.3*, f. 338v.) Quaestio 23 of *Optice* differed from the later English in the paragraph that asserted the necessity of active forces. The English as we know it says that we meet with very little motion in the world beside that caused by the active principles. The earlier Latin added another possible source: 'Nam admodum paullum Motus in mundo invenimus, praeterquam quod vel ex his Principiis actuosis, vel ex imperio *Voluntatis*, manifesto oritur.' (*Optice*, p. 343.) Second, he associated efforts to explain forces by means of material media with the concept of a self-contained material order, that is, with materialistic atheism. Query 28 makes this clear enough. The ancient atomists, he said, assigned gravity to the atoms, tacitly attributing it to some cause other than dense matter. Later philosophers banish the consideration of such a cause from natural philosophy, feigning hypotheses to explain all things mechanically and referring other causes to metaphysics. However, Newton continued, the main business of natural philosophy is to argue from effects to causes until we come to the first cause which is not mechanical – that is, which is God. In a draft of *Quaestio* 23, Newton was even more explicit. In the passage, he was asserting that all reasoning must start from experience. 'Even arguments for a Deity if not taken from Phaenomena are slippery & serve only for ostentation. An Atheist will allow that there is a Being absolutely perfect, necessarily existing, & the author of mankind & call it Nature: & if you talk of infinite wisdom or of any perfection more then he allows to be in nature heel reccon it a chimaera & tell you that you have the notion of *finite* or *limited wisdom* from what you find in yor self & are able of your self to prefix ye word *not* or *more* y^n to any *verb* or *adjective* without the existence of *wisdome not limited* or *wisdome more then finite* to understand the meaning of the phrase as easily as Mathematicians understand what is meant by an infinite line or an infinite area. And hee may tell you further that ye Author of mankind was destitute of wisdome & designe because there are no final causes & that matter is space & therefore necessarily existing & having always the same

quantity of motion, would in infinite time run through all variety of forms one of wch is that of a man Metaphysical arguments are intricate & understood by few The argument wch all men are capable of understanding & by wch the belief of a Deity has hitherto subsisted in the world is taken from Phaenomena. We see the effects of a Deity in the creation & thence gather the Cause & therefore the proof of a Deity & what are his properties belong to experimental Philosophy. Tis the business of this Philosophy to argue from the effects to their causes till we come at ye first cause & not to argue from any cause to the effect till the cause as to its being & quality is sufficiently discovered.' (University Library Cambridge, *Add. MS. 3970.9*, ff. 619–19v.) On Newton's concern to relate Deity to the creation, cf. David Kubrin, 'Newton and the Cyclical Cosmos: Providence and the Mechanical Philosophy,' *Journal of the History of Ideas*, **28** (1967), 325–46.

182. University Library Cambridge, *Add. MS. 3970.9*, f. 619. Other drafts of essentially the same material are found in ibid., f. 620v, and *3970.3*, f. 252v. Cf. J. E. McGuire and P. M. Rattansi, 'Newton and the "Pipes of Pan",' *Notes and Records of the Royal Society of London*, **21** (1966), 108–43, and J. E. McGuire, 'Force, Active Principles, and Newton's Invisible Realm,' *Ambix*, **15** (1968), 154–208. In the latter work especially, McGuire takes a somewhat different position from that which I am advancing.

183. Cf. a draft of material for the Queries apparently from about 1715: 'Qu. 17 Is there not something diffused through all space in & through wch bodies move without resistance & by means of wch they act upon one another at a distance in harmonical proportions of their distances.' (University Library Cambridge, *Add. MS. 3970.3*, f. 234v.) In his fourth reply, Clarke referred to the world, 'to which he [God] is present throughout, and acts upon it as he pleases, without being acted upon by it' (*Leibniz-Clarke Correspondence*, p. 50).

184. 'Epicurei naturam totam in corpus et inane distinguentes Deum pernegabant: at absurde nimis. Nam Planetae duo ab invicem longe vacui intervallo distantes non petent se mutuo vi aliqua gravitatis neq; ullo modo agent in se invicem nisi mediante principio aliquo activo quod utrumq; intercedat, et per quod vis ab utroq; in alterum propagetur. [Hoc medium ex menti veterum non erat corporeum cum corpora universa ex essentia sua gravia esse dicerent, atq; atomos ipsos vi aeterna naturae suae absq; aliorum corporum impulsu per spatia vacua in terram cadere.] Ideoq; Veteres qui mysticam Philosophiam rectius tenuere, docebant spiritum quendam infinitum spatia omnia pervadere & mundum universum continere & vivificare; et hic spiritum supremum fuit eorum numen, juxta Poetam ab Apostola citatum: In eo vivimus et movemur et sumus. Unde Deus omnipraesens agnoscitur et a Judaeis Locus dicitur. Mysticis autem Philosophis Pan erat numen illud supremum. . . . Hoc Symbolo Philosophi materiam in spiritu illo infinito moveri docebant et ab eodem agitari non inconstanter sed harmonice seu secundum rationes harmonicas ut modo explicui.' (University Library Cambridge, *Add. MS. 3965.6*, f. 269.) In a draft of the General Scholium Newton used a sentence about God that did not find its way into any published work: 'Vivit sine corde et sanguine, praesens praesentia sentit et intelligit sine organis sensuum et sine cerebro, agit sine manibus, et corpore minime vestitus videri non potest sed Deus est prorsus invisibilis.' (*3965.12*, f. 361.) Cf. a memo-

randum by David Gregory of 20 February 1697/8: Christopher Wren, he reported, 'smiles at Mr. Newton's belief that it [gravity] does not occur by mechanical means, but was introduced originally by the Creator' (*Correspondence*, **4**, 267). There is another piece of information deriving from Gregory. In a note dated 21 December 1705, Gregory mentioned the new Queries that Newton was adding to the Latin edition of the *Opticks*. 'His Doubt was whether he should put the last Quaere thus. *What the space that is empty of body is filled with*. The plain truth is, that he believes God to be omnipresent in the literal sense; And that as we are sensible of Objects when their Images are brought home within the brain, so God must be sensible of every thing, being intimately present with every thing: for he supposes that as God is present in space where there is no body, he is present in space where a body is also present. But if this way of proposing this his notion be too bold, he thinks of doing it thus. *What Cause did the Ancients assign of Gravity*. He believes that they reckoned God the Cause of it, nothing els, that is no body being the cause; since every body is heavy.' (*David Gregory, Isaac Newton and their Circle*. Extracts from David Gregory's Memoranda 1677–1708, ed. W. G. Hiscock, (Oxford, 1937), p. 30.) Passages in Whiston's *Astronomical Principles of Religion, Natural and Reveal'd* express the same conception of the ultimate source of apparent attractions in the physical world. He contended that gravity is 'an entirely immechanical Power, and beyond the Abilities of all material Agents whatsoever.' Mechanical causes act only on the surface, gravity on the very inward substantial parts. Gravity acts the same whether a body is a rest or in violent motion, whereas mechanical causes act only with the excess of their velocity. 'By this Power Bodies act upon other Bodies *at a Distance*, nay at all Distances whatsoever; that is, they *act where they are not*: Which is not only impossible for Bodies Mechanically to do, but indeed is impossible for all Beings whatsoever to do, either Mechanically or Immechanically, it being just as good Sense to say, an Agent can act *when* he is not in Being, as *where* he is not present. Whence, by the way, as we shall see hereafter, it will appear, that this Power of Gravity is not only Immechanical, or does not arise from Corporeal Contract [sic] or Impulse, but is not, strictly speaking, any Power belonging to Body or Matter at all; tho' for ease of Conception and Calculation we usually so speak; but is a Power of a superior Agent, ever moving all Bodies after such a manner, as if every Body did Attract, and were Attracted by every other Body in the Universe, and no otherwise.' (pp. 45–6.) Later Whiston argued that the cause of Gravity must be a Being that constantly knows the sizes and distance of bodies and moves them always with such velocities as those factors demand. 'Since Power can be exerted no where but where the Being which exerts that Power is actually present; and since it is certain, as has been shewn, that this Power is constantly exerted all over the Universe, 'tis certain that the Author of the Power of Gravity is present at all Times in all Places of the Universe also. Since this Power has been demonstrated to be Immechanical, and beyond the Abilities of all Material Agents; 'tis certain that the Author of this Power is an Immaterial or Spiritual Being, present in, and penetrating the whole Universe.' (p. 89.) Pierre Coste, the translator into French of both the *Opticks* and of Locke's *Essay Concerning Human Understanding*, reported a conversation with Newton on a passage in Locke's *Essay*: 'On pourrait, dit'il, se former en quelque manière une idée de la création de la matière en supposant que Dieu eût empêché par sa puissance que rien ne pût entrer dans une certaine portion de l'espace pur, qui de sa nature est pénétrable, éternel, nécessaire, infini, car dès la cette portion d'espace aurait l'impénétrabilité, l'une des qualités essentielles à la matière; et comme l'espace pur est absolument uniforme, on n'a qu'à supposer que Dieu aurait communiqué cette espèce d'impénétrabilité à autre pareille portion de l'espace, et cela nous donnerait en quelque sorte une idée de la mobilité de la matière, autre qualité qui lui est aussi très essentielle.' (Cited in Alexandre Koyré, *Newtonian Studies*, (Cambridge, Mass., 1965), p. 92 fn, reprinted by permission of the Harvard University Press and Chapman & Hall Ltd.) According to Coste's account, Newton said that the idea came into his mind one day when he was in conversation with Locke. Newton was not being entirely candid; although it may have come into his mind during the conversation with Locke, it had originally occurred to him long before. It is exactly the position of *De gravitatione* and, according to my argument, virtually identical, in its conception of the relation of God to physical phenomena, to his ultimate position as well. Cf. also various passages from Clarke's papers in the *Leibniz-Clarke Correspondence*: 'But God is present to the world, not as a part, but as a governor; acting upon all things, himself acted upon by nothing.' (p. 24.) 'Space void of body, is the property of an incorporeal substance. ... In all void space, God is certainly present, and possibly many other substances which are not matter; being neither tangible, nor objects of any of our senses.' (p. 47.) 'To suppose that in spontaneous animal-motion, the soul gives no new motion or impression to matter; but that all spontaneous animal-motion is performed by mechanical impulse of matter; is reducing all things to mere fate and necessity. God's acting in the world upon every thing, after what manner he pleases, without any union, and without being acted upon by any thing; shows plainly the difference between an omnipresent governor, and an imaginary soul of the world.' (p. 51.) For other discussions, see Henry Guerlac, *Newton et Epicure*, Conférence donnée au Palais de la Découverte le 2 Mars 1963, (Paris, 1963); A. R. Hall and Marie Boas, 'Newton's Theory of Matter,' *Isis*, **51** (1960), 131–44; A. J. Snow, *Matter & Gravity in Newton's Physical Philosophy*, (London, 1926); and the articles by Rattansi and McGuire and by McGuire cited above in fn. 182.

185. *Principia*, p. 544.
186. University Library Cambridge, *Add. MS.* 3965.6, f. 266v. Cf. a MS. dating roughly from the time of the first edition of the *Principia*: 'An Prophetae rectius Deum locis omnibus absolute praesentem dixerint et corpora contenta secundum leges mathematicas constanter agitans nisi ubi leges illas violare bonum est.' (*3965.13*, f. 542.) In his fourth reply, Clarke made the same point to Leibniz: 'That one body should attract another without any intermediate means, is indeed not a miracle, but a contradiction: for 'tis supposing something to act where it is not. But the means by which two bodies attract each other, may be invisible and intangible, and of a different nature from mechanism: and yet, *acting regularly and constantly*, may well be called natural . . .' (*Leibniz-Clarke Correspondence*, p. 53, my italics.)

187. *Optice*, pp. 320–1. In a MS. dating from about 1710, Newton carried out essentially the same computation, comparing the motion of a ray of light to the motion of the earth in its orbit. He concluded that 'the gravity of our earth towards the Sun in proportion to the quantity of its matter is above ten hundred millions of millions of millions of millions of times less then the force by wch a ray of light in entring into glass or crystall is drawn or impelled towards the refracting body.' All of the calculations are found on the sheet (University Library Cambridge, *Add. MS. 3070.9*, f. 621).

188. *Opticks*, pp. 392–4. The most detailed consideration of this experiment is found in a MS. from about 1710: Si vitra erant 20 digitos longa, Gutta olei ad distantiam quatuor digitorum a concursu vitrorum stabat in aequilibrio ubi vitrum inferius in clinabatur ad horizontem in angulo graduum plus minus sex. Est autem Radius 10 000 ad sinum anguli 6^{gr} vizt 1 045 ut totum guttae pondus, quod dicatur P, ad vim ponderio juxta planum vitri inferioris $\frac{1\,054}{10\,000} P$, cui vis attractionis versus concursum vitrorum aequalis est. Haec autem vis est ad vim attractionis versus plana vitrorum ut sinus semissus anguli quem vitra continent ad Radium id est ut tricessima secunda pars digiti ad viginti digitos: ideoq; vis attractionis versus plana vitrorum est $\frac{20 \times 32 \times 1\,045}{10\,000} P$ & vis attractionis versus planum vitri alterutrius est $\frac{10 \times 32 \times 1\,045}{10\,000} P$ seu 33, 44.P [Sit P pondus grani unius et vis attractionis in planum alterutrum aequalis erit ponderi granorum 33] Distantia vitrorum ad guttam seu crassitudo altitudo [*sic*] guttae erat octesima pars digiti ideoq; pondus cylindri olei cujus diameter eadam est cum diametro guttae & altitudo est digiti unius aequat 80 P & hoc pondus est ad vim qua gutta attrahitur in vitrum alterutrum ut 80 ad 33, 44 ideoq; vis attractionis aequatur ponderi cylindri cujus altitudo est 33, 44 pars digiti. Haec ita se habent ubi distantia guttae a concursu vitrorum est digitorum quatuor: in alijs distantijs vis attractionis prodit per experimentum reciproce ut distantia guttae a concursu vitrorum quam proxime id est reciproce ut crassitudo guttae: & propterea vis attractionis aequalis est ponderi cylindri olei cujus basis eadem est cum basi guttae et altitudo est ad $\frac{33,44}{80}$ dig ut $\frac{1}{80}$ ad crassitiem guttae. Sit crassitudo guttae $\frac{1}{10\,000\,000}$ pars digiti et altitudo cylindri erit $\frac{3\,344\,000}{64}$ dig seu 52 250 dig. id est 871 passuum. et pondus ejus aequabitur vi attractionis Tanta autem vis ad cohaesionem partium corporis abunde sufficit.' (University Library Cambridge, *Add. MS. 3970.3*, f. 428v.) Another consideration of the same experiment, noting that the attraction varies reciprocally as the distance between the glasses for the same quantity of surface, went on to conclude that if the glasses were separated by one ten millionth of an inch, the attraction in a circle one inch in diameter would sustain a cylinder of water of the same diameter over a mile long. Again he concluded that there are agents in nature whereby particles can cohere. (*3970.9*, f. 622.)

Chapter Eight

Newtonian Dynamics

In contrast to the continual development of Newton's speculations on nature from their origin in the *Questiones quaedam Philosophiae*, his work in mechanics, begun so fruitfully in the *Waste Book* and continued into *De gravitatione*, was interrupted and broken off. During roughly fifteen years, he attended to mechanics only when external stimuli caused him to – and then only as long as the stimuli endured. Only with the redaction of the tract *De motu* late in 1684 did he pick up again the thread of discourse he had laid down. From that point on, starting from approximately the level he had reached in *De gravitatione*, he rapidly extended his command in an investigation that was identical, for all practical purposes, with the composition of the *Principia*. One of the enduring questions of Newtonian scholarship is his delay of twenty years between the ideas on celestial dynamics in the 1660's and the publication of the *Principia*. In one sense, it begs the question to cite the level of his mechanics, since there is no obvious reason why he could not have continued in the 1660's to develop his initial insights as well as he did in the 1680's. Nevertheless, it is relevant to the question to insist that he did not, to insist on the fact that before 1685 Newton did not command a dynamics adequate to express the *Principia*.

During the intervening years at least three occasions summoned Newton to the display of his command of mechanics. The surviving papers help to illustrate the level of understanding he had reached and the possibilities that were open to him, but in none of the cases was he prompted seriously to examine the problems that remained inherent in his dynamics. Upon the publication of Huygens' *Horologium Oscillatorium*, apparently, Newton explored the properties of isochronous oscillations in a cycloid.[1] In comparison to Huygens' kinematics, Newton followed his usual dynamical approach and arrived immediately at the conclusion that he later included in the *Principia* – that at any point along the cycloid the effective component of gravity is proportional to the displacement along the curve from the lowest point, the point at which the pendulum hangs at rest. What makes the paper of interest in the history of Newton's dynamics is its implicit demonstration of the ease with which the fundamental equation of dynamics emerged from a frankly dynamical approach to Galileo's kinematics, in this place his kinematics of inclined planes. Without hesitation or even a word of justification, he set 'the efficacy of gravity' at any point proportional to 'the acceleration of descent . . .'[2] If we limit ourselves to what the paper contains, we seem to be face to face with the essentials of modern dynamics. The problem, of course, is what the paper does not contain. As far as the treatment of gravity is concerned, Newton had said virtually as much in his papers on circular motion, and the examination of the cycloid added nothing new. It did omit much that had been present in earlier papers. For example, it did not explicitly announce any principles of dynamics whatever. It did not generalise at all. It did not employ the word 'force' or a substitute for it, and did not suggest that the 'efficacy of gravity' is representative of a more general dynamical principle. Above all, it did not consider the other model of dynamic action, impact, and hence made no effort to choose between the two models. That is, in sum, the paper did nothing to resolve or to eliminate the contradictions that remained in his early dynamics.

Late in 1679, Newton received a letter from Robert Hooke inviting him to resume his correspondence with the Royal Society that had broken off with the death of Henry Oldenburg. The exchange of letters that followed is well known. Part of its interest lies in Hooke's letters, which furnished the basis for his later claim of plagiary. Newton himself acknowledged that the correspondence stimulated him to demonstrate that when a body follows an elliptical orbit about a centre of attraction located at one focus, the force of attraction must vary inversely as the square of distance. The demonstration did not get into the correspondence, but some of the concepts of dynamics worked out ten years earlier did appear in a form that presaged the demonstration.

In his first reply to Hooke, Newton proposed an experiment to demonstrate that the earth turns on its axis. If a body is dropped from

FORCE IN NEWTON'S PHYSICS

a tower, the greater tangential velocity of the top of the tower should cause the body to move somewhat ahead of the tower as it falls, and hence to land slightly to the east. In a diagram that he drew, Newton showed the trajectory as part of a spiral ending at the earth's centre (see Fig. 38). Here was a slip that Hooke could not bring himself to ignore. Suppose that the body is let fall at the equator and that the earth is

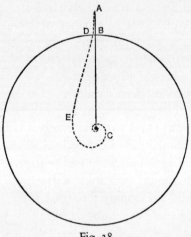

Fig. 38

separated into two halves with a vacuum between them while the gravitation to the centre remains unchanged. In such a case, a body that started to move with the tangential velocity of the surface of the earth would follow a path that resembles an ellipse (see Fig. 39). In effect, Hooke had converted the problem of fall into the problem of orbital motion. The described orbit followed, he said, from his 'Theory of Circular motions compounded by a Direct motion and an attractive one to a Center.'[3]

In the mechanics of circular motion, Hooke's suggestion was of capital importance, slicing away as it did the confusion inherent in the idea of centrifugal force and exposing the basic dynamic factors with striking clarity. In contrast, Newton's response suggests that his comprehension of circular motion, despite his successful quantitative analysis, was at that time less clear. If the gravity of a body let fall as Hooke imagined remains constant, 'it will not descend in a spiral to ye very center but circulate wth an alternate ascent & descent made by its *vis centrifuga* & gravity alternately over ballancing one another.'[4] The

426

NEWTONIAN DYNAMICS

comment is significant for the light it casts on Newton's understanding of his earlier quantitative analysis of circular motion. Whereas Hooke spoke of a central attraction continually diverting a rectilinear tangential motion into a closed orbit, Newton appears still to have viewed circular motion as an equilibrium of opposing forces, and elliptical motion as their alternating disequilibrium. In the continuing debate on

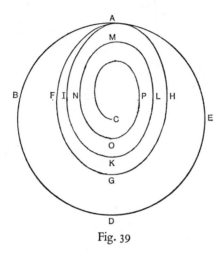

Fig. 39

the relative priorities of Newton and Hooke, it is necessary to insist that Hooke was the one who in fact set upright the crucial problem of orbital motion, which had been conceptualised, as it were, upside down. Hooke rather than Newton stated the mechanical elements of orbital motion in terms adequate to the concept of inertia. There is no good reason to doubt that in this matter Hooke was Newton's mentor, teaching him to conceptualise the problem in those terms by which alone the dynamics of orbital motion could be brought into harmony with his general dynamics.

In the very letter that employed the image of the alternating disequilibrium of opposing forces, a second passage, incapable of reconciliation with the first, foreshadowed the fruitful application of Newton's dynamics to Hooke's conceptualisation of orbital motion. It suggested as well the line of attack whereby a concept of force developed for problems of impact could be brought to bear on a problem that appears radically different. Hooke contended that in the problem he set, the body would follow a path similar to an ellipse. Newton did not agree. He argued that when the body crossed the diameter AD, its motion would not be directed toward *N* (as it would have to be if it traced an ellipse) but rather somewhere between *N* and *D* (see Fig. 40).

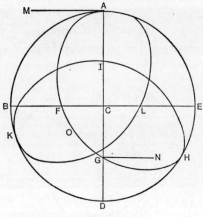

Fig. 40

For yᵉ motion of ye body at *G* is compounded of ye motion it had at *A* towards *M* and of all ye innumerable converging motions successively generated by ye impresses of gravity in every moment of its passage from *A* to *G*: The motion from *A* to *M* being in a parallel to *GN* inclines not ye body to verge from ye line *GN*. The innumerable & infinitely little motions (for I here consider motion according to ye method of indivisibles) continually generated by gravity in its passage from *A* to *F* incline it to verge from *GN* towards *D*, & ye like motions generated in its passage from *F* to *G* incline it to verge from *GN* towards *C*. But these motions are proportional to ye time they are generated in, & the time of passing from *A* to *F* (by reason of ye longer journey & slower motion) is greater than ye time of passing from *F* to *G*. And therefore ye motions generated in *AF* shall exceed those generated in *FG* & so make ye body verge from *GN* to some coast between *N* & *D*.[5]

In his reply, Hooke countered with the assertion that he did not believe gravity to be constant; rather it decreases in proportion to the square of the distance. Hooke later based his claim of plagiary directly upon this letter. The claim was made in ignorance of Newton's earlier derivation of the inverse square relation, of which Hooke could not have had any knowledge, but the claim also overrated the importance of a bright idea in comparison to a demonstrated theory, of which he should have understood the significance. Despite the fact that he treated gravity as constant in the letter, Newton revealed his command of a rudimentary, relatively systematised quantitative dynamics, something Hooke never had at his disposal. Ignoring his own idea of *conatus* from the centre, and accepting Hooke's conceptualisation in terms of an external force operating to change the direction of motion, Newton employed indivisibles to adapt the idea of force developed for impact to the problem. Gravity became a series of discrete impulses, like separate impacts, each generating a discrete change of motion. The total change of motion during a period of time, proportional to the total force exerted, would be the sum of the discrete impulses during that interval. The analysis as given can be faulted. It ignores the direction of motion at the time of each impulse and assumes in each a constant component normal to the diameter *BE*. Whatever its faults, however, it was pregnant with future possibilities.

The offspring was delivered almost at once. In a paper to which he alluded many times in the future and which has recently been identified,[6] Newton extended the earlier analysis to the problem Hooke had defined. He began by stating three hypotheses.

> Hypoth. 1. Bodies move uniformly in straight lines unless so far as they are retarded by the resistance of the Medium or disturbed by some other force.

A casual statement of the principle of inertia, the hypothesis repeats Hooke's definition of the problem. Since Newton's last explicit consideration of motion in *De gravitatione* had abandoned inertia with Cartesian relativity, and since *De motu*, composed in 1684, began from the position of *De gravitatione*, Hypothesis 1 poses something of a riddle in the history of Newtonian mechanics. It appears to me that Newton merely adopted the premises Hooke had set.

> Hyp. 2. The alteration of motion is ever proportional to the force by which it is altered.

To this restatement of the *Waste Book's* concept of force Newton added, as the third hypothesis, a statement of the parallelogram of motions as the device by which the motions mentioned in hypotheses one and two are combined.

In the first of three propositions, Newton applied these concepts to demonstrate Kepler's law of areas. Let *A* be a centre to one side toward which a body is attracted as it moves inertially, 'and suppose the attraction acts not continually but by discontinued impressions made

FORCE IN NEWTON'S PHYSICS

at equal intervalls of time which intervalls we will consider as physical moments.' By means of the parallelogram and some simple geometry, he demonstrated that the areas swept out in successive moments are equal. He then allowed the moments to diminish in length and to increase in number *in infinitum* until the polygon approaches a curve and the impulses of force become continual; the areas swept out are thus proportional to the times. Five years later, *De motu* began with the same proposition demonstrated in substantially the same way, and Proposition 1 of the *Principia* repeated *De motu*.

In 1679, Newton's strategy was to apply the law of areas to the calculation of forces in an ellipse. Given a body attracted to either focus with a force sufficient to make it orbit in the ellipse, such a force must vary inversely as the square of the distance from the focus. In the second proposition, he attacked the problem at the most accessible points, the two ends of the ellipse where the two curvatures are equal (see Fig. 41).

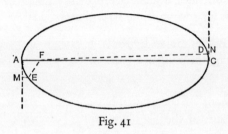

Fig. 41

In equal times the areas *AFE* and *CFD* are equal. When the arcs are short enough to approximate right lines, then AF/FC equals CD/AE. If there were no attraction, the bodies would move along the tangents *AM* and *CN*. 'Tis by the force of the attractions that the bodies are drawn out of the tangents from *M* to *E* and from *N* to *D* . . .' and therefore *ME* and *ND* are proportional to the attractions. The statement marks a significant step in Newton's comprehension of the mechanics of curvilinear motion. Under the tutelage of Hooke, he bid centrifugal force adieu and turned down the path that would lead him eventually to the *Principia*. Meanwhile, since the curvatures at the ends are equal, the geometry of the circle led immediately to the conclusion that at those two points the forces of attraction are inversely as the squares of the distance. It is noteworthy in passing that Proposition 2 did not employ the discontinuous force of Proposition 1, but rather one understood from the beginning to be continuous.

430

Proposition 3, which generalised the second, however, returned to physical moments and discrete impulses. With the aid of three lemmas that establish properties of the ellipse, and with the law of areas applied initially to individual 'physical moments,' he was able to demonstrate that the forces in such a polygon described around an ellipse must vary inversely as the square of the distance. Hence, when the sides of the polygon increase in number *in infinitum*, the force necessary to hold a body in an elliptical orbit must also vary as the square of the distance from the focus.[7] As in Proposition 1, Newton began the demonstration with a concept of force drawn from the model of impact, and in the passage to the limit slid unobtrusively into a concept of continuous force based on the model of gravity and free fall. Such a procedure served merely to obscure the problem of force, not to solve it.

The paper of 1679 contains Newton's original demonstration of two propositions that were later basic to *De motu* and to the *Principia*. It is perhaps more important in its demonstration, despite its logical problems, of the potential powers of a dynamics that applied the *Waste Book's* concept of force to the principle of inertia. What appears obvious to us was not necessarily so to Newton, however. Although he utilised the principle of inertia at this time, he had not determined yet to embrace it. In a word, the paper demonstrated two fundamental propositions but did nothing to resolve the problems internal to Newton's dynamics.

Approximately a year after the exchange with Hooke, Newton found himself in correspondence with Flamsteed about the great comet (or, as everyone except Flamsteed believed, comets) of 1680–1.[8] Newton's expertise in dynamics came into play in criticism of the fantastic mechanics by which Flamsteed derived an orbit that reversed itself without passing around the sun. The criticisms demanded no more than a qualitative discussion of orbital motion, and hence it did not carry into a new consideration of the dynamics question.

*

That new consideration began only in 1684. The circumstances of Halley's visit to Newton in August of that year are well enough known to require no repetition here. For my purposes, it is enough to recall that Newton claimed to be unable to find the paper of 1679 in which he had demonstrated the connection of the inverse square relation with elliptical orbits.⁹ Before Halley left, however, he agreed to demonstrate the proposition anew and to send it to Halley when it was done. As a result, a paper was sent to London in the autumn of 1684, a paper commonly referred to as *De motu*.¹⁰ It was registered with the Royal Society to insure Newton's rights of priority; and under the Royal Society's encouragement, Newton extended the investigation of orbital motion into the *Principia*, which was published in June 1687. As far as his dynamics is concerned, the development of its central concepts into their permanent form was associated with the revision and expansion of the material that composed the introductory sections of Book I. This development appears to have been completed during roughly the six months following the redaction of the original version of *De motu*.¹¹

In the original version of *De motu*, Newton derived the law of areas, the connection of the inverse square relation with elliptical orbits, and basic propositions concerning motion through a uniformly resisting medium from a dynamics less satisfactory than that he had employed in 1679, a dynamics inconsistent with itself and utterly inadequate to the load of demonstration it was asked to carry.

He started with two definitions. The first of them utilised the lesson of 1679 to introduce a new word into the language of mechanics.

> I call that by which a body is impelled or attracted toward some point which is regarded as a center centripetal force.¹²

Later, Newton explained that he had coined the word 'centripetal' in conscious imitation of Huygens' word 'centrifugal'. The change of word was symbolic of the more fundamental change in the comprehension of circular motion. Whereas Huygens' word, which means literally 'fleeing the centre', expressed the view that the crucial aspect of circular motion is the endeavour of bodies so moving to recede from the centre, Newton's force 'seeking the centre', grasped the realisation that, given the concept of rectilinear inertia, circular motion is possible only when a force continually diverts a body from its rectilinear path. This he had stated in 1679, and now, in Theorem 2 of *De motu*, he stated again that centripetal forces 'continually pull bodies [revolving in circles] back from the tangents to the circumferences.'¹³

What Newton recorded was his permanent debt – and that of the science of mechanics – to the conceptual insight of Robert Hooke. Let no one consider the debt insignificant. Established conceptualisations of problems are endowed with extraordinary powers of persistence. Hooke's insight freed Newton from the inhibiting concept of an equilibrium in circular motion. It opened the door to the supreme act of imagination in the construction of modern dynamics, the recognition that uniform circular motion is dynamically equivalent to its apparent opposite, uniformly accelerated motion in a right line. All curvilinear motion could now be subjected to the principles of a generalised dynamics adequate to every change of an inertial state. Without this insight, the *Principia* would have been impossible.

Beyond the comprehension of centripetal force, however, the first version of *De motu* ignored the paper of 1679 and returned to the dynamics of *De gravitatione*. Definition two defined a second type of force.

> And [I call that] by which it indeavors to persevere in its motion in a straight line the force of a body or the force inherent in a body [*vim . . . corpori insitam*].[14]

Lest the meaning of the definition should be unclear, he added an Hypothesis 2 which placed its interpretation beyond doubt.

> Every body by its inherent force alone proceeds uniformly in a right line to infinity unless something extrinsic hinders it.[15]

Perhaps nothing illustrates more clearly the intent of *De gravitatione* when it announced that something other than mere translation is essential to motion. The nearest approach to the principle of inertia in *De motu* treated uniform rectilinear motion as the product of a uniform force, the *vis insita* of the moving body.

In the MS. in which it has survived, the first version of *De motu* contains two stages, an original form and a set of emendations and additions. The original form started with the two definitions cited above and two hypotheses. Hypothesis one said that bodies are not impeded by a medium or by other external causes which would prevent them from exactly obeying the inherent and centripetal forces.[16] That is, *De motu* intended originally to confine itself to orbital motion treated as the resultant of the combined actions of inherent and centripetal forces. After he had completed this treatment, apparently, he hit upon the means of mathematically describing motion through a resisting medium. At that point, he inserted a third definition – of resistance as the force of a medium uniformly impeding motion – and altered hypothesis one to say that resistance is assumed to be zero in the first nine propositions and proportional to the velocity of the body and the density of the medium in the remaining ones.[17] Newton's success in subjecting resistance to the same mathematical treatment as other forces was an event of some significance in the history of mechanics. As far as the central problem of orbital motion to which *De motu* directed itself was concerned, however, resistance was no factor. Orbital motion continued to be seen as the resultant of the combined actions of inherent and centripetal forces. Sometime after the original redaction, and probably at the same time as the additions mentioned above, he entered a third hypothesis in the margin and the heading 'Hyp 4,' although he did not write in the fourth hypothesis itself. The third one stated explicitly the dynamics already employed in the following propositions.

> Hyp. 3. When it is acted upon by [two] forces simultaneously, a body is carried in a given time to that place to which it would be carried by the forces acting separately in succession during equal times.[18]

Newton had already employed the parallelogram of forces in Theorem 1. By the action of its inherent force alone, a body moves in a straight line, while impulses of centripetal force at equal intervals of time divert it into new paths. Using the parallelogram of forces and some simple geometry, Newton was able to demonstrate that Kepler's law of areas must be satisfied under such conditions.[19] Thus *De motu* embodied the ideal of mechanics that *De gravitatione* had presented, and sought to submerge kinematics in a universal dynamics.

By returning to *De gravitatione*, Newton plunged the dynamics of *De motu* into a series of contradictions to the resolution of which the work of the next six months was largely devoted. The definitions, first two and then a third, posited three kinds of force – inherent force, centripetal force, and resistance. The parallelogram of force was introduced as the means to determine the result when two of the forces act conjointly. Could they be compared in this way? In fact, the word 'force' [*vis*] was the only factor common to concepts that were otherwise utterly disparate. The 'inherent force' of definition two was conceived to maintain a uniform rectilinear motion, the force introduced originally to evade Cartesian relativism. 'Centripetal force' and 'resistance,' on the other hand, repeated the concept of force developed in the *Waste Book*, that of force external to a body which acts to alter its uniform rectilinear motion.

Nothing illustrates the incompatibility of the two concepts of force more clearly than the fourth hypothesis, if we may assume that the statement Halley entered into version two of *De motu*, after he had copied the imperfect version one, was the one Newton had in mind when he wrote 'Hyp 4' in the margin. In Halley's version, the space that a body, 'under the action of any centripetal force whatever, describes in the very beginning of motion is proportional to the square of the time.'[20] In version three of *De motu*, Newton offered a demonstration of the proposition (now called Lemma 2) which is interesting for what it assumes. When we read the proposition, we are apt to think that its point is the proportionality of distance to the square of time, so that a uniform force is seen as the cause of a uniform acceleration. In fact, this is what Newton assumed – referring it to Galileo's exposition. The point lay rather with the word 'whatever'. He was concerned to demonstrate that with a non-uniform centripetal force, as well as with a uniform one, the distance described at the very beginning of motion is proportional to the square of the time.[21] Whereas 'force' as inherent force causes a uniform motion ($F = mv$), 'force' as centripetal force causes a uniform acceleration ($F = ma$). Newton's parallelogram of forces was an adaptation of Galileo's parallelogram of motions, an adaptation which expressed his principle that force is more basic than motion. The two forces in question here, however, have utterly different relations to motion. Newton's parallelogram assumed that they are identical in their relations to motion.[22]

Nor was this the only problem that beset the conception of force in *De motu*. As far as inherent force is concerned, we can overlook the minor ambiguity arising from the fact that *De motu* did not contain a definition of mass, and say that inherent force is measurable by mv. What is the measure of external forces that alter uniform motion? In so far as *De motu* defined them, it appears that such forces are measured by ma. Hypothesis 4 explicitly assumed that a uniform (centripetal) force generates a uniform acceleration. Even before he set out to add the fourth hypothesis, he had assumed the same relation in Theorem 3 which, like Proposition 6 of the ultimate *Principia*, established a geometric expression for the centripetal force necessary for motion in any curve. He stated that the short line QR 'is proportional to the centripetal force when the time is given, to the square of the time when the force is given, and to the centripetal force and the square of the time conjointly when neither is given . . .'[23] A note in the margin, added later, referred to Lemma 2, an indication of the revision in version three. Theorem 2, which investigated the quantity of centripetal force in uniform circular motion, measured it in terms of uniform centripetal acceleration. And in the final two problems on motion through a resisting medium, added at the revision, the dynamic principle basic to the solutions held that 'the decrease in the speed is proportional to the resistance . . .'[24] In the context, the statement means instantaneous decrease of speed, that is, acceleration. With the exception of the last case, then, the dynamics of *De motu* seems to rest squarely on Galileo's kinematics of free fall. Lemma 2 of version three even referred the proportionality of distance to the square of time explicitly to him. Newton's solution of the difficult problem of motion through a resisting medium, with its constant change of velocity and resistance, rested as well on his perception of the analogy of the instantaneous decrease of velocity to the uniform acceleration of idealised free fall – the analogy, that is, of resistance to other forces that change uniform rectilinear motion.

Unfortunately, however, another measure of such forces also appears in *De motu*, another measure which returns to the other model of force, the phenomena of impact. In his letter to Hooke in 1679, Newton had imagined a series of discrete impulses acting in successive instants. Theorem 1 of *De motu* employed the same conception.

> Let the time be divided into equal intervals, and in the first interval of time suppose that the body describes the right line *AB* by reason of its inherent force. Likewise in the second interval of time, if nothing were to impede it, it would proceed in the right line to *c* describing the line *Bc* equal to *AB* . . . In fact, when the body arrives at *B*, let the centripetal force act with a single but great impulse, and make the body turn from the line *Bc* and proceed in the line *BC*.[25]

Similarly, the centripetal force is conceived to act at points *C*, *D*, and *E*, removed from each other by equal intervals of time, turning the body aside on each occasion into a new rectilinear path. Force conceived in these terms, as a 'single but great impulse,' can only be measured, of course, by Δmv. In fact, such a measurement had much to offer Newton. Although it is no more commensurable in rigorous terms with inherent force than a uniformly acting accelerative force is, it is amenable to treatment by the parallelogram in a way that a continuous accelerative force cannot be. For example, in Theorem 1, one side of the parallelogram (or triangle) corresponds to the motion generated by the inherent force. Another side, also rectilinear motion, corresponds to that generated by the instantaneous impulse, and the third side is the resultant. Given the mathematical device Newton chose to employ in Theorem 1, so that equality of areas in successive intervals of time was demonstrated from the geometry of triangles, he was encouraged to employ a definition of force compatible with it. Although *De motu* does not contain a formal exposition of the method of nascent and ultimate ratios which he employed in the *Principia*, it does make use of the method a number of times. The method involves the construction of small triangles and rectangles through two points of a curve which are then imagined to approach each other more closely than any assignable distance. A nascent or ultimate ratio of two sides of such a figure, when the two points approach each other, is a mathematical conception similar to the differential though expressed in terms of variables not commonly employed. Whenever one side of such a figure represents a force, as one side of the little triangle in Theorem 1 implicitly does, the initial statement of the problem encourages a definition of force as Δmv. Thus the mathematics of *De motu*, which in some cases regards curves as the limiting figures of polygons, helped to direct Newton toward one definition of force whereas his physical intuition of uniformly accelerated motion directed him toward another. Apart from the issue of inherent force, the dynamics of *De motu* in its original form rested primarily on the concept, derived from Galileo's analysis of free fall, that a uniform force generates a uniformly accelerated motion. The mathematical devices of *De motu* were pressing toward the model that impact offered. In the following revisions a variety of considerations helped the model of impact to assume greater importance, until in the *Principia* the second law of motion stated that the 'change of motion is proportional to the motive force impressed . . .'[26]

De motu also suggests that Newton did not consider the two definitions to be incompatible. As I have been arguing, dynamics in the seventeenth century was bedevilled by dimensionally incompatible definitions of force. However great his contributions to dynamics, Newton did not succeed in escaping from the same dilemma, and already in *De motu* he indicated the device by which he would obscure the problem from his own perception. In Theorem 1, discrete impulses of force act at successive intervals of time, each impulse generating an increment of motion proportional to it. The triangular areas corresponding to individual intervals are shown to be equal. Assume now, Newton continued, that 'these triangles [are] infinite in number and infinitely small so that single triangles correspond to single moments of time...' The phrase is somewhat ambiguous. By itself it might be taken to suggest an ultimate discontinuity of time and of force in which the measure of force could only be a discrete increment of motion. The letter to Hooke in 1679 had seemed to embrace such a conception. From this point of view there would be only one measure of force, Δmv, and the concept of total force that he employed in the *Waste Book* would correspond to a simple summation both of impulses and of the motion generated by a similar summation of increments. In fact, however, this was not Newton's view. The same sentence in Theorem 1 spoke of 'centripetal force acting without intermission' in the limiting situation.[27] Theorems 2 and 3 employed an explicit passage to the limit which assumed time to be a continuum infinitely divisible rather than a series of discrete instants. Indeed, the definitions of the problems considered in those two theorems employed a concept of continuous force proportional to *ma* even before the passage to the limit. From every appearance, Newton understood the 'centripetal force' of Theorem 1 to be identical to the 'centripetal force' in Theorems 2 and 3.

How was this possible? Although a dimension of time was not present in the definition of force in Theorem 1, it was present in the definition of the problem. Separate impulses act at equal intervals of time. In the passage to the limit, the succession of intervals approaches the continuous flow of time as a limit, and the force concentrated in separate impulses is spread out evenly, as it were, over the entire period. In terms of Newton's earlier phrase, the 'total force' before and after the passage to the limit remains unchanged. Total force after passage to the limit refers to the integral $\int F dt$. In 1684, Newton had long been mathematically equipped to state such a concept. Nevertheless he did not. In a later revision of *De motu*, he did reconcile the two uses of 'force' in a definition of the concept of moment, a definition that looked forward to the ultimate Lemma II of Book II in the *Principia*.

> The moments of quantities are the principles from which they are generated or altered by a continual flux: as present time [is the moment] of the past and future, present motion of past and future motion, centri-

petal or any other instantaneous force whatever of impetus, a point of a line, a line of a surface, a surface of a solid, and an angle of contact of a rectilinear angle.[28]

Despite the distinction indicated here, he continued to employ the word 'force' indiscriminately in two cases that were dimensionally incompatible with each other. To this extent, he remained entangled in the century's confusion as to what quantity to single out as the dynamic unit. Already in *De motu* the confusion had been sharply limited, however. In *De motu*, as in the *Principia* that followed, force as Δmv was employed solely in contexts that assumed a succession of impulses at equal intervals of time. If the dimensional incompatibility remained, the confusion of thought did not, and Newton's failure to distinguish force as Δmv from force as ma did not hobble his dynamics as similar failures had earlier hobbled the dynamics of others.

★

As far as Newton's ultimate dynamics was concerned, the central problem that the first version of *De motu* left for solution was the resolution of the contradiction posed by his acceptance of two radically different concepts of force, inherent force which maintains uniform motion, and centripetal force which alters it. To the solution of the problem, version two, which was little more than a fair copy of the amended form of version one, added nothing. Version three, in contrast, marked a further stage in Newton's thought. Hypotheses 3 and 4 of the earlier versions were moved to the status of Lemmas, enhanced by demonstrations, but not otherwise altered. To the remaining two hypotheses Newton added three more, decided that the word 'Hypothesis' did not express his meaning, struck it out in all five cases, and replaced it with a new word, '*Lex.*'

In their original incarnation, then, the laws of motion numbered five. The first remained substantially unchanged from version one, still asserting that by its inherent force alone a body proceeds with a uniform motion in a right line. Law two, however, underwent a significant change. The word 'impressed' replaced the word 'centripetal,' not only increasing the generality of the law, but also expressing the dichotomy that Newton was seeking to exploit. The inherent or internal force of a body maintains it in uniform motion. The force external to a body, impressed on it from without, alters its uniform motion.

> The change of motion is proportional to the impressed force and is made in the direction of the right line in which that force is impressed.

With the exception of one word (the adjective 'motive' modifying impressed force), the law is identical to that which appeared in the *Principia*. At this time, however, Newton was not satisfied with the wording. He changed the initial phrase to read: 'The motion generated or the change of motion . . .,' and then changed it a second time to the form it retained in version three: 'The change in the state of moving or of remaining at rest is proportional to the impressed force . . .'[29] Two things are worthy of note in the definition. His insistence on naming both rest and motion in the definition was connected with the form of definition one in which the absolute motion of a body is determined by its absolute force. As long as rest was not identical to uniform motion, the law of force needed to include changes in both states. Second, the statement of the law he now adopted was the one based on the model of impact, this despite the fact that Lemma 2 and divers theorems continued to embody a conception of force that can only be measured by acceleration.

Of the remaining three hypotheses – or laws – added in this version, the last of them formalised what he had discovered in version one about the resistance of a medium. It asserted that the resistance of a medium is proportional to its density, to the surface of the body moved, and to the velocity, all three taken conjointly. This peculiar 'law' experienced a peculiar fate. In the next revision, it suffered the indignity of an additional note in which Newton confessed that he did not affirm the law to be exact; it was sufficient that it be approximate. Whereupon he recognised the absurdity of what he was saying, and with a stroke of the pen removed it forever, as it deserved, from the laws of motion.[30] The other two laws of version three were less anomalous but in the end more troublesome. One asserted that the 'relative motions of bodies contained in a given space are the same whether the space in question rests or moves perpetually and uniformly in a straight line without circular motion.' The other complemented it by affirming that the common centre of gravity of a system of bodies does not change its state of motion or rest because of the mutual actions of the bodies on each other.[31] In a word, these two laws returned to the insights of the *Waste Book* and its realisation that in an inertial system two bodies isolated from outside influences can be considered as one concentrated at their common centre of gravity. Thus he referred the second of them to Law 2, the force law, for confirmation. Law 2 itself implied the concept of inertia in its renewed assertion of the linearity of all changes of motion. In version three of *De motu*, then, three new laws of motion marked the beginning of Newton's return to inertial mechanics. Law 1, on the other hand, contradicted them directly in its assertion that uniform motion is distinguishable by the internal force that sustains it. As if to insure that the contradiction not be lost, Newton also inserted a Scholium after Problem 5 in which he referred to 'the immense and truly immobile space of the heavens . . .'[32] Thus two contradictory currents of thought appear to have governed the amendments that went into version three, as though Newton felt the need to assert the reality of absolute space with more vehemence the more he undermined its operative significance as he clarified his dynamics.

In the papers revising version three that stood between it and the so-called *Lectiones de motu*, the divergent tendencies become fully manifest. A greatly expanded set of definitions began with absolute time, relative time, absolute space, and relative space. After definitions of body, centre and axis of a body, place, and rest, Newton continued with a definition of motion which came immediately to grips with his central concern, the distinction of absolute from relative motion. In circular motions, the endeavour to recede from the centre enables us to determine an absolute rotation. In general, the distinguishing factor is force.

> Furthermore, that motion and rest absolutely speaking do not depend on the situation and relation of bodies between themselves is evident from the fact that these are never changed except by a force impressed on the body moved or at rest, and by such a force, however, are always changed; but relative motion and rest can be changed by a force impressed only on the other bodies to which the motion and rest are related and is not changed by a force impressed on both such that their relative situation is preserved.[33]

Both in their form and in their content, the definitions seem to return to the essay *De gravitatione* with its impassioned rejection of Cartesian relativism and its conviction that relativism can be escaped via dynamics. The dichotomy of relative and absolute motion was equally the dichotomy of kinematics and dynamics. As he had said in *De gravitatione*, 'physical and absolute motion is to be defined from other considerations than translation, such translation being designated as merely external.'[34]

And exactly here was the dilemma. Further down the page, the definition of the inherent force of a body, that which is not merely an external relation – what in this paper, as though consciously recalling the earlier assertion, he first called the 'force . . . inherent and innate in a body . . .' [*Vis corporis seu corpori insita et innata* . . .], and then the 'inherent, innate and essential force of a body . . .' [*Corporis vis insita, innata et essentialis* . . .][35] – was in process of a revision which would render it useless in the determination of absolute motion. In its first three versions, *De motu* defined the inherent force of a body as that by which it endeavours to persevere in uniform motion in a right line, and the first law of motion had reaffirmed what the definition asserted. 'The inherent, innate and essential force of a body,' he now stated, 'is the power by which it perseveres in its state of resting or of moving uniformly in a right line, and is proportional to the quantity of the body; it is actually exerted proportionally to the change of state. . . .'[36] The first law of motion underwent a corresponding change, in effect abandoning the concept that a uniform motion is the product of a uniform force, converting inherent force into a resistance to change in a body's state, and tacitly embracing the principle of inertia.

In light of the direction in which his dynamics was tending, we cannot avoid asking whether Newton's insistence on absolute motion had any practical consequences for his mechanics. Although the discussion of absolute space, time, and motion was to be greatly expanded in 1685 and fixed in substantially its final form in the *Lectiones de motu*, all of the essential ideas were present in the passage under consideration. Did he in fact have any criteria by which he might identify what in the *Principia* he called the true motions of bodies? Remember that absolute motions – not absolute space, but absolute space as a frame of reference for absolute motions – were the object of Newton's inquiry. In *De gravitatione*, he proclaimed as the ultimate absurdity of Cartesian relativism that in its terms 'a moving body has no determinate velocity and no definite line in which it moves. . . . On the contrary, there cannot be motion since there can be no motion without a certain velocity and determination.'[37] In the *Principia*, he gave the example of a sailor walking east with a velocity of one on a ship sailing west with a velocity of ten. If that part of the earth on which the ship is located is truly moved to the east with a velocity of 10 010, then the sailor will truly move east with a velocity of 10 001. It is exactly this true and absolute motion that Newton now found himself asserting while he denied himself at the same time the means to identify it. As Law 4 on the sheet of revisions asserted, the 'relative motion of bodies enclosed in a given space is the same whether that space rests absolutely or moves perpetually and uniformly in a right line without circular motion.'[38]

In the *Principia*, Newton detailed three criteria by which absolute and relative motions can be distinguished – their properties, causes, and effects. It is a property of rest that bodies truly at rest are at rest in respect to each other. This is the criterion of translation or kinematics alone; and true to his unchanged conviction, Newton pointed out the impossibility of ever determining if a given body that might serve as a reference point is truly at rest. The causes of true motion are forces impressed on bodies to alter it. This criterion is no better than the last. It speaks solely of changes in motion, and is helpless to determine absolute motion itself. The effects of true motion are forces of recession from axes of circular motion. Here in fact was the gravamen of Newton's argument.

By way of illustration, he gave two examples. A bucket of water hangs on a twisted rope. When it is set free to turn, the bucket begins to rotate while the water in it remains for a time at rest. As the motion of the bucket communicates itself to the water, the water begins to rotate as well. When the bucket and water are in relative motion, the flat surface of the water demonstrates that it is truly at rest. When the bucket and water are rotating together and are therefore relatively at rest, the depressed surface caused by the water's centrifugal tendency demonstrates that it is truly in motion.[39] Perhaps the argument proved too much. When Newton referred to the flat surface of the water, he was speaking of a liquid body which he believed to be, not at rest, but in motion, and indeed in circular motion on a rotating earth. Is there any way to determine with assurance that in fact the earth does rotate on its axis? Newton was convinced that there is. If we connected two globes together with a cord and set them rotating about their common centre of gravity 'even in an immense vacuum, where there was nothing external or sensible with which the globes could be compared,' we could determine that they were indeed in motion by the tension in the cord. By impressing equal forces on alternate faces of the globes and measuring the effect by means of the tension, we could determine the direction of the rotation.

But now, if in that space some remote bodies were placed that kept always a given position one to another, as the fixed stars do in our regions, we could not indeed determine from the relative translation of the globes among those bodies, whether the motion did belong to the globes or to the bodies. But if we observed the cord, and found its tension was that very tension which the motions of the globes required, we might conclude the motion to be in the globes, and the bodies to be at rest; and then, lastly, from the translation of the globes among the bodies, we should find the determination of their motions.[40]

Usually, though not always, Newton's discussions of absolute space implicity identified it with the space of the solar system. The *Principia*, for example, explicitly introduced the hypothesis that the centre of gravity of the solar system is at rest.[41] Did he in fact have a criterion by which to establish that in the solar system the earth truly rotates, as the passage above so clearly sought to do? What his two examples establish is a criterion, within the framework of an inertial mechanics, for absolute rotation about a given axis. The criterion can be and has been criticised. Drawn from a mechanics that was derived from the universe as given, it assumes the outcome of a thought experiment in which all the rest of the universe is imagined not to exist. Since my purpose here is not philosophic analysis but the reconstruction of Newton's thought in its historical context, let me accept the criterion as valid and examine its consequences. Performed at Cambridge, the experiment with the two globes would have been difficult to interpret in 1685 even if measurements sufficiently fine could have been made; performed at the north or south pole, it could in principle have demonstrated that the earth rotates on its axis. Fortunately nature has undertaken to perform equivalent experiments on her own. Newton was convinced that the non-spherical shape of the earth and the measured acceleration of gravity both provide concrete evidence of the absolute rotation of the earth. The measured lengths of seconds pendulums at different latitudes had suggested the oblate shape of the earth in Newton's day. If the conclusion remained in doubt, it was confirmed to the general satisfaction of the scientific community within a generation. Newton's correlation of the acceleration of gravity with the centripetal acceleration of the moon, a calculation in which the centrifugal effect due to the earth's rotation was inserted as one correction, was utterly misleading. The correlation depended ultimately on the distance of the moon, and the range of possible values from available parallactic observations introduced a margin of error considerably larger than the centrifugal correction. Nevertheless, Newton's use of the measured acceleration of gravity to detect centrifugal effect was correct in principle.

Beyond this point, however, he could not advance. According to the heliocentric system, which he accepted as beyond question, the earth also revolves around the sun. In *De gravitatione*, in the definition of absolute motion connected with *De motu*, and in the *Principia*, he objected, now with more, now with restrained vehemence, that Cartesian relativism denied the motion of the planets about the sun despite their endeavours to recede. At this point certainly, Mach's criticism of Newton's argument becomes overwhelming. 'All our principles of mechanics are . . . experimental knowledge concerning the relative positions and motions of bodies. . . . No one is warranted in extending these principles beyond the boundaries of experience.'[42] Indeed how did Newton know that the planets endeavour to recede from the sun? No confining bucket, no constraining cord renders the endeavour observable. It can only be inferred from the curvilinear motions in the context of an inertial mechanics. Newton's principle was exactly the reverse – forces were supposed to reveal motions, not motions forces.

At best, then, Newton advanced a criterion able to distinguish a narrowly confined set of absolute rotations. Since the revision of the concept of inherent force left him with no criterion at all for absolute translations, his purpose – to determine absolute motions – was beyond fulfilment in any case. As I mentioned, Newton tended to equate absolute space with the space of the solar system. Nevertheless, at least once he acknowledged explicitly the possibility that the solar system as a whole moves inertially when he raised the question, in the *Principia*, of whether the so-called fixed stars are indeed at rest.[43] Meanwhile, as he continued to assert, all of the relative motions remain the same whether or not the space in which they take place is moving with a uniform rectilinear motion. With the revisions of the third version of *De motu*, his dynamics became what dynamics has remained ever since, the science of the causes, not of motions, but of changes of motion. To this science, even as it is presented in the *Principia*, the concept of absolute motion is utterly without consequence.

Why then did Newton assert it? Possibly germane to the question is the increasing stridency of his assertions, which grew, both in vehemence and in length, in exact proportion as the development of his dynamics rendered the concept operationally meaningless. In the first version of *De motu*, the inherent force of a body provided a criterion to identify absolute motion, which was referred to only by implication. Step by step, as he modified the concept of inherent force toward reconciliation with the principle of inertia, he introduced the idea of absolute motion and explained it in ever increasing detail, though not with additional criteria. To understand Newton's motives, we need to return to *De gravitatione* where absolute space expressed his revulsion from the absolute insecurity of a world in which no guidelines and reference points were present. 'The eternal silence of these infinite spaces fills me with fear.' Pascal's *cri de coeur* found its echo in Newton's refusal to set sail on the shoreless sea of relativity. By vehemence alone, when all else failed, he would refuse the manifest conclusion to which his own dynamics led him. His assertion of absolute motion has all the appearance of an act of defiance hurled in the face of the very current of thought on which his dynamics itself was borne inexorably toward its ultimate form.

Perhaps another phrase in *De gravitatione* also sheds some light on Newton's motives. The absurdity of Cartesian relativism, he pronounced, stands fully disclosed in the consequence that in its terms a moving body can have no determinate velocity and no definite direction. 'And, what is worse, that the velocity of a body moving without resistance cannot be said to be uniform, nor the line said to be straight in which its motion is accomplished.'[44] The relativity inherent in inertial mechanics had brought Newton face to face with the ultimate paradox of modern mechanics – that to assert the principle of inertia is to renounce in the same breath any criterion by which its truth may be demonstrated. The relativity associated with inertia renders it impossible to ascertain even that a motion is either uniform or rectilinear. In part, at least, we should perhaps understand Newton's assertion of absolute motion as his refusal to embrace that paradox. If so, we must realise as well that whether or not he embraced it he could not escape it, and his inability to escape it was the price he paid to create the science of dynamics.

If this analysis is correct, if Newton's assertion of absolute motion went hand in hand with his abandonment of a dynamics in which absolute motion did not have to be proclaimed because it was implicit, then the motives behind the other change are relevant as well. Again Newton did not announce them. Nevertheless, since we can see how the change eliminated difficulties internal to the dynamics as it stood, we are not left to speculate wholly without foundation. The original dynamics of *De motu*, following the line of *De gravitatione*, saw the inherent force of a body as the sum total of the forces that had acted on it. What was one to make of circular motion under these circumstances? Here Newton confronted another anomalous riddle in which so much of seventeenth-century science revolved. Already in his early paper, 'The Lawes of Motion,' he had seen the fact that in certain impacts in which there are changes of direction, quantity of motion is not conserved.[45] In uniform circular motion as he now conceived it, the constant exercise of centripetal force fails to alter the force of the moving body at all. In a revision of version three, he inserted a definition of the 'force of motion or the force which a body possesses accidentally because of its motion,' which is usually called impetus, which is proportional to the quantity of motion, and which is, according to the motion, either absolute or relative.[46] This definition he crossed out, and in the final revision of the definitions that preceded the *Lectiones de motu*, he added a sentence to the definition of impressed force that became a permanent part of it, continuing unchanged through all the editions of the *Principia*. 'This force consists in the action only, and remains no longer in the body when the action is completed.'[47] Newton's effort wholly to absorb the science of motion into dynamics had foundered on the unavoidable shoal where every seventeenth-century effort to build a dynamics on the model of impact had been wrecked. The inherent force of a body, the internalised sum of the forces that have acted on it, is a concept incapable by its nature of vectorial expression. Given the idea of rectilinear motion, the inherent force of a body can only express its force to press on in its present direction. Given the composition of forces in the parallelogram, the inherent force of a body after an oblique force has changed the direction of motion cannot be the sum of the forces that have acted on it. When Descartes considered the force of a body's motion, the necessity implicit in the concept led him to assert that no action is required for a change of direction. Newton faced the same problem convinced that the principle of inertia places changes of direction on the same plane as changes of speed, and in the end he altered his concept of inherent force to one totally different from the original, one not inextricably bound to the model of impact. From this time, he was free to pursue the central insight of the *Waste Book* and to build a dynamics, different from that of *De motu*, on the principle of inertia and the concept of force proportional to the change of inertial motion it causes. Once this essential alteration was made, the concept of absolute motion ceased to have any operative function in Newton's mechanics.

*

The changes in Newton's dynamics introduced in the revisions of version three of *De motu* extended beyond the rejection of his original notion of inherent force. It was in this manuscript that Newton consciously moved outside the limited range of celestial dynamics and attempted systematically to examine the basic concepts of dynamics as a whole. In its original form, *De motu* had been far more confined. To solve the mechanics of planetary motion, Newton had seized on a minimal number of dynamic concepts ready at hand but unexamined. The original version contained two definitions only, and version three contained only four. At one time or another as it underwent continual emendation, the revision of version three proposed eighteen definitions of quantities possibly required for a systematic dynamics. With the redaction of this paper, Newton's dynamics assumed a consciously generalised formulation and began to approach its ultimate shape.

One of the quantities that struggled toward precise definition was quantity of matter or mass. Mass had not appeared originally in *De motu* beyond its implicit presence in the idea of quantity of motion, a concept which had also not been defined. The paper of revisions now defined quantity of motion as the product of velocity and the quantity of the body in motion. 'However the quantity of a body is measured by the bulk of corporeal matter, which is usually proportional to its gravity.'[48] In the later reworking of the definitions, he moved quantity of matter to the head of the list.

> *Quantity of matter* is that which arises from its density and the magnitude [of the body] conjointly. A body twice as dense in twice the space is quadruple in quantity. This quantity I designate by the name body or mass.[49]

Thus the word 'mass' entered the vocabulary of mechanics. Mach has criticised the definition on the grounds of circularity. Whereas Newton defined mass in terms of density, we can define density only in terms of mass and thus are left with two mutually dependent but otherwise undefined terms.[50] From a positivist point of view, the criticism is undoubtedly valid, but it does nothing whatever to illuminate Newton's process of thought. From the mechanical philosophers he accepted without question an idea of matter in terms of which density evoked an image requiring no definition. Matter as such was homogeneous extended stuff, proportional always to its extension and differentiated only by quantity. Though atomists rejected Descartes' identification of matter with extension, they accepted his view of the homogeneity of matter in proportion to extension. To explain differences in density, Gassendi employed the image of a bushel of wheat which can be shaken to cause the kernels to fit more closely together. Descartes employed essentially the same figure, the sponge, though of course the pores in this case were not voids but were filled with subtle matter. Because of the unquestioned assumption of the homogeneity of matter, Newton could regard density as a measure merely of the quantity of matter in a given volume and define mass in its terms. Far from an operational term, density was an intuitive term evoked by Gassendi's image of the bushel of wheat, the proportion of solid matter in a volume partly composed of voids. Of course, Mach is correct in the contention that Newton had no way to measure mass and density independently of each other.

In reality, the definition of quantity of matter furnished only half of Newton's concept of mass. The other half came from his revised idea of inherent force.

> The inherent, innate, and essential force of a body is the power by which it perseveres in its state of resting or of moving uniformly in a right line, and is proportional to the quantity of the body. It is actually exerted proportionally to the change of state, and in so far as it is exerted, it can be called the exerted force of a body.[51]

As he thought about it, Newton liked the idea of exerted force and added a definition of it. Exerted force is that by which a body strives to preserve 'that part of its state of moving or of resting which it loses in single moments and is proportional to the change of that state or to the part lost in single moments. . . .'[52] Significantly, in a later revision of these definitions, Newton altered that of inherent force so that it read, not 'inherent force of a body,' but 'inherent force of matter,' as it remained in the *Principia*. In any given body, he continued, the inherent force of matter is proportional to the mass and differs in no way from the inactivity of the mass.[53]

Contrary to Newton's assertion, the inactivity of matter as conceived by the prevailing mechanical philosophy was exactly what inherent force, as he now defined it, did differ from. If matter is endowed with an inherent force by which it resists efforts to change its state, it cannot be wholly inactive, or in the classic phrase of the seventeenth century, wholly indifferent to motion. Indeed, Newton's formulation of the laws of mechanics implied a conception of matter at once inert and active, unable to initiate any action itself, passively dominated by external forces, but endowed with a power to resist their actions. Although Leibniz had arrived at a similar view, the *Principia* was the primary vehicle by which this strange conception of matter passed into the mainstream of modern science, where familiarity coupled with the power of the dynamics that employs it has dulled our perception to its ultimate paradox. In the *Principia*, Newton himself summarised the paradox in another anomalous phrase, *vis inertiae*, which we might translate freely as 'the activity of inactivity', or perhaps 'the ertness of inertness'.

Given the other factors already established in the science of mechanics, it is difficult to see how a workable dynamics could have developed without some such idea of matter. The indifference of matter to motion expressed admirably Galileo's basic insight that a body in motion will continue in motion. Meanwhile it was apparent to everyone that matter cannot be wholly indifferent to motion since unequal amounts of effort are required to cause equal changes of velocity in unequal bodies. Struggling to express this observation, Descartes had said that a body persists, as much as in it lies, in its motion in a straight line. With its connotations of activity applied to inert matter, the verb 'persist' foreshadowed the paradox of Newton's *vis inertiae*. Similarly, the analysis of impact in the *Waste Book* had stated as a basic proposition that equal forces applied to unequal bodies generate motions inversely proportional to the sizes of the bodies. Facing the same problem of impact, Mariotte was to arrive at the same idea of an internal resistance, and in the century's most consciously analytical examination of the question, Leibniz concluded that if matter were indifferent to motion then any force would be able to generate any motion in any body.[54] Not unlike Newton, Leibniz reformulated the conception of matter as we have seen, and defined its essence as force rather than extension. Newton did not go that far. To him, *vis inertiae* represented one of the universal properties of matter, not displacing extension, but standing equally beside it together with hardness and impenetrability. Inevitably, he set it proportional to the quantity of matter. As he revised *De motu* into the *Principia*, he returned specifically to Descartes' discussion of motion which had influenced the passage in the *Waste Book*. The inherent force of matter is a power of resisting by which 'any body perseveres, as much as in it lies, in its state of resting or of moving uniformly in a right line . . .'[55] Because of the inherent force of matter, there is a constant proportion between the force applied and the change it generates in a body's state of motion. Without some such concept, a quantitative dynamics would have been impossible.

Newton devoted part of the re-examination of basic terms that contributed to the expansion of the number of definitions in *De motu* to the concept of force itself. *De motu* had started with only inherent force and centripetal force, to which he added resistance not much later. In the revision of version three, he now defined six kinds of force – inherent force, the force of motion, exerted force, impressed force, centripetal force, and resistance. 'There are also other forces,' he added, 'arising from the elasticity, softness, tenacity, etc., of bodies, which I do not consider here.'[56] I have already discussed inherent force, in the revised form in which it now appeared. The force of motion represented a last effort to save the idea that had been eliminated from the revised definition of inherent force. Exerted force, on the other hand, supplemented the new concept of inherent force. Whereas inherent force is proportional to mass, exerted force is proportional to the impressed force and thus to the change of the body's state of motion or rest, and is not improperly called the reluctance or resistance of the body. One species of exerted force is the centrifugal force of revolving bodies.[57] Obviously the idea of exerted force was a groping step toward the insight expressed in the third law, first announced later in this same paper. If a body resists the actions seeking to change its state of motion, it exerts in reaction a force on whatever acts on it.

The other three forces, impressed force, centripetal force, and resistance, were in fact identical in Newton's opinion; centripetal force and resistance were merely specific forms of impressed force. The phrase 'impressed force' had appeared in the wording of the second law of motion in version three of *De motu*, although the definitions, which included centripetal force and resistance, had not included it. Now he defined impressed force as a general term subsuming the other forces that act on a body from outside to alter its state of motion or rest.

> The force brought to bear or impressed on a body is that by which the body is urged to change its state of moving or resting and is of divers kinds such as the pulse or pressure of percussion, continuous pressure, centripetal force, resistance of a medium, etc.[58]

When Newton proceeded to cancel his definitions of the force of motion and exerted force, two kinds of force remained as the foundation of his dynamics – the inherent force of matter, by which a body resists efforts to change its state of motion or rest, and impressed force, any action arising outside a body that attempts to change its state of motion or rest. Inherent force, which is proportional to the quantity of matter or the mass, establishes a constant proportion between an impressed force and the change of motion it produces.

The suggestion has been advanced that Newton's use of the adjective 'impressed' was consciously intended to distinguish mere 'force' from 'impressed force'. The point of the suggestion rests on the specific form in which Newton expressed Law II.

> The change of motion is proportional to the motive force impressed...[59]

We are accustomed to word the second law of motion somewhat differently so that it refers, not to change of motion (Δmv), but to rate of change of motion (ma). The distinction between force and impressed force proposes to resolve this dilemma by the argument that Newton understood 'impressed force' to mean the application of force over a period of time ($\int F dt$).[60] I have already presented my interpretation and resolution of the manifest discrepancy between Newton's statement of the second law and his frequent use of it in the other sense, and I shall need to return to it in connection with the *Principia*. Meanwhile, the development of the definitions in *De motu* seems clearly to argue that he intended no distinction between force and impressed force. The fact that he built a dynamics on divers concepts of force but not on force unmodified, which he never undertook to define, is surely decisive. The distinction is not between 'force' and 'impressed force', but rather between 'inherent force' and 'impressed force', *vis insita* and *vis impressa*. The distinction was expressed most aptly by the contrasting adjectives 'internal' and 'external', which he used in *De gravitatione*. The extra adjectives that he added in the revision of version three, even if they later dropped out of sight, also helped to express the distinction – on the one hand 'the inherent, innate, and essential force of a body' [*corporis vis insita, innata et essentialis*], and on the other hand 'the force brought to bear and impressed on a body' [*vis corpori illata et impressa*].[61]

Although the revisions of version three of *De motu* did not include a statement of the principle of inertia, they implied it, and they set the enduring pattern of Newton's dynamics – the establishment of equilibrium between inherent and impressed forces by means of changes in inertial motion. In an unexpected way, a concept drawn from statics reappeared at the very heart of the system that established the point at which dynamics ceases to be statics. Surprisingly as well, we are forced to recognise a central role for the model of impact in Newton's vision of dynamics. On the one hand, it furnished his definition of impressed force, although I have argued that Newton understood the definition not to stand in conflict with a definition drawn from the model of free fall. More importantly, impact supplied the insight formalised in Law III in a way that free fall could not have done. And Law III, in establishing the relations of inherent and impressed force, provided the capstone of his dynamics.

Another related interpretation of Newton's mechanics seeks to find a precisely defined meaning in the term 'action', as he employed it in the third law. Like the interpretation of impressed force mentioned above, 'action' is taken to refer to the product of force and the time during which it acts, or better to the effect of a force acting over a period of time.[62] Again, the revisions of *De motu* argue against such an interpretation. To use 'action' in such a way would have been to import a new term into mechanics, and Newton's definitions were intended to establish the terms he considered new. He offered no definition of 'action'. And why should he have done so? 'Action' and 'passion', to 'act' and to 'suffer' (*agere* and *pati*) were terms of daily philosophic commerce and as such scarcely in need of definition. Significantly, the first statement of the third law did not employ the abstract nouns 'action' and 'reaction', but rather the verbs 'to act' and 'to suffer'.[63] 'Action' (with its various cognates) appeared repeatedly in Descartes' writings, and Newton employed it as a commonly understood word to express the activity of force. Neither in the revisions of *De motu* nor in any other place that I have found, did he attempt to define 'action' as a quantitative term in his dynamics.[64]

Newton undertook two successive revisions of the definitions in version three of *De motu*. In the first of the two, the definitions approached their final form; the second fixed their wording substantially as they later appeared in the *Principia* without altering their intent beyond what I have indicated above. Whereas the second revision limited itself to definitions, the first went on to state six laws of motion.

> Law 1. By its innate force every body perseveres in its state of resting or of moving uniformly in a right line unless it is compelled to change that state by impressed forces. Moreover this uniform motion is of two sorts, progressive motion in a right line which the body describes with its center which is borne uniformly, and circular motion about any one of its axes which either rests or remains ever parallel to its prior position as it is carried with a uniform motion.
> Law 2. The change of motion is proportional to the impressed force and is made in the right line in which that force is impressed. By means of these two laws, which were already well known, Galileo discovered that in a medium offering no resistance projectiles describe parabolic trajectories under gravity acting uniformly along parallel lines. And experience confirms this conclusion except in so far as the motion of projectiles is retarded a little by the resistance of the air.
> Law 3. As much as any body acts on another, so much it suffers in reaction. Whatever presses or pulls another body is pressed or pulled equally by it. If a bladder filled with air presses or strikes another similar to it, both yield equally inward. If a body striking another changes the motion of the other by means of its force, its own motion as well is changed as much by the force of the other (because of the equality of their mutual pressure). If a magnet attracts a piece of iron, it is itself attracted as much in return, and so also in other cases. Indeed this law follows from Definitions 12 [inherent force] and 14 [impressed force] in so far as the force of a body exerted to conserve its state is identical to the force impressed on the other body to change its state, and the change of state of the first body is proportional to the first force, and that of the second to the second force.[65]

Laws 4 and 5, substantially identical to Laws 3 and 4 in version three of *De motu*, stated that the relative motions of bodies in a given space are unchanged by the inertial motion of the space and that the mutual actions of bodies do not affect the inertial state of their common centre of gravity. In the *Principia*, Newton demoted the two 'laws' to the status of Corollaries V and IV to the laws of motion. A stroke of the pen removed Law 6, on resistance of media, from the status of a law; and before the *Principia* was completed, an expanded investigation of motion through media fundamentally altered Newton's view of how they resist. Two Lemmas stated the same propositions as Lemmas 1 and 2 in version three and Hypotheses 3 and 4 in versions one and two – the parallelogram of force (still improperly stated as it was to be until edition two of the *Principia*), and the proportionality of distance to the square of time at the beginning of motion under the action of any force. They appeared in the *Principia* as Corollary I to the laws of motion and Lemma X of Book I.

The first three laws can be clearly recognised as the three laws of motion we still receive. The wording of Law 3 received a substantial revision in the so-called *Lectiones de motu*, but its content was already set. Law 2 had already received, not only its central idea, but also its final wording with the exception of the adjective '*motrici*' added between the words '*vi impressae*.' Law 1, however, presents a different picture. In at least two ways it appears strange. On the one hand, it refers to uniform rotation as an inertial motion. In this, Newton repeated his early insight into the conservation of angular momentum in a different context, one to which it was inappropriate. By the *Lectiones de motu*, he had seen its inappropriateness. It disappeared from the first law, and the principle itself of the conservation of angular momentum, despite its importance, played no role in his dynamics. The other anomaly is the continued presence of the phrase '*vi insita*.' Aside from its apparent attribution of inertial motion to inherent force, however, the first sentence of Law 1 is identical to its ultimate form in the *Principia*. Is the attribution in fact only apparent? The crucial change from version three was the insertion of the word 'resting [*quiescendi*].' With that word present, inherent force can no longer be considered as the internal cause of uniform motion. It has been altered to its final form, resistance to changes of a body's state of motion, and in fact the statement of Law 1 differs in no essential way from its final form. Certainly the later elimination of the reference to inherent force from a context where it could only confuse was a victory for clarity.

In every essential way, the revisions of version three of *De motu*, which apparently took place early in 1685, fixed the elements of Newton's dynamics in their permanent form. When the scope of version three of *De motu* is compared with the *Lectiones de motu*, composed initially later in 1685, the importance of the revisions becomes manifest. With the inherently unsatisfactory dynamics of *De motu*, Newton had solved the central problems of orbital dynamics and had worked out his basic approach to motion through resisting media. In neither case could his principles have withstood systematic criticism, and it is difficult to believe that they could have supported the immensely expanded investigation into mechanics celestial and mundane that Newton launched in 1685 and recorded in the *Lectiones*. With the new dynamics formulated in the revisions, however, far more was possible. Indeed, the *Principia* and the law of universal gravitation were possible – a satisfactory return perhaps on the effort invested.

*

During the first half of 1685, Newton overcame the main problems that stood in his path and formulated the central features of a quantitatively consistent dynamics based on the concept of force. The ontological status of force is a question distinct from its status in a quantitative dynamics. What relation, if any, did Newton's speculations on natural philosophy have to his dynamics? In the *Principia*, he asserted more than once that his use of the word attraction and his subjection of it to exact quantitative treatment were not to be understood as attempts to define the causal processes at work. Whatever the nature of physical reality, force was merely a precise term in a mathematical dynamics. On a logical plane, Newton's contention can be sustained. As a commentary on the actual development of his thought, however, it is wholly misleading. As late as February 1679, when he wrote the famous letter to Boyle, Newton still approached natural philosophy in categories that were primarily mechanical. As numerous examples including his own *Waste Book* demonstrate, the mechanical philosophy was not ideally adapted to encourage the elaboration of a sophisticated quantitative dynamics. Late in the same year 1679, Newton received the letter from Hooke that proposed a celestial dynamics based on the concept of attraction, and the resultant demonstration of the connection of elliptical orbits with the inverse square relation suggests how liberating the concept of attraction could be. Whatever the definition of force employed, the idea of attraction fostered an exact quantitative treatment of the kind that causal mechanisms eternally obstructed. But Hooke's proposal was able to take root in Newton's mind and flourish because it fell on fertile soil. The year 1679 was the crucial year in Newton's intellectual life, the time when problems inherent in his philosophy of nature were leading him to restructure it in terms of attractions and repulsions between particles of matter. In December 1679, Newton was ready to receive the suggestion that the sun attracts the planets as he would not have been prepared before.

Hooke did more than merely restate an idea which had already appeared in Newton's mind, however. Newton's speculations had been concerned primarily with microscopic phenomena, above all with chemical reactions. The attractions between particles that he saw as the ultimate reality of chemical reactions were subject in principle to quantitative description, but in fact the data available to the seventeenth-century scientist confined their treatment to the verbal plane. What Hooke did was to turn Newton's attention from the microscopic to the macroscopic realm; he suggested the application of the concept of attraction to celestial dynamics, the area of inquiry uniquely open in the seventeenth century to extended quantitative treatment. It was here that the idea of attraction could furnish the central conceptual member in the construction of a quantitative dynamics, and it was here that a quantitative dynamics could demonstrate its power by its success.

To suggest an attraction of the sun on the planets as a principle of celestial dynamics was not, however, equivalent to suggesting the concept of universal gravitation. As I have argued, Hooke never arrived at such a concept. Some element of specificity expressing the unique congruity of the sun with the planets circling it remained embedded in his celestial dynamics. Newton too had arrived at the concept of attraction via the path of specificity. The 'secret principle of sociability,' which owed so much to Hooke, had translated itself into attractions between particles that could not, by their nature, be general. The man who believed that gold dissolves in *aqua regia* because the acid attracts it, but does not dissolve in *aqua fortis* for lack of an attraction was not likely immediately to see the idea of universal gravitation in the concept of celestial attraction. Perhaps the context of the correspondence with Hooke, connecting as it did planetary dynamics with the fall of heavy bodies on the earth, involved a step in the direction of universal gravitation, but neither in his letters to Hooke nor in the paper that apparently contains his subsequent demonstration that elliptical orbits entail an inverse square attraction did Newton mention universal gravitation. What Newton completed in 1679 was nothing more than a sketch of planetary dynamics based on the concept of an undesignated centripetal attraction to the sun.

There was, moreover, at least one major obstacle that made a concept of universal gravitation difficult if not impossible in 1679. Something over a decade earlier, as we have seen, Newton had substituted his quantitative expression of centrifugal endeavour into Kepler's third law and derived the inverse square relation for the solar system. In the same paper, he had compared the centrifugal endeavour of the moon with the measured acceleration of gravity and found, not an exact correlation, but only an approximation. According to Pemberton, Newton therefore concluded that 'some other cause must at least join with the action of the power of gravity on the moon.'[66] Both DeMoivre and Whiston identified the 'other cause' as an effect of the vortex,[67] and so it may well have been in 1666. Knowing what we do about the evolution of Newton's thought, we can scarcely believe that he still referred the cause to the vortex in 1679, and there are also a number of reasons to believe that he did not correct the correlation before 1685. In the concept of specific attractions, however, the concept that developed from his chemical speculations and was never eliminated from them, he had an alternative explanation ready at hand. The idea of gravities specific to the planets had been advanced by Roberval among others. Hooke had applied it to the solar system. It is easier to imagine Newton applying to celestial dynamics an idea already part of his speculations than to imagine him seizing on a notion of uniform universal attraction which was not present elsewhere in his thought. The gravitational attraction of planets to the sun, in this construction, would have been related but not identical to the gravitational attraction of terrestrial bodies, including the moon, to the earth.

The correspondence that Newton held with Flamsteed between December 1680 and April 1681 about the comet of 1680–1 suggests strongly that he did in fact hold such a view. In the history of cometary theory, the correspondence is notable for Flamsteed's suggestion that the two comets observed in the winter of 1680–1 were in fact one comet which nearly reversed its path in the vicinity of the sun. It was a radical suggestion, one which flew in the face of the unanimous opinion of received tradition, and it was Flamsteed's misfortune to express it in terms of a shallow mechanics of magnetic attraction and repulsion all too easy to refute.[68] Nevertheless, Newton received it a year after he had solved the mechanics of planetary motion, and to anyone who held a concept of universal gravitation Flamsteed's proposed cometary orbits would seem to have offered a golden opportunity for confirmation by a simple extension of the mechanics worked out for planets. Hence the significance of Newton's response – steadfastly he refused seriously to entertain the suggestion.[69] Although he criticised Flamsteed's mechanics, which was certainly open to criticism, the ultimate gravamen of his argument fell elsewhere. 'But whatever there be in these difficulties,' he stated, 'this sways most with me that to make ye Comets of November & December but one is to make that one paradoxical. Did it go in such a bent line other comets would do ye like & yet no such thing was ever observed in them but rather the contrary.'[70] Despite his reference to observations, we must not mistakenly understand the objection as empirical. Only four years later, the same observations no longer presented any obstacle. The crucial point is that in 1681 Newton could not conceive that cometary orbits are governed by the same dynamic factors as planetary orbits. The entire tradition of scientific thought had been united in treating comets as foreign bodies which do not conform to the planetary system. Hooke's theory of comets, for example, had been explicit on that score. Newton's letters to Flamsteed in 1681 are only comprehensible if we apply a similar conception to him.

Less than four years later, when he composed the first version of *De motu*, he included comets under the orbital mechanics the tract proposed; and in a letter to Flamsteed, he stated his intention to determine the paths of the comets of 1664 and 1680 according to the principles of motion observed by the planets.[71] A manuscript containing sixteen points on cometary theory, which appears to stem from the period between the correspondence with Flamsteed and the composition of *De motu*, records the significance of Newton's change of opinion.

> 5 There is gravitation toward the center both of the sun and of each planet: and that toward the center of the sun is by far the greatest.
> 6 That gravitation in things more remote from the surface of the sun or the planet decreases in duplicate ratio of the distance from the center of the sun or planet.
> 7 The motion of a comet is accelerated until it reaches perihelion and afterwards is retarded.
> 8 A comet is not carried in a right line but in some curve the greatest curvature of which is at its minimum distance from the sun, the concave side of which is toward the sun, and the plane of which passes through the sun, and the sun is approximately at its focus.[72]

The use of the word 'gravitation' may again mislead. Before the *Principia*, '*gravitas*' and '*gravitatio*' referred to the tendency of systems to remain united without implying that the gravitation of one system (perhaps the earth) was identical to that of another (such as Mars).[73]

Nevertheless, the realisation that cometary orbits are determined by the same attraction to the sun that determines planetary orbits, the realisation, that is, that the matter of comets is identical to that of planets, constituted a major step in the direction of universal gravitation.

Although his thought was tending strongly in that direction, Newton had not yet arrived at the concept of universal gravitation when he began to compose *De motu* late in 1684. In its original version, *De motu* was a tract on celestial dynamics based on the concept of centripetal attraction, but referring to terrestrial phenomena only to the extent that its last proposition sought to define the motion of bodies falling in a right line under the action of a force that varies inversely as the square of the distance. It is true that Newton initially employed the word 'gravity' without qualifying adjectives, instead of the neutral term, 'centripetal force', to describe the attraction to the sun.[74] The original wording of Problem 5, on bodies falling vertically, also implied a correlation of terrestrial gravity with the centripetal force operating in the solar system.

> Given that gravity is inversely proportional to the square of the distance from the center of the earth, find the distances described in given times by heavy bodies [*gravia*] in falling.[75]

Nevertheless, Newton did not suggest a correlation of the centripetal force holding the moon in its orbit with the measured acceleration of gravity on the surface of the earth, and the law of universal gravitation was to rest ultimately squarely on that correlation. Only in the third revision did he venture to insert the correlation, and even then his use of the adverb 'very nearly [*quamproxime*],' together with the past tense of the relevant verb, suggests that he was still citing the earlier, not entirely satisfactory, correlation.[76] In a revision of version one, Newton systematically replaced the word 'gravity' by the phrase 'centripetal force', and confined himself to remarking in a Scholium to Problem 5 that gravity is one species of centripetal force.[77] Apparently the revision of wording accompanied the addition of further propositions in which he first undertook the quantitative treatment of motion through a resisting medium. The more general phrase, 'centripetal force', was consistent with the generalisation of his dynamics which began at that point, but the elimination of the word 'gravity' is ironic, since the generalisation of his dynamics was the necessary condition to a rigorous derivation of the law of universal gravitation.

What stood between Newton and the concept of universal gravitation at the end of 1684? I have already suggested the still unresolved failure of the correlation of the moon's centripetal acceleration with that on the surface of the earth. The first satisfactory correlation appeared only in the original version of the final book, which he composed later in 1685.[78] There is evidence to suggest further that for a time Newton had no incentive to check the correlation because of theoretical problems which seemed to exist. How far was the famous apple from the earth? At first blush, its distance appears to have been about ten feet, whereas the inverse square relation demands four thousand miles. In this context, Newton's vehement outburst to Halley in 1686, in reply to Hooke's claim of the inverse square law, is revealing. 'I never extended ye duplicate proportion lower then to ye superficies of ye earth,' he asserted, '& before a certain demonstration I found ye last year have suspected it did not reach accurately enough down so low . . .' Further on in the same letter, he expressed himself with still more heat.

> There is so strong an objection against ye accuratness of this proposition, yt without my Demonstrations, to wch Mr Hook is yet a stranger, it cannot be beleived by a judicious Philosopher to be any where accurate.[79]

That is to say, before he achieved the demonstration that an homogeneous sphere, composed of particles that attract with inverse square forces, itself attracts any body external to it with a force inversely proportional to the square of the distance from its centre, Newton had no reason carefully to recompute the correlation of the centripetal acceleration of the moon with the measured acceleration of gravity on the surface of the earth. And until that correlation was demonstrated, he cannot be said to have discovered the law of universal gravitation.

Universal gravitation, as a demonstrated conclusion rather than a mere idea, rested primarily on two foundation stones – the demonstration that elliptical orbits entail an inverse square force, and the demonstration of the attraction of homogeneous spheres, which allows the correlation of celestial dynamics with terrestrial gravity. By his own testimony, Newton did not arrive at the second demonstration before 1685. That demonstration in turn, with its suggestion that the attractions of finite bodies are the sums of the attractions of their infinitesimal parts, supported the concept of universal gravitation in more ways than one.

Is it possible that Newton's hesitation in extending the inverse square relation to the surface of the earth also sprang from other sources? I have already argued that the concept of universal gravitation, an attraction pertaining uniformly to all matter as matter, stood in contradiction to the concept of specific attractions born of his chemical studies. Nothing in the letter about the inverse square law sent to Halley in 1686 rules out the possibility that a residue of specificity also inhibited his general application of the inverse square law. At least two pieces of evidence directly support it. Late in December of 1684, Newton wrote to Flamsteed requesting information on the motion of Saturn when it is in conjunction with Jupiter.[80] As Flamsteed perceived, the request implied that Newton thought the two planets might attract each other and influence each other's motion. The idea itself is another significant step toward universal gravitation. The letter Newton wrote in response to the data Flamsteed sent presents a more complicated picture, however.

> Your information about ye error of Keplers tables for ♃ [Jupiter] & ♄ [Saturn] has eased me of several scruples. I was apt to suspect there might be some cause or other unknown to me, wch might disturb ye sesquialtera proportion. For ye influences of ye Planets one upon another seemed not great enough tho I imagined ♃'s influence greater than Your numbers determin it. It would ad to my satisfaction if you would be pleased to let me know the long diameters of ye orbits of ♃ & ♄ assigned by your self & Mr Halley in your new tables, that I may see how the sesquiplicate proportion fills ye heavens together wth another small proportion wch must be allowed for.[81]

The final sentence of the passage is perplexing, but the simplest interpretation is to accept it as a reference, rather ineptly phrased, to the disturbance that the mutual attractions of planets introduce into the ideal orbits determined by the attraction of the sun alone. Even if this interpretation is accepted, however, the letter testifies that before he received Flamsteed's data Newton did not believe that universal gravitation by itself would account for all the phenomena.

The second piece of evidence derives from the revisions of the third version of *De motu*. Twenty years earlier, in the *Waste Book*, Newton had asserted that quantity of matter is proportional to weight. Now, apparently, he was not sure. In defining quantity of motion, Newton found it necessary to define the quantity of a body as well. It is measured, he said, 'by the bulk of corporeal matter which is usually proportional to its gravity. Using equal pendulums, let the oscillations of two bodies of the same weight be counted, and the bulk of matter in each will be reciprocally as the number of oscillations made in the same time.' Having defined the operation, he proceeded to try it, and a blank space opposite the original statement records an altered conclusion.

> By weight I understand the quantity or bulk of matter to be moved, abstracting consideration of gravitation whenever there is no question of gravitating bodies. Of course the weight of gravitating bodies is proportional to their quantity of matter, and it is proper for analogous quantities to stand for and designate each other. The analogy can in fact be determined in the following way. Using equal pendulums, let the oscillations of two bodies of the same weight be counted, and the bulk of matter in each will be reciprocally as the number of oscillations made in the same time. When experiments were carefully made with gold, silver, lead, glass, sand, common salt, water, wood, and wheat, however, they resulted always in the same number of oscillations.[82]

In composing and revising *De motu*, Newton had set *vis inertiae* proportional to the quantity of matter. The concept of universal gravitation demands that weight also be proportional to the quantity of matter, and until Newton recorded that conclusion in the unmistakable terms quoted above, he cannot be said fully to have embraced the concept.

Two decades before the *Principia*, Newton had perceived the analogy of terrestrial gravity with the dynamic factors that determine the motions of the solar system. The same analogy had been implicit in the correspondence with Hooke in 1679 and explicit in *De motu*. The analogy is by no means equivalent to the law of universal gravitation, however, and every indication suggests powerfully that the concept of universal gravitation, far from dominating the composition of *De motu*, gradually dawned on Newton's consciousness during the revision of the tract. The subjection of cometary orbits to the principles of planetary dynamics, the demonstration of the attraction of spheres, the influence of Jupiter on Saturn, the experiment proving that mass is proportional to weight, the exact correlation of the centripetal acceleration of the moon with the acceleration of gravity – except for the first, the crucial pieces of evidence fell into place some time in 1685. The immense expansion of scope of the so-called *Lectiones de motu* in comparison to version three of *De motu* may record Newton's reaction to the discovery as its realisation swept over him.

What is of particular importance, the realisation came upon him at the very time when he was perfecting his dynamics. Hence he found himself at one and the same time in possession both of the concept of universal gravitation and of a quantitative dynamics sophisticated enough to demonstrate the validity of the concept. The attraction of spheres, the proportionality of mass to weight, the exact correlation of the moon's motion with the acceleration of gravity – as rigorous conclusions, all of these drew on the dynamics he was engaged in perfecting. Once the concept had reached the full light of consciousness, moreover, the same dynamics was able to furnish innumerable pieces of evidence to confirm it. Experiments with pendulums prove that the mass of terrestrial bodies is proportional to their weight. Nature has been kind enough to perform analogous experiments in the heavens. If the planets move around the sun according to Kepler's third law, it follows from the quantitative expression of centripetal force and the inverse square relation that all the planets would move in the same orbit if they were located at the same distance from the sun. This in turn would only be possible if the attractions on unequal planets were proportional to their quantities of matter.[83] The satellites of Jupiter offer a somewhat more complicated case. In so far as they circle Jupiter according to Kepler's third law, an argument identical to that for the planets shows that they gravitate toward Jupiter with forces proportional to their masses. At the same time, both they and Jupiter are attracted by the sun. What if the satellites and Jupiter were not homogeneous in respect to the sun, so that the attraction of the sun for Jupiter were not identical in proportion to mass to its attraction for Jupiter's satellites? Deploying his analysis of perturbations due to a third body, Newton was able to show that if the attraction of the sun for the satellites in proportion to mass varied by only one thousandth of the whole from its attraction for Jupiter, the centre of the satellites' orbits would be displaced by a distance equal to one-fifth the radius of the outer satellites' orbit, an eccentricity readily observable from the earth.[84]

Hence the *Principia* could rise to its climax in the assertion of which *De motu* had not even dreamed.

> That there is a power of gravity pertaining to all bodies, proportional to the several quantities of matter which they contain.[85]

In the original version of the final book, he had expressed himself more forcibly when he asserted that the attractive forces of bodies 'arise from the universal nature of matter.'[86] In consequence, he insisted that we must understand the attractions to be directed toward material bodies and not toward imaginary points. In an early version of his demonstration that parts of the earth attract each other, he imagined one section of the earth to be cut off and moved some distance away. What would happen? Obviously the two parts would attract each other. We cannot imagine the gravitation of either section to be directed to a place distinct from the other. Space as such is uniform; it contains no point to which gravitation might be directed in preference to another. If the earth were moved entire to another place, there can be no doubt that its parts would continue to gravitate toward its centre and not toward the centre of its former location.

> For the affections and operations of bodies depend on the bodies, and hence they will not remain in the spaces from which the bodies move but will accompany the bodies. Magnetic force seeks the magnet, electric force the electric body, centripetal force the planet.[87]

It would be impossible, I believe, to reconcile such statements with Newton's repeated assertions that his word 'attraction' implied no assumption about the ultimate nature of the force. Newton's critics saw as much and did not neglect to insist on it. In principle, of course, it was always possible to imagine an *ad hoc* mechanism to account for any phenomenon. When the phenomenon was the universal attraction of every particle of matter in the universe for every other, however, an aethereal mechanism of whatever sort would become so unwieldly as to be utterly unbelievable. In Newton's mind, at least, the law of universal gravitation was the ultimate confirmation of the concept of attraction at which he had arrived by another route.

In its development to full consciousness, universal gravitation was equally the product of his concept of attraction. If it was possible, once he had arrived at the conclusion, for him to say of the cause of gravity that he framed no hypotheses, and if it was logically defensible to separate universal gravitation as a quantitatively demonstrated phenomenon from its cause, it is inconceivable that he could have formulated the idea without the prior concept of attraction. Forces as real entities and a quantitative dynamics, the Siamese twins of Newtonian science – from the complex play of their mutual interaction emerged the *Principia* and the law of universal gravitation.

In one respect of overwhelming importance, the law of universal gravitation differed from the other forces that appeared in Newton's speculations. The concept of attractions and repulsions between particles, derived from the hermetic tradition of Renaissance Naturalism, threatened always, like that tradition, to dissolve the unity of nature into a multiplicity of specific agents. In contrast, universal gravitation embodied the assertion of generality over specificity. In the never ending struggle of the hermetic with the mechanical tradition, which was one of the basic factors in Newton's scientific thought, the formulation of the law of universal gravitation represented that point at which the mechanical tradition, apparently in full retreat before the swelling tide of hermeticism, reasserted its role. Although the idea of attraction as such remained an abomination to mechanists of orthodox persuasion, Newton attached it unmistakably to the mechanist conception of matter when he made it universal. To the uniform matter of the mechanical tradition, his dynamics had already attributed a uniform *vis insita* which resists all changes of motion or rest. To the same uniform matter, he now attributed as well a uniform attraction that pertains to matter as matter. Perhaps it was not entirely by accident that in the first edition of the *Principia* he suppressed the projected discussion of specific forces and inserted the hypothesis proclaiming the universal convertibility of all bodies. When he omitted the hypothesis in later editions and published his speculations on forces in the *Opticks*, the pendulum of his thought swung back in the direction of hermeticism. Once reached, however, the conclusion of universal gravitation was never relinquished. It remained the unbroken bond attaching the concept of force to the mechanical philosophy.

Both to his dynamics and to the mechanical philosophy of nature, universal gravitation had much to offer. To his dynamics, it offered a unique opportunity. Although the forces of Newton's speculations were all quantifiable in principle, in fact the level of scientific knowledge at the end of the seventeenth century confined their discussion to a verbal plane. Gravitation, and gravitation almost alone, offered a field in which the body of data available allowed a quantified dynamics both to elaborate its concepts and to display its power. To the mechanical philosophy, universal gravitation showed how the concept of force applied to particles in motion could lead it out of the choking morass of imaginary mechanisms and on to the solid ground of quantitative demonstration. The concept of force as the central element in a quantitative dynamics was the heart of Newton's contribution to science.

NEWTONIAN DYNAMICS

The *Principia* may be seen as the extension and deployment of the dynamics elaborated during the early months of 1685.

★

True to his original insight recorded in the *Waste Book*, Newton always regarded force – that is, force impressed on a body by an agent external to it – as an action to change its state of motion or rest. His study of mechanics had begun with the principle of inertia, and to it he returned after a temporary aberration. It is both historically accurate and symbolically representative of the structure of his dynamics that the first law of motion should state the principle of inertia.

> Every body continues in its state of rest, or of uniform motion in a right line, unless it is compelled to change that state by forces impressed upon it.[88]

Matter is inert. Bodies are 'perfectly indifferent to the receiving of all impressions.'[89] They are incapable of changing their state of motion or rest by any action of their own. Consider a perfect sphere set rotating in free space by an impulse on its surface.

> Since this globe is perfectly indifferent to all the axes that pass through its centre, nor has a greater propensity to one axis or to one situation of the axis than to any other, it is manifest that by its own force it will never change its axis, or the inclination of its axis. Let now this globe be impelled obliquely by a new impulse in the same part of its surface as before; and since the effect of an impulse is not at all changed by its coming sooner or later, it is manifest that these two impulses, successively impressed, will produce the same motion, as if they had been impressed at the same time; that is, the same motion, as if the globe had been impelled by a simple force compounded of them both (by Cor. II of the Laws), that is, a simple motion about an axis of a given inclination.[90]

Matter is not wholly inert, however. If bodies are indifferent to any state of motion, they are not indifferent to leaving whatever state they are in. The case above of the rotating globe holds only for perfect globes.

> But let there be added anywhere between the pole and the equator a heap of new matter like a mountain, and this, by its continual endeavour to recede from the centre of its motion, will disturb the motion of the globe, and cause its poles to wander about its surface describing circles about themselves and the points opposite to them.[91]

Although the example above is complicated by the issue of circular motion, it embodies the inherent ambiguity of the Newtonian conception of matter. Matter is at once inert and active. Inert, it is dominated in its motion by external forces impressed upon it. Active, it endeavours to persevere in its present state. Because matter is inert, every external force impressed upon a body will generate a new motion in the body if the force is not balanced by an equal and opposite one. Because matter is active, bodies resist external forces and in so doing insure a constant proportionality between force and the motion it generates.

The possibility of a quantitative dynamics rests on the immutable uniformity of that proportionality. Hence the second law of motion:

> The change of motion is proportional to the motive force impressed; and is made in the direction of the right line in which that force is impressed.[92]

The brief discussion of the law emphasises the extent to which it assumes the principle of inertia. Since the motion a force generates is directed in the same line as the force, 'if the body moved before, [the new motion] is added to or subtracted from the former motion, according as they directly conspire with or are directly contrary to each other; or obliquely joined, when they are oblique, so as to produce a new motion compounded from the determination of both.'[93] Because of the principle of inertia, every force can be regarded as finding the body on which it acts in a state of rest, and the uniformity of matter and force transposes itself into the uniform linearity of changes of motion. The principle of inertia almost demanded to be complemented with a concept of force as the cause and measure of changes of motion. With its formulation, a quantitative dynamics prepared to take its place beside the quantitative kinematics already existing.

One of the immediate by-products of the concept of force was a clear perception of the mutual relations of statics and dynamics, which solved, once and for all, a perplexed conundrum on which projected dynamics had repeatedly come to grief during the seventeenth century. With a satisfactorily dynamic conception of force, Newton found himself liberated from the tyranny that the simple machines had exercised over dynamic thought, and statics represented itself to him as merely a special case in which the forces impressed on a body are in equilibrium.

> The power and use of machines consist only in this, that by diminishing the velocity we may augment the force, and the contrary; from whence, in all sorts of proper machines, we have the solution of this problem: *To move a given weight with a given power*, or with a given force to overcome any other given resistance. For if machines are so contrived that the velocities of the agent and resistant ['estimated according to the determination of the forces'] are inversely as their forces, the agent will just sustain the resistant, but with a greater disparity of velocity will overcome it. So that if the disparity of velocities is so great as to overcome all that resistance which commonly arises either from the friction of contiguous bodies as they slide by one another, or from the cohesion of continuous bodies that are to be separated, or from the weights of bodies to be raised, the excess of the force remaining, after all those resistances are overcome, will produce an acceleration of motion proportional thereto, as well in the parts of the machine as in the resisting body.[94]

The clear differentiation of statics and dynamics also appeared in his analysis of pendulums. Newton resolved the force acting on the bob into two components, a radial one (CX) parallel to the thread PT, and a tangential one (TX), 'of which CX impelling the body directly from P stretches the thread PT, and by the resistance the thread makes to it is totally employed, producing no other effect; but the other part TX, impelling the body transversely or towards X, directly accelerates the motion in the cycloid.'[95] As he remarked in regard to a similar problem: 'That whole force, therefore, will be spent in producing this effect.'[96] Galileo's realisation that one heavy body cannot press on another when both are falling freely was thereby generalised as one of the first conclusions of dynamics.

The relation of statics to dynamics was susceptible to a different formulation in the context of Newton's mechanics. The passage on simple machines appeared in an extended discussion justifying the third law, and Newton went on to show how machines obey the third law when they are not in equilibrium.

> For if we estimate the action of the agent from the product of its force and velocity, and likewise the reaction of the impediment from the product of the velocities of its several parts, and the forces of resistance arising from the friction, cohesion, weight, and acceleration of those parts, the action and reaction in the use of all sorts of machines will be found always equal to one another.[97]

From the point of view of the second law, an unbalanced force expends itself in generating a new motion. From the point of view of the third law, an unbalanced force is an impossibility. Dynamics is the science of equilibrium quite as much as statics. In the one case, the equilibrium of force has as its result the maintenance of an inertial state. In the other case, changes in the inertial state re-establish equilibrium, between the *vis inertiae* and the impressed force. From Newton's point of view, indeed, the perspective of the third law was the more fundamental. His dynamics concerned itself primarily with the relation of two forces, internal and external force, *vis insita* and *vis impressa*, and the third law stated their relation in terms of the ancient tradition of static equilibrium.

If the second law, the law of impressed force, can be seen as a generalisation of Galileo's insight into the dynamics of free fall, Newton's perception of the relation of statics and dynamics may also be seen to have generalised Galileo's analysis of the role of the medium. In any dynamic problem, the first step is to establish the effective force impressed, by a process of addition and subtraction. The questions considered in the *Principia* are such that the best examples of this procedure are found in problems concerned with motion through resisting media. Unlike Galileo, however, Newton did not confine his conception of resistance to their buoyant effect, and his treatment of their resistance is perfectly generalisable to resistance of any sort. When a body is thrown upward through a resisting medium, the effective force operating to decelerate it at any instant is its weight plus the resistance of the medium at that instant; when it falls through the same medium, the effective force operating to accelerate it at any instant is its weight minus the resistance of the medium at that instant.[98]

At this point, Newton achieved a decisive step beyond Galileo. In reforming the understanding of free fall, Galileo had insisted that the resistance of the medium be subtracted from the weight rather than divided into it. In generalising his conclusion, Newton saw that the process of subtracting (or adding) is confined to the calculation of the net force, but that the determination of its effect requires exactly the division of force by resistance that Galileo's conclusion had rejected. The resistance was displaced of course from the medium to the moved body itself – a step reminiscent of the displacement of motive force from the medium to the projectile that had furnished the heart of impetus mechanics. Called *vis inertiae* now, or mass, it became the factor that established the proportion between a force and the effect it produces. Hence the basic equation of Newtonian dynamics,

$$a = F/m.$$

In a form yet more reminiscent of an earlier dynamics, Newton was also prepared to write the equation as

$$\Delta v = F/m.$$

In either form, the term 'F' repeats Galileo's analysis of the medium's function. F is established by adding and subtracting to find the effective force.

A basic requirement of a quantitative dynamics was an exact definition of the effect that a force produces. As I have just indicated, the *Principia* proposed not one, but two definitions, two definitions which are mutually incompatible with each other. Indeed there are passages in which the two appear in the same sentence.

> When a body is falling, the uniform force of its gravity acting equally, impresses, in equal intervals of time, equal forces upon that body, and therefore generates equal velocities; and in the whole time impresses a whole force, and generates a whole velocity proportional to the time.[99]

Is it in fact possible to speak of Newton as the founder of quantitative dynamics if the fundamental term in his dynamics remained ambiguous?

The second law of motion defined the effect of impressed force in unmistakable terms. 'The change of motion is proportional to the motive force impressed . . .' There is only one way to represent this statement in an equation:

$$F = \Delta mv.$$

Moreover, there can be no doubt that Newton intended the words to mean what they say. In a paper from the early 1690's, he attempted no less than eight alternative statements of the second law. A diagram of the parallelogram of motion on the sheet shows a curved diagonal, indicating that he understood himself to be dealing with a continuous force such as gravity altering the inertial motion of a body. Nevertheless, all eight statements agreed in setting the change of motion proportional to the impressed force. In the statement that the paper apparently accepted, the second law went as follows:

> Every new motion by which the state of a body is changed is proportional to the motive force impressed and is made from the place which the body would otherwise occupy in the direction which the impressed force pursues.[100]

The measurement of force by Δmv appears repeatedly throughout the *Principia*. As I have mentioned before, Proposition 1, which demonstrates Kepler's law of areas for any centripetal force, continues to repeat the demonstration in *De motu* where the force is applied as a series of discrete impulses. In explaining how the precession of the equinoxes indicates the shape of the earth, in a passage cited above, Newton imagined a perfect sphere set in motion by two oblique 'impulses' [both *impetus* and *impulsus*]. The globe will move, he asserted, with a single rotation as though it 'had been impelled by a simple force [*vi simplici*] compounded of them both.'[101] His treatment of the resistance of media derived its major component from the impacts of material particles, each one of which impacts has to be measured by Δmv. When he asserted that Wallis, Huygens, and Wren had employed his three laws of motion in their treatments of impact, he implied the same measure of force again, whatever the historical accuracy of the contention.[102] There can be no doubt that the *Principia* employs the formula Δmv as a measure of force.

Equally, there can be no doubt that the *Principia* employs the formula ma as the measure of force. Numerous passages explicitly set acceleration proportional to force. In demonstrating the validity of the third law in the case of two bodies attracting each other, he imagined B to attract A more strongly than A attracts B. If that were the case, there would be an unbalanced force in the system of two bodies, and the system would move in the line AB '*in infinitum* with a motion perpetually accelerated'[103] – a situation contradicting the first law. To demonstrate that the weight of bodies toward any gravitating centre is proportional to their mass, he cited the evidence of pendulums and satellites with the reasoning that 'forces which equally accelerate unequal bodies must be as those bodies . . .'[104] When he examined the motion of a body through a medium that resists in proportion to the velocity squared, he utilised a rectangular hyperbola, with the y axis representing velocity, and the x axis time (see Fig. 43). The area under the hyperbola between two ordinates AB and DG is proportional to the distance traversed in the time AD. Newton went on to remark that the resistance of the medium is given by making it equal at the beginning

NEWTONIAN DYNAMICS

of the motion to the uniform force which would generate the velocity *AB* in the time *CA* in the absence of all resistance.

> For if *BT* be drawn touching the hyperbola in *B*, and meeting the asymptote in *T*, the right line *AT* will be equal to *AC*, and will express the time in which the first resistance, uniformly continued, may take away the whole velocity *AB*.[105]

The triangle *BAT* in Newton's diagram is the exact dynamic equivalent to the triangle by which Galileo expressed the kinematics of uniformly

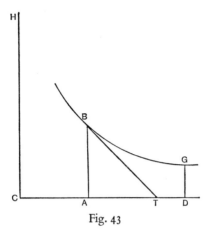

Fig. 43

accelerated motion. It is intelligible only if Newton intended to set a uniform force proportional to a uniform acceleration.

More important in the *Principia* than forces that alter the linear velocity of bodies are centripetal forces which alter the direction of their motion. In Proposition I, Newton employed the device of discrete impulses to represent such forces, and after setting up the demonstration, imagined a passage to the limit in which the series of impulses becomes a continuous force. In most of the other propositions, he started with a continuous force and established its effect by Lemma X of Book I:

> The spaces which a body describes by any finite force urging it, whether that force is determined and immutable, or is continually augmented or continually diminished, are in the very beginning of the motion to each other as the squares of the times.[106]

The substance of the Lemma had appeared in the third version of *De motu* as Lemma 2. I have discussed it already and shall confine myself to repeating that it assumes a uniform acceleration as the product of a uniform force and demonstrates that in the limit spaces are proportional to the squares of time at the very beginning of motion for non-uniform forces as well as for uniform ones. This measure of force is employed throughout the *Principia* wherever the curvature of an orbit allows centripetal force to be represented by the deviation from the tangent. Clearly it is not restricted to centripetal forces, and when the occasion demanded Newton generalised it without obvious difficulty. In discussing a spiral path in a resisting medium, for example, he laid out an increment of arc, *PQ*, described in a moment of time and another, *PR*, described in twice the time. The 'decrements of those arcs arising from the resistance, or their differences from the arcs which would be described in a nonresisting medium in the same times,' he remarked, 'will be to each other as the squares of the times in which they are generated . . .'[107] When the problem required it, he also represented centripetal force in a circular orbit by an expression equivalent to our mv^2/r, which is equivalent, of course, to employing ma as the measure of force.[108]

Finally, there can also be no doubt that Newton continued to consider Δmv and ma, as the measures of force, to be identical. The basic dynamics of *De motu* continued unchanged through all the editions of the *Principia*. The discussion itself of the second law asserted the identity. 'If any force generates a motion, a double force will generate double the motion, a triple force triple the motion, whether that force be impressed altogether and at once, or gradually and successively.'[109] Already in the analysis of circular motion carried out in his undergraduate days, Newton had employed the concept of total force, and such a concept remained implicit in his mature dynamics, so obvious apparently as not to require explicit discussion. Where the distinction of force and total force (F and $\int F dt$) implies dimensional incompatibility to us, the 'force' common to both implied identity to Newton. We must constantly remember that the concept of force itself, defined as Newton defined it, was original with him, his contribution to dynamics. Following the principle of inertia in Law I, Law II identified force as the cause of changes in the inertial state of a body. 'An impressed force,' Definition IV stated, 'is an action exerted upon a body, in order to change its state, either of rest, or of uniform motion in a right line.'[110] Total force falls under the definition quite as well as instantaneous force.

Newton's acceptance of the continuity of force and total force expressed itself repeatedly in the *Principia*. It is present in the very discussion of Law II; and in Proposition I, the interval between successive impulses is allowed to decrease *in infinitum* until 'the centripetal force, by which the body is continually drawn back from the tangent of this curve will act continually . . .'[111] The propositions following the first suggest that he might have avoided the use of impulses in that demonstration, but in no way could his investigation of the source of resistance in fluids have dispensed with it. That resistance arises from the impact of the moving body with discrete particles of the fluid. The 'forces' of the individual impacts 'are as the velocities and the magnitudes and the densities of the corresponding parts conjointly . . .' A body moving through a fluid and striking its particles impresses a quantity of motion on them and suffers in return a reaction which is the resistance.[112] To compute the resistance that a sphere encounters as it moves through a fluid, Newton first compared the resistance a sphere suffers to that of a cylinder of the same material and of equal diameter. He imagined that particles moving with equal velocities in parallel paths strike every point on the flat surface of the cylinder and every point on one side of the sphere. Because the particles do not strike the sphere perpendicularly, the force of each impact varies with the obliquity; and because each force is normal to the surface of the sphere, only a component of it is effective to move the sphere in the direction the particles are moving. Summing their effects, Newton demonstrated that 'the whole force of the medium upon the globe is half the entire force of the same upon the cylinder.' The 'whole force' in this case, of course, is an integral over an area and not over time. It represents, as it were, the impact of one particle on every point of a surface. Now imagine that the medium is at rest and the cylinder moves through it. If we assume perfect elasticity, it will communicate a velocity double its own to each particle it meets. It follows therefore 'that the cylinder in moving forwards uniformly half the length of its axis, will communicate a motion to the particles which is to the whole motion of the cylinder as the density of the medium to the density of the cylinder.' The resistance to the globe is half as great but its mass is only two-thirds as large. Therefore in the time required for the sphere to move two-thirds the length of its diameter it will communicate a motion to the particles of the medium that is to its whole motion as the density of the medium is to the density of the sphere. At this point,

Newton slid unobtrusively from the sum of discrete impacts ($\Sigma F = \Delta mv$) to the integral of force over time ($\int F dt = \Delta mv$). 'And therefore the globe meets with a resistance which is to the force by which its whole motion may be either taken away or generated in the time in which it describes two-thirds of its diameter moving uniformly forwards, as the density of the medium is to the density of the globe.'[113] In solving the problem, Newton assumed that the particles of the fluid are evenly spaced and that individual impacts therefore follow each other at regular intervals of time. As in Proposition 1, time is in the definition of the problem if it is not in the definition of force, and his solution of the problem involved no essential error. Nevertheless, he was caught in the same dimensional inconsistency that harassed the attempted establishment of a quantitative dynamics throughout the seventeenth century, and it remained for the eighteenth century to define the concept of force with adequate rigour.

To account for the deficiency at the heart of Newton's dynamics, we must recall the process by which his dynamics developed. The inconsistency that remained marked the extent to which his concept of force was unable to emancipate itself from its own history. He had formulated the concept initially in the analysis of impact, and he never wholly separated his dynamics from that model even when his philosophy of nature had proposed the model of attraction and repulsion, force acting continuously. In 1679, and in 1684 when he began *De motu* before he reformulated his dynamics, he adopted an approach to orbital mechanics that was dictated by the impact model of force, and that approach, once adopted, continued to dominate his procedure despite the subsequent refinement of his dynamics. It rested on the mathematical concept that a polygon approaches a continuous curve as the number of its sides increases without limit. In 1679, Newton imagined the attraction of the sun to act as discrete impulses at the beginning of each moment of time. In 1684, he employed the same procedure in the first theorem of *De motu* and institutionalised it, as it were, in the parallelogram of forces, the ultimate device by which he attempted to combine the disparate forces of that treatise in a unified dynamics. The parallelogram was deceptively simple. As a mechanical concept, it transposed the parallelograms of statics and kinematics into dynamic terms. As a mathematical concept, it adapted nicely to the representation of a polygon inside or outside a curve. Its apparent transparency concealed the fact that in attempting to represent both the path and the forces, it embodies the very ambiguity that was thereby crystallised in Newton's conception of force.

Already in its initial statement in *De motu*, Newton's parallelogram was a perplexing device.

> When it is acted upon by [two] forces simultaneously, a body is carried in a given time to that place to which it would be carried by the forces acting separately in succession during equal times.[114]

As we have seen, the forces Newton intended thus to combine were utterly disparate. As far as inherent force, still conceived as a uniform force producing a uniform velocity when he composed Hypothesis 3, and impressed force, conceived as an impulse, were concerned, the parallelogram involved no problems beyond conceptual untidiness. Each of the two sides could represent three proportional quantities – force, motion (or velocity, since a constant mass is involved), and distance traversed in unit time (since the motions generated by each of the two forces are uniform). The problem arises with the third force, impressed force as a continuous force producing a uniform acceleration. Even in this case the issue is deceptive. For unit time, each side of the parallelogram can still represent force, motion, and space traversed. The diagonal of the parallelogram as a straight line, however, cannot represent the uniform relation of all three quantities during the unit of time. What is the same problem in other terms, if we compare two equal and similar parallelograms in which corresponding legs represent equal impressed forces, in one parallelogram an impulse and in the other a continuous force (for, as the second law was later to state, the change of motion is the same whether the force be impressed altogether and at once, or gradually and successively) the same corresponding legs cannot also represent the distances traversed. If we wish force as impulse to be identical to continuous force, the same parallelogram cannot serve both to represent the forces and the paths. But orbital motion lent itself readily to geometric representation, and the parallelogram was such a convenient device for thinking about force that Newton even continued to refer to bodies moving by their inherent force after he had ceased to conceive of uniform motion in such terms.[115] In the concept of the ultimate ratio, which is similar to that of the differential, he found an apparent means to establish the continuity both of path and of force as the polygon approached the curve in the limit. That is, he continued to employ the parallelogram both for force and for

FORCE IN NEWTON'S PHYSICS

path, and in so doing maintained the inherent ambiguity in his concept of force.

Proposition I of Book I illustrates how smoothly one could slide from one use of the parallelogram to the other. In essentially the form of 1679, Proposition I employs a conception of force as discrete impulses to demonstrate Kepler's law of areas. Let time be divided into equal parts, and in the first part let the body describe the right line AB by its innate force alone (see Fig. 44). If nothing acted on it, the body

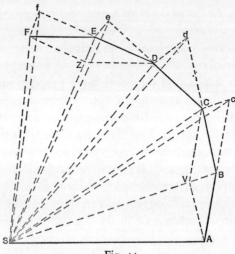

Fig. 44

would describe the line Bc ($=AB$) in the second part of time, and by simple geometry Newton showed that the triangle ABS is equal to the triangle BcS. But suppose that a centripetal force 'acts [on the body] at once with a great impulse' when it arrives at B, turning it aside into the line BC. Again, it was a simple matter to prove that $\triangle BCS = \triangle BcS = \triangle ABS$; and by extension of the same procedure, all successive triangles swept out in equal periods of time are also equal. Until this point, Newton had employed the parallelogram explicitly to represent the path and implicitly to represent the forces. 'Now let the number of those triangles be augmented, and their breadth diminished *in infinitum*; and (by Cor. IV, Lem. III) their ultimate perimeter ADF will be a curved line: and therefore the centripetal force, by which the body is continually drawn back from the tangent of this curve, will act continually; and any described areas *SADS, SAFS,* which are always proportional to the times of description, will, in this case also, be proportional to the times.' In the third and fourth corollaries, the implied dual representation of the parallelogram both in the initial problem and in the limit, is made explicit.

> Cor. III If the chords *AB*, *BC*, and *DE*, *EF*, of arcs described in equal times, in spaces void of resistance, are completed into the parallelograms *ABCV*, *DEFZ*, the forces in *B* and *E* are one to the other in the ultimate ratio of the diagonals *BV*, *EZ*, when those arcs are diminished *in infinitum*. For the motions *BC* and *EF* of the body (by Cor. I of the Laws [the parallelogram]) are compounded of the motions *Bc*, *BV*, and *Ef*, *EZ*; but *BV* and *EZ*, which are equal to *Cc* and *Ff*, in the demonstration of this Proposition, were generated by the impulses of the centripetal force in *B* and *E*, and are therefore proportional to those impulses.
> Cor. IV. The forces by which bodies, in spaces void of resistance, are drawn back from rectilinear motions, and turned into curvilinear orbits, are to each other as the versed sines of arcs described in equal times; which versed sines tend to the centre of force, and bisect the chords when those arcs are diminished to infinity. For such versed sines are the halves of the diagonals mentioned in Cor. III.[116]

The initial form of the parallelogram of forces, as it was formulated in *De motu*, appeared without substantial alteration in the first edition of the *Principia*. Its implication that a uniform force produces a uniform motion was obviously in conflict with the shape his dynamics had assumed, and in the 1690's Newton undertook to alter it that it might account for accelerated motions as well as uniform ones.

> Cas. 2. By the same argument, if in a given time a body be carried with a uniform motion from *A* to *B* by the force *M* alone, impressed apart in the place *A*, and be carried with an accelerated motion in the right line *AC* from *A* to *C* by the force *N* alone, not altogether and at once but continuously impressed, complete the parallelogram *ABDC* & the body will be carried by both forces from *A* to *D* in the same time. For it will be found in the line *BD* as well as in the line *CD* at the end of the time and therefore

A heavily belaboured revision of that statement ended in the following form:

> And therefore if the forces *M* & *N* be impressed altogether and at once in the lines *AB* and *AC* as though they were separately generating uniform motions in those lines *AB* and *AC*: the body will proceed from *A*

479

to D in that given time in the rectilinear diagonal AD with a uniform motion arising from both forces. But if those forces M and N in the same lines AB and AC be applied gradually and continuously as though they were separately generating accelerated motion in those lines AB and AC or in either one of them: the body will proceed in the same time from A to D in the curvilinear diagonal AD with an accelerated motion arising from both forces.[117]

In so far as the second statement continued to assert that quantitatively equal forces, one impressed as an impulse and one as a continuous force, will both generate equal motions and cause equal distances to be traversed, he was still caught up in the same ambiguity.

At about the same time, Newton attacked what was basically the same problem in a projected revision of the second law. The statement of the law itself continued to assert the proportionality of force to the change in motion it generates. After the discussion that appears in the *Principia*, stating the linear relationship of force and motion, the identity of impulse and continuous application, and the rules of composition, however, he continued with a passage that was never published:

If a body A were moving before the force is impressed and with the uniformly continued motion that it had in A could have described the distance AB in a given time and meanwhile if it be impelled in a given direction by the impressed force: it will be necessary to consider that the place in which the body is relatively at rest is moved together with the body from A to B and that the body is driven from that moving place by the impressed force and leaves it in the direction of that impressed force with a motion which is proportional to the force. And therefore if the force is determined, for example, in the direction of the right line AC and in that given time the body initially at rest could have been driven from the unmoved place A to the place C, construct BD parallel and equal to AC, and according to the meaning of this Law, the same force will drive that body in the same time from its moving place B to the new place D. Therefore the body will be moved in some line AD with a motion which arises from the motion of its relative place from A to B and the motion of the body from that place to another place D, that is, from the motion AB in which the body participated before the impressed force and the motion BD which the impressed force generates according to this Law. And the motion of the body in the line AD will arise from these two motions conjoined according to their determinations.[118]

Further down the page, a revision substituted the word 'translation' [*translatio*] for the word 'motion'. The change was indicative of the whole. By abandoning any effort to represent either force or quantity of motion on the diagram, Newton was able accurately to present the path a body follows when a continuous force operates to change its inertial motion. Clearly, this treatment corresponds to the *Principia's* concept of force as the parallelogram of force never could. It achieved this goal, indeed, by abandoning the parallelogram of force – which may be the reason it never appeared in print.[119] In fact the diagram – though not the discussion of it – does not differ in any significant way from Galileo's parabolic trajectory based on a kinematic composition. What Newton inserted in the second edition of the *Principia* was instead a revision of the parallelogram which avoided reference to a uniform force causing a uniform motion by adopting the device quoted above – the forces M and N are 'impressed apart in the place A [*in loco A impressa*].' That is, the parallelogram as amended deals with two impulses, a situation wholly irrelevant to the problems in the *Principia* to the solution of which it is applied.

Because of the ambiguity inherent in the parallelogram, Newton's dynamics appears most modern, that is, most 'Newtonian,' in those problems in which the geometric representation of the path of motion, if it appears at all, is separated entirely from the diagrammatic representation of force. In such problems, in which Newton treated force in a purely functional relationship to the change in motion it causes, he escaped from the complexity of his definition of force and without hesitation set it proportional to acceleration. It is here, rather than in the formal statement of the second law, that we find the relation known to physics today as the second law of motion:

$$F = ma, \quad \text{or} \quad F = \frac{d}{dt} mv.$$

Significantly, almost none of these propositions were part of *De motu*. Nearly all of them were composed after his dynamics was revised and formulated in its mature form.

The analysis of pendular motion and of other similar motions such as vibrations in an elastic medium offers the readiest example. Newton was familiar with the dynamics of the pendulum more than ten years before the *Principia*, of course. The pendulum did not figure in *De motu*, however, and the analysis in the *Principia* set isochronous vibrations in the context of a generalised dynamics. The diagram traces the cycloidal path of the pendulum, it is true, but the force that Newton was concerned to investigate has nothing to do with the shape of the path, which is determined by the string vibrating between cycloidal cheeks. The isochronous motion of the pendulum cannot appear on the diagram (see Fig. 45). The analysis of gravity into its normal and tangential components at each point along the path is equivalent to establishing a functional relation between force and displacement. From the

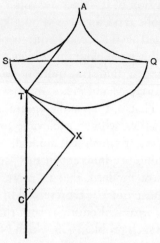

Fig. 45

properties of the cycloid, he proved that the tangential component is proportional to the displacement from the vertical measured along the cycloid. 'Then it is plain,' he concluded, 'that the acceleration of the body, proportional to this accelerating force, will be every moment as the length $TX\ldots$'[120] By a sort of verbal integration, he proceeded to demonstrate that acceleration, velocity, distance traversed, distance to the vertical remaining to be traversed, and time consumed are equal for all vibrations in the same cycloid, and therefore that all vibrations are isochronous.

Motion through resisting media presented a more complicated problem. In the *Principia*, Newton dealt with three cases, in which bodies are resisted in proportion to their velocity, to the square of their velocity, and partly to their velocity and partly to the square of their velocity. In order to get the required exponential function, he employed the known quadrature of the rectangular hyperbola, $xy = k$ (see Fig. 46). The area under the curve between x_1 and x_2 is equal to the integral

$$A = \int_{x_1}^{x_2} y\,dx = k \int_{x_1}^{x_2} \frac{dx}{x} = k(\ln x_2 - \ln x_1) = k \ln \frac{x_2}{x_1}.$$

If the ratio, $\frac{x_2}{x_1}$, is constant, that is, if successive values of x are taken in geometric ratio, then the increments of area will be equal to each other. By varying the quantities that x, y, and A represent as the

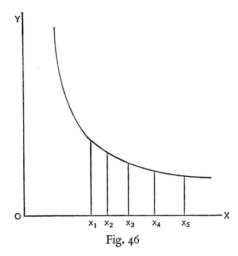

Fig. 46

problem demanded, Newton adapted this mathematical device to all of the problems of resistance.

The simplest case is that in which bodies are resisted in proportion to their velocity. Let time be divided into equal intervals, and suppose the resistance acts at the beginning of each interval 'with one single impulse which is as the velocity . . .' Since the decrease in velocity is proportional to the resistance, the decrease will be proportional to the velocity also, and the velocities at the beginning of successive intervals of time will be continually proportional, as will the velocities at the beginning of longer periods formed from an equal number of intervals. 'Let those equal intervals of time be diminished, and their number increased *in infinitum*, so that the impulse of resistance may become continual,' and the velocities at the beginning of equal intervals of time will be in geometric progression.[121] To present the problem, Newton used an hyperbola in the second quadrant, $xy = -k$ (see Fig. 47). Distance along the x axis $(-x)$ represents both velocity and resistance, and equal increments of area represent time. The diagram is the logical extension of Galileo's kinematic triangle into a complex problem in dynamics. Its primary point is not to picture the path traversed, although the distance does appear in this case as $x_2 - x_1$. Rather, the diagram presents the functional relation of velocity, resistance, and

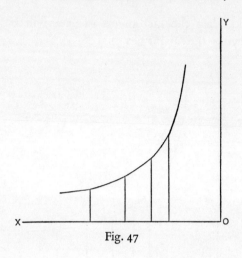

Fig. 47

time. Although it employs a series of impulses, the mathematical representation of resistance has nothing to do with the physical conception of multiple discrete impacts. Hence the problem effectively sets resistance proportional to $\Delta v/\Delta t$, even though it adopts the same apparent definition of force that is used in Proposition 1 of Book 1; and in this case the passage to the limit leads to the true differential, dv/dt. In Proposition 1, the attempt to represent both distance and motion leaves the signification of the ultimate ratio utterly confused.

Since the revision of the original version of *De motu* added the proposition on motion through a medium that resists in proportion to the velocity, it is impossible to maintain that the more abstract, functional treatment of force belonged entirely to the period after the development and maturation of Newton's basic dynamics. Such treatment was vastly expanded in that period, however. To *De motu's* propositions on resistance proportional to velocity, the *Principia* added the consideration of resistance proportional to velocity squared, and proportional partly to velocity and partly to velocity squared. Faced with the increasing complexity of such problems, Newton solved them by uninhibited recourse to the functional representation of force. He found it necessary to introduce the concept of the moment of a *genitum*. A *genitum* he defined as any quantity, such as a product, a quotient, or a root, which increases and decreases 'by a continual motion or flux,' and by 'moments' he designated the 'momentary increments or decrements' of such quantities. Moments, he continued, are not to be

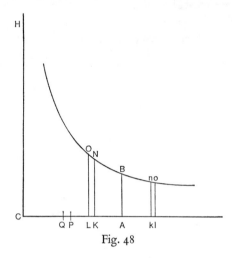

Fig. 48

considered as finite; finite quantities are not moments but quantities generated by moments. In contrast, moments are to be conceived 'as the just nascent principles of finite magnitudes,' nor are their magnitudes to be considered but rather their first proportion as they are nascent.[122] The language of Lemma II, Book II, is virtually identical to that which introduced the concept of nascent and ultimate ratios in Book I. Whereas the ultimate ratios are applied to the geometric representation of orbits, however, moments are applied to functional relations. They are identical to what we call differentials of the quantities in question with respect to time.

Proposition VIII of Book II examines vertical motion through a medium that resists in proportion to the square of velocity. Employing an hyperbola in the first quadrant, it sets gravity proportional to a given quantity AC and resistance proportional to the variable quantity AK (or Ak for upward motion), so that the absolute force operating to accelerate the body in motion is KC (or kC), equal to $AC - AK$ (or $AC + Ak$) (see Fig. 48). The line AP, the mean proportional between AK and AC (and therefore proportional to the square root of the resistance AK, since AC is a fixed quantity) represents the velocity. The short line KL is the increment of resistance generated in a momentary interval of time, and the other short line PQ the contemporaneous decrement of velocity. Because AK is proportional to AP^2, their moments KL and $2AP.PQ$ are also proportional, and by substitution KL is proportional to $AP.KC$ 'for the increment PQ of the velocity is (by Law II) proportional to the generating force KC.'[123] This is the crucial step in the demonstration. In a context explicitly dealing with differentials, he set force proportional to the differential of velocity dv/dt. Where the complexity of the problem required Newton to introduce both the functional representation of force, velocity, and distance and the fluxional calculus, he stated unequivocally that the meaning of the second law of motion is

$$F = \frac{d}{dt} mv.$$

From this step, it was a matter of simple geometry to demonstrate that the effective forces (AC, IC, KC) are in geometrical progression for equal increments of space traversed. Unlike the earlier problem, when resistance is proportional to the square of velocity, nothing on the diagram represents time. To find the time elapsed, Newton had to construct a separate diagram in which the area of an incremental sector (of a circle in ascent, of an hyperbola in descent) is 'directly as the increment of the velocity, and inversely as the force generating the increment; and therefore as the interval of the time answering to the increment.'[124] That is,

$$\Delta t = m\frac{\Delta v}{F}, \text{ or } F = m\frac{\Delta v}{\Delta t},$$

and since he was dealing explicitly with increments which pass into differentials,

$$F = \frac{d}{dt} mv.$$

Quite a different problem led Newton to a similar functional treatment of force. Proposition XXXIX of Book I set the general problem of finding at any point the velocity of a body falling vertically under the influence of a force that varies in any arbitrary way. In the seven preceding propositions, Newton had applied his analysis of orbital dynamics to the solution of specific problems when the force varies inversely as the square of the distance or directly as the distance. By allowing conic sections to shrink and approach their axes as limits, he

NEWTONIAN DYNAMICS

ingeniously extended previous conclusions on velocity in orbit to determine velocity in the limiting situation. Now he wanted a general solution for any force. Every indication suggests that Proposition XXXIX was one of the last additions to Book I. Whereas the treatment of motion in a vertical line as the limiting case of motion in a conic section dated back to *De motu*, a comparison of the numbers assigned to propositions in the surviving sections of the so-called *Lectiones de motu* with those in the *Principia* indicates that Proposition XXXIX was probably absent from the *Lectiones* together with Sections IV and V.

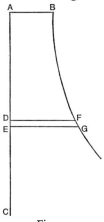

Fig. 49

Hence it may not have been composed before the winter of 1685–6, well after Newton's dynamics had assumed its final form. As the construction of the proposition indicates, its appeal to nascent ratios is an appeal to the moments or differentials of Lemma II, and its treatment of force is similar to that applied to motion under resistance.

The construction of Proposition XXXIX is peculiar. A problem, it announces the solution instead of deriving it, and then demonstrates that the announced solution is correct. The diagram is built on the vertical line of fall, *AC* (see Fig. 49). Erected perpendicular to *AC* are the lines *AB, DF, EG*, representing at every point the force tending to the centre *C*, so that the curve *BFG* expresses force as a function of distance from the centre *C*. Suppose that a body falls from *A*. The velocity at any place *E* will be as a right line, the square of which is equal to the area *ABGE*. What Newton stated is the mathematical equivalent of the work-energy equation, expressed indeed in a more general form than Leibniz ever achieved. The area *ABGE* is the integral $\int F ds$. The problem specifically stated that the quadrature of curves is necessary to its solution. Proportional to the integral is the square of velocity. Since a single body is assumed and it starts from rest, the conclusion is entirely equivalent to the equation

$$\int F ds = \Delta(\tfrac{1}{2}mv^2).$$

Corollary III, in which the body was allowed to start with a known velocity, makes the equivalence more clear.

How did Newton arrive at the conclusion? His demonstration takes DE as an increment of path. If the velocity in D is v, then the velocity in E will be $v + \Delta v$, and according to the announced solution

$$\frac{DFGE}{DE}\left(=\frac{F\Delta s}{\Delta s}\right) = \frac{v^2 + 2v\Delta v + (\Delta v)^2 - v^2}{\Delta s}.$$

In the limit, when the area is just nascent,

$$DF = \frac{2v dv}{ds}, \quad \text{or} \quad F \propto v\frac{dv}{ds}.$$

To establish the significance of this relation, Newton started again from known principles. The time in which the body describes DE is proportional to the distance DE and inversely proportional to the velocity:

$$\Delta t = \frac{\Delta s}{v}.$$

The force is 'directly as the increment I of the velocity and inversely as the time; and therefore if we take the first ratios when those quantities are just nascent, as $\dfrac{I.V}{DE}$, that is, as the length DF.'[125] In other words,

$$F \propto \frac{\Delta v}{\Delta t} \propto \frac{\Delta v}{\Delta s/v} \propto \frac{v \Delta v}{\Delta s},$$

and in the limit $Fds \propto vdv$, the relation derived from the diagram. In light of his own exposition, it appears obvious that Newton arrived at this relation, integrated, and with the result in hand composed the inverse demonstration that the problem expounds. Crucial to it, as to the demonstrations on resistance, is the proportionality of force to the differential of velocity with respect to time. We might note in passing that the quantities represented by the equations had no significance whatever in Newton's eyes. Exploiting a dynamics built on the equa-

NEWTONIAN DYNAMICS

tion $F = ma$, he was unable to grasp the importance of the work-energy equation he had implicitly derived. Above all, the quantity mv^2 held no meaning for him. This was Leibniz's conception of force. It was not Newton's. In the first corollary, he arrived at a result identical to the equation

$$\frac{W}{\Delta W} = \frac{v^2}{2v\Delta v}.$$

His immediate impulse, which he followed, was to simplify the right hand side to $\frac{\frac{1}{2}v}{\Delta v}$.

The same mathematical relations, equivalent to the work-energy equation, appeared again in Newton's examination of the effect of

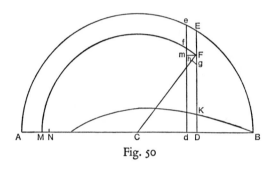

Fig. 50

resistance (proportional to the square of velocity) on the motion of a pendulum. He concluded that the difference between the arc in descent and the arc in ascent, multiplied by the sum of the two arcs, is equal to the area under the curve that represents the resistance at each point along the arc. The area under the curve is obviously equivalent to the integral $\int R ds$; and if S_1 be taken as the arc of descent and S_2 the arc of ascent, the integral is set proportional to $S_1^2 - S_2^2$, what we would call the loss of potential energy. To achieve this result, Newton constructed a diagram to show the effective component of gravity (CD), the resistance (DK), the velocity at D given the loss already incurred via the resistance (DF), the moment of velocity that it lost at D by the resistance of the medium while it describes a least space (Fg), and the decrement of arc due to that loss of velocity (MN) (see Fig. 50). 'Draw Fm perpendicular to df, and the decrement Fg of the velocity DF generated by the resistance DK will be to the increment fm of the same velocity, generated by the force *CD*, as the generating force *DK* to the generating force *CD*.' That is,

$$\frac{-\Delta v}{\Delta v} = \frac{R}{F}, \quad \text{or} \quad \frac{F}{\Delta v} = \frac{R}{-\Delta v},$$

the crucial proportionality of force to change of velocity which, in the limit (and he did speak of moments) becomes the proportionality of force to acceleration. By the similarity of triangles, he established that *fm* is to *Fm* or *Dd* as *CD* is to *DF*, which is equivalent to

$$\frac{\Delta v}{\Delta s} = \frac{F}{v}, \quad \text{or} \quad F\Delta s = v\Delta v,$$

the work-energy equation. Also *Fh* or *MN* is to *Fg* as *DF* to *CF*,

$$\frac{-\Delta S}{-\Delta v} = \frac{v}{S}, \quad \text{or} \quad -S\Delta S = -v\Delta v.$$

By multiplication of corresponding terms, *MN* is to *Dd* as *CF* to *CM*:

$$\frac{-\Delta S}{\Delta s} = \frac{R}{S}, \quad \text{or} \quad R\Delta s = -S\Delta S.$$

Therefore 'the sum of all the *MN.CM* will be equal to the sum of all *Dd.DK*.'[126] Which is to say, in our terms,

$$-\int S dS = \int R ds,$$

or the loss of potential energy due to resistance is equal to the work done against the resistance. Again Newton's understanding of force in a functional relation to distance and his manipulation of moments had led him to an important expression of the work-energy relation. Again his lack of interest in any equation except that relating force to change of velocity led him to ignore the significance of the equations he implicitly wrote.

Judged in the perspective of seventeenth-century dynamics as a whole, Newton had taken an immense stride in the direction of precision and clarity in the definition of force. After a fruitless flirtation with an absolutistic dynamics, in which the force of a body would express its absolute motion, he abandoned completely the conception of force as *mv*, the force of a body's motion. Phrases about bodies moving by their inherent force alone, artifacts left behind in the historical development of his dynamics, should not be allowed to mislead us on that score. Another seventeenth-century measure of force, Fs or $\int Fds$, what we call work, never figured as 'force' in his dynamics, even though the quantity appeared in the contexts of given problems. As far as the other two measures of force are concerned, however, ma and Δmv, he never freed himself completely from the ambiguity of using both. Nevertheless, as I have indicated, the ambiguity that remained, like his ultimate ratios, had been reduced to the vanishing point. If it appeared disturbing in the axiomatic structure of his dynamics, spreading a faint haze of conceptual unclarity over the whole, the fact that he almost always employed Δmv in a context of periodically repeating impulses, which implied the factor Δt that was absent from the definition of force, eliminated virtually all the disruptive quality from the ambiguity. Especially in those problems in which force appeared as a function of distance or velocity and the fluxional calculus was employed, the confusion inherent in his parallelogram of force dropped away, and the second law appeared without ambiguity in the form we try to read into Newton's own statement of the second law,

$$F = \frac{d}{dt} mv.$$

I have argued that the model of impact played a major role in Newton's dynamics, both in the development and exposition of his concept of force, and in providing the insight recorded in the third law. At the same time, it is impossible to miss the ultimate correlation of Newton's dynamics with Galileo's kinematics. The successful quantification of kinematics had based itself on two fundamental concepts, uniform motion and uniformly accelerated motion. Newton's first two laws of motion stated their dynamic conditions. Whatever the dimensional problems in his definition of force, Newton's second law embodied the dynamic conditions for the model of free fall, uniformly accelerated motion. With this foundation finally and securely in place, the science of mechanics could proceed to erect a quantitative dynamics as the natural complement of the quantitative kinematics already constructed.

*

To the science of dynamics Newton contributed not only a concept of force quantitatively defined, but a concept of force successfully generalised to cover every dynamic situation in which an inertial state is changed. Indebted more in this respect to the tradition of mechanics during the seventeenth century, which had consistently sought to reduce all phenomena of motion to a common ontological plane, Newton applied his mathematical gifts to that tradition and succeeded in reducing to quantitative treatment a host of problems that had seemed at the beginning of the century beyond the limits of scientific investigation. The paradigm problems of impact and free fall were simple to state in dynamic terms once the key concepts had been grasped. A science of dynamics, however, could be said to exist only when the paradigm cases had been successfully generalised to a host of problems infinitely more complex. In that direction also, the *Principia* took gigantic strides.

First in importance was the treatment of curvilinear motion. In so far as celestial dynamics is the central problem of the *Principia*, it can be said that the work rests on Newton's recognition that change of direction and change of speed are identical from the point of view of a dynamics that starts with the principle of inertia. The ultimate triumph of the *Principia* was its reduction of centripetal acceleration to the paradigm of a uniformly accelerated motion, thus bringing to a successful conclusion the century's struggle with the perplexities of circular motion. There is no need to repeat the details of Newton's quantitative treatment of centripetal force. It is interesting to note that even the man who coined the term 'centripetal force' and defined it quantitatively could not wholly exorcise the ghost of centrifugal force. To establish that the terrestrial force of gravity is identical to the force that holds the moon in its orbit, he imagined a satellite circling the earth at mountain height. If such a satellite 'be deserted by its centrifugal force that carries it through its orbit, and be disabled from going onward therein,' it will fall to the earth.[127] Another deposit left behind by the movement of his thought, the phrase does not mean that Newton continued to think of circular motion as an equilibrium of opposed forces. Rather the passage recalls the demonstration that a body moving under the influence of a centripetal force will be held in a closed orbit only if the velocity of the body is below a limit that can be calculated for a given intensity of force. If the velocity is above that limit, the body will be diverted into an hyperbolic path but not into a closed orbit. Conversely, if the velocity is too low, the orbit will intersect the central attracting body, and the moving body, like a projectile on the earth, will strike it and come to rest. The phrase, 'centrifugal force', was a legacy from the past that evoked the orbital velocity of the moving body. The elements of orbital dynamics are a tangential motion and a centripetal force sufficiently strong continually to divert the body from its inertial path into a closed orbit. 'For every body that moves in a curved line is (by Law I) turned aside from its rectilinear course by the action of some force that impels it.'[128] Not only did he apply the concept to the problems of celestial dynamics, he employed it as well to derive the sine law of refraction, on the assumption that light is corpuscular.[129]

If the complexities of circular motion had baffled the science of mechanics in the seventeenth century, the problem of resistance had seemed frankly beyond the possibility of scientific treatment. Galileo had contended that resistance is not subject to 'fixed laws and exact description' because of the innumerable ways in which it disturbs the motion of bodies according to their variations in shape, weight, and velocity. 'Of these accidents of weight, of velocity, and also of shape, infinite in number, it is not possible to give any exact description; hence, in order to handle this matter in a scientific way, it is necessary to cut loose from these difficulties; and having discovered and demonstrated the theorems, in the case of no resistance, to use them and apply them with such limitations as experience will teach.'[130] Galileo had therefore confined himself to ideal motions from which resistance had been eliminated by definition. Almost without precedent, Newton created the scientific treatment of motion under conditions of resistance, that is, of motion as it is actually found in the world. Recognising the exponential nature of such motion, he seized on a means of expressing it mathematically by means of the rectangular hyperbola. With this device, he was able to derive the consequences for various assumed functional relations of velocity and resistance.

Newton was not satisfied, however, with a purely hypothetical treatment of motion under resistance. He was convinced that the principal component in the resistance of a medium derives from the inertia of the medium itself which must be set in motion in order for a body to pass. This component must be proportional to the density of the medium, the square of the velocity of the moving body, and the area of the body's cross-section (or the square of its diameter). By experiments with bodies falling through air and through water and with pendulums swinging in various fluids, he convinced himself that the reality of resistance is as his theory predicts. The climax of his consideration of motion through resisting media is found in Corollary 3 to Proposition XXXVIII, Book II.

> If there be given both the density of the globe and its velocity at the beginning of the motion, and the density of the compressed quiescent fluid in which the globe moves, there is given at any time both the velocity of the globe and its resistance, and the space described by it.[131]

This was to do for motion through a resisting medium exactly what Galileo had done for motion under idealised conditions from which resistance is eliminated. Inevitably there were shortcomings in Newton's treatment, as there were in the whole of Book II. Especially his conception of the source of a fluid's resistance, a conception which embodied the entities basic to the whole of Newtonian dynamics, had to be revised in the eighteenth century before a satisfactory treatment of it could emerge. The shortcomings cannot conceal the achievement. By generalising his dynamics to include resistance, whatever its physical source, he created almost *ex nihilo* a new branch of the science of mechanics; and whatever his errors, he pointed out unmistakably the path that would need to be followed in order to correct them.

Closely analogous to the question of resistance was that of perturbations of orbital motion due to the presence of third bodies. The three body problem is an immensely complex one which admits no general analytic solution. Newton's treatment of it was to be replaced before the eighteenth century had passed. The significance of his attempt lay again in the generalisation of his dynamics to cover a new body of phenomena and in his successful indication that they could be removed from the domain of qualitative discussion into the realm of quantitative science, even though he did not carry their quantitative treatment to a high level of precision.

His basic device consisted in analysing the force by which the third body S attracts the orbiting satellite P into two components, LM and SM, both of which must vary, of course, during the orbit of P around T (see Fig. 51). The component LM is parallel to the centripetal force by which T attracts P and holds it in orbit. Its effect is to decrease the centripetal force, and that in varying amounts according to P's position, thus causing the force to diverge from the inverse square relation necessary for a stable elliptical orbit. The other component, SM, which is neither proportional to $1/PT^2$ nor parallel to PT, disturbs the motion of the satellite still more. Its effect is largely offset, however, by the attraction of S on T, one of the basic insights from which, as we have seen, the law of universal gravitation was drawn. Since the distances of P and T from S are not always equal, the intensities or accelerative quantities of the two attractions are not equal, and again they vary

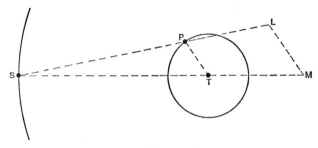

Fig. 51

according to the position of P, the component SM now exceeding the attraction on T when P is between S and T, now falling short when P is beyond T.

With immense ingenuity, Newton made the one analysis serve four problems. Initially and overtly, it applied to the influence of one planet on the motion of another in an inferior orbit. Since the sizes of S and T can be varied at will to pose different problems, and the construction remains unchanged whether S orbits T or T orbits S, it can be employed to examine the effect of the sun on the moon's orbit. The choice of the letters S and T (*Sol* and *Terra*) indicates that this was the central purpose of the analysis. The radius PT of the orbit can also be varied at will without changing the central features of the analysis. Let P become a belt of matter encircling the earth and contiguous to it. As a liquid, able to move independently of the body it surrounds, it represents the sea, and the analysis can be applied to the tides. As solid matter attached to the body and able to move only by moving the body with it, it represents the matter around the equator by which the earth departs from perfect sphericity, and the analysis offers a grasp on the precession of the equinoxes.[132]

In the case of orbital perturbations, Newton was explicitly dissatisfied with his results and confined himself to a verbal consideration of the major inequalities that ought to appear. In the case of tides, the body of observations available was limited and crude and the complications offered by irregular shorelines immense. Within those limits, his analysis offered a satisfactory explanation of the major phenomena in their roughly known quantities. In the case of precession, he allowed himself to create an illusion of precision that was grossly misleading. In the first edition of the *Principia*, the treatment of precession included as a necessary element a defective proposition which was corrected in the second edition.[133] In both cases, however, the derivation of precession in its observed quantity depended on the subsequent application of the ratio of the moon's attraction on the earth to that of the sun. This ratio was derived from the observed phenomena of the tides. In the first edition, the ratio that the observations presented to Newton was $6\frac{1}{3}$ to 1. In the second edition, where the corrected proposition led to different quantities, the ratio presented by the same observations was 4·4815 to 1. In fact, he was able to make the demonstration conclude with whatever degree of precision he wished by manipulating the ratio of the attractions of moon and sun.

It is unnecessary to insist at great length on the deficiencies of Newton's treatment, which were inevitable in a pioneering attempt to bring a whole new range of phenomena within the purview of quantitative dynamics. It is unnecessary as well to insist that the immense increase in sophistication both in rational mechanics and in the calculus during the eighteenth century, which combined to leave his effort far behind, drew their strength primarily from the beginning he provided.

Newton also applied his dynamics to the motion of pulses or waves through fluid media. The inverse square law is associated with his name so intimately that it is frequently overlooked that the *Principia* also contains the first satisfactory analysis of simple harmonic motion, though he did not employ that name. The basic dynamic condition of simple harmonic motion, that force varies directly in proportion to displacement, is equivalent, of course, to Hooke's Law; and before Newton, Hooke had seized intuitively on the dynamic identity of pendulums and vibrating springs. As we have seen, both his mathematics and his dynamics were inadequate for the task of exact analysis. Sophisticated mathematics and dynamics were precisely the tools that Newton brought to it. From the consideration of orbital motion when the centripetal force varies directly as the distance he moved to pendulums, from pendulums to columns of water vibrating in U-shaped tubes, and from vibrating columns to waves on a surface of water, finding in all cases that vibrations of whatever amplitude are isochronous because the restoring force increases in direct proportion to the displacement. He was able as well to determine correctly the functional relation of acceleration, velocity, displacement, and time during a single vibration.[134]

From waves on a surface of water, he proceeded to waves propagated through an elastic medium. With supreme irony, the champion of the corpuscular theory of light provided the rival theory with the wave mechanics its own champions had been unable to develop for themselves. Newton recognised that Boyle's Law implies the dynamic condition of simple harmonic motion. Examining the force exerted on a differential segment EG of the medium when it is displaced to $\varepsilon \gamma$, he demonstrated that 'the accelerative force of the physical short line $\varepsilon \gamma$ is as its distance from Ω, the middle place of the vibration.'[135] Hence vibrations propagated through an elastic medium are isochronous, and the velocity of propagation in a given medium is constant for all frequencies.

To demonstrate that his theory of wave motion was correct, Newton calculated the velocity of sound in air. According to his theory, the velocity should vary directly as the square root of the elastic force and inversely as the square root of the density. Boyle's Law yielded the first factor, and measurements of the air's density in comparison to water the second. When carried through, the calculation yielded a velocity of 968 feet per second. His own experiments at Trinity College had placed the velocity of sound between 920 and 1085 feet per second, and in the first edition of the *Principia* he considered the calculation to be confirmed by the measurements. Alas, measurements are made in order that they may be improved. The improvements moved the permissible velocity upward to a range between 984 and 1109 feet per second. The lower figure was above Newton's calculation; and as he prepared the second edition of the *Principia*, he began to fudge. He decided that sound moves instantaneously through the solid parts of the air, which means, in view of the measured density of air, that roughly one-ninth to one-tenth the path in any direction will be through solid matter. The calculation assumed that water is solid matter (the cube root of 850, the ratio of the densities of water and air that he used, falls between 9 and 10); in view of Newton's conception of matter, this was an egregious fraud. Nevertheless, it raised the calculated velocity of sound roughly 100 feet per second, and Newton was prepared to let it stand. Then the news arrived that a French scientist,

Sauveur, had determined the velocity of sound by measuring the length of pipe in which air vibrates at a known frequency. The new figure was 1142 feet per second, and Newton did not question it. After all, further adjustments were available. He corrected the density of the air. In the first edition the ratio of densities was 850 to 1. As he amended the passage, he wrote '850 or 870,' and in the second edition took 870 straightaway. The lowered density of air gained 11 feet per minute. To this he added the correction for the solid parts, and because he wanted the answer to be precise he experimented with figures again, trying a number of corrections between 8 and 10. Finally, he wrote that the ratio of the diameters of particles of air to the space between them is 1 to 9 or 10, used nine in the calculation and added another 109 feet per second. The calculated velocity was now 1088 feet per second, still 54 feet short. Fortunately, an adjustment for water vapour still remained. It appeared that water vapour, for reasons of its own, does not vibrate, and if the atmosphere contains 10 parts of air to one of vapour, the velocity of sound will be increased by the ratio of the square root of 11/10, which is very nearly 21/20, which in turn, *mirabile dictu*, increases the calculated velocity of sound to 1142 feet per second.[136]

Like the precession of the equinoxes and the correlation of the acceleration of gravity on the earth and at the moon, the calculation of the velocity of sound illustrates the growing sense of infallibility that kept pace with Newton's years. The obvious fraud in the pretended degree of precision does not destroy the underlying importance of the theory, however. The correction he needed was one of which he could not have known, a correction of the elasticity equal to the ratio of the specific heat of air at constant pressure and its specific heat at constant volume. With that correction, Laplace used Newton's theory to calculate the velocity of sound correctly. The theory of the propagation of waves, together with the analysis of simple harmonic motion on which it rests, is one of the triumphant achievements of the *Principia*.

★

Newton also extended his principles into the area of fluid dynamics. Neither the subject as a whole nor the problems he undertook were original with him, nor for that matter were his efforts in fluid dynamics blessed with compelling success. Indeed fluid dynamics seems almost to have been invented to demonstrate the limitation of Newtonian dynamics when not enhanced by additional principles. As its clarification of the concept of force implied, his dynamics devoted itself primarily and with greatest success to problems in which discrete bodies experience accelerations. Via the model of impact, Newton attempted to apply this vision of dynamics to the resistance of fluids – an effort that the eighteenth century had to abandon before it could bring the treatment of resistance to a satisfactory conclusion. Fluid dynamics was intimately related to the problem of resistance, and again Newton's spontaneous effort to reduce problems of continuous media in steady states to solid bodies subjected to accelerating forces imposed sharp limitations on his success.

Perhaps the judgment is too severe in regard to vortical motion, for Newton successfully introduced quantitative dynamics into a problem that verbal images had hitherto dominated. His very success, however, was proportional to the degree in which he treated fluids as solids. The basic element of his analysis was a laminar shell concentric with the body central to the vortex (a cylinder in one case, a sphere in the other), and the possibility of dividing the vortex into such shells derived, of course, from its fluidity. Once he had postulated the shells, however, the success of his analysis lay in his treatment of them as solids, and the shortcomings of the analysis in turn were related to the fact that fluid shells cannot remain intact under the conditions he set. To examine the motion of the laminar shells, Newton proceeded on two assumptions which are implicit in the concept of viscosity accepted today – that the friction between shells is proportional to their relative velocity, and that it is proportional to the area of surface in contact. Each shell, of course, is subject to two frictional forces. The shell inside, turning with a motion ultimately derived from the central body generating the vortex, seeks by its friction continually to accelerate the rotation of the shell in question; the shell outside, turning only to the extent that the shell in question has moved, seeks by its friction continually to retard it. 'Any part will continue in that motion in which its attrition on one side retards it just as much as its attrition on the other side accelerates it.'[137] Since the surface areas are a funtion of the radii, it is a simple matter to calculate angular velocities as a function of radius, and Newton was able to demonstrate that under the assumptions he made the periods of rotation in a spherical vortex (the only vortex of interest to natural philosophy) vary directly as the square of the distance from its centre. In the first instance, the analysis applied to the matter located along the equator or ecliptic of the shell. By dividing each shell into rings, and arguing that friction between the rings (which unlike the shells do not form a series of unlimited extension) would operate to speed up those nearer the poles should they turn more slowly, he concluded that the whole of each shell must turn with the same angular motion.[138]

What Newton sought to establish in his analysis of the vortex were the conditions of equilibrium in which each shell would continue in its uniform rotation. The peculiar nature of the dynamics he brought to the problem and the peculiar approach to it that he took, led to the postulation of an anomalous equilibrium in which the uniform motion of each shell assumes the constant transfer of motion through it to matter on beyond that is not yet moved.

> Because the inward parts of the vortex are by reason of their greater velocity continually pressing upon and driving forwards the external parts, and by that action are continually communicating motion to them, and at the same time those exterior parts communicate the same quantity of motion to those that lie still beyond them, and by this action preserve the quantity of their motion continually unchanged, it is plain that the motion is continually transferred from the centre to the circumference of the vortex, till it is quite swallowed up and lost in the boundless extent of that circumference. The matter between any two spherical surfaces concentric to the vortex will never be accelerated; because that matter will be always transferring the motion it receives from the matter nearer the center to that matter which lies nearer the circumference.[139]

Equilibrium as such could never be reached. Motion would be communicated outward *in infinitum*, and equilibrium would be approached as the communication proceeds. In a general sense, the constant transfer of motion out from the central body to be 'swallowed up and lost' in infinite space does service for the conversion of kinetic energy to heat, a concept which was not available to Newton, of course. Although his vortex was generated by friction between the layers, he was not in a position to consider the heat that would be generated thereby and would constantly undermine the steady state his analysis imagined. In so far as his investigation of vortices was directed against Descartes, he would not have been upset to learn that even the assumption of a steady state grants more than the case will bear.

Motion is not all that is communicated outward, and heat is not all that upsets the steady state. Although he did not take centrifugal force into account as a condition of his analysis, he did recognise that in a given shell it is less near the pole than at the equator. As a consequence, he agreed, there will be a circulation of matter in meridional planes, outward at the equator and inward at the poles.[140] Beyond mentioning it, however, Newton did nothing to calculate its effect – which appears to be nothing less than the constant dissolution of the laminar shells on which the whole analysis rests.

Vortical motion has not remained a central question of modern physics, as other problems examined in the *Principia* have. In the eighteenth century, the legacy of Descartes contained enough vitality to lead to improvements on Newton's analysis, which certainly needed them. It is worth noting its positive features as well. Before him, the vortex was discussed entirely in qualitative terms. Newton insisted that vortices, as matter in motion, must be subject to the same dynamic conditions that govern the motions of all matter. In the concept of the laminar shell, a differential segment of fluid matter, he imagined the device whereby dynamic considerations could be brought to bear on fluids. To the extent that he could only treat the differential segment as a solid body, his principles were inadequate for a wholly successful fluid dynamics. Nevertheless, he took at least one step in the direction of bringing this area of inquiry within the realm of quantitative mechanics.

In the other major problem in fluid dynamics to which he devoted serious attention, the rate of flow of water from a hole in the bottom of a container, which he examined in Proposition XXXVI of Book II, his dynamics broke down completely. In the first edition (in which it was Proposition XXXVII), Newton set out in a straightforward manner to reduce the problem to that of a body falling under the uniform force of gravity. If we imagine the hole in the vessel to be plugged, the plug will obviously sustain the weight of the water above it. It follows that when we remove the plug, 'the motion of all the water flowing out is that which the weight of the water perpendicularly above the hole is able to generate.' The situation is somewhat ambiguous, however. In the dynamics of discrete masses, the effect of a force is registered in the acceleration of a mass. In the case of a fluid flowing from a hole, the vessel presents a steady state in which water flows uniformly through the hole, and the problem is to calculate that one velocity to which every particle of water has been accelerated as it emerges. To equate the two cases, Newton imagined the column of water above the hole to fall from rest under the force of its own weight. Let F be the area of the hole and A the height of the column, S the space it would describe in falling freely for time T, and V its velocity at the end of time T. Its 'acquired motion $AF \times V$ will be equal to the motion of all the water that flows out in the same time.' Let the velocity of that water be to V as d is to e. With velocity V it would describe $2S$ in time T; with its actual velocity it would describe $2\frac{d}{e}S$. Hence a column of water that long – $2\frac{d}{e}SF$ – flows out in time T and its motion is $2\frac{d^2}{e^2}SFV$ 'which is all the motion generated in the time of its efflux.' Newton then proceeded to set the two motions equal since both are generated by the same force. Hence,

$$\frac{d}{e}\left(=\frac{v}{V}\right) = \sqrt{\frac{A}{2S}}.$$

The velocity of the effluent stream is that of a body falling through half the depth of the water, and if the stream were diverted upward it would rise half way to the surface.[141]

The later development of this proposition suggests that basic to Newton's solution was an experiment in which he measured the quantity of water flowing from a hole of known cross-section in a given time. The experiment indicated that the velocity of flow must be approximately that which a body would acquire in falling through a distance equal to half the depth of the water above the hole. Years later, in connection with the calculus controversy, Newton attempted to explain the errors that had crept into the *Principia*, and among them this problem.

> By measuring the quantity of water w[ch] ran out of a vessel in a given time through a given round hole in the bottom of the vessel: I found that the velocity of the water in the hole was that w[ch] a body would acquire in falling half the height of the water stagnating in the vessel. And by other experiments I found afterwards that the water accelerated after it was out of the vessel untill it arrived at a distance from the vessel equal to the diameter of the hole; & by accelerating acquired a velocity equal to that w[ch] a body would acquire in falling the whole height of the water stagnating in the vessel, or thereabouts.[142]

If we can trust the testimony of later editions, the other experiments were those in which the stream of water was directed horizontally and vertically upward. When horizontal, it traced a parabolic trajectory demanding the higher velocity; when vertical, it rose virtually to the level of water inside the vessel. Already by 1690, these experiments had been pressed on Newton's attention, apparently by Fatio de Duillier, and by March of that year he had amended the proposition. A confusing passage asserted, among other things that appear directly in

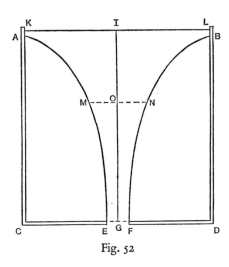

Fig. 52

contradiction, that 'the speed with which the water issues from the hole is the same as if it fell from the top of the water in the vessel . . .' He concluded, moreover, that 'the quantity of water flowing out in the time in which a body in falling can fall through the height of water in the vessel above the tube, is equal to a column of water whose diameter is the same as the tube's and whose length is twice as great as the height of water in the vessel above the tube.'[143] Thus apparently, according to the dynamics he had employed in the first edition, the effective force had to be increased by the square root of two since the velocity of water flowing out was increased by that amount.

How was he to account for the increased force? In so far as I can reconstruct the progress of his thought, he hit upon the idea of the contracting cataract that appeared in the second and subsequent editions (see Fig. 52). That part of the water in the vessel which is effectively involved in the efflux is a cataract as broad as the vessel at the surface of the water and narrowing to the cross-section of the hole at its base. One can imagine all the rest of the water in the vessel to be frozen and to play no part whatever in the dynamics of flow. The 'velocity of the falling water, when it reaches any surface *MN* parallel to the horizon, will be that which the water can acquire in falling and by its fall describing the altitude *IO* . . .' The imagined cataract represents a column of water falling freely, its cross-section continually narrowing in proportion to its increase in velocity whereby a given quantity can pass through a smaller hole in the same time. Hence 'the velocity of the water when in falling it reaches the hole *EF* will be that which the water can acquire in describing the altitude *IG* in its fall.'[144] The quantity of matter in the cataract equals that in the column twice the height of water above the hole, and apparently Newton had this in mind when he asserted in the final revised form that 'the whole weight of the column of water *ABNFEM* will be taken up as before in forcing out the water . . .'[145] The same dynamic principles employed in the first edition were in play, but the quantity of force had been increased.

The revision took care of the experiments that showed the higher velocity. What was to be done, however, with the original experiment measuring the quantity of water that flows out? Either the experiment had been wrong, or it had to be accounted for. A manuscript exists in which Newton mentioned having Cotes perform the original experiment. Cotes confirmed the original result.[146] From the pressure of the conflicting experiments emerged the strange proposition which first appeared in edition two, an extended exercise in obfuscation that talked around the problem, announced results that satisfied all the evidence, but offered nothing whatever which might be called a demonstration. It included the cataract but not the principle of freely falling water which had evoked it in the first place. It asserted indeed the basic conclusion of the first edition, that the 'running water in passing through the hole itself has a velocity downwards nearly equal to that which a heavy body would acquire in falling through half the height of the stagnant water in the vessel.'[147] So much for the original experiment. To satisfy the other experiments, he now discovered that the water is accelerated for a short distance after it passes through the hole. Because most of the particles of water approach the hole obliquely, they force the stream to contract – a sort of inverse application of his cataract. By measurement, the diameter of the stream has contracted by about one-sixth (he used the ratio 25/21) when it reaches a point below the hole that is one diameter distant. If the stream contracts, the water must accelerate, and that by the ratio $25^2/21^2$ which is virtually identical to $\frac{\sqrt{2}}{1}$, not too surprisingly the exact ratio of acceleration that he required.

The revision of Proposition XXXVI is an extended demonstration that even Newton could not have his cake and eat it too. If he apparently succeeded in reconciling the conflicting evidence, he did nothing to enhance his scientific reputation in the process. Fundamental to the revision was his inability to find new principles with which to attack the problem. In contrast, Huygens approached it from the outset via the principle that under ideal conditions the velocity of a moving body is always able to restore it to the height from which it fell in acquiring that velocity – an implicit principle of the conservation of energy, if you will. In one sentence he obtained the answer that Newton extracted by brute force after torturing the acceleration of freely falling bodies through five pages.

It should be obvious that Newton's dynamics, as published in the *Principia*, was not an unalloyed success. It should equally be obvious that the extent of its failure, in proportion to its achievement in synthesising and completing earlier tentative gropings toward a quantitative dynamics, and thus in adding a new dimension to rational mechanics, was extraordinarily small. If in nothing else, in this proportion alone – the ratio of its failure to its success – the genius of the *Principia*'s author is readily assessed. In asserting that Newton's dynamics added a new dimension to rational mechanics, I do not intend in any way to derogate from the achievement of Leibniz or to open again at this late date the barren debate on their relative attainments. Leibniz's contributions to science in general and to dynamics in particular cannot be called into question. The principles of his dynamics, moreover, were exactly those required to make good the inadequacies of Newton's. As far as seventeenth-century mechanics was concerned, it does appear to me that Newton's conception of force and Newton's laws of motion were more immediately relevant to the solution of problems being faced. Certainly the scope of his dynamics, as measured by the problems he solved, exceeded that of Leibniz's. Historically speaking, Newtonian dynamics, based on the Newtonian concept of force, had to be incorporated into mechanics and thoroughly digested before Leibnizian dynamics with its different idea of force, rebaptised later as kinetic energy, could be utilised to the full. In the *Principia*,

Newton both stated the principles of his dynamics and applied them with general success to most of the problems current in mechanics.

*

One of the features of the *Principia* – and one of the criteria by which the power and utility of its dynamics are readily measured – is its constant effort to generalise problems to all possible variations of force with distance. Newton examined orbits and trajectories that would result from any functional relation of force and distance chosen arbitrarily. He calculated the attractions of finite bodies composed of particles that attract according to various proportions. He investigated the variation of density of atmosphere around a globe that attracts by any force, and the relation of pressure and volume in a gas composed of particles that repel by any force. Rational mechanics, he stated in the preface to the *Principia*, is 'the science of motions resulting from any forces whatsoever, and of the forces required to produce any motions, accurately proposed and demonstrated.'[148] And again, in the original version of Book III, he declared that his purpose was to demonstrate from phenomena the quantity and properties of the force that must act to hold planets in their closed orbits and to apply the conclusions as principles by which effects in more complicated cases can be estimated in a mathematical way. 'We said, *in a mathematical way*, to avoid all questions about the nature or quality of this force, which we would not be understood to determine by any hypothesis . . .'[149] The possibility of proceeding in a mathematical way was exactly what Newton's conception of force offered him, and in the *Principia* he followed the mathematical way both to generalise his conception of force by applying it to every change of an inertial state and to abstract it from specific physical problems.

Despite what he said, however, force was never a mere mathematical abstraction to him. The form in which he stated Proposition XXIV of Book II, which establishes the proportionality of mass to weight, so fundamental to universal gravitation, unexpectedly expresses Newton's ultimate conviction.

> The quantities of matter in pendulous bodies, whose centres of oscillation are equally distant from the center of suspension, are in a ratio compounded of the ratio of the weights and the squared ratio of the times of the oscillations in a vacuum.[150]

In this proposition, force is the given and mass the derived term. The demonstration of the proposition assumes the second law of motion. Had Newton been pressed to reduce his dynamics to three ultimate terms, as we now know dynamics can be reduced, he would certainly have chosen, in addition to length and time, force rather than mass. Force was more than a convenient term to express the product of mass times the rate of change of translation. Force was an entity ontologically existent in the universe.

Moreover, although he generalised the problems he treated for any relation of force to distance, two particular relations riveted his attention – the direct proportionality of force to distance, the inverse proportionality of force to the square of distance. 'I have now explained the two principal cases of attractions,' he declared in a scholium following his demonstrations of the attractions of homogeneous spheres;

> to wit, when the centripetal forces decrease as the square of the ratio of the distances, or increase in a simple ratio of the distances, causing the bodies in both cases to revolve in conic sections, and composing spherical bodies whose centripetal forces observe the same law of increase and decrease in the recess from the centre as the forces of the particles themselves do; which is very remarkable.[151]

It was indeed remarkable in Newton's eyes. The entire *Principia* shows how remarkable. The book started with the examination of orbits, and discovered that bodies will circle attracting centres in elliptical orbits if the force is directly proportional to distance or inversely proportional to the square of distance. In the one case, the attracting body is located at the centre of the ellipse; in the other case, at one focus. When he investigated the stability of orbits in space, he discovered that bodies can orbit attracting centres if the force does not decrease at a ratio greater than the third power of the distance, but that the orbit itself remains fixed and unmoving in space only for two relations of force to distance, if we exclude from consideration such unlikely possibilities as the direct proportionality of force to the sixth power of distance. The two relations yielding orbits that are stable in space are, of course, direct proportionality and inverse square.[152] Most remarkable of all, as the quotation above indicates, were the attracting properties of spheres composed of attracting particles. In two cases, homogeneous spheres were found to attract by exactly the same relation as the corpuscles that compose them, and again the two relations were those of direct and inverse square proportionality. One subtle aspect of this remarkable property does not emerge from the statement quoted above. The strength of the attraction is not dependent on the distance of attracted bodies from the surface of such spheres. In other cases, such as spheres composed of particles attracting with an inverse cube or inverse quadruple force, the attraction increases to infinity on contact with the surface. On a globe composed of particles that attract with an inverse square force, the weight of bodies is effectively constant near the surface; if the particles composing the globe attracted with an inverse cube force, weight would vary radically and nothing could lift a body that had once touched the earth.[153]

In one case, attraction directly proportional to distance reveals itself to possess an inherent advantage over the inverse square force. The presence of other bodies also attracting with forces that vary directly as distance does not disturb the perfect ellipticity of orbits but merely increases the velocity of motion.[154] For any other relation of force to distance, the presence of a third body distorts the perfect geometrical shape of orbits, and the further the force relation departs from direct proportionality the greater the perturbations become. If the inverse square force cannot avoid the disturbance of orbits by third bodies, perturbations can be minimised by certain conditions, so that orbits at least approach perfect ellipticity. When a number of satellites orbit one central body, the larger the central body in relation to the others, so that its centre more nearly coincides with the whole system's centre of gravity, the less the smaller bodies disturb each other's motion. When a system of two or more bodies, such as the earth with its moon or Jupiter with its satellites, orbits another body, the greater the distance of the other body, so that the attractions on the whole system follow lines more nearly parallel, the less the perturbation. The perturbation is also minimised when the attraction of the external body follows the same proportionality to distance and mass for all bodies in the system, and when the bodies in the system attract the external body in return.[155] That is, the perturbations are minimised for a system such as our universe is in fact.

'When I wrote my treatise about our Systeme,' Newton declared to Bentley, 'I had an eye upon such Principles as might work wth considering men for the beleife of a Deity . . .'[156] No principle was more directly under his eye than the inverse square law of attraction. Small wonder that he resented Hooke's claim to the inverse square relation,

NEWTONIAN DYNAMICS

when Hooke had done no more than glimpse it obscurely from a distance. Newton had contemplated its very essence. Far more than a bare empirical fact in Newton's view, the inverse square relation embodied an inherent rationality whereby it alone could support an architecture suitable to a cosmos. In the end, the direct proportionality of force to distance, with its consequent infinite accelerations and infinite velocities in an infinite universe, did not provide a suitable foundation. As though in compensation, moreover, a universe of solid spherical bodies composed of particles that attract by the inverse square relation, contains the relation of direct proportionality as well, between the centres of the spheres and their surfaces; and Newton specifically remarked that what he had demonstrated about the motions of bodies under the influence of such forces takes place when the bodies are moved within homogeneous spheres.[157] At best it is a perplexing statement, but it may be pertinent to note that his demonstration of isochronous vibrations in a cycloid avoids the assumptions that gravity is uniform and its lines of action parallel by utilising a cycloid described inside a circle and assuming that the force of attraction to the centre varies directly as the distance.[158] Although motion in the vulgar cycloid, as he called it, is effectively identical, the demonstration is wholly rigorous under his conditions alone. Be that as it may, the direct proportionality of force to distance plays its role in nature in connection with vibratory motions. For cosmic architecture, only the inverse square relation is suitable. Nothing reveals the omniscience of God more clearly than His decision to construct a universe composed of particles that attract according to it.

The inverse square relation exhibits anew the dual aspect of Newton's conception of force. On the one hand, as long as we confine ourselves to the scientific level of discourse, attractions between particles of matter are real entities existing in nature. God has created a universe composed of particles which attract and repel each other at a distance. Along with other specific forces, he has endowed each particle of matter with the power to attract every other particle with a force directly proportional to the product of their masses and inversely proportional to the square of the distance between them. If there were any doubt as to the ontological status of force, the rationality of the inverse square attraction and its inherent suitability to construct a stable and harmonious universe must remove it. On the other hand, forces act in accordance with precise mathematical laws such that they fall within the competence of a quantitative dynamics. If force is not a mere mathematical abstraction, it is not for that less mathematically exact. The essential task of natural philosophy is to demonstrate quantitatively how particles of matter, in motion and endowed with the forces experience shows them to possess, produce the observed phenomena of nature.

If Book I of the *Principia* may be said to have devoted itself to that task, Book II was primarily concerned to demonstrate that the prevailing mechanical philosophy – the kinetic mechanical philosophy which admitted moving particles alone and rejected attractions and repulsions as occult – could not similarly account for the same phenomena in exact quantitative terms. 'The hypothesis of vortices is pressed with many difficulties,' the General Scholium stated in its opening sentence.[159] So indeed it was, as the *Principia* demonstrated, and with the hypothesis of vortices the entire mechanical philosophy. According to the law of universal gravitation, the weights of bodies are universally proportional to their masses. The proportionality can be demonstrated experimentally if the three laws of motion are granted. Although in principle it was possible to devise an aethereal mechanism to account for the proportionality, such a mechanism would require a host of *ad hoc* assumptions, as Newton well knew. To explain the universal gravitation of matter, moreover, it would be necessary to surrender a basic premise of the mechanical philosophy by assuming the existence of a material medium that is not convertible into ordinary bodies. By examining motion through resisting media, he further demonstrated to his own satisfaction that the density of the medium, and not the subtlety of its parts, is the major factor in the resistance it offers. Hence the resistance can never be removed. In a medium as dense as the body in motion – that is, in the matter of Descartes' vortex – the body will lose half its motion before it travels a distance equal to twice its diameter. In a medium no denser than our atmosphere, Jupiter, which can be shown to have a density roughly equal to that of water, would lose a tenth of its motion in thirty days. An orbiting planet subject to any resistance whatever must ultimately spiral in to the centre of attraction. Should Cartesians object that planets move with the vortex and not through it, the motions of comets, even the tails of which, tenuous exhalations though they be, are observed to move freely through the heavens at every angle, raise the issue again in a form which Cartesians could not avoid. And finally, the hypothesis of vortices was pressed with the difficulty of Newton's dynamical analysis of their motion. Since the analysis rested on certain assumptions, Cartesians might seek to invoke other assumptions that would yield a vortex compatible with Kepler's third law. What they could never evade, however, were two other conclusions independent of Newton's assumptions. On the one hand, the variation of velocity with distance in the orbit of an individual planet, as demanded by Kepler's second law, could never be reconciled to the variation of velocity with distance among the planets demanded by his third. On the other hand, and in the end most compelling of all, a vortex cannot be a self-sustaining system. In order for it to continue, 'some active principle is required from which the globe may receive continually the same quantity of motion which it is always communicating to the matter of the vortex.'[160] The hypothesis of vortices, he concluded, 'is utterly irreconcilable with astronomical phenomena, and rather serves to perplex than explain the heavenly motions.'[161] *Mutatis mutandis*, the same could be said of the mechanical philosophy.

Rather, the same could be said of the prevailing kinetic mechanical philosophy. In Newton's view, the concept of force carried the prospect, not of the mechanical philosophy's demise, but of its salvation. In his philosophy, particles of matter in motion still produce the observed phenomena of nature. The particles have been endowed now with the further property of attractive and repulsive forces, to be sure, but such forces are never separated from material particles and act only on other material particles. They are subject, moreover, to exact quantitative discussion as the shapes and sizes and even motions of the kinetic philosophy never were. The same forces that emerged from his speculations found their quantitative expression in his dynamics. In the *Principia*, he declared, he derived the force of gravity from celestial phenomena; and from the force derived, he deduced the motions of the planets, of comets, of the moon, and of the seas.

> I wish we could derive the rest of the phenomena of Nature by the same kind of reasoning from mechanical principles, for I am induced by many reasons to suspect that they may all depend upon certain forces by which the particles of bodies, by some causes hitherto unknown, are either mutually impelled towards one another, and cohere in regular figures, or are repelled and recede from one another. These forces being unknown, philosophers have hitherto attempted the search of Nature in vain, but I hope the principles here laid down will afford some light either to this or some truer method of philosophy.[162]

Through the concept of force, Newton effected a synthesis of the two dominant traditions of seventeenth-century science – the Democritean or corpuscular-mechanical, and the Pythagorean or mathematical. On the foundation thus established, the structure of modern science has been raised.

★

The discoveries of one generation, however profound and whatever the labour attending their birth, have a way of becoming commonplace truisms for the following generation, and for their descendants vulgar errors. By the time P. G. Tait published a work suitably titled *Dynamics* in 1895, mechanics had assimilated the concept of force so thoroughly that the immensity of the achievement it embodied was beyond recognition. The conservation of energy now dominated the stage, and beside energy the concept of force was little short of contemptible.

> But in all methods and systems which involve the idea of force there is a leaven of artificiality. The true foundations of the subject, based entirely on experiments of the most extensive and most varied kinds, are to be found in the inertia of matter, and the conservation and transformation of energy. With the help of kinematical ideas, it is easy to base the whole science of dynamics on these principles; and there is no necessity for the introduction of the word 'force' nor of the sense-suggested ideas on which it was originally based.[163]

Sic transit gloria mundi.

NOTES

1. The paper is published in Hall and Hall, *Unpublished Papers*, pp. 170–80, and in Herivel, *Background*, pp. 198–207. If we confine ourselves to the evidence of the MS. itself, which survives without a date (University Library Cambridge, Add. MS. 3958.5, ff. 90–91v.), the suggestion that Huygens' *Horologium* occasioned the paper is irresistible. Not only did Newton demonstrate the isochronism of cycloidal oscillations, but he demonstrated as well that a pendulum swinging between cycloidal cheeks moves in a cycloidal path. Moreover, Newton's correspondence establishes his knowledge of Huygens' work (Newton to Oldenburg, 23 June 1673; *Correspondence*, **1**, 290). Opposed to the internal evidence is the statement of David Gregory, after a visit to Cambridge in 1694, that he saw a MS. written in 1669 in which Newton demonstrated 'the principle of equal times of a pendulum suspended between cycloids, before the publication of Huygens' *Horologium Oscillatorium*' (ibid., **1**, 301). Newton himself must have

supplied the date. Perhaps his word is suspect because he showed Gregory the paper that arrived at the inverse square proportion for the centrifugal force of planets at the same time and as part of the same MS. Since he was probably showing that paper in refutation of Hooke's claims, it could be argued that he let himself be carried away and claimed a priority over Huygens as well. The question of centrifugal force, also set forth in the *Horologium*, was involved in the paper that arrived at the inverse square relation. The problem with this argument is that no one seriously doubts the approximate dating of the paper with the inverse square relation; the hand seems clearly to place it before 1679, the year crucial to Hooke's charge. If we accept Newton's date for the one, why not accept his date for the other? There can be no doubt that in 1669 he was technically equipped to do the paper on the cycloid. I know of no evidence by which the issue can be resolved definitively. I continue to feel that the internal evidence suggests Huygens' stimulus. Nothing in my exposition of the development of Newton's dynamics hinges on the date of this paper, however, and I shall not pretend that the date I accept is conclusively established. Herivel prefers to accept 1669 (p. 192). The Halls take 1673 (pp. 170-80).

2. Herivel, *Background*, p. 203.
3. Hooke to Newton, 9 December 1679; *Correspondence*, **2**, 305-6.
4. Newton to Hooke, 13 December 1679; ibid., **2**, 307. Cf. Newton to Crompton for Flamsteed, *c*. April 1681: In rejecting Flamsteed's notion of a sharply curved orbit for a comet that turns short of the sun, Newton proposed the possibility that the comet could follow a path that embraces the sun, 'the *vis centrifuga* at C [perihelion] overpow'ring the attraction & forcing the Comet there notwithstanding the attraction, to begin to recede from ye Sun' (ibid., **2**, 361). Cf. D. T. Whiteside, 'Newton's Early Thoughts on Planetary Motion: A Fresh Look,' *British Journal for the History of Science*, **2** (1964), 117-37; and 'Before the *Principia*: the Maturing of Newton's Thoughts on Dynamical Astronomy, 1664-1684,' *Journal for the History of Astronomy*, **1** (1970), 5-19.
5. Newton to Hooke, 13 December 1679; ibid., **2**, 307-8.
6. *Add. MS.* 3965.1, ff. 1-3. A somewhat different version of it exists among Locke's papers. Both versions have been published a number of times. The version among Newton's papers appeared most recently in Herivel, *Background*, pp. 246-54, and in Hall and Hall, *Unpublished Papers*, pp. 293-301. The version among Locke's papers was published in Lord King, *The Life and Letters of John Locke*, 2nd ed. (London, 1830 – in the new edition of 1858 the paper is found on pp. 210-16), and recently in Newton's *Correspondence*, **3**, 71-6.

In some ways Newton's testimony to his demonstration in 1679 is puzzling. In a number of memoranda, written after 1715 in connection with the calculus controversy, he said that he composed propositions I and XI of Book I (of the *Principia*) in 1679. (*Add. MS.* 3968.9, ff. 101, 106. Brewster printed the memorandum on f. 106, which is a later version of that on f. 101; David Brewster, *Memoirs of the Life, Writings, and Discoveries of Sir Isaac Newton*, 2 vols. (Edinburgh, 1855), **1**, 471.) DeMoivre's memorandum on Newton, composed after Newton's death and presumably based on information from the last years of his life, gives the same account (*Add. MS.* 4007, f. 706). Proposition I is the demonstration of Kepler's law of areas for motion around any source of centripetal force. Proposition XI demonstrates that in elliptical orbits the centripetal force directed to one focus varies inversely as the square of the distance. Proposition XI does not depend for its demonstration on Proposition I. It does depend on Proposition VI, which establishes a general geometrical expression for centripetal force in any curve. Newton said that he composed Proposition VI, which is quite a different matter from the formula in terms of mass, velocity, and radius for centripetal force in uniform circular motion, in 1684. The paper in question here demonstrates the equivalents of Propositions I and XI, deriving XI from the law of areas in I, just as Newton's memoranda state. Herivel, who originally identified the paper with 1679-80, argues at some length that the paper was composed either at that time or in 1684, basing his position on a careful analysis of the level of Newton's dynamic thought as the paper presents it (*Background*, pp. 108-17). Newton's memoranda support 1679-80. Even stronger support for that date, as I shall indicate in the text, is the evident identity of the procedure he followed in the paper with that of his letter to Hooke of 13 December 1679. Also supporting a date earlier than *De motu* of 1684 are the labelling of the diagrams, which differ from that he established once and for all in 1684, and the absence of the word 'centripetal', which he had coined by the time of the first draft of *De motu*. For a fuller discussion of this paper, cf. Richard S. Westfall, 'A Note on Newton's Demonstration of Motion in Ellipses,' *Archives internationales d'histoire des sciences*, **22** (1969), 51-60.

7. Herivel, *Background*, pp. 246-54.
8. The exchange began with a letter of 15 December 1680 from Flamsteed to James Crompton, an extract of which passed on to Newton (*Correspondence*, **2**, 315). After a number of letters via Crompton, it was concluded with a letter from Newton to Flamsteed on 16 April 1681 (ibid., **2**, 363-7).
9. Newton always found it difficult to part with papers, fearing that they were still imperfect and might, in their imperfection, subject him to ridicule. In the 1670's, delays and excuses preceded every paper on colours. In light of this, I am inclined to doubt that Newton's problem in August 1684 was his inability to find the paper of 1679. I wonder instead whether he was not unwilling to show it before it was thoroughly revised. If Herivel's identification of the paper of 1679-80 is correct, as I have argued it is, its survival among Newton's MSS. supports this hypothesis.
10. Three distinct versions of *De motu* exist. The earliest (*Add. MS.* 3965.7, ff. 55-62; published with a translation in Herivel, *Background*, pp. 257-89) carries the title *De motu corporum in gyrum*. The second (3965.7, ff. 63-70; its content is compared with version I by Herivel, pp. 292-4; it is published in W. W. Rouse Ball, *An Essay on Newton's 'Principia,'* (London, 1893), pp. 35-51) has no title. Version II is identical to the copy in the register of the Royal Society if I may judge, without having examined the copy myself, from its publication by Rouse Ball, who took it from that source. It is said to be identical as well to a copy in the Macclesfield collection which I have also not seen. The third version (3965.7, ff. 40-54; the passages in which it departs from version II are published by Herivel, pp. 294-303) is called *De motu sphaericorum Corporum in fluidis*. There exist also papers

that contain two revisions of version III. The first of these (*Add. MS.* 3965.5a, ff. 23–6; *Background*, pp. 304–14) is entitled *De motu corporum in mediis regulariter cedentibus*), and the second (*Add. MS.* 3965.5, ff. 21–2; *Background*, pp. 315–20) merely *De motu corporum*. The two revisions of version III were preparatory to the so-called *Lectiones de motu* (University Library Cambridge, *Dd.* 9. 46; parts that differ from the corresponding passages in the *Principia* published in *Background*, pp. 321–5), which were not lectures at all but either a vastly expanded version of *De motu*, or a first draft of the *Principia*, or both. They bore the title, *De motu corporum Liber primus*, and were completed by another MS. (*Add. MS.* 3990) with the title *De motu Corporum, Liber secundus* (published in Isaac Newton, *Opera quae exstant omnia*, ed. Samuel Horsely, 5 vols. (London, 1779–85), **3**, 179–242; English translation in the Cajori edition of the *Principia*, pp. 549–626) which was the original version of Book III. Like *De motu*, the so-called *Lectiones* went through at least one major revision which can be traced in the MS.

An extensive literature on *De motu* exists, attempting among other things to identify which (if any) of the three versions was the paper Edward Paget carried to London in November 1684, presumably fulfilling the pledge Newton made to Halley in August. To me, at least, it appears that this paper can only have been version I. As it was originally written, version I had two 'Hypotheses.' In a revision, Newton added a third in the margin and entered a heading 'Hyp. 4' with nothing after it (see the reproduction in Herivel, *Background*, Plate 5 between pp. 292 and 293). Version II is in Halley's hand. When it was written, a heading and place for Hypothesis 4 were left, but the hypothesis itself was added later in a different ink, as were a couple of other minor additions also indicated in version I. I can only explain this by assuming that Halley copied Paget's paper, which letters testify to have been in great demand, and carried it with him on his trip to Cambridge before December 10. There he entered the hypothesis in the place he had left for it. The completed paper must then have been copied into the register of the Royal Society and dated December 10, the day Halley reported the paper to the Society. Version III contains an interesting amendment that deletes a reference to the satellites of Saturn. A letter from Flamsteed on 27 December questioned their existence. Newton's letter of 30 December asked specifically about them. Thus version III must have been completed around the end of December. In a paragraph added in the third version to the end of Theorem 4, Newton raised the question of the mutual action of the planets on each other and said that he was not then in a position to deal with it. The paragraph certainly reminds one of his correspondence with Flamsteed at the same time (cf. his letters of 30 December 1684 and 12 January 1684/5; *Correspondence*, **2**, 407, 413). The second of the two papers that revise version III is on a sheet which contains, on its reverse side, a list of 'Stellarum Longitudines et Latitudines desumptis Nominibus ex Bayero.' This list appears to be related to Flamsteed's letter of 27 December 1684, which compared locations of certain stars with those in Bayer. This is the only evidence of which I know by which one might presume to date the two papers, and primarily on its basis I assume that they date from early 1685.

11. Most of the conceptual difficulties central to Newton's dynamics were solved, as

I shall argue, in the two revisions of version III, which I have dated with a considerable degree of uncertainty to early 1685. The final form of the so-called *Lectiones de motu* is virtually identical in its early parts, which expound the principles of dynamics, with the first edition of the *Principia*. Since the original version of these passages has not been found, it is impossible to state with certainty how much the revision of the *Lectiones* affected Newton's dynamics, but in the light of the revisions of version III of *De motu* it is difficult to imagine that the dynamics could have undergone much further change. When were the *Lectiones* originally composed? Once again it is impossible to say with certainty. The dates later added in the margins, suggesting that the content of the papers was delivered as lectures beginning in October 1684, have no meaning. Newton was accustomed to fulfill the legal obligation to deposit his lectures by sending over any handy papers at a later time. Other supposed lectures have been shown to differ from what was delivered, and there is no reason to think these papers, which have no semblance to lectures, differ from the others. In fact, the lecture numbers run consecutively over a first draft and a later revision of the same material. Herivel has broken the misarranged folios of the *Lectiones* into four groups, α, β_1, β_2, and γ. β_2 is a revision of β_1, a revision that consists primarily of adding what became Sections IV and V of Book I. Newton's correspondence suggests that this was done over the winter of 1685-6, as he sought a method to deal with comets. At the same time apparently, he also revised the first section, what Herivel calls α, inserting an extra twelve folios, as altered numbers on the pages testify. Alterations in headings further suggest that this revision coincided with the decision to divide the material of the original Book I into Books I and II, although it is true that Newton claimed, in a letter to Halley, to have completed Book II in 1685. Elsewhere in the Portsmouth Papers (*Add. MS.* 3965.3, ff. 7-14) is another of the gatherings of eight folios characteristic of the *Lectiones*, with the numbers of the original version in the upper right hand corner. In these folios is found the demonstration of what became Proposition LXXI, that an homogeneous spherical shell, composed of particles that attract with a force decreasing as the square of the distance, attracts bodies external to it with a force inversely proportional to the square of the distance from its centre. Newton's letter to Halley of 20 June 1686 (*Correspondence*, **2**, 435) placed this demonstration in the year 1685, but no more accurately. All in all, it appears likely to me that the first version of the so-called *Lectiones* was composed fairly early in 1685, and that it contained a dynamics that remained thereafter substantially unchanged.

12. Herivel, *Background*, p. 257 and p. 277. I cite the location both of the original Latin and of the translation. It is necessary to add that my translations do not always agree with Herivel's. I had already translated the passages in question before I saw his. Although I have generally published his, and cite their location, I have ventured to introduce some alterations (as in this case), where, no doubt mistakenly, I could not bring myself to relinquish my own phrases.
13. Ibid., p. 259 and p. 279.
14. Ibid., p. 257 and p. 277.
15. Ibid., p. 258 and p. 277.

16. 'Corpora nec medio impediri nec alijs causis externis quin minus viribus insitae et centripetae exquisite cedant.' (*Add. MS.* 3965.7, f. 55.) I believe that Herivel has transcribed this original version incorrectly (p. 274).
17. Herivel, *Background*, p. 257 and p. 277.
18. Ibid., p. 258 and p. 278.
19. Ibid., pp. 258–9 and p. 278. This theorem, together with a couple of others, remained substantially unchanged through all the revisions of *De motu*; hence propositions in Section I of Book I and propositions on motion through resisting media in Book II of the *Principia* contain phrases about a body moving in a straight line *sola vi insita*, even though his dynamics had meanwhile left that concept behind.
20. Ibid., p. 258 and p. 278.
21. Ibid., pp. 295–6 and p. 300. Cf. Theorem 3, which establishes a general expression for the centripetal force holding a body in a curved orbit in terms of the radius vector SP from the centre of force S to any point P arbitrarily chosen on the curve, the line QT that is perpendicular to the radius vector, and the line QR (the subtense in Newton's terminology) (see Fig. 42). In the indefinitely small figure $QRPT$, he argued, the short line QR, which represents the deflection of

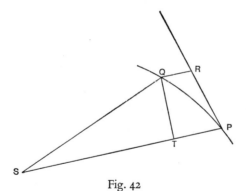

Fig. 42

the body from its rectilinear path, is proportional to the square of time for a given force (ibid., p. 260 and pp. 279–80). Although Lemma 2 employs the phrase 'centripetal force', which Newton was prepared to apply to gravity as a force attracting to the centre of the earth whatever the path of motion, its demonstration deals explicitly only with rectilinear accelerations. The applicability of the same variation of motion with time to accelerations that change the direction of motion depends on the property of the circle implicitly called upon in Theorem 3. Since PR is tangent to the curve, equal increments of distance along it can represent equal increments of time in the uniform motion that a body would follow if no force diverted it. RQ, of course, is the new motion generated by the centripetal force. For a circle, which the segment PQ of the curve approaches in the limit, RQ is proportional to PR^2 for small values of the angle PSQ. Hence the space described at the very beginning of the motion is proportional to the square of the time.

22. Although version I had no diagram of the parallelogram, version III did. 'When it is acted upon by [two] forces simultaneously, a body describes the diagonal of a parallelogram in the same time in which it would describe the sides under the action of the forces separately.' (Ibid., p. 295 and p. 299.) In fact, his use of the parallelogram in the propositions of version I assumed the same thing, that the body describes the rectilinear diagonal.

It may be remarked in passing that this version of Lemma 2 appeared, with only a minor change of wording, as Corollary I to the Laws of Motion in the first edition of the *Principia* (Corpus viribus conjunctis diagonalem parallelogrammi eodem tempore describere, quo latera separatis). Only in the second edition did he eliminate the paradox that such a statement presents as a Corollary to Law I, which states the principle of inertia. In fact he did not alter the wording of the Corollary, but where edition one referred to a body being 'carried' from A to B by force M and from A to C by force N, edition two spoke of the forces being 'impressed' on the body at point A. The parallelogram returned then to its original Galilean form, a parallelogram of motions, in this case of motions set proportional to the (total) forces which generate them. Even in its amended form, it remains a curious artifact that marks the path Newton's dynamics followed in its development.

23. Ibid., p. 260 and pp. 279-80.
24. Ibid., p. 270 and p. 287.
25. Ibid., pp. 258 and 278.
26. *Principia*, p. 13. Except where I explicitly state otherwise, references to the *Principia* are references to Cajori's edition of the English translation, which, for all its faults, is the recognised standard edition.
27. Herivel, *Background*, p. 259 and p. 278.
28. Ibid., p. 306 and pp. 311-12.
29. *Add. MS.* 3965.7, f. 40. The original version read as follows: 'Mutationem motus proportionalem esse vi impressae et fieri secundum lineam rectam qua vis illa imprimitur.' Newton altered it to read: 'Motum genitum vel mutationem motus . . .' A second revision then left it in its final form: 'Mutationem status movendi vel quiescendi proportionalem esse . . .' The last is the form that Herivel has published (p. 294 and p. 299).
30. Ibid., p. 295 and p. 299; p. 308 and p. 313.
31. Ibid., p. 294 and p. 299.
32. Ibid., p. 298 and p. 302.
33. Ibid., p. 305 and p. 310.
34. Hall and Hall, *Unpublished Papers*, p. 128.
35. *Add. MS.* 3965.5a, f. 26. Herivel, *Background*, p. 306 and p. 311.
36. Ibid., p. 306 and p. 311.
37. Hall and Hall, *Unpublished Papers*, p. 129.
38. Herivel, *Background*, p. 307 and p. 313.
39. *Principia*, pp. 10-11.
40. Ibid., p. 12.
41. Ibid., p. 419. Cf. the 'system of the World,' the first version of Book III: Arguing against the Ptolemaic system, Newton stated that the stationary points and

retrogressions of the planets are only apparent and that their 'absolute motions' are nearly uniform. Moreover, since the planets describe equal areas around the sun in equal times, it follows that the sun itself is not being moved by any significant force unless it is one that affects all the planets as well in parallel lines so that the entire system is translated rectilinearly by it. 'Reject that translation of the whole system,' he continued, 'and the sun will be almost quiescent in the centre thereof.' (Ibid., pp. 573–4.) Earlier in the same piece, he had also considered the possibility that the entire solar system is being acted upon by some external force which moves all the bodies in parallel lines. From such a force there would be no relative change among the planets or any sensible effect whatever. 'Let us, therefore, neglect every such force as imaginary and precarious, and of no use in the phenomena of the heavens . . .' (Ibid., p. 558.)

42. Ernst Mach, *The Science of Mechanics*, trans. Thomas J. McCormack, 6th ed. (Lasalle, Illinois, 1960), p. 280.
43. *Principia*, p. 9.
44. Hall and Hall, *Unpublished Papers*, p. 129.
45. Herivel, *Background*, pp. 214–15.
46. Ibid., p. 306 and p. 311.
47. Ibid., pp. 315–16 and p. 318.
48. Ibid., p. 306 and p. 311.
49. Ibid., p. 315 and p. 317.
50. Mach, *Mechanics*, p. 237.
51. Herivel, *Background*, p. 306 and p. 311.
52. Ibid., p. 317 and p. 320.
53. Ibid., p. 315 and p. 318.
54. Somewhat ironically, Samuel Clarke read Leibniz a lesson on the meaning of *vis inertiae* which made the same point: *vis inertiae*, he said, is 'that passive force, not by which (as Mr. Leibnitz from Kepler understands it,) matter resists motion; but by which it equally resists any change from the state 'tis in, either of rest or motion . . . Without this *vis*, the least force would give any velocity to the greatest quantity of matter at rest; and the greatest quantity of matter in any velocity of motion, would be stopped by the least force, without any the least shock at all. So that, properly and indeed, all force in matter either at rest or in motion, all its action and reaction, all impulse and all resistance, is nothing but this *vis inertiae* in different circumstances.' (*Leibniz-Clarke Correspondence*, pp. 111–12.) Newton's role in the Clarke-Leibniz correspondence is well known, and that passage can be taken as a statement of his own conception of the function of *vis inertiae* in a quantitative dynamics.
55. Herivel, *Background*, p. 315 and p. 318. Cf. I. Bernard Cohen, '"Quantum in se est": Newton's Concept of Inertia in Relation to Descartes and Lucretius,' *Notes and Records of the Royal Society of London*, **19** (1964), 131–55.
56. Herivel, *Background*, p. 306 and p. 311.
57. Ibid., p. 317 and p. 320. Herivel has placed this definition in the second revision. It is written on the blank space opposite definitions 12 and 13 of the first revision, and was in my opinion clearly intended as part of that paper when Newton wrote it. He later crossed it out.

58. Ibid., p. 306 and p. 311.
59. *Principia*, p. 13.
60. R. G. A. Dolby, 'A Note on Dijksterhuis' Criticism of Newton's Axiomatization of Mechanics,' *Isis*, **57** (1966), 108–15; I. Bernard Cohen, 'Newton's Second Law and the Concept of Force in the *Principia*,' *Texas Quarterly*, **10** (1967), No. 3, 127–57. Cf. E. J. Dijksterhuis, *The Mechanization of the World Picture*, trans. C. Dikshoorn (Oxford, 1961), pp. 464–77.
61. Herivel, *Background*, p. 306 and p. 311. See Appendix F.
62. Dolby, 'A Note on Dijksterhuis' Criticism,' and Cohen, 'Newton's Second Law.' Cf. note 60.
63. 'Corpus omne tantum pati reactione quantum agit in alterum.' (Herivel, *Background*, p. 307.)
64. See Appendix G.
65. Herivel, *Background*, p. 307 and pp. 312–13.
66. Pemberton, *View of Newton's Philosophy*, Preface.
67. *Add. MS.* 4007, f. 706v; Whiston, *Memoirs*, p. 37.
68. Flamsteed to Crompton for Newton, 15 December 1680; Newton, *Correspondence*, **2**, 315. Flamsteed to Halley, 17 February 1680/1; ibid., **2**, 336–9. Flamsteed to Crompton for Newton, 7 March 1681; ibid., **2**, 348–53.
69. Newton to Crompton for Flamsteed, 28 February 1680/1; ibid., **2**, 340–7. Newton to Flamsteed, 16 April 1681; ibid., **2**, 363–7.
70. Newton to Flamsteed, 16 April 1681; ibid., **2**, 364.
71. Herivel, *Background*, pp. 267–8 and p. 285. Newton to Flamsteed, 12 January 1684/5; *Correspondence*, **2**, 412–13.
72. *Add. MS.* 3965.14, f. 613.
73. Cf. Flamsteed to Crompton for Newton, 7 March 1681: 'Mr. Halley thinkes the Comet to be a body that has lost its principle of gravitation, & yet I perceive would have it attracted by ye sun which I cannot assent to, for then I see no reason why its masse should not dissipate & the atoms composeing it separate themselves & scatter over ye wide aether.' (*Correspondence*, **2**, 351.) In this passage, the comet's gravitation refers to the mutual cohesion of its parts whereby it remains a united body.
74. Herivel, *Background*, pp. 275–6.
75. *Add. MS.* 3965.7, f. 61. After he altered it, the final wording of Problem 5 in version I stated: 'Given that the centripetal force is inversely proportional to the square of the distance from the centre, find the distances that a body describes in falling perpendicularly in given times.' (Herivel, *Background*, p. 269 and p. 286.)
76. Ibid., p. 298 and p. 302.
77. Ibid., p. 270 and p. 286.
78. *Principia*, p. 560.
79. Newton to Halley, 20 June 1686; *Correspondence*, **2**, 435, 437. Cf. Proposition VIII of Book III: 'After I had found that the force of gravity towards a whole planet did arise from and was compounded of the forces of gravity towards all its parts, and towards every one part was in the inverse proportion of the squares of the distances from the part, I was yet in doubt whether that proportion inversely as the square of the distance did accurately hold, or but nearly so, in the

total force compounded of so many partial ones; for it might be that the proportion which accurately enough took place in greater distances should be wide of the truth near the surface of the planet, where the distances of the particles are unequal, and their situation dissimilar.' (*Principia*, pp. 415–16.)

80. Flamsteed to Newton, 27 December 1684; *Correspondence*, **2**, 403–4; Newton to Flamsteed, 30 December 1684; ibid., **2**, 406–7. Flamsteed saw the point of Newton's request, and in his letter of 5 January 1684/5 he stated his opinion that the spheres of action of bodies so small could not possibly extend far enough for them to influence each other. At their closest, Jupiter and Saturn are separated from each other by a distance equal to four radii of the earth's orbit, a distance much greater in proportion to their sizes than one hundred yards in proportion to the size of a magnet, but the spheres of activity of magnets do not extend one hundred yards (ibid., **2**, 408–9). Flamsteed's objection is a fascinating reflection of the conceptual problem with which the idea of universal gravitation confronted the scientific community of the late seventeenth century, even those who were prepared to accept the concept of attraction at a distance.
81. Newton to Flamsteed, 12 January 1684/5; ibid., **2**, 413.
82. Herivel, *Background*, pp. 306, 316–17 and pp. 311, 319. I believe that Herivel is mistaken in placing this correction of the first revision with the second revision.
83. *Principia*, p. 567.
84. Ibid., p. 412.
85. Ibid., p. 414. For a similar discussion of the late development of the concept of universal gravitation, cf. Curtis A. Wilson, 'From Kepler's Laws, So-called, to Universal Gravitation: Empirical Factors,' *Archive for History of Exact Sciences*, **6** (1970), 89–170.
86. Ibid., p. 571.
87. *Add. MS.* 3990, f. 15v. The passage was later crossed out.
88. *Principia*, p. 13.
89. Ibid., p. 186.
90. Ibid., p. 188.
91. Ibid., pp. 188–9.
92. Ibid., p. 13.
93. Ibid., p. 13.
94. Ibid., p. 27.
95. Ibid., p. 154.
96. Ibid., p. 128.
97. Ibid., p. 28.
98. Ibid., pp. 237–9, 252–3. Cf. Corollary 2 of Proposition XXXVIII, Book II, which establishes the terminal velocity that a body can attain in falling through a given medium. If the body has this velocity, 'the force of weight will be equal to the force of resistance, and therefore cannot accelerate the globe' (ibid., p. 352).
99. Ibid., p. 21. The translation has unintentionally contributed to my point. The first line of the Latin original refers, not to 'the uniform force of its gravity,' but merely to 'uniform gravity.' [*Corpore cadente gravitas uniformis, singulis temporis particulis aequalibus aequaliter agendo imprimit vires aequales* . . .] Nevertheless, Newton always treated gravity as an impressed force and referred to it re-

peatedly in those express terms. The essential ambiguity of meaning attached to the words in the translation is in fact present in the passage and indeed throughout the *Principia*.

100. 'Motum in spatio vel immobili vel mobili genitum proportionalem esse vi motrici impressae & fieri secundum lineam rectam qua vis illa imprimitur. [This statement was amended by deletion of the phrase, 'in spatio vel immobili vel mobili.']

'Motum a loco quem corpus alias occuparet, vi motrici impressae proportionalem esse et in plagam ejus dirigi.

'Motum omnem novum quo status corporis mutatur vi motrici impressae proportionalem esse & fieri a loco quem corpus alias [occup]aret in plagam quam vis impressa petit.

'Vis omnis in corpus liberum impressa motum sibi proportionalem a loco quem corpus alias occuparet in plagam propriam generat. [This statement was amended by deletion of the phrase 'in corpus liberum.']

'Vis omnis impressa motum sibi proportionalem g[enerat]

'Motum genitum vi motrici impressae proportionalem esse & a loco quam corpus alias occuparet in plagam vis illius fieri.'

'Vis impressa motum sibi proportionalem a loco quem corpus alias occuparet in plagam propriam generat.

'Mutationem motus quo status corporis mutatur proportionalem esse vi motrici impressae & fieri a loco quem corpus alias occuparet in plagam qua vis imprimitur.' [This statement was amended by deletion of the phrase 'Mutationem motus' and replacement of it by the phrase 'Motum omnem novum.']
(*Add. MS.* 3965.6, ff. 274–74v.) Newton crossed out all of the statements after the third ('Motum omnem novum . . .') and thus appears to have found it the most acceptable at the time he was working on this sheet.

101. *Principia*, p. 188.
102. Ibid., p. 22.
103. Ibid., p. 25.
104. Ibid., p. 412.
105. Ibid., p. 246. Cf. Proposition XXXVII, Book II, in which he demonstrated that the resistance to a cylinder moving parallel to its axis through a medium of the same density as the cylinder will, if it is continued unchanged, destroy the motion of the cylinder in the time it takes to describe four times the length of its axis. If the length of the cylinder is increased or diminished, both its motion and the time in which it describes four times its length will be increased or diminished in the same proportion, 'and therefore the force by which the same motion, so increased or diminished, may be destroyed or generated, will continue the same . . .' (ibid., p. 347).
106. Ibid., pp. 34–5.
107. Ibid., p. 283.
108. The expression is $(rk)^2/2kC$, in which rk is half a chord equal to half its arc when the arc is just nascent, and kC is the radius of the circle in question. He uses the relation to express 'the force with which a body may revolve with a circular motion from R to K . . .' in an equal circle (ibid., p. 138).

109. Ibid., p. 13.
110. Ibid., p. 2.
111. Ibid., p. 41.
112. Ibid., pp. 329–31.
113. Ibid., pp. 331–4.
114. Herivel, *Background*, p. 258 and p. 278.
115. Cf. Proposition I, Book I (*Principia*, p. 40.) and Proposition I, Book II (ibid., p. 235). The same phrase appears in Propositions V and XI, Book II (ibid., pp. 245, 272), which treat motion through media that resist in proportion to different powers of velocity; although neither proposition was in *De motu*, they are analogous to Proposition I, Book II. The phrase was capable, however, of inserting itself insidiously into propositions composed after *De motu* that had no relation to that tract. Cf. Corollary 6, Proposition XLIV, Book I (on the stability or motion of the line of apsides): 'For the body P, by its inertia alone, no other force impelling it, will proceed uniformly in the right line VP.' (Ibid., p. 140.) 'Nam corpus P, per vim inertiae, nulla alia vi urgente, uniformiter progredi potest in recta VP.' (1st ed., p. 137.)
116. *Principia*, pp. 40–2.
117. 'Cas. 2. Eodem argumento si corpus dato tempore vi sola M in loco A impressa ferretur uniformi cum motu ab A ad B & vi sola N non simul & semel sed perpetuo impressa ferretur accelerato cum motu in recta AC ab A ad C compleatur parallelogrammum $ABDC$ & corpus vi utraq; ferretur eodem tempore ab A ad D. Nam reperietur in fine temporis tam in linea CD quam in linea BD et propterea'
    'Ideoq; si vires M & N sedundum lineas AB et AC simul et semel imprimantur, sic ut motus seorsim generarent in lineis istis AB et AC uniformes: corpus dato illo tempore perget ab A ad D in diagonali rectilinea AD uniformi cum motu ex vi utraq; oriundo. Sin vires istae M et N secundum lineas easdem AB et AC imprimantur paulatim & perpetuo sic ut motus seorsim generarent in lineis istis AB et AC vel in earum alterutra acceleratos: corpus perget eodem tempore ab A ad D in diagonali curvilinea AD cum motu accelerato ex vi utraq; oriundo.' (*Add. MS.* 3965.6, f. 86.)
118. 'Si corpus A ante vim impressam movebatur & motu quem habuit in A uniformiter continuato distantiam AB dato tempore describere posset et interea a vi impressa in datam plagam urgeatur: cogitandum erit quod locus in quo corpus relative quiescit movetur una cum corpore ab A ad B quodq; corpus per vim impressam deturbatur de loco hocce mobili et ab eo migrat in plagam vis illius impressae cum motu qui vi eidem proportionalis est. Ideoq; si vis determinatur, verbi gratia, in plagam rectae AC ac dato illo tempore corpus motu omni privatum impellere posset a loco immobili A ad locum C, age BD ipsi AC parallelam et aequalem & vis eadem eodem tempore ex mente Legis hujus impellet corpus idem a loco suo mobili B ad locum novum D. Corpus igitur movebitur in linea aliqua AD cum motu qui oritur ex motu loci sui relativi ab A ad B & motu corporis ab hoc loco ad locum alium D, id est ex motu AB quem corpus ante vim impressam participabat et motu BD quem vis impressa per hanc Legem generat. Ex hisce duobus motibus secundum determinationes suas conjunctis orietur motus corporis [in] linea AD.' (*Add. MS.* 3965.6, f. 274.)

119. In the third edition, Newton added a somewhat similar passage deriving the parabolic trajectory to the Scholium to the laws of motion. It differs from the passage above in being still more purely kinematic (*Principia*, pp. 21–2).
120. Ibid., pp. 154–5.
121. Ibid., p. 236.
122. Ibid., p. 249.
123. Ibid., pp. 252–3.
124. Ibid., p. 255.
125. Ibid., p. 126.
126. Ibid., pp. 312–13.
127. Ibid., p. 409. Cf. Corollary 20, Proposition LXVI, Book I (his construction to explain the tides). Newton applied his general solution of the three body problem, shrinking the orbit of the satellite circling the earth until it coincides with the earth's surface, and replacing the satellite with a ring of water girdling the earth in a channel. The water, he said, 'is no longer sustained and kept in its orbit by its centrifugal force, but by the channel in which it flows.' (Ibid., p. 187.)
128. Ibid., p. 42.
129. Ibid., pp. 226–7.
130. Galileo, *Discourses*, pp. 252–3.
131. *Principia*, p. 352.
132. Ibid., pp. 173–89.
133. Lemma I, Book III, in 1st ed. (pp. 467–9); Lemmas I and II in later eds (*Principia*, pp. 485–8).
134. Ibid., p. 124.
135. Ibid., p. 378.
136. 1st ed., p. 370. University Library Cambridge, *Adv. b.* 39.1 (an interleaved copy of the 1st ed. with additions and emendations on the added sheets), f. 370A. *Principia*, pp. 382–3.
137. Ibid., p. 391.
138. Ibid., pp. 388–9.
139. Ibid., p. 390.
140. Ibid., p. 389.
141. 1st ed., pp. 330–2.
142. *Add. MS.* 3968.9, f. 102v.
143. A correction entered by Newton in his copy of the 1st ed.; quoted in *Correspondence*, **3**, pp. 39–40 fn. Fatio transcribed the passage on 13 March 1689/90.
144. *Add. MS.* 3965.12, f. 205. This sheet contains a revision of the proposition intended for the 2nd ed.
145. *Principia*, p. 338.
146. *Add. MS.* 3965.12, f. 232v.
147. *Principia*, p. 339.
148. Ibid., p. xvii.
149. Ibid., p. 550.
150. Ibid., p. 303.
151. Ibid., pp. 202–3.

152. Ibid., p. 145.
153. Ibid., pp. 214–15.
154. Ibid., pp. 169–71.
155. Ibid., pp. 171–2, 190.
156. Newton to Bentley, 10 December 1692; *Correspondence*, **3**, 233.
157. *Principia*, p. 198.
158. Ibid., pp. 154–5.
159. Ibid., p. 543.
160. Ibid., p. 390.
161. Ibid., p. 396.
162. Ibid., p. xviii.
163. P. G. Tait, *Dynamics* (London, 1895), p. 361.

Appendix A
Galileo's Usage of Force

GALILEO used the word 'force' (*forza*) in a variety of ways. Easily his most consistent usage occurs in discussions of simple machines; the '*forza*' applied to one end of a lever (and we should remember that Galileo reduced all the simple machines to the case of the lever) balances and overcomes the resistance at the other end. (Cf. *Discourses*, pp. 110–15, 137–8; *Mechanics*; *Motion and Mechanics*, trans. I. E. Drabkin and Stillman Drake, (Madison, Wisconsin, 1960), p. 158. Although I quote from the English translations, I have checked these cases and all of those in which I cite the Italian word against the *Opere*.) Since Galileo's analysis of percussion rests on its analogy to the lever, as I shall discuss below, the phrase *forza della percossa* is a natural extension of this usage; the velocity of a moving body multiplies its weight just as levers and screws multiply the '*forza*' applied to them (cf. *Discourses*, p. 271). *Forza* derives from the Latin *fortis*, and when it is used with the lever it suggests the physical effort applied to raise a weight. In other contexts, Galileo used *forza* as a synonym for physical strength. Thus he referred to the '*forza*' of a draught animal turning a capstan (one of the simple machines he analysed) and to simply the '*forza*' of a horse; in drawing wire, workers pull it through the draw-plane with great '*forza*'; a thread held between the fingers does not slip even when pulled with considerable '*forza*' (*Mechanics: Motion and Mechanics*, pp. 161, 150; *Discourses*, pp. 53, 9). An extension of the meaning, analogous to that in English, allowed him to speak of the '*forza*' of a discourse and the '*forza*' of truth (ibid., pp. 165, 164). Sometimes *forza* in the sense of strength took on an abstracted and generalised, though not very precise, meaning. In the *Dialogue*, Galileo solved Tycho's conundrum of the cannon fired east and west on a moving earth by applying the

APPENDIX A

results of a thought experiment with crossbows fired from a moving carriage. He argued that if the strength (not *forza* at this point) of the bow is the same for shots made forward and backward, the arrows will fall equal distances from the carriage; so also with cannon shots east and west on an earth turning from west to east when they are made with the same '*forza*' (*Dialogue*, p. 171). Without any effort at precision, the strength of the bow and the *forza* of a shot from a cannon were set equivalent to each other and given an implicit quantitative definition.

A body in motion can also manifest *forza*, as the reference to the *forza* of percussion has suggested. In the *Dialogue*, Simplicio declared that a man throwing a stone moves his arm with speed and '*forza*' so that the resulting impetus carries the stone along, and Sagredo asked why a hoop thrown with a cord goes farther and consequently with more '*forza*' than one thrown by hand (*Dialogue*, pp. 151, 158). In fragments associated with the so-called sixth day of the *Discourses*, Galileo further explored the idea that the *forza* of the projector is conserved in the projectile. Thus large bells must be set swinging by repeated pulls, each one of which adds '*forza*' to that acquired from the earlier pulls; the larger the bell the more '*forza*' it acquires (*Opere*, **8**, 346, cf. also p. 345). Galileo's analysis of the lever by means of the principle of virtual velocities suggested the connection of *forza* applied to the lever with the *forza* of motion. His word '*momento*', from which both our 'moment' and 'momentum' derive (and the word '*impeto*' which he used interchangeably with *momento*), expressed the *forza* of a body's motion. One of his standard devices to measure *momenti* was to compare *forze* of percussion, and at several places he used *forza* to mean the impulse which can generate a given *momento* (*Discourses*, p. 286, 273–5, cf. *Opere*, **8**, 338). Just to confuse the issue further, he also stated in the *Discourses* that even if the '*forza*' of a moving body is small, its velocity can overcome the great resistance of a slowly moving body when the ratio of velocities is greater than the ratio of the resistance to the '*forza*' – a return to *forza* as the measure of static weight (*Discourses*, p. 291). The concept of *momento* or *impeto* is central to the question of force in Galileo's mechanics; I shall discuss it at some length below.

One other use of *forza* requires comment. In Aristotelian terms, Galileo continued to think of forced or violent motion as the opposite of natural motion. A ball rolls down a slope spontaneously, and '*forza*' is necessary to keep it at rest; in order for it to move up the slope, more '*forza*' is needed (cf. *Dialogue*, pp. 147, 264). The '*forza*' that holds a body outside its natural place, he said in *De motu*, must be equal to its weight (*On Motion: Motion and Mechanics*, p. 98) – a condition like that of equilibrium of a balance. The concept here was similar to the use of *forza* to overcome resistance with a lever, an idea that he continued to use throughout his life. When a rope is 'violated' by enough '*forza*', it breaks (*Discourses*, pp. 121–2). A similar use of *forza* appeared in discussions of projectile motion. Birds in flight were treated as a special case of such motion, and Salviati declared that the citation of birds in flight as an argument against the motion of the earth depended on their being animate and able to use '*forza*' at will against the original inherent motion of terrestrial objects (*Dialogue*, p. 186). Similarly he argued that the motion of a ship is not identical to the diurnal rotation of terrestrial bodies; while the latter is natural, the motion of the ship is accidental, conferred on it by the '*forza*' of the oars (*Dialogue*, p. 142).

Appendix B

Descartes' Usage of Force

THE word *'force'* appears in Descartes' writings in great profusion; probably it was his frequent use which gave it general currency and led to its ultimate selection as the name of the concept central to dynamics. Sometimes he saw it as a definite quantified term; *'force, motion,* and percussion, etc., are kinds of quantities,' he wrote to Mersenne (11 March 1640; *Oeuvres,* **3**, 36). Sometimes the quantity even seemed to measure change of motion, as we expect to find in a dynamics built on discrete impacts. Thus, in a letter to Mersenne, his concept of quantity of motion led him to an erroneous statement that is intelligible only with such a definition of force – 'the largest bodies, being pushed by the same *force,* as the largest boats by the same wind, always move more slowly than the others.' (December 1638; ibid., **2**, 467). 'In general,' he later wrote, 'the larger bodies are, the more slowly they ought to move when they are pushed by the same *force,*' (Descartes to Mersenne, 25 December 1639; ibid., **2**, 627). Sometimes he used *'force'* to mean simply that which will put a body in motion. When a perfect sphere lies on a perfect horizontal plane and there is no air resistance, *'la moindre force'* will move it (Descartes to Mersenne, 3 May 1632; ibid., **1**, 247). A hard body, however large, can be set in motion by *'la moindre force'* when it is surrounded by a fluid (Descartes to Cavendish, 2 November 1646; ibid., **4**, 562). His correspondence with Henry More referred to a *'vis movens'* and a *'vis corpus movendi,'* which, in the context, meant forces that put bodies in motion (August 1649; ibid., **5**, 403–4). He also asserted, however, that a body in motion has the *'force'* to continue its motion, and in rest, the *'force'* to remain at rest (Descartes to Mersenne, 28 October 1642; ibid., **3**, 213).

By far the most common usage of the word was in phrases such as 'the *force* of its [a ball's] motion' (cf. *Dioptrique*; ibid., **6**, 99). In part, the force of a body's motion was measured by its velocity; thus, when water takes away half the speed of a ball that enters it from the air, it takes away as much of its '*force*' (*Dioptrique*; ibid., **6**, 98). Apparently size and solidity also add to the force of a body's motion; large and heavily laden boats have more '*force*' to continue their motion than the river down which they float (*Le monde*; ibid., **11**, 58). Although the similarity of this concept to our concept of momentum gives it a misleading aura of precision, in fact Descartes' use of it was often closer to a vague meaning of strength, reminding us of the original derivation of the word. For example, as glass cools, its particles become interlaced until the air no longer has the '*force*' to maintain their agitation (*Principles*, **iv**, 127; ibid., **9**, 268). With a different sentence pattern, the adjective '*fort*' can replace the noun '*force*'. If the fire is not '*fort*' enough to turn iron into droplets, steel cannot be made (*Principles*, **iv**, 142; ibid., **9**, 277). When air is heated, if the agitation of its particles is '*plus forte*' than the weight of the atmosphere, the particles separate and the air rarefies (Descartes to Reneri, 2 June 1631; ibid., **1**, 207). Descartes could move back and forth from the nominal to the adjectival or adverbial form in passing from one clause or sentence to the next. The third law of motion spoke of a body meeting another either '*plus fort*' or '*plus foible*' than itself, while the discussion following said that the first body '*a moins de force*' or '*a plus de force*' than the one it meets (*Principles*, **II**, 40; ibid., **9**, 86–7). When sound travels through air, air must be moved a certain distance, time is consumed, and the motion therefore loses part of its '*force*'; when it travels the same distance down a solid beam, both ends move at the same time, and it is heard '*plus fort*' (Descartes to Mersenne, 3 May 1632; ibid., **1**, 246). With a vice such that the '*force*' travels one hundred feet while the load travels one, a man can press '*aussy fort*' as a hundred men can press without the vice (Descartes to Huygens, 5 October 1637; ibid., **1**, 441–2). Equally the antonym of '*fort*' could be used; sometimes a flame is '*si foible et si debile*' that it does not have the '*force*' to kindle the wick (*Principles*, **IV**, 116; ibid., **9**, 263). In the discussion of the magnet all of these usages were stirred together into a rare and wonderful hash. Steel receives a magnetic virtue '*plus forte*' than that which iron can receive; the virtue of the earth is not '*si forte*' as that of most loadstones. The '*force*' of a magnet can be destroyed by fire or by placing it for a long time in opposition to the earth's field. When a magnet is not lined up with the

APPENDIX B

earth's field, the channelled pieces, by the '*force*' they have to continue their motion in a straight line, push it into alignment, unless it is held by other bodies '*plus forts.*' The sphere of activity of a long magnet extends further than that of a short one because the channelled pieces in passing through its pores have time to acquire the '*force*' to move further through the air in a straight line; hence the virtue of long magnets extends further even though the virtue may be '*plus foible.*' The '*force*' of an armature consists solely in the way it touches pieces of iron the magnet attracts. The '*force*' of a loadstone to sustain a weight of iron is increased by the presence of another magnet. The more numerous and the more agitated are the channelled pieces that pass through a magnet, the more '*force*' it has. Thus he can speak of the '*plus fort*' magnet and the '*plus foible*' (*Principles*, **IV**, 145–79; ibid., **9**, 281–303). Descartes' use of '*force*' with the meaning of strength shaded imperceptibly from the seemingly precise to the wholly analogous. In the case of pendulums, he spoke of the '*force*' of their agitation (Descartes to Cavendish, 30 March 1646; ibid., **4**, 384). He discussed the '*force*' with which water flows from a hole, (Descartes to Huygens, 18 February 1643; ibid., **3**, 627) the '*force*' that a blow has, (Descartes to Mersenne, 11 March 1640; ibid., **3**, 42) the '*force*' of a spring (Descartes to Mersenne, 9 February 1639; ibid., **2**, 505). Heated air bursts its container with great '*force*' (*Principles*, **IV**, 47; *Ibid.*, **9**, 226). Even light has '*force*' (*Principles*, **III**, 63, 135; ibid., **9**, 135, 187). One can even speak of the '*force*' of a medicine (Descartes to Wilhelm, 24 June 1640; ibid., **3**, 93).

Closely allied to the concept of the force of a body's motion was a use of '*force*' to mean virtually what we call energy. In discussing salt water, Descartes described the particles of salt as straight rods around which the pliable particles of water are rolled. The '*force*' of the subtle matter which agitates the particles of water is employed solely in making them turn around the saline particles and move from one to another. In the case of fresh water, however, in which the pliable particles are interlaced like worms in a can, the '*force*' of the subtle matter must be employed as well in bending them and disengaging them from each other and hence cannot make them move so easily (*Météores*; ibid., **6**, 251–2).

Quite another use of 'force' appears in Descartes' discussions of problems in statics. Here the same tradition from which Galileo drew had prepared a clear conception of static force as that which is applied, often to a simple machine, in overcoming a resistance. In analysing the inclined plane, he spoke of the '*force*' necessary to hold a body in equilibrium on it (Descartes to Mersenne, 18 November 1640; ibid., **3**, 245). Of a given device to raise water quickly, he said that one loses as much '*force*' as one gains in time (Descartes to Mersenne, 29 January 1640; ibid., **3**, 13). He also spoke of the '*forces*' necessary to break different cylinders projecting from walls (Descartes to Mersenne, December 1638; ibid., **2**, 465). Somewhat analogously, since circular motion continued to be seen as a form of equilibrium, he could speak of the '*force*' of a body to move away from the centre of a circular motion (*Le monde*: ibid., **11**, 76–7. *Principles*, **III**, 60, 83; ibid., **9**, 133, 149). The opposite was the '*force*' with which a body tends to descend (Descartes to Mersenne, 13 July 1638; ibid., **2**, 226), which, of course, is identical to its weight in static problems. As with Galileo, he could employ the word '*force*' to indicate the resistance to be overcome. In the case of a beam projecting horizontally from a wall, he tried to calculate the '*force*' with which it is attached and the '*force à resister*' of the wall to the pressure on it (Descartes to Mersenne, September 1647; ibid., **5**, 74–6). In this context, he defined the absolute weight of a body as 'the *force* with which it tends to descend in a straight line when it is in our ordinary air at a certain distance from the centre of the earth, being neither pushed nor supported by another body, and finally having not yet begun to move' (Descartes to Mersenne, 13 July 1638; ibid., **2**, 226–7). In the seventeenth century, the concept of static force tended always to creep into dynamic contexts where it was likely to confuse, and Descartes fell into the same trap when he commented on the '*force*' necessary to distort the shape of a ball of lead, where impact rather than pressure was employed (Descartes to Mersenne, 9 January 1639; ibid., **2**, 483). Further confusion arose from reference to the '*force*' of a lever and to the '*force*' of a pulley, where he could only have meant their mechanical advantages (Descartes to Mersenne, September 1647 and 30 September 1640; ibid., **5**, 74 and **3**, 186). Finally, Descartes' analysis of simple machines also made use of '*force*' as a concept explicitly meaning what we now call work. To Mersenne, he wrote that by '*force*' he did not mean 'the power called the *force* of a man when we say that such a man has more *force* than another' (15 November 1638; ibid., **2**, 432). Unfortunately, he did frequently use '*force*' with exactly that meaning, as the examples above illustrate. To his credit, he distinguished explicitly between one dimensional force (our static force) and two dimensional force (our work), but his use of the same word for both was a constant invitation to confusion.

Beyond the usages above, Descartes also employed '*force*' in a vague, intuitive way to signify any power. He denied that the sun has '*quelque force*' by which it attracts vapours causing them to rise (*Météores*; ibid., **6**, 239). With Mersenne, he discussed where the greatest '*force*' of a sword is, at its point or its centre of gravity or elsewhere. (15 September 1640 and 26 April 1643; ibid., **3**, 180, 658.) To More, he wrote that he understood souls as 'certain virtues or forces' (*virtutes aut vires quasdam*) which apply themselves to material things but are not extended (5 February 1649; ibid., **5**, 270).

Now and then, he also fell into the old usage that connected force with violence opposed to the natural order, even though his philosophy did not allow such a notion. In explaining elasticity, he said that the pores of a distorted spring have '*une figure forcée*' (Descartes to Mersenne, 29 January 1640; ibid., **3**, 8–9). Again, he mentioned the '*force ou violence*' of the motion of aqueous particles in a vapour (Descartes to Mersenne, 19 January 1642; ibid., **3**, 482) and said of a cannon ball that it moves with the '*force*' impressed on it and is pushed '*avec grande violence*' (Descartes to Mersenne, 17 November 1642; ibid., **3**, 592).

It should be evident that the word '*force*', for all its frequent use, had not yet acquired a precise definition except perhaps when it was used in problems of statics. The chaos of incompatible meanings which attached to the word can be seen, among other places, in Descartes' letter to Mersenne of 25 December 1639. He discussed why a spring loses its '*force*' when it is stretched for a long time. Taking up percussion, he asserted that larger bodies go more slowly when pushed by the same '*force*.' It is not true that the parts of the body are not moved when pressed '*assez fort*' under water; the whole body can be compressed and thus moved. When a man jumps, he rises into the air because the '*force*' with which he presses the earth with his feet is reflected. The '*force*' of percussion depends on the velocity of motion. A hammer of 100 pounds with a velocity of 1 unit presses an anvil with the '*force*' which that velocity gives to 100 pounds; a hammer of 1 pound with 100 degrees of velocity presses the anvil '*aussy fort.*' A hammer swung by a hand may have ten thousand times '*plus de force*' than one at rest. Archimedes' screw is the best instrument to raise water; in the pump there is too much '*force*' lost. Not only the subtle matter in our bodies, but the animal spirits as well give '*force*' to our movements (ibid., **2**, 626–35. Cf. Descartes to Mersenne, 11 March 1640; ibid., **3**, 37. Descartes to Mersenne, 23 February 1643; ibid., **3**, 635. *Principles* **III**, 114–16; ibid., **9**, 169–70).

Appendix C
Gassendi's Usage of Force

GASSENDI'S use of the word 'force' (usually *vis* – Charleton translated '*vis*' by 'force' with great consistency, and further inserted the word 'force' into discussions from which Gassendi had omitted '*vis*'), as the usage of Galileo and Descartes, reveals the lack of precision associated with the term. His most frequent use was analogous to Descartes' 'force of a body's motion'. He called the gravity of atoms their 'native force [*ipsa nativa Atomorum vis*]' and their 'motive force [*motrix sua vis*]'. (*Syntagma philosophicum*; *Opera*, I, 343.) In considering the problem of cannons firing east and west on a turning earth, he spoke of them firing their balls 'with equal forces [*paribus viribus*]'; that shot from the west has an 'added force from the earth [*superaddita a Terra vis*]'. In a similar situation on a moving ship, cannons at the prow and the poop impress 'equal force [*par vis*]' on two balls. (*Institutio astronomica iuxta hypothesis tam veterum, quam Copernici, et Tychonis*, (Paris, 1647), pp. 199–200.) His derivation of the parabolic trajectory of a projectile used 'force' rather than motion. On his diagram of a ball thrown straight up from a moving ship, *B* is the summit of the parabola *GBH*, and *GR* is a vertical line reaching to the same height. Do not imagine that a greater 'force [*virtus*]' is required on the part of the thrower to send the ball along *GB* than along *GR*. True, if the ship were at rest, the thrower would have to apply greater 'force [*virtus*]'. When the ship moves, however, what is lacking from the 'force [*virtus*]' applied by the projector is supplied by the 'force of motion added by the ship [*vis translatitia adiecta a navi*]'. As the motion of the ball is composed of the forward motion of the ship and the upward motion from the thrower, so also the 'force of projection [*vis proiectionis*]' is composed from the 'force due to the projector [*vis proicientis propria*]' and the 'force impressed by the ship [*vis impressa a navi*]'. If a stone is dropped from the top of the mast, no 'force [*vis*]' from the projector is required; the perpendicular component is supplied by the gravity of the stone, the horizontal by the 'force of motion' of the ship [*a vi, seu a motu ipsius navi*]. Moreover, neither of these 'forces [*neuter harum virium*]', whether of the projector upward or of gravity downward, destroys or diminishes the other, horizontal one. The stone does not tend upward or forward any less than two separate stones projected in those two directions 'by separated forces [*viribus separatis*]'. (*Epistolae tres. De motu impresso: Opera*, **3**, 484.) Two other concepts of force have crept into this passage; implicitly Gassendi has set the force of projection (Δmv) equal to the force of the projectile's motion (mv), and he has further substituted the force of gravity (perhaps ma) for the force of the projector. (Cf. also *Epistolae tres. De motu impresso*; ibid., **3**, 498; and *Syntagma philosophicum*; ibid., **1**, 388.) In so far as 'force' referred to the force of motion, he also used the word '*impetus*' as a synonym. (Cf. *Epsitolae tres. De motu impresso*; ibid., **3**, 488–99.) Although he denied that anything except motion is impressed, he continued to use the word 'force' in this context. (Cf. *Syntagma philosophicum*; ibid., **1**, 354.) Examples of similar usage in Charleton are legion; cf. *Physiologia*, pp. 199, 279–80.

As with other seventeenth century writers, Gassendi also commanded a concept of static force. For example, his discussion of the barometer employed the concept of pressure. The air above the surface of mercury is compressed by 'external Force' due to the air above it on the one hand and the gravity of the mercury on the other. If the external force is removed, the particles of air recover their natural contexture 'with united forces.' (I use Charleton's passage; ibid., pp. 55–6.)

'Force' applied as well to attractions, such as gravity. Thus Gassendi said that the earth, as every body, has an 'innate force [*insita vis*]' of conserving itself by which it resists the separation of any part. (*Epistolae tres. De motu impresso: Opera*, **3**, 491.)

The association of 'force' with violence still continued. Charleton cited the definition of a poison, '*quod in corpus ingressum, vim infert Naturae, illamque vincit*,' which he translated as 'That which being admitted into the body, offers violence to Nature, and conquers it.' (*Physiologia*, p. 377.)

As in the case of others, the word 'force' frequently took on a meaning merely of strength. When Charleton stated that the 'force' of a loadstone is sufficient to overcome the gravity of the iron it picks up, the word hovered between the precise and the figurative (ibid., p. 395). When he said that magnetic effluvia become 'less vigorous' at a distance, and at a great distance are 'languid and of no force at all . . .' (ibid., p. 283), he was using the word only as a synonym for 'strength'. With a similar meaning he spoke of the 'force' of fire (ibid., p. 429). Wholly figurative was its use in the statement that by the 'force of Nature [*vis Naturae*]' two bodies cannot be in the same place at the same time (*Syntagma philosophicum, Opera*, **I**, 392).

Appendix D
Huygens' Usage of Force

In the end, Newton resolved the question of terminology by imposing a precise definition (or better, an apparently precise definition) on the word 'force' (or '*vis*'). As we shall see, Newton also did not wholly escape the ambiguity of terms. Meanwhile, Huygens' use of the word 'force' ('*force*' in French and '*vis*' in Latin) suggests the wisdom of his move to import into dynamics a new term that was not compromised by conflicting usages. It is proper to stress that Huygens' usage achieved a consistency well beyond that realised by any earlier student of mechanics. In the great majority of cases, the word 'force' clearly referred to a static context consistent with the established terminology of the simple machines – for example, the 'force of gravity' [*vis gravitatis*], referring to the tendency of bodies to descend, a tendency measured by their weight. This usage corresponded to the minimal dynamic assumption behind his kinematics of descent. It was not his only use of 'force', however, and the fact that a man who endeavoured so consistently to deal in kinematic terms was unable to avoid repeating most of the standard contradictory usages is further testimony to the terminological problem in seventeenth-century dynamics.

One of the common uses of 'force' during the century – one which raised no serious problem beyond a vague fuzziness of meaning similar to that attached to analogical uses of 'force' that exist in English today – recalled the origin of the word in the concept of strength applied to a lever to move a weight. In discussing the motion of the pendulum of a clock, for example, he said that it is maintained 'by the force of the wheels pulled by the weight . . .' (*Oeuvres*, **18**, 95.) In his clock with a conical pendulum, the bob describes circles that are greater or less according as the axis is turned by the wheels 'with a greater or lesser force...' (Ibid., **18**, 363.) In a similar vein, he spoke of the 'force' of gunpowder and the 'force' of springs which derives from their elasticity (ibid., **21**, 461). In these cases, 'force' is less a precise quantitative term than a synonym for strength. The same meaning in its original context appeared in a paper that laid out a programme for the systematic treatment of mechanics. He proposed to start with 'moving forces,' which are those of animals (such as horses or men), weights, water, wind, or springs. Here we can see both the original image of strength applied to some instrument or machine to move heavy bodies, and the use of weight (as Huygens explicitly stated) to provide an exact quantitative measure of strength – the movement of thought from 'force' as strength to 'force' as an exact scientific concept (ibid., **19**, 27).

The idea of 'moving forces' introduced one of the prime sources of ambiguity in the concept of force. The simple machines existed to move heavy bodies, but the analysis of them applied to the conditions of equilibrium. They were further complicated by the issue of mechanical advantage which tended to be confused with force. In a paper read to the *Académie royale*, Huygens stated that knowledge of 'moving forces' is useful in the construction of mills. For example, if the quantity and speed of the water available is known, we can calculate the 'force' of horses or men to which the mill will be equal, and in the case of windmills we can calculate the size of the sails necessary to make their effect equal to a given 'force.' In a problem with a windmill, he then proceeded to calculate that its 'force' was equivalent to a weight of 1646 pounds on a lever arm of 16 feet. In the context, there was no room for confusion, but the word 'force' was applied to a concept we have distinguished as moment (ibid., **19**, 140–1). Another manuscript on the simple machines (which used the word *'potentia'* as well as the word *'vis'*) stressed that a 'force' of one pound can move one hundred pounds if it moves one hundred times as far. So far so good – but a question in the margin beside the analysis of the block and tackle asked 'How much force is required to impress a certain speed on the body' (ibid., **19**, 31–3).

The simple machines helped to maintain the confusion of force and motion even among those who had left Aristotelian mechanics far behind. In discussing the physical pendulum, Huygens spoke of bodies with a certain motion 'by the force of which' they can ascend a certain distance. The pendulum itself ascends from the lowest point in its swing 'by the force of its motion' (ibid., **16**, 417). Liquids that flow from a hole in a vessel 'have the force' to rise again to the level of the surface of the liquid in the container (ibid., **19**, 166–7). Here was the concept of the 'force of a body's motion', which to most students of mechanics in the seventeenth century meant a quantity analogous to mv. Whether or not Huygens had a precise quantity in mind when he wrote the passages above, his analysis of impact and of pendulums left no doubt that the force of a body's motion could only be measured by mv^2. For the most part, his applications of the word 'force' to this quantity derive from the later years of his life. In the year 1690, he referred to 'the law . . . that bodies conserve the force which causes their centre of gravity to rise to the height from which it descended' (ibid., **9**, 456). A remarkable paper from the same period analysed the two motions, linear and rotational, that a ball acquires in rolling down an inclined plane and stated that its 'forces' are divided into two equal parts (ibid., **18**, 433–6). The quantity mv^2 was always associated with a product of weight times height, to which 'force' was also sometimes set equal. In analysing the resistance of water, Huygens said that the 'force' of water in regard to its ability to move other bodies must be found – 'that is to say, how much weight a certain amount of water running with a certain speed can raise a certain height in a given time.' In our terms, force was to be measured here in terms of work. Before the page was completed, he was measuring the 'force' of the same water in terms of weight alone (ibid., **19**, 122). In the passage quoted above (as fn. 48), he stated that in all movements 'no force is lost or disappears without producing a subsequent effect for the production of which the same amount of force is needed as that which has been lost. By force I mean the power of raising a weight.' (Ibid., **18**, 477.) Here force (as mv^2) was set directly proportional to the work (in our sense of the word) it can perform. At least once, he spoke of the 'force' required to impress a certain velocity on a body – a usage that can only be measured by Δmv (ibid., **19**, 482). The same measure of force was implicit in the concept of *vis collisionis* employed in his early papers on impact. As I have indicated already, Huygens' usage of 'force' was more consistent than that of earlier physicists, and it is somewhat misleading to single out examples of conflicting usage. Nevertheless, the examples illustrate the continued presence of ambiguous terminology and the problems it entailed.

Appendix E
Borelli's Dynamical Terminology

IT is equally true that the chaos of terminology nowhere reached a higher peak than it did in Borelli's work. He offers a striking example of the confusion inherent in the ready application of intuitive concepts of force to problems in quantitative dynamics. His confusion of terminology, in which the entire range of available terms was applied equably and incontinently to the entire range of intuitive concepts, compounded his confusion of concepts. No other example makes the role of conceptual definition and terminological precision in the history of mechanics so clear.

A passage in which he discussed projectile motion epitomises the chaos. He considered the possibility that a projectile set in motion acquires a 'force and motive faculty' [*vis, & facultas motiva*] or a 'cause and motive force [*causa, & vis motiva*]' by which the motion continues. In disproving the theory that antiperistasis continues the motion, he declared that the 'moment and force of resistance [*momentum, & vis resistentiae*]' of the fluid that must be moved out of the way is composed of the 'force of density [*vis densitatis*]' of the fluid and the 'force of velocity [*vis velocitatis*]' with which it must be moved. In the same way the 'moment of the virtue [*momentum virtutis*]' with which the fluid fills the space behind the projectile is composed of its density and of the velocity with which it must move. Since the fluid and its density are the same in both cases and the velocities of the 'same strength [*idem robur*],' the two moments are equal, and therefore the resistance of the fluid to being moved out of the way is equal to the 'returning force of the fluid [*vis recursus fluidi*]' in filling the space behind. In the continuing discussion, '*vis*' and '*facultas*' are used interchangeably with 'moment' of resistance and return, and with 'moment of impulse [*momentum*

541

impulsus],' 'impulsive force [*vis impulsiva*],' 'impulsive virtue [*impulsiva virtus*],' and 'moment of impelling virtue [*momentum virtutis impellentis*].' The same discussion contains reference to the 'force and energy [*vis, & energia*]' by which compressed air seeks to expand and the 'force [*vis*]' which rarefied air exercises in being compressed. (*De vis percussionis*, pp. 9–18.) That is, 'force' (*vis, facultas, momentum, virtus, energia*) might refer to any quantity associated with some apparent activity.

Borelli's most common dynamic term was 'motive force [*vis motiva*].' Needless to say, its usages were legion. Particularly (though not solely) in *De motu animalium*, it referred to static force (despite the adjective 'motive') and especially to the context of the simple machines. Since the lever was the basic concept of *De motu animalium*, it is not surprising that 'motive force' appeared consistently on one end of the lever, moving a 'resistance' on the other – the established terminology of the simple machines. (*De motu animalium*, **1**, 10–34, *passim*.) Any number of other terms replaced 'motive force' in the same context without evident pause or embarrassment – 'force' [*vis*], 'contractive force' [*vis contractiva*], 'virtue' [*virtus*], 'motive virtue' [*virtus motiva*], 'power' [*potentia*], 'energy' [*energia*], 'strength' [*robur, validitas*], and various combinations such as 'force of a power' [*vis potentiae*] (ibid., **1**, 10–113, *passim*.) Following again a well established pattern, Borelli saw that weight could provide a quantitative measure of static force and that in any static problem a weight on a line can replace a force of another kind. Thus one finds phrases such as 'downward motive force in heavy bodies [*vis motiva deorsum in gravibus corporibus*]' (ibid., **2**, 9) and 'force or conatus of heaviness [*vis seu conatus gravitatis*].' (*De motionibus naturalibus*, p. 2.) Clear as the idea of static force seemed to be, it kept getting confused with other quantities. In comparing the strength of muscles, he concluded from his analysis of them that the weight a muscle can support is proportional to its cross-section; the height it can lift a weight, however, is proportional to its length; and its 'power [*potentia*]' is therefore proportional to the product of its cross-section and its length. (*De motu animalium*, **1**, 210–11.) An analysis of the bobbing motion of a log in water used 'force [*vis*]' to refer both to the resultant force on the log (its weight minus the upward thrust of the displaced water) and to the integral of the resultant forces over a period of time. (*Mediceorum*, pp. 70–1.) From the examples above, it is clear that 'force' in Borelli's usage was more an intuitive term, which referred to any apparent source of activity, than a precise technical one.

APPENDIX E

Much of the time it is virtually impossible to distinguish it from 'strength'; which would be an acceptable translation of many of his Latin words, including '*vis*' of course. Thus he said that 'great force [*magna vis*]' is required for drops of water to insinuate themselves among the fibres of a rope. (*De motu animalium*, **2**, 48–9.) Muscles contract with 'immense force [*ingens vis*]' because the particles of cuneiform shape that insert themselves in the muscular fibres expand the space by a 'percussive force [*vis percussiva*],' an action which is similar to the 'force and action of a wedge [*vis, & actio cunei*].' (Ibid., **2**, 29–30.) In a similar vein, the preface to the same work set forth two aims: first he would reveal the fabric of the muscles and show by how much 'motive force [*vis motiva*]' and by what mechanical organs the parts of animals are moved; second, he would examine the 'motive force [*vis motiva*]' diffused through the nerves, by which the muscles are moved (ibid., **1**, Preface).

Frequently, 'motive force' carried another meaning which is distinct though often difficult to distinguish – a meaning virtually identical to the Medieval concept of impetus. A body in motion is not 'inert [*iners*];' it is endowed with a 'motive power [*potentia motiva*]' or 'motive virtue [*virtus motiva*],' or 'motive force and energy [*vis, & energia motiva*].' (*Mediceorum*, pp. 57–61.) When a piece of wood rises in water, it acquires an impetus, and 'the persevering impetus necessarily produces its effect of velocity . . . [*impetus perseverans ex sui natura suum effectum velocitatis producit . . .*]' (*De motionibus naturalibus*, p. 303.) A light pendulum stops swinging more quickly than a heavy one because it is carried by a smaller 'motive force [*vis motiva*].' (*De motionibus naturalibus*, pp. 273–4.) When two unequal bodies are impelled by equal 'motive forces [*vires motivae*],' their velocities are reciprocally as their sizes. (*De vi percussionis*, p. 40.) Birds stay aloft in the air, not because they have lost their 'force of gravity [*vis gravitatis*],' but because of the 'force and acquired impetus [*vis, & impetus acquisitus*]' of their motion. (*De motu animalium*, **1**, 320–1.) Again, an indefinable chaos of usage existed, however. On occasion, 'force' appears to refer primarily to velocity. Thus the 'motive force [*vis motiva*]' of a body (more than just velocity here) is composed 'from its vehement or intensive force [*ex vehementia seu intensiva eius vi*]' and its quantity of matter, so that every particle is moved with equal velocity by the 'same impetus [*idem impetus*]' (*De vi percussionis*, p. 196). In the very common case of a lever, the same 'power [*potentia*]' requires 'less force [*minus vis*]' to move a resistance that is closer to the fulcrum, and in the case both of the power and of the resistance we must multiply by the velocity to get the 'moment or energy [*momentum, sive energia*]' by which it acts (*Mediceorum*, pp. 63–4). Also, he referred to the 'energy composed from the motive power P, and the velocity AB . . . [*energia composita ex potentia motiva P, & velocitate AB . . .*]' of a body, in a context in which P refers to its size (*Mediceorum*, p. 59). Just to be sure that no possibility was left untried, he also asserted that a 'force not only weak but endowed with an extremely slow motion [*vis nedum debilis, sed tardissimo motu praedita*]' can impress a huge speed, even a thousand times greater than its own, on a body (ibid., p. 62). It is already clear that the usual range of terms could be applied to 'force' in its general sense of impetus. Obviously, '*impetus*' itself was the most common word. Frequently, he apparently did not understand 'impetus' to differ in any way from velocity as such, and the words '*velocitas*' or '*celeritas*' or even '*motus*' could substitute for it, and the word '*momentum*', alone or in phrases (e.g., *momentum potentiae*), expressed the same concept.

Another prominent concept of force was associated with the phrase, 'force of percussion', which also appeared in a considerable range of synonymous terms. (For example, cf. *De vi percussionis*, pp. 63–78.) Most frequently, Borelli did not distinguish the concept from force as impetus, so that it was virtually identical to our momentum. On occasion, however, he did wish to distinguish the two, and in such places the 'force of percussion' was implicitly set proportional to the change of motion the impact causes, Δmv. Thus, on one occasion, he imagined a ship at rest with a spring attached to it. When the spring is released, it impels the ship with 'contrary motions [*contrariis motibus*]' caused by 'equal force [*equali vi*],' so that the ship continues at rest (ibid., p. 167). A similar use in a different context found him setting the 'motive forces [*vires motivae*]' by which equal bodies are projected upward proportional to the square roots of the distances they can rise, that is, proportional to Δmv (*De motu animalium*, **1**, 277).

Finally, it is worthy of note that in certain contexts at least, especially when associated with an external and therefore unnatural action, force continued to bear the connotation of violence. When drops of water insinuate themselves into the pores of wood causing it to swell, they must overcome a large resistance. Therefore they must 'have the force to exercise such great violence . . . [*vim habeant exercendi tam grandem violentiam . . .*]' (*De motu animalium*, **2**, 49). When air is in its natural state 'with all violence removed from it [*remota omni violentia*]' it occupies two thousand times as much space as when it is 'in its state of maximum violent constriction [*in statu maximae ejus violente constrictionis*],' and if it is freed from the latter state it spontaneously restores itself 'to its pristine native rarity . . . [*ad pristinam nativam raritatem . . .*" (*De motionibus naturalibus*, p. 161).

Appendix F

Newton's Usage of Impressed Force

Herivel, *Background*, p. 306 and p. 311.

I AM unable to see that Newton intended the phrase 'impressed force' to be a precise term in rational mechanics. Certainly his usages of the adjective 'impressed' and of the associated verb do not suggest that he had such an end in mind. A paper from the early '90's revising the laws of motion contains two successive drafts of the corollary on the parallelogram. In the first, a body is carried in a given time '*vi sola* M *in loco* A *impressa*' from *A* to *B* with a uniform motion, and '*vi sola* N *non simul & semel sed pertetuo impressa*' from *A* to *C* with an accelerated motion. In the second draft he shifted to the verb, and the forces *M* and *N* '*simul et semel imprimantur* . . .' (*Add. MS.* 3965.6, f. 86.) He was prepared to speak of motion being impressed as well as force. Thus equal bodies moving with equal velocities through a medium of given density will strike an equal quantity of matter in an equal time whatever the subtlety of the particles of the medium and therefore 'impress [*imprimant*] on that matter an equal quantity of motion . . .' (*Principia*, p. 331). If motion can be impressed, there can also be 'impressed motion,' (ibid., p. 584) a phrase which was present already in the paper of 1679 that connected ellipses with the inverse square law (Herivel, *Background*, p. 246). Finally, the same stem could yield a noun, 'impression', which he also employed. A sphere is 'perfectly indifferent to the receiving of all impressions [*impressiones*].' When it is surrounded by a ring of matter and the ring endeavours to diminish its inclination, the endeavour 'impresses a motion [*motum imprimit*]' on the globe, and the globe 'retains this motion impressed [*motum impressum*]' (*Principia*, p. 186). His analysis of vortices also spoke of 'impressions [*impressiones*]' made on the two surfaces of a laminar shell (ibid., p. 385). With the exception of the last case, in which it is difficult to translate the word 'impression' into anything other than 'force' as we use it today, it might be possible to find a common quantity (of impressed force or of motion generated by it) in all of the passages above. The question, however, is whether he understood the specific phrase 'impressed force' as a precise term in a quantitative dynamics, and in my opinion the range of usage indicated above is impossible to reconcile with that contention.

Appendix G

Newton's Usage of Action

NEWTON's characteristic use of 'action' recalls his phrase 'active principle' far more than it suggests a precisely defined term in a quantitative mechanics. Newton stated, for example, that he was considering forces mathematically and not physically, and therefore the reader should not imagine that when he used the words attraction, impulse, or propensity towards a centre, he was trying to define 'the kind, or the manner of any action, the causes or the physical reason thereof,' or that he attributed 'forces, in a true and physical sense,' to any centre (*Principia*, pp. 5–6). In comparing the attractive forces of the sun and the earth on the moon, he concluded that the 'action of the sun, attracting the moon from the earth' is to the 'force' exerted by the earth as 1/179 (ibid., p. 407). The planets would move in perfect ellipses according to Kepler's laws if they did not act [*agerent*] on each other. The 'actions' on each other are so very small, however, that they can be neglected, except in the case of the 'action' of Jupiter on Saturn (ibid., pp. 420–2). Although 'the mutual actions' of two planets may be distinguished and considered as two, there is only one operation. 'It is not one action by which the sun attracts Jupiter and another by which Jupiter attracts the sun; but it is one action by which the sun and Jupiter mutually endeavour to approach each other.' (Ibid., p. 569.) In a similar way, Newton denied that light could consist in 'action alone' since it does not spread into a shadow (ibid., p. 382). As one example cited above indicates, the verb '*agere*' could serve quite as well in similar contexts, reinforcing the conclusion that Newton's intent was to emphasise the activity. For example, he said that when a body falls, its gravity 'acting [*agendo*] equally' impresses equal forces on the body in equal times and generates equal velocities (ibid., p. 21). Other verbs

APPENDIX G

could replace '*agere*'. Newton maintained that the only comets that approach the sun closely are small, in order that they do not 'agitate' [*agitent*] it by their attraction (ibid., p. 532). If several globes rotate in a fluid, they will propagate their motion throughout the fluid so that every part of it will be agitated' [*agitetur*] with a motion resulting from the 'actions' of all the globes (ibid., p. 391). A motion is propagated through a medium because the parts of the medium near the source of the motion 'disturb and agitate [*urgent commoventque*]' those which lie beyond them (ibid., p. 371).

If Newton frequently used 'action' in a way that can be transposed accurately into a quantity of motion proportional to the integral $\int F dt$, he also used the word in other ways that require other integrals. Consider the following cases. 'Because the action of the centripetal force upon the planets decreases inversely as the square of the distance and the periodic time increases as the 3/2th power of the distance, it is evident that the actions of the centripetal force, and therefore the periodic times, would be equal in equal planets at equal distances from the sun; and in equal distances of unequal planets the total actions of the centripetal force would be as the bodies of the planets ...' (Ibid., p. 567.) Here 'action' is the total attraction summed up over all the matter in a body. He said the same thing in discussing his definitions of force: 'For the sum of the actions of the accelerative force upon the several particles of the body, is the motive force of the whole.'(Ibid., p. 5.) When he calculated the pressure of a fluid pressing in from all sides on a central attracting sphere, he divided the fluid into concentric orbs and imagined the force of gravity 'to act [*agere*]' only on the upper surface of each orb. He then allowed the number of orbs to increase and their thickness to decrease *in infinitum* 'so that the action of gravity from the lowest surface to the uppermost may become continued ...' (ibid., pp. 292–3). In this case, the 'action' of gravity that becomes continuous is extended over space rather than time. In his brief discussion of simple machines, Newton employed 'action' in yet another way. 'For if we estimate the action of the agent from the product of its force and velocity, and likewise the reaction of the impediment from the product of the velocities of its several parts, and the forces of resistance arising from the friction, cohesion, weight, and acceleration of those parts, the action and reaction in the use of all sorts of machines will be found always equal to one another.' (Ibid., p. 28.)

In all of the cases cited above (and in others too numerous to include), the English word 'action' stands in place of '*actio*' in the original Latin. To me, at least, the instances reveal an intuitive use of 'action' without any effort to impose an exact quantitative meaning on it.

Bibliography

Literature on Mechanics in the Seventeenth Century

Agassi, Joseph. 'Leibniz's Place in the History of Physics,' *J. Hist. Ideas*, **30** (1969), 331–44.

Aiton, E. J. 'The Cartesian Theory of Gravity,' *Ann. Sci.*, **15** (1959), 27–50.

'The Celestial Mechanics of Leibniz,' *Ann. Sci.*, **16** (1960), 65–82.

'The Celestial Mechanics of Leibniz: a New Interpretation,' *Ann. Sci.*, **20** (1964), 111–23.

'The Celestial Mechanics of Leibniz in the Light of Newtonian Criticism,' *Ann. Sci.*, **18** (1962), 31–41.

'An Imaginary Error in the Celestial Mechanics of Leibniz,' *Ann. Sci.*, **21** (1965), 169–73.

'Newton and the Cartesians,' *School Sci. Rev.*, **40** (1959), 406–13.

'The Vortex Theory of the Planetary Motions – I, II, III,' *Ann. Sci.*, **13** (1957), 249–64; **14** (1958), 132–47; **14** (1958), 157–72.

Andrade, E. N. da C. *Isaac Newton*. London, 1950.

'Newton and the Science of his Age,' *Nature*, **150** (1942), 700–6.

'Newton, considérations sur l'homme et son oeuvre,' *Rev. hist. sci.*, **6** (1953), 289–307.

Sir Isaac Newton, London, 1954.

Armitage, A. ' "Borell's Hypothesis" and the Rise of Celestial Mechanics,' *Ann. Sci.*, **6** (1950), 268–82.

'The Deviation of Falling Bodies,' *Ann. Sci.*, **5** (1941–7), 342–51.

Arons, A. B. and A. M. Bork. 'Newton's Laws of Motion and the 17th Century Laws of Impact,' *Am. J. Phys.*, **32** (1964), 313–17.

Auger, Leon. *Un savant méconnu: Giles Personne de Roberval (1602–1675)*. Paris, 1962.

Ball, W. W. Rouse. *An Essay on Newton's Principia*. London, 1893.

'A Newtonian Fragment, relating to Centripetal Forces,' *Proc. London Math. Soc.*, **23** (1892), 226–31.

Belaval, Yvon. 'La rise de la geomitrisation de l'univers dans la philosophie des lumières,' *Revue int. phil.*, **6** (1952), 337–55.
Leibniz critique de Descartes. Paris, 1960.
'Premières animadversions de Leibniz sur les *Principes* de Descartes,' *Mélanges Alexandre Koyré, II. L'aventure de l'espiit,* (Paris, 1964), pp. 29–56.
Bell, A. E. *Christian Huygens and the Development of Science in the Seventeenth Century.* London, 1947.
'Hypotheses non fingo,' *Nature*, **149** (1942), 238–40.
Newtonian Science. London, 1961.
Blackwell, Richard J. 'Descartes' Laws of Motion,' *Isis*, **57** (1966), 220–34.
Blake, R. M. 'Newton's Theory of Scientific Method,' *Phil. Rev.*, **42** (1933), 453–86.
'Isaac Newton and the Hypothetico-Deductive Method,' *Theories of Scientific Method: the Renaissance through the Nineteenth Century,* by Ralph M. Blake, Curt J. Ducasse, and Edward H. Madden, (Seattle, Wash., 1961), pp. 119–43.
Bloch, L. 'La mécanique de Newton et la mécanique moderne,' *Rev. sci.*, **23** (1908), 705–12.
La philosophie de Newton. Paris, 1908.
'Les théories newtoniennes et la physique moderne,' *Rev. met. et mor.*, **35** (1928), 41–54.
Bork, Alfred M. 'Logical Structure of the First Three Sections of Newton's *Principia*,' *Am. J. Phys.*, **35** (1967), 342–4.
Boutroux, Pierre. 'L'enseignement de la mécanique en France au XVIIe siecle,' *Isis*, **4** (1921), 276–94.
'L'histoire des principes de la dynamique avant Newton,' *Rev. met. et mor.*, **28** (1921), 657–88.
Brewster, David. *Memoirs of the Life, Writings, and Discoveries of Sir Isaac Newton,* 2 vols, Edinburgh, 1855.
Broad, C. D. *Sir Isaac Newton.* Annual lecture on a master mind. Henriette Hertz Trust of the British Academy. London, 1927.
Brodetsky, S. *Sir Isaac Newton.* London, 1927.
Brougham, Lord Henry and Routh, E. J. *Analytical View of Sir Isaac Newton's Principia.* London, 1855.
Brown, G. Burniston. 'Gravitational and Inertial Mass,' *Am. J. Phys.*, **28** (1960), 475–83.
Buchdahl, G. 'Science and Logic: Some Thoughts on Newton's

Second Law of Motion in Classical Mechanics,' *Brit. J. Phil. Sci.*, **2** (1951), 217–35.
Burke, H. R. 'Sir Isaac Newton's Formal Conception of Scientific Method,' *New Scholast.*, **10** (1936), 93–115.
Burtt, E. A. *The Metaphysical Foundations of Modern Physical Science*, 2nd ed. London, 1932.
Cajori, Florian. 'Newton and the Law of Gravitation,' *Arch. storia sci.*, **3** (1922), 201–4.
— 'Newton's Twenty Years' Delay in Announcing the Law of Gravitation,' *Sir Isaac Newton, 1727–1927. A Bicentenary Evaluation of His Work*, F. E. Brasch, ed. (Baltimore, 1928), pp. 127–88.
— 'Sir Isaac Newton on Gravitation,' *Scientific Monthly*, **27** (1928), 47–53.
Carrington, Hereward, 'Earlier Theories on Gravitation,' *The Monist*, **23** (1913), 445–58.
Carteron, H. 'L'idée de la force mécanique dans le système de Descartes,' *Rev. phil.*, **94** (1922), 243–77, 483–511.
Cassirer, Ernst. *Leibniz System in seinen wissenschaftlichen Grundlagen*. Marburg, 1902.
— 'Mathematical Mysticism and Mathematical Science,' *Galileo, Man of Science*, ed. Ernan McMullan, (New York, 1967), pp. 338–51.
— 'Newton and Leibniz,' *Phil. Rev.*, **52** (1943), 366–91.
Centre internationale de synthèse. *Galilée, aspects de sa vie et de son oeuvre*. Paris, 1968.
— *Pierre Gassendi, 1592–1655. Sa vie et son oeuvre*. Paris, 1955.
Clagett, Marshall. *The Science of Mechanics in the Middle Ages*. Madison, 1959.
Clark, Joseph T. 'Pierre Gassendi and the Physics of Galileo,' *Isis*, **54** (1963), 352–70.
Clavelin, Maurice. *La philosophie naturelle de Galilée*. Paris, 1968.
Comité du tricentenaire de Gassendi. *Actes du Congrès du tricentenaire de Pierre Gassendi*. Paris, 1957.
Cohen, I. B. *The Birth of a New Physics*. Garden City, 1960.
— 'The Dynamics of the Galileo-"Plato Problem" – its Relation to Newton's *Principia*,' *Actes du IXᵉ congrès international d'histoire des sciences*. (Barcelona, 1959), pp. 187–96.
— 'Dynamics: the Key to the "New Science" of the Seventeenth Century,' *Acta hist. rerum nat. tech.*, **3** (1967), 78–114.
— *Franklin and Newton*. (Philadelphia, 1956.)

'Galileo's Rejection of the Possibility of Velocity Changing Uniformly with Respect to Distance,' *Isis*, **47** (1956), 231–5.

'Hypotheses in Newton's Philosophy,' *Physis*, **8** (1966), 163–84.

Isaac Newton. The Creative Scientific Mind at Work. The Wiles Lectures, 1966. Mimeographed.

Isaac Newton's Papers & Letters on Natural Philosophy and Related Documents. Cambridge, Mass., 1958.

'Leibniz on Elliptical Orbits: as seen in his Correspondence with the Académie Royale des Sciences in 1700,' *J. Hist. Med.*, **17** (1962), 72–82.

'Newton's Attribution of the First Two Laws of Motion to Galileo,' *Atti del simposio su 'Galileo Galilei nella storia e nella filosofia della scienza'.* (Firenze-Pisa, 14–16 Settembre 1964), pp. XXIII–XLII.

'Newton's "Electric and Elastic Spirit",' *Isis*, **51** (1960), 337.

'Newton's Second Law and the Concept of Force in the *Principia*,' *Texas Q.*, **10** (1967), 127–57.

'Newton's Use of "Force," or Cajori versus Newton: A Note on Translations of the *Principia*,' *Isis*, **58** (1967), 226–30.

'"Quantum in se est": Newton's Concept of Inertia in Relation to Descartes and Lucretius,' *Notes Rec. R. Soc. Lond.*, **19** (1964), 131–55.

Cohen, I. B. and Koyré, A. 'The Case of the Missing "Tanquam": Leibniz, Newton and Clarke,' *Isis*, **52** (1961), 555–66.

'Newton and the Leibniz-Clarke Correspondence, with Notes on Newton, Conti and Des Maizeaux, *Arch. int. hist. sci.*, **15** (1962), 63–126.

Costabel, Pierre. 'Essai critique sur quelques concepts de la mécanique cartésienne, *Arch. int. hist. sci.*, **20** (1967), 235–52.

Leibniz et la dynamique: les textes de 1692. Paris, 1960.

'Newton's and Leibniz' Dynamics,' trans. J. M. Briggs, Jr., *Texas Q.*, **10** (1967), 119–26.

Costello, William T. *The Scholastic Curriculum at Early Seventeenth-Century Cambridge.* Cambridge, Mass., 1958.

Crombie, A. C. 'Newton's Conception of Scientific Method,' *Bull. Inst. Phys.*, **8** (1957), 350–62.

Dijksterhuis, E. J. 'Christiaan Huygens,' *Centaurus*, **2** (1951–3), 265–82.

The Mechanization of the World Picture, trans. C. Dikshoorn, Oxford, 1961.

'The Origins of Classical Mechanics from Aristotle to Newton,' *Critical Problems in the History of Science*, ed. Marshall Clagett, (Madison, 1962), pp. 163–84.

Dolby, R. G. A. 'A Note on Dijksterhuis' Criticism of Newton's Axiomatization of Mechanics,' *Isis*, **57** (1966), 108–15.
Drake, Stillman. 'The Concept of Inertia,' *Saggi su Galileo Galilei*, (Firenze, 1967), pp. 3–14.
— 'Free Fall in Galileo's *Dialogue*,' *Isis*, **57** (1966), 269–71.
— 'Galileo Gleanings. V. The Earliest Version of Galileo's *Mechanics*,' *Osiris*, **13** (1958), 262–90.
— 'Galileo Gleanings XVI. Semicircular Fall in the "Dialogue",' *Physis*, **10** (1968), 89–100.
— 'Galileo and the Law of Inertia,' *Am. J. Phys.*, **32** (1964), 601–8.
— 'Galileo's 1604 Fragment on Falling Bodies,' *Br. J. Hist. Sci.*, **4** (1968–9), 340–58.
— 'Uniform Acceleration, Space, and Time,' *Br. J. Hist. Sci.*, **5** (1970), 21–43.
Dubarle, D. 'Sur la notion cartésienne de quantité de mouvement,' *Mélanges Alexandre Koyré, II, L'aventure de l'esprit*, (Paris, 1964), pp. 118–28.
Dugas, Rene. 'De Descartes à Newton par l'école anglaise,' *Les conferences du Palais de la Découverte*, Sèrie D No. 16. Paris, 1953.
— *Histoire de la mécanique*. Neuchatel, 1950.
— *La mécanique au XVIIᵉ siècle*. Neuchatel, 1954.
— 'Sur le cartésianisme de Huygens, *Rev. hist. sci.*, **7** (1954), 22–33.
Duhem, P. 'De l'accélération produite par une force constante,' *Congrès international de philosophie*, IIᵉ session, Geneve, 1905.
— *Les origines de la statique*, 2 vols. Paris, 1905–6.
Ellis, Brian D. 'Newton's Concept of Motive Force,' *J. Hist. Ideas*, **23** (1962), 273–8.
— 'The Origin and Nature of Newton's Laws of Motion,' *Beyond the Edge of Certainty*, ed. Robert G. Colodny, (Englewood Cliffs, N.J., 1965), pp. 29–68.
Fierz, Markus. 'Über den Ursprung und Bedeutung von Newtons Lehre vom absoluten Raum,' *Gesnerus*, **11** (1954), 62–120.
François, Charles. 'La théorie de la chute des graves. Evolution historique du problème,' *Ciel et Terre*, **34** (1913), 135–7, 167–9, 261–73.
Gagnebin, Bernard. 'De la cause de la pesanteur. Mémoire de Nicholas Fatio de Duillier présenté à la Royal Society le 26 Février 1690,' *Notes Rec. R. Soc. Lond.*, **6** (1949), 106–60.
Giacomelli, R. *Galileo Galilei giovane e il suo 'De motu'*. Pisa, 1949.

Gerhardt, C. I. 'Leibniz über den Begriff der Bewegung,' *Arch. Gesch. Phil.*, **1** (1888), 211–15.

Glansdorff, Maxime. 'La philosophie de Newton,' *Synthese*, **2** (1947), 25–39.

Goldbeck, Ernst. *Die Gravitationshypothese bei Galilei und Borelli.* Berlin, 1897.

Kepler's Lehre von der Gravitation. Halle a. S., 1896.

Grant, Edward. 'Aristotle, Philoponus, Avempace, and Galileo's Pisan Dynamics,' *Centaurus*, **11** (1965–7), 79–95.

'Bradwardine and Galileo: Equality of Velocities in the Void,' *Arch. Hist. Exact. Sci*, **2** (1962–6), 344–64.

Greenhill, George, 'Definitions and Laws of Motion in the *Principia*,' *Nature*, **111** (1923), 224–6, 395–6.

Gregory, Joshua C. 'The Newtonian Hierarchic System of Particles,' *Arch. int. hist. sci.*, **33** (1954), 243–47.

Gregory, Tullio. *Scetticismo ed empirismo. Studio su Gassendi.* Bari, 1961.

Grigoryan, A. T. 'Appraisal of Newton's Mechanics and of Einstein's "Autobiography",' *Arch. int. hist. sci.*, **14** (1961), 13–22.

Guerlac, Henry. 'Francis Hauksbee: expérimenteur au profit de Newton,' *Arch. int. hist. sci.*, **16** (1963), 113–28.

Newton et Epicure. Conférence donnée au Palais de la Découverte le 2 Mars 1963. Paris, 1963.

'Newton's Optical Aether,' *Notes Rec. R. Soc. Lond.*, **22** (1967), 45–57.

'Sir Isaac and the Ingenious Mr. Hauksbee,' *Mélange Alexandre Koyré I, L'aventure de la science.* (Paris, 1964), pp. 228–53.

Guéroult, Martial. 'La constitution de la substance chez Leibniz,' *Rev. met. et mor.*, **52** (1947), 55–78.

Dynamique et métaphysique Leibniziennes. Paris, 1934.

'Métaphysique et physique de la force chez Descartes et Malebranche,' *Rev. met. et mor.*, **59** (1954), 1–37, 113–34.

Guzzo, A. 'Meccanica e cosmologia newtoniane,' *Filosofia*, **5** (1954), 229–66.

'Ottica e atomistica newtoniane,' *Filosofia*, **5** (1954), 383–419.

Haas, A. E. *Die Entwicklungsgeschichte des Satzes von der Erhaltung der Kraft.* Vienna, 1909.

Hall, A. R. *Ballistics in the Seventeenth Century.* Cambridge, 1952.

'Cartesian Dynamics,' *Arch. Hist. Exact Sci.*, **1** (1961), 172–8.

'Correcting the Principia,' *Osiris*, **13** (1958), 291–326.

From Galileo to Newton. New York, 1963.

'Galileo and the Science of Motion,' *Br. J. Hist. Sci.*, **2** (1964–5), 185–99.

'Mechanics and the Royal Society, 1668–1670,' *Br. J. Hist. Sci.*, **3** (1966), 24–38.

'Newton on the Calculation of Central Forces,' *Ann. Sci.*, **13** (1957), 62–71.

The Scientific Revolution. London, 1954.

Hall, A. R. and Marie Boas. 'Clarke and Newton,' *Isis*, **52** (1961), 583–5.

Hall, A. R. and M. B. 'The Date of "On Motion in Ellipses",' *Ach. int. hist. sci.*, **16** (1963), 23–8.

'Newton and the Theory of Matter,' *Texas Q*, **10** (1967), 54–68.

Hall, A. R. and Marie Boas. 'Newton's Chemical Experiments,' *Arch. int. hist. sci.*, **11** (1958), 113–52.

Hall, A. R. and M. B. 'Newton's Electric Spirit: Four Oddities,' *Isis*, **50** (1959), 473–6.

Hall, A. R. and Marie Boas. 'Newton's "Mechanical Principles",' *J. Hist. Ideas*, **20** (1959), 167–78.

Hall, A. R. and M. B. 'Newton's Theory of Matter,' *Isis*, **51** (1960), 131–44.

eds. *Unpublished Scientific Papers of Isaac Newton.* Cambridge, 1962.

Hankins, Thomas L. 'Eighteenth-Century Attempts to Resolve the *Vis viva* Controversy,' *Isis*, **56** (1965), 281–97.

'The Reception of Newton's Second Law of Motion in the Eighteenth Century,' *Arch. int. hist. sci.*, **20** (1967), 43–65.

Hanson, N. R. 'Galileo's Discoveries in Dynamics,' *Science*, **147** (1965), 471–8.

'Newton's First Law: a Philosopher's Door into Natural Philosophy,' *Beyond the Edge of Certainty*, ed. Robert G. Colodny, (Englewood Cliffs, N.J., 1965), pp. 6–28.

Harré, R. *Matter and Method.* London, 1964.

Hawes, Joan L. 'Newton and the "Electrical Attraction Unexcited",' *Ann. Sci.*, **24** (1968), 121–30.

'Newton's Revival of the Aether Hypothesis and the Explanation of Gravitational Attraction,' *Notes Rec. R. Soc. Lond.*, **23** (1968), 200–12.

Hay, W. H. 'On the Nature of Newton's First Law of Motion,' *Phil. Rev.*, **66** (1956), 95–102.

Henrici, J. *Die Erforschung der Schwere durch Galilei, Huygens, Newton als Grundlage der rationallen Kinematik und Dynamik historisch-didactisch dargestellt.* Leipzig, 1885.

Herivel, J. W. *The Background to Newton's 'Principia'.* Oxford, 1965.

'Early Newton Dynamical MSS.' *Arch. int. hist. sci.*, **15** (1962), 149-50.

'Galileo's Influence on Newton in Dynamics,' *Mélanges Alexandre Koyré, I. L'aventure de la science.* (Paris, 1964), pp. 294-302.

'The Growth of Newton's Concept of Force,' *Proc. 10th Int. Cong. Hist. Sci.*, 1962, **2**, 711-13.

'Halley's First Visit to Newton,' 'On the Date of Composition of the First Version of Newton's Tract de Motu,' 'Suggested Identification of the Missing Original of a Celebrated Communication of Newton's to the Royal Society,' *Arch. int. hist. sci.*, **13** (1960), 63-6, 67-70, 71-8.

'Newton on Rotating Bodies,' *Isis*, **53** (1962), 212-18.

'Newtonian Studies III. The Originals of the Two Propositions Discovered by Newton in December 1679?' *Arch. int. hist. sci.*, **14** (1961), 23-34.

'Newtonian Studies IV,' *Arch. int. hist. sci.*, **16** (1963), 13-22.

'Newton's Achievements in Dynamics,' *Texas Q*, **10** (1967), 103-18.

'Newton's Discovery of the Law of Centrifugal Force,' *Isis*, **51** (1960), 546-53.

'Newton's First Solution of the Problem of Kepler Motion,' *Br. J. Hist. Sci.*, **2** (1965), 350-4.

'Newton's Test of the Inverse Square Law Against the Moon's Motion,' *Arch. int. hist. sci.*, **14** (1961), 93-7.

'Sur les premières recherches de Newton en dynamique,' *Rev. hist. sci.*, **15** (1962), 105-40.

Hesse, Mary B. 'Action at a Distance in Classical Physics,' *Isis*, **46** (1955), 337-53.

'Hooke's Vibration Theory and the Isochrony of Springs,' *Isis*, **57** (1966), 433-41.

Hiebert, Erwin N. *Historical Roots of the Principle of Conservation of Energy.* Madison, 1962.

Home, Roderick W. 'The Third Law in Newton's Mechanics,' *Br. J. Hist. Sci.*, **4** (1968), 39-51.

Hooykaas, R. *Das Verhältnis von Physik und Mechanik in historischer Hinsicht.* Wiesbaden, 1963.

Humphreys, W. C. 'Galileo, Falling Bodies and Inclined Planes. An Attempt at Reconstructing Galileo's Discovery of the Law of Squares,' *Br. J. Hist. Sci.*, **3** (1967), 225–44.

Jammer, Max. *Concepts of Force*. Cambridge, Mass., 1957.
Concepts of Mass. Cambridge, Mass., 1961.
Concepts of Space. Cambridge, Mass., 1954.

Jourdain, Philip E. B. 'Elliptic Orbits and the Growth of the Third Law with Newton,' 'Newton's Theorems on the Attraction of Spheres,' *The Monist*, **30** (1920), 183–98, 199–202.
'Galileo and Newton,' *The Monist*, **28** (1918), 629–33.
'Newton's Hypothesis of Ether and of Gravitation,' *The Monist*, **25** (1915), 79–106, 233–54, 418–40.
'The Principles of Mechanics with Newton, from 1666 to 1679,' *The Monist*, **24** (1914), 187–224.
'The Principles of Mechanics with Newton, from 1679 to 1687,' *The Monist*, **24** (1914), 515–64.
'Robert Hooke as a Precursor of Newton,' *The Monist*, **23** (1913), 353–84.

Kargon, Robert. *Atomism in England from Hariot to Newton*. Oxford, 1966.
'Newton, Barrow and the Hypothetical Physics,' *Centaurus*, **11** (1965), 45–56.

Knudson, Ole. 'A Note on Newton's Concept of Force,' *Centaurus*, **9** (1963–4), 266–71.

Knudson, Ole and Kurt Pedersen. 'The Link Between "Determination" and Conservation of Motion in Descartes' Dynamics,' *Centaurus*, **13** (1968), 183–6.

Koyré, Alexandre. 'Le *De Motu Gravium* de Galilée. De l'expérience imaginaire et de son abus,' *Rev. hist. sci.*, **13** (1960), 197–245.
'De motu gravium naturaliter cadentium in hypothesi terrae motae,' *Trans. Am. Phil. Soc.*, **45** (1955), 329–55.
A Documentary History of the Problem of Fall from Kepler to Newton. Philadelphia, 1955.
Entretiens sur Descartes. New York and Paris, 1944.
Etudes galiléennes. Paris, 1939.
'Etudes newtoniennes III. Attraction, Newton et Cotes,' *Arch. int. hist. sci.*, **14** (1961), 225–36.
'An Experiment in Measurement,' *Proc. Am. Phil. Soc.*, **97** (1953), 222–37.

From the Closed World to the Infinite Universe. Baltimore, 1957.
'Galileo and Plato,' *J. Hist. Ideas*, **4** (1943), 400–28.
'Galileo and the Scientific Revolution of the Seventeenth Century,' *Phil. Rev.*, **52** (1943), 333–48.
'La gravitation universelle de Kepler à Newton,' *Arch. int. hist. sci.*, **4** (1951), 638–53.
'L'hypothèse et l'expérience chez Newton,' *Bull. soc. fran. phil.*, **50** (1956), 59–79.
'La mécanique céleste de J. A. Borelli,' *Rev. hist. sci.*, **5** (1952), 101–38.
'Newton, Galilée et Platon,' *Annales*, **6** (1960), 1041–59.
Newtonian Studies. Cambridge, Mass., 1965.
'Les regulae philosophandi.' 'Les Queries de l'Optique,' *Arch. int. hist. sci.*, **13** (1960), 3–14, 15–29.
La révolution astronomique. Copernic, Kepler, Borelli. Paris, 1961.
'The Significance of the Newtonian Synthesis,' *Arch. int. hist. sci.*, **11** (1950), 291–311.
'An Unpublished Letter of Robert Hooke to Isaac Newton,' *Isis*, **43** (1952), 312–37.
Kubrin, David. 'Newton and the Cyclical Cosmos: Providence and the Mechanical Philosophy,' *J. Hist. Ideas*, **28** (1967), 325–46.
Kuhn, T. S. 'The Independence of Density and Pore-size in Newton's Theory of Matter,' *Isis*, **43** (1952), 364–5.
'Newton's 31st Query and the Degradation of Gold,' *Isis.*, **42** (1951), 296–8.
Lenoble, Robert. *Mersenne ou la naissance du mécanisme.* Paris, 1943.
Lenzen, V. F. 'Newton's Third Law of Motion,' *Isis*, **27** (1937), 258–60.
'Newton's Third Law,' *Science*, **87** (1938), 508.
Lindberg, David C. 'Galileo's Experiments on Falling Bodies,' *Isis*, **56** (1965), 352–4.
Lohne, Johs. 'Hooke *versus* Newton,' *Centaurus*, **7** (1960), 6–52.
Losee, John. 'Drake, Galileo, and the Law of Inertia,' *Am. J. Phys.*, **34** (1966), 430–2.
Macauley, W. H. 'Newton's Theory of Kinetics,' *Bull. Am. Math. Soc.*, **3** (1896–7), 363–71.
McGuire, J. E. 'Atoms and the "Analogy of Nature": Newton's Third Rule of Philosophizing,' *Studies Hist. Phil. Sci.*, **1** (1970), 3–58.
'Body and Void in Newton's De Mundi Systemate: Some New Sources,' *Arch. Hist. Exact. Sci.*, **3** (1966), 206–48.

'Force, Active Principles and Newton's Invisible Realm,' *Ambix*, **15** (1968), 154–208.

'The Origin of Newton's Doctrine of Essential Qualities,' *Centaurus*, **12** (1968), 233–60.

'Transmutation and Immutability: Newton's Doctrine of Physical Qualities,' *Ambix*, **14** (1967), 69–95.

McGuire, J. E. and P. M. Rattansi. 'Newton and the "Pipes of Pan",' *Notes Rec. R. Soc. Lond.*, **21** (1966), 108–43.

McMullin, Ernan. 'Galileo, Man of Science,' *Galileo, Man of Science*, ed. Ernan McMullin, (New York, 1967), pp. 3–51.

Mach, Ernst. *The Science of Mechanics: A Critical and Historical Account of its Development*, trans. Thomas J. McCormack, 6th ed., Lasalle, Ill., 1960.

Meldrum, Andrew Norman. 'The Development of the Atomic Theory: (3) Newton's Theory, and its Influence in the Eighteenth Century,' *Mem. and Proc. Manchester Lit. and Phil. Soc.*, **55** (1910), 1–15.

Metzger, Hélène. *Attraction universelle et religion naturelle chez quelques commentateurs anglais de Newton*. Paris, 1938.

Newton, Stahl, Boerhaave et la doctrine chimique. Paris, 1930.

Milhaud, Gaston. *Descartes savant*. Paris, 1921.

Moody, Ernest A. 'Galileo and Avempace: The Dynamics of the Leaning Tower Experiment,' *J. Hist. Ideas*, **12** (1951), 163–93, 375–422.

'Galileo and His Precursors,' *Galileo Reappraised*, ed. Carlo L. Golino, (Berkeley and Los Angeles, 1966), pp. 23–43.

More, L. T. *Isaac Newton, a Biography*. New York, 1934.

'Newton's Philosophy of Nature,' *Sci. Monthly*, **56** (1943), 491–504.

Moscovici, S. *L'expérience du mouvement*. Paris, 1967.

'Recherches de Giovanni-Battista Baliani sur le choc des corps élastiques,' *Actes du symposium international des sciences physiques et mathematique dans la première moitié du XVIIe siècle*, (Paris, 1960), pp. 98–115.

'Rémarques sur le dialogue de Galilée "De la force de la percussion",' *Rev. hist. sci.*, **16** (1963), 97–137.

'Torricelli's *Lezioni Academiche* and Galileo's Theory of Percussion,' *Galileo, Man of Science*, ed. Ernan McMullin, (New York, 1967), pp. 432–48.

Mouy, P. *Le développement de la physique cartésienne, 1646–1712.* Paris, 1934.
Les lois du choc des crops d'après Malebranche. Paris, 1927.
'Malebranche et Newton,' *Rev. met. et mor.*, **45** (1938), 411–35.
Natucci, Alpinolo. 'Il concetto di lavoro meccanico in Galileo e Cartesio,' *Actes du symposium international des sciences physiques et mathematiques dans la première moitié du XVIIe siècle.* Paris, 1960.
'I meriti di Evangelista Torricelli come fisico,' *Convegno di studi torricelliani* (Faenza, 1959), pp. 77–91.
Neményi, P. F. 'The Main concepts and Ideas of Fluid Dynamics in their Historical Development,' *Arch. Hist. Exact Sci.*, **2** (1962–6), 52–86.
Patterson, Louise D. 'Hooke's Gravitation Theory and its Influence on Newton,' *Isis*, **40** (1949), 327–41, **41** (1950), 32–45.
'A Reply to Professor Koyré's Note on Robert Hooke,' *Isis*, **41** (1950), 304–5.
Pav, Peter Anton. 'Gassendi's Statement of the Principle of Inertia,' *Isis*, **57** (1966), 24–34.
Perl, Marguta R. 'Newton's Justification of the Laws of Motion,' *J. Hist. Ideas*, **27** (1966), 585–92.
Pla, Cortès. *Isaac Newton.* Buenos Aires, 1945.
Reichenbach, Hans. 'The Theory of Motion According to Newton, Leibniz, and Huygens,' *Modern Philosophy of Science.* (London, 1959), pp. 46–66.
Rigaud, S. P. *Historical Essay on the First Publication of Sir Isaac Newton's Principia.* Oxford, 1838.
Rochot, B. 'Beeckman, Gassendi et le principe d'inertie,' *Arch. int. hist. sci.*, **31** (1952), 282–9.
'Gassendi et l'expérience,' *Mélanges Alexandre Koyré, II. L'aventure de l'esprit*, (Paris, 1964), pp. 411–22.
'Sur les notions de temps et d'espace chez quelques auteurs du XVIIe siècle, notamment Gassendi et Barrow,' *Rev. hist. sci.*, **9** (1956), 97–104.
Les travaux de Gassendi sur Epicure et sur l'atomisme, 1619–1658. Paris, 1944.
Ronchi, Vasco. 'Il dubbii di Isaaco Newton circa la universalitá della legge dell'attrazione,' *Arch. int. hist. sci.*, **13** (1960), 31–7.
Rosenberger, Ferdinand. *Isaac Newton und seine physikalischen Principien.* Leipzig, 1895.

Rosenfeld, L. 'Newton and the Law of Gravitation,' *Arch. Hist. Exact Sci.*, **2** (1962–6), 365–86.

'Newton's Views on Aether and Gravitation,' *Arch. Hist. Exact Sci.*, **6** (1969), 29–37.

Russell, Bertrand. *A Critical Exposition of the Philosophy of Leibniz*, new ed. London, 1937.

Russell, J. L. 'Kepler's Laws of Planetary Motion,' *Br. J. Hist. Sci.*, **2** (1964–5), 1–24.

Sabra, S. I. *Theories of Light from Descartes to Newton*. London, 1967.

Schimank, Hans. 'Die geschichtliche Entwicklung der Kraftbegriffs bis zum Aufkommen der Energetik,' *Robert Mayer und das Energieprinzip*, ed. H. Schimank and E. Pietsch, (Berlin, 1942).

Scott, J. F. *The Scientific Work of René Descartes (1596–1650)*. London, 1952.

Scott, Wilson. *Conflict Between Atomism and Conservation Theory, 1644–1860*. London, 1970.

Serrus, Ch. 'La mécanique de J. A. Borelli et la notion d'attraction,' *Rev. hist. sci.*, **1** (1947–48), 9–25.

Settle, Thomas B. 'Galileo's Use of Experiment as a Tool of Investigation,' *Galileo, Man of Science*, ed. Ernan McMullin, (New York, 1967), pp. 315–37.

Shapere, Dudley, 'The Philosophical Significance of Newton's Science,' *Texas Q*, **10** (1967), 201–15.

Snow, A. J. *Matter and Gravity in Newton's Physical Philosophy*. London, 1926.

'The Role of Mathematics and Hypothesis in Newton's Physics,' *Scientia*, **42** (1927), 1–10.

Sortais, R. P. Gaston. *Pierre Gassendi, sa vie et son oeuvre*. Paris, 1955.

Stein, Howard. 'Newtonian Space-Time,' *Texas Q*, **10** (1967), 174–200.

Strong, E. W. 'Hypotheses non fingo,' *Men and Moments in the History of Science*, ed. Herberg M. Evans, (Seattle, 1959), pp. 162–76.

'Newton and God,' *J. Hist. Ideas*, **13** (1952), 147–67.

'Newtonian Explications of Natural Philosophy,' *J. Hist. Ideas*, **18** (1957), 49–83.

'Newton's "Mathematical Way",' *J. Hist. Ideas*, **12** (1951), 90–110.

Sullivan, J. W. N. *Isaac Newton, 1642–1727*. New York, 1938.

Tait, P. G. 'Note on a Singular Passage in the Principia,' *Proc. R. Soc. Edin.*, **13** (1886), 72–8.

Taliaferro, R. C. *The Concept of Matter in Descartes and Leibniz.* Notre Dame, 1964.
Tannery, P. 'Galilée et les principes de la dynamique,' *Mémoires scientifiques,* **6** (Paris, 1926), pp. 387–413.
Tarozzi, G. 'I infinito cosmico e la mecanica celeste di Newton,' *Arch. st. filos.,* **1** (1932), 5–22.
Thackray, Arnold. *Atoms and Powers. An Essay on Newtonian Matter-theory and the Development of Chemistry.* Cambridge, Mass., 1970.
   ' "Matter in a Nut-Shell": Newton's *Opticks* and Eighteenth Century Chemistry,' *Ambix,* **15** (1968), 29–63.
Toulmin, Stephen. 'Criticism in the History of Science: Newton on Absolute Space, Time, and Motion,' *Phil. Rev.,* **68** (1959), 1–29, 203–27.
Truesdell, C. 'A Program toward Rediscovering the Rational Mechanics of the Age of Reason,' *Arch. Hist. Exact Sci.,* **1** (1960–2), 1–36.
   'Rational Fluid Mechanics, 1687–1765,' editor's introduction to vol. II, 12 *Euleri Opera Omnia,* (Zurich, 1954), IX–CXXV.
   'Reactions of Late Baroque Mechanics to Success, Conjecture, Error, and Failure in Newton's *Principia,*' *Texas Q,* **10** (1967), 238–58.
Vavilov, S. I. *Isaac Newton.* Trans. Josef Grun, Wien, 1948.
Weisheipl, James A. 'The Principle *Omne quod movetur ab alio movetur* in Medieval Physics,' *Isis,* **56** (1965), 26–45.
Westfall, R. S. 'The Foundations of Newton's Philosophy of Nature,' *Br. J. Hist. Sci.,* **1** (1962), 171–82.
   'Hooke and the Law of Universal Gravitation: a Reappraisal of a Reappraisal,' *Br. J. Hist. Sci.,* **3** (1967), 245–61.
   'Newton and Absolute Space,' *Arch. int. hist. sci.,* **17** (1964), 121–32.
   'Newton and Order,' *The Concept of Order,* ed. Paul G. Kuntz, (Seattle, 1968), pp. 77–88.
   'A Note on Newton's Demonstration of Motion in Ellipses,' *Arch. int. hist. sci.,* **22** (1969), 51–60.
   'The Problem of Force in Galileo's Physics,' *Galileo Reappraised,* ed. Carlo L. Golino, (Berkeley and Los Angeles, 1966), pp. 67–95.
Whiteside, D. T. 'Before the *Principia*: The Maturing of Newton's Thoughts on Dynamical Astronomy, 1664–84,' *J. Hist. Astro.,* **1** (1970), 5–19.
   'Newtonian Dynamics,' *History of Science,* **5** (1966), 104–17.

'Newton's Early Thoughts on Planetary Motion: a Fresh Look,' *Br. J. Hist. Sci.*, **2** (1964), 117–37.

Wiener, Philip Paul. 'The Tradition Behind Galileo's Methodology,' *Osiris*, **1** (1936), 733–46.

Wilson, Curtis A. 'From Kepler's Laws, So-called, to Universal Gravitation: Empirical Factors,' *Arch. Hist. Exact Sci.*, **6** (1970), 89–170.

Wohlwill, Emil. 'Die Entdeckung der Beharrunggesetzes,' *Zeitschrift fur Völkerpsychologie und Sprachwissenschaft*, **14** (1883), 365–410; **15** (1884), 70–135, 337–87.

Wohlwill, E. 'Die Entdeckung der Parabelform der Wurflinie,' *Abh. Gesch. Math.*, **9** (1889), 577–624.

Index

Accademia del cimento, 258
Académie royale des sciences, 146, 186, 191–3, 243, 265–6, 268
Acta eruditorum, 284
Archimedes, 13, 15, 16, 99, 153, 248
Aristotle, 326
   Aristotelian mechanics, 3, 5, 13–16, 20–1, 34, 39, 57, 118, 235, 342
   Aristotelian philosophy, 325, 414
Arnauld, Antoine, 319
Atwood's machine, 52

Bacon, Francis, 266
Baliani, Giovanni Battista, 115–17
   De motu gravium solidorum, 115
   free fall, model of, 115
   mass, concept of, 115
Bayer, Johann, 515
Beeckman, Isaac, 56, 104
Benedetti, Giambattista, 15
Bentley, Richard, 396, 508
Bernoulli, Jean, 301–2
Borelli, Giovanni Alfonso, 136, 213–31, 234, 243–4, 258–60, 264–5, 276–8, 281
   circular motion. See *orbital mechanics*
   Copernican system, 218–19
   De motionibus naturalibus, 213
   De motu animalium, 213, 217
   De vi percussionis, 213, 222, 228
   force. See *motive force*
   'force', use of word, 541–5
   free fall, model of, 228, 230
   gravity, cause of, 228–9
   impact, model of, 222–3, 230, 265
   impact, study of, 215–16, 222–7
      elastic bodies, 226–7
      force of percussion, 222–5, 228
      perfectly hard bodies, 224–6
   mass, concept of, 223, 277
   mathematics, in relation to mechanics, 257–9, 262
   measure of force, 229
   mechanical philosophy, 213–14, 264–5
      rejection of attractions, 264, 281
   motion, conception of:
      inertia, principle of, 216, 276
      relativity, 226
   motion through resisting media, 257–8, 281
   motive force, concept of, 214–18, 220, 223–30
   occult leap, concept of, 217–18, 276
   orbital mechanics, 218–21, 307
      effort to recede, 219, 221–2
      natural instinct to unite with sun, 219, 221–2
   simple harmonic motion, 221, 258–9, 261
   simple machines, 215–18, 220–4, 226–8, 230
   static force, concept of, 214–15
   statics and dynamics, 215, 217–18, 220, 224, 227, 230
   Theoricae mediceorum planetarum, 213, 218
   uniformly accelerated motion, 228–30
Boulliau, Ismael, 214
Boyle, Robert, 326, 329, 334–5, 366, 368–9, 371, 374–5, 385, 387, 408–9, 411

INDEX

Boyle—*contd.*
  Boyle's law, 246, 497
  Mechanical Origin of Qualities, 369–70, 408–9
  New Experiments Physico-Mechanical, 403
  Origin of Forms and Qualities, 407
Brahe, Tycho, 105, 218
Brouncker, William, Lord:
  cycloid, 192

Cambridge platonists, 341
centrifugal force. See Galileo, Huygens, Leibniz, Newton: *circular motion*
Charleton, Walter, 101, 103–6, 108, 326–7, 400, 535–7
  impressed force, concept of, 139
  Physiologia, 325
Christina, Queen of Sweden, 57
circular motion, 82, 168, 174, 256, 310, 447. *See also* Borelli, Descartes, Galileo, Gassendi, Hooke, Huygens, Leibniz, Newton, Torricelli
Clarke, Samuel:
  Clarke–Leibniz correspondence, 415, 420, 422, 519
Clerselier, Claude, 71
Cohen, I. Bernard, 402, 414
Colbert, Jean Baptiste, 146
Copernican system, 171–2. *See also* Borelli, Galileo, Huygens, Newton, Roberval
Coste, Pierre, 421–2
Cotes, Rogers, 504

Dante, 42
DeBeaune, Florimond, 88
De Chales, Claude-François Milliet, 189, 200–3, 273–5
  Cursus seu mundus mathematicus, 200
  impact, study of, 201–2
  impetus, concept of, 200–1, 273–4
  mathematics, in relation to mechanics, 274–5
  measure of force, 201–2
  '*momentum*', usage of, 274
  simple machines, 201–2
  uniformly accelerated motion, 200, 202, 273–4
  '*vis*', usage of, 274

Democritean tradition, 90, 99, 146, 148, 315–16, 398, 512
DeMoivre, Abraham, 406, 458, 513
Descartes, René, 43–4, 56–99, 104, 109, 112, 114, 136, 139, 147, 149–50, 155, 182, 184–5, 196, 212, 216, 218, 254, 256, 265, 268, 276, 284–5, 288–9, 303, 315, 320–1, 326, 328–9, 335, 337, 414, 500
  action, concept of, 61–3, 65–8, 73, 75–8, 91, 93, 113, 453
  agitation, concept of, 61–2, 90–1
  circular motion, 59, 78, 80–2, 93–4, 219
    conatus to recede, 78–81, 93–5, 167, 172–3, 210, 304–5, 334, 351–2, 360
    necessity in a plenum, 82, 94
  composition of motions, 89, 93–4
  Correspondence, Clerselier edition, 403
  criticism of Galileo, 89, 97, 187
  deception by the senses, 85
  Dioptrics, 91
  Discourse on Method, 57, 86
  dynamics, philosophical foundations of, 60
  epistemology, 85
  force of a body's motion, 63–71, 75, 81, 104, 112, 114, 118, 137, 148, 150, 167, 285–6, 288, 297, 316, 344, 347, 447
    distinct from determination, 64–9, 71, 78, 83, 91, 94, 156, 291
  'force', use of word, 529–34
  free fall, model of, 44
  gravity, cause of, 78, 80–1, 89, 92, 173, 178, 185–7, 281, 305, 330
  immutability of God, 60, 68, 82
  impact, laws of, 68, 70–2, 82–4, 89, 92, 106, 123, 147–8, 151, 153, 289–91, 343–4
    force of percussion, 82
    instantaneous, 318
  impact, model of, 60–1, 63–4, 72, 81, 83
  impetus, 60
  light, conception of, 62, 65–6, 78, 81
    law of reflection, 65–7, 92
    law of refraction, 65–6, 92, 366
  matter, conception of, 100, 397
    extension, 63, 69–70, 89, 339, 449
    indifference to motion, 56, 69–70
    inert, 60, 63, 69, 70, 72, 94

INDEX

natural inertia, 70–2, 93, 182, 195
uniformity of matter, 388
measure of force, 74–8, 93
mechanical philosophy, 56–7, 60, 63, 68, 72, 79, 81–8, 90, 99, 101, 147, 173, 327–9, 331–2, 334, 337, 387
    causal explanations, 86–9, 95, 97
    particle as causal agent, 68, 81, 83
    plenum, 60, 78, 328, 385
    rejection of attractions, 86–8
    rejection of occult qualities, 86
    transparency of nature to reason, 86, 339
Meditations on First Philosophy, 57
metaphysics, 325
Le monde, 58, 79, 94–5
motion, conception of, 72, 84–5, 289, 343, 450
    changes of direction, 66–8, 78, 80, 91, 447
    changes of motion, 60, 63–4, 67–8, 72, 85
    finite initial velocity, 69, 92
    God the source of motion, 60
    identical to rest, 58
    identity of all motions, 345
    inertia, principle of, 52, 57–9, 67–8, 70, 78, 80–1, 93–4, 96–7, 99, 172, 347, 451
    rejection of violent motion, 56, 59, 69, 90
    relativity, 58, 61, 82, 93, 151, 337–8, 347, 361–2, 441–2, 445–6
motion, laws of, 58–9, 66–7, 78, 80, 106, 319
nature, laws of. See *motion, laws of*
Principles of Philosophy, 57, 79–80, 83, 86, 93, 97, 185, 319
quantity of motion, 70, 156
    quantity of motion, conservation of, 60–1, 64, 68, 70, 82–4, 106, 151, 157, 292, 390
rational mechanics, 60, 84, 87–9, 96–7, 148, 398
Rules for the Direction of the mind, 86
simple machines, 72–8, 93, 95, 285–6, 319
static force, concept of, 75–6
statics and dynamics, 75, 297
vortices, 396, 406, 501, 510–11
work, concept of, 73, 75–7

Digby, Kenelm, 326
    gravity, cause of, 330

Endeavour to recede. See Borelli, Descartes: *circular motion*
Epicurus, 99–100, 103–4
Euclid, 12
Eustachius of St Paul, 325

Fatio de Duillier, Nicolas, 184, 503
Flamsteed, John, 431, 458–9, 462, 513–15, 521
force, conception of, 346. See also Borelli, Descartes, Galileo, Hobbes, Hooke, Leibniz, Mariotte, Newton, Wallis, Wren
force of percussion, 348. See also Borelli, Descartes, Galileo, Hobbes, Huygens, Pardies, Torricelli, Wallis: *impact*
free fall, model of, 21, 64, 88, 118, 120, 245, 256, 310. See also Baliani, Borelli, Descartes, Galileo, Gassendi, Hobbes, Hooke, Huygens, Leibniz, Marci, Mariotte, Newton, Torricelli, Wallis
Frénicle de Bessy, Bernard, 186
gravity, conception of, 265

Galilei, Galileo, 1–56, 61, 63, 69, 72, 75, 88–90, 97, 99, 102–5, 109, 114–17, 119, 125–9, 143, 146, 154–5, 162–5, 213, 215–16, 243–4, 246, 248, 254, 257, 494
The Assayer, 6
circular motion, 19, 22, 42, 44, 49–50, 82, 171, 174, 303
    centrifugal force, 49–50
    comparison of gravity to centrifugal force, 355, 405
    composition of motions, 4, 42–3, 52, 54, 435, 481, 518
    parabolic trajectory, 4, 20, 30, 42, 54
Copernican system, 3, 9, 18–19, 22, 30, 355.
De motu, 2, 12–23, 25–6, 29–33, 35, 43, 45, 119
Dialogue, 3–5, 10, 18–19, 42, 45, 326, 355, 405

569

INDEX

Galilei—*contd.*
  Discourse on Bodies in Water, 27, 34
  Discourses, 3-4, 7-8, 18-19, 23, 26, 28-9, 31-2, 34, 36-7, 42, 45, 114-17, 127, 153, 161-2, 402, 413
  dynamical terminology, 7
  force, conception of, 7-8, 11-12, 16-17, 25, 35, 39-41, 43-4
  force of percussion, 24-5, 28-30, 36-8, 50, 123, 129, 147, 222
  'force', use of word, 526-8
  free fall, model of, 7-9, 22, 30, 37-40, 43-4, 46, 53
  geometry, in relation to mechanics, 40-1
  gravity, 55
    acceleration of gravity, 356-7
  impact. See *force of percussion*
  impact, model of, 11-12, 44, 64
  impeto. See *momento*
  impetus, 49-51, 346
  impressed force, concept of, 12-13, 15-16, 18-19, 21, 23, 25-6
  incompatibility, concept of, 55, 368, 413
  Inquisition, 3, 58, 218, 337
  mass, concept of, 25, 32, 35, 38, 44, 54
  mathematical description of motion. See *rational mechanics*
  mathematics. See *geometry*
  matter, conception of:
    all bodies heavy, 6, 12-13, 21
    indifference to motion, 4, 6, 35, 39-40, 42, 56, 219, 224, 450
    inert, 6
    uniformity of matter, 21
  measure of force, 26-30, 44, 50-1
  mechanical philosophy, 42, 44, 46, 55
  On Mechanics, 2, 28
  momento or *impeto*, concept of, 11-12, 25-9, 38
  motion, conception of, 102, 289
    inertia, principle of, 2, 4-5, 9, 11, 18, 21-2, 30, 35, 44, 99, 117, 127
    motion in a void, 15
    natural motions, 8-10, 12, 17, 22-3, 33, 35, 41-6
    uniform horizontal motion, 9-11, 18-20, 23, 43
    uniform motion identical to rest, 4, 19, 30

  motion through medium, 53, 230, 470-1, 493
  resistance of medium, 14-15, 31-5, 39
  terminal velocity, 32-4, 41
  nature, conception of, 6, 19, 41-2, 45, 82
  Newton's second law, 2-3, 7-8, 31-2, 35
  pendulums, 26, 41, 53, 281
    period of pendulum, 158
  rational mechanics, 2, 12, 18-19, 47, 49, 99, 148
  simple machines, 17-18, 25, 27-8, 31, 35-7, 44, 50-1, 74
  specific gravity, concept of, 14-16, 21, 23, 33, 35
  static force. See *weight as the measure of force*
  statics and dynamics, 16, 18, 20, 23, 27, 30, 36-9, 44, 51, 469
  telescope, 3
  uniformity accelerated motion, 5, 9, 20-3, 30, 32, 35, 41, 43, 46, 89, 99, 114-15, 118, 120, 130, 154, 157-9, 161-4, 168, 177, 186-7, 200-1, 206-8, 212, 214, 217, 244, 253-4, 262, 275-6, 280, 285, 289, 292, 295, 302-3, 315, 333, 355, 357, 398, 425, 435-7, 470, 473, 484, 491
    dynamic function of weight, 6-7
    equal acceleration of all bodies, 5, 31, 33, 39, 333, 401
    free fall, 5, 7-8, 45
    free fall as standard of measurement, 23-5, 30, 37, 43
    inclined planes, 26, 31-2, 160
    naturally accelerated motion, 5
  vibrating springs, 53-4
  weight as the measure of force, 17, 199
Gassendi, Pierre, 99-109, 114, 326-7, 329, 335, 337, 400-1
  absolute space, 339
  atomism, 99-100, 104, 325, 327-9, 400, 449
    gravity of atoms, 103-6
  circular motion, 103
  Epistola tres. De motu impresso, 107
  force of a body's motion, 105-7

570

'force', use of word, 535–7
free fall, model of, 107–9
gravity, cause of, 101, 103, 107, 109, 330, 401
impact, 105–7
impact, model of, 104–5, 107–8
mass, concept of, 109
measure of force, 108
mechanical philosophy, 99–100, 103–4, 329
  causal explanations, 100–1, 109
  rejection of occult qualities, 100–1
motion, conception of, 100, 102–5
  distinction of natural and violent motions, 102
  inertia, principle of, 101–3, 108, 138
Opera omnia, 400
rational mechanics, 104, 109
scepticism, 339
Syntagma philosophicum, 100
Gilbert, William, 214, 392
Glanville, Joseph:
  Vanity of Dogmatizing, 326
gravity. *See* Borelli, Descartes, Digby, Frénicle, Galileo, Gassendi, Hobbes, Hooke, Huygens, Kepler, Leibniz, Marci, Mariotte, Newton, Roberval, Wallis
Gregory, David, 358, 421, 512–13

Hall, A. R. and M. B., 513
Halley, Edmond, 364, 431–2, 435, 461–2, 515–16, 520
Harris, John:
  Lexicon Technicum, 378
Hauksbee, Francis, 384, 392, 399
Helmont, Johannes Baptista van, 367, 369, 389
Herivel, John, 93–4, 355, 404, 406, 513–14, 516
hermetic philosophy, 265–8, 331, 367–8, 387, 389
Hobbes, Thomas, 109–14, 326, 335
  action, concept of, 112–13
  De corpore, 109–10
  endeavour, concept of, 110–14, 119, 139, 317
  force, definition of, 112
  force of a body's motion, 112
  free fall, model of, 114
  gravity, cause of, 111
  impact:
    force of percussion, 113
    impact, model of, 114
  impetus, definition of, 110
  light, conception of, 112
    laws of reflection and refraction, 112
  mechanical philosophy, 109–11, 114
    rejection of occult qualities, 110
  rational mechanics, 114
  simple machines, 111
  statics and dynamics, 110–11, 113
  uniformly accelerated motion:
    finite initial velocity, 134
    velocity of the whole motion, 140
Hooke, Robert, 206–13, 260–1, 275–6, 280–1, 384, 412, 418, 425–6, 429, 457
  An Attempt for Explication of Phaenomena, 412
  Attempt to Prove Motion of Earth, 269–70
  attraction, concept of, 209–11
  charge of plagiary against Newton, 211, 358, 364, 425, 428, 461, 508–9, 513
  circular motion. *See orbital mechanics*
  cometa, 270
  congruity, concept of, 268–72, 332, 368, 457–9
  Cutlerian lectures, 207, 210
  De potentia restitutiva, 275
  Discourse of the Nature of Comets, 271
  force, conception of, 272
    inverse square force, 428
  force of a body's motion, 206–7
  free fall, model of, 206, 212
  gravity, conception of, 269–72
  Hooke's law, 133, 206, 211, 260, 496
  influence on Newton, 272
  mathematics, in relation to mechanics, 260–1
  measure of force, 207–8, 212, 275
    general rule of mechanics, 207, 260
  mechanical philosophy, 269
  Micrographia, 269, 408, 412
  motion, conception of:
    inertia, principle of, 209–10, 270, 427

INDEX

Hooke—*contd.*
  motion through resisting medium, 280
  orbital mechanics, 209–11, 270, 272, 426–7, 430, 432–3
    force toward centre, 352
  rational mechanics, 268, 272
  Simple harmonic motion, 211–12, 260–1, 303, 496
  simple machines, 206, 212, 275–6
  statics and dynamics, 210, 276
Huygens, Christiaan, 24, 36, 114–15, 146–94, 196, 204–5, 213, 239, 243, 248, 306, 372, 412
  Académie royale des sciences, 146, 186, 191–3
  circular motion, treatment of, 80, 158, 167–77, 179
    centrifugal force, 81, 167–77, 181, 188, 191–3, 351, 405, 432
    dynamic status of change of direction, 172, 179
  composition of motions, 161
  conservation of mv^2, 167, 284, 292, 296, 303, 312
    centre of gravity unable to rise, 163, 165–6, 189, 505
    perpetual motion impossible, 155, 162, 166–7
  Copernican system, 171–2, 191–2
  De descensu gravium, 160
  De motu corporum ex percussione, 152–3
  De vi centrifuga, 167, 177–8, 184, 193
  Discourse on the Cause of Gravity, 186–7
  endeavour, concept of, 168–9, 173, 175–80
  force of a body's motion, 148–9, 157
  'force', use of word, 538–40
  free fall, model of, 158, 164, 167, 169, 177–81, 188
  gravity, cause of, 159, 163, 173, 181–2, 185–8, 191–2, 265
    acceleration of, 165, 171–2, 191
  Horologium oscillatorium, 160, 163, 167, 424, 512–13
  impact, model of, 149, 157–8, 174
  impact, study of, 146–57, 159–60, 165, 167, 172, 177, 182–4, 189–90, 203, 205, 245, 247, 289–90, 294, 344, 361, 472
    common centre of gravity, 152–6, 247, 291–2, 312
    concept of momentum, 156
    conservation of mv^2, 157–8
    criticism of Descartes, 147–9, 151–2, 155
    force of percussion, 149, 152, 180, 182
    instantaneous change of direction, 155–6, 318
    perfectly hard bodies, 148, 151, 155–6, 190, 240, 246
    quantity of motion not conserved, 151–2, 156–7, 190
  incitation, concept of, 177–81, 184, 188, 195
  mass, concept of, 157, 181–4, 191, 193
  matter, conception of:
    inert, 154
  mechanical philosophy, 147, 154, 156, 163, 173–4, 178, 184–6, 188, 268
    rejection of attractions, 184–8, 302
  conception of motion:
    distinction of quantity and determination of motion, 291
    inertia, principle of, 152–4, 156, 159, 161, 168, 172, 177, 180, 183
    relativity, 149–52, 154, 156–7, 160
  pendulums:
    isochronous property of cycloid, 159, 164–5
    period of pendulum, 158–9, 165, 405
    pes horarius, 164, 191
    physical pendulum, 165–7, 181, 183
  quantity of motion, 156
  rational mechanics, 148, 153, 178, 187–8
  simple harmonic motion, 180–1
    vibration of strings, 180–1, 183
  simple machines, 163, 169, 177, 179–80
  statics and dynamics, 169, 176, 180
  Treatise of Light, 173, 186
  uniformly accelerated motion, 154, 158–62, 164–5, 167, 171, 177–8, 184, 186–8
    endeavour of gravity, 168, 171
    free fall, 169, 171
    inclined planes, 160, 162–4
Huygens, Constantijn, 58, 72, 147

impact, model of, 118, 125, 156, 206, 245, 256, 310, 323. *See also* Descartes, Galileo, Gassendi, Hobbes, Huygens, Leibniz, Marci, Mariotte, Mersenne, Newton, Pardies, Torricelli
impetus, concept of, 13–14, 25, 470. *See also* de Chales, Descartes, Galileo, Hobbes, Marci, Mersenne
indifference to motion, 196. *See also* Descartes, Galileo, Leibniz, Newton, Pardies: *matter*
inertia, principle of, 310, 319, 323, 346. *See also* Borelli, Descartes, Galileo, Gassendi, Hooke, Huygens, Kepler, Leibniz, Mariotte, Newton, Pardies, Torricelli, Wallis: *motion*
Inquisition, 218–19, 337

Jordanus, 3
Journal des scavans, 189–90

Kepler, Johannes:
 Astronomia nova, 213
 celestial mechanics, 214, 218–19, 305, 307
 gravity, conception of, 265
 inertia, concept of, 70, 318, 519
 law of velocities, 211, 221, 272, 306
 laws of planetary motion, 90, 187, 358, 398, 429, 457, 464, 472, 478, 511, 514

Laplace, Pierre Simon, Marquis de, 498
Leibniz, Gottfried Wilhelm, 29, 64, 75, 136, 146, 167, 213, 256, 283–322, 415, 519
 Brevis demonstratio, 284, 287
 calculus, 284, 296, 298–300, 308
 circular motion, treatment of, 303–10
  centrifugal force, 304–5, 307–8
 continuity, principle of, 290–1, 294, 318, 320
 dead force, concept of, 286–7, 297–301, 303, 319
 Discourse on Metaphysics, 321
 'dynamics', coinage of word, 284, 287
 Essay de dynamique, 284, 291, 302
 force, conception of. See *dead force* and *vis viva*

force in relation to time, 301, 303, 310, 320
free fall, model of, 296, 302
gravity, cause of, 305
Hypothesis physica nova, 284, 289, 317
impact, model of, 296, 310, 323
impact, treatment of, 290–5
 general equations, 291–3
 imperfectly elastic bodies, 295
 perfectly elastic bodies, 242, 291–2, 294–5, 300, 302–3, 312–13, 321
 vectorial nature of momentum 291–2
mass, concept of, 298
mathematics, in relation to mechanics. See *calculus*
matter, conception of, 316, 450, 519
 indifference to motion rejected, 316–18
 passive force, 316–19
measure of force by effect, 285–6, 288, 296, 301–2, 319
Mechanical philosophy, 288, 305, 311, 314, 323
 rejection of attractions, 302, 315
 solidity, cause of, 305
metaphysics:
 autonomy of substances, 301, 312–14, 321
 pre-established harmony, 311, 314, 316
 primitive and derived forces, 314
 reality of force, 315, 323
 substance, conception of, 311–12, 314, 316, 318–19
motion, conception of:
 inertia, principle of, 303, 311, 318, 320
moving action, concept of, 289, 292, 294, 296, 302–3, 320
 formal and violent effects, 293
orbital mechanics, 305–10
 gravity, role of, 307
 harmonic circulation, 306–8
 paracentric motion, 306–10
rational mechanics, 296, 316, 318, 398, 505
simple machines, 286–7
Specimen dynamicum, 299
static force. See *dead force*
statics and dynamics, 287, 297–8

INDEX

Leibniz—*contd*.
   Tentamen de motuum coelestium causis, 305–7
   Uniformly accelerated motion, 297, 301–3, 308
   vis viva, concept of, 29, 149, 158, 284–7, 289, 295–302, 308, 310, 312, 319, 323, 489, 505
      conservation of, 287–9, 291–6, 298, 302–3, 312–14, 318, 487
      perpetual motion impossible, 288
lever, law of the. *See* Borelli, de Chales, Descartes, Galileo, Hobbes, Hooke, Huygens, Leibniz, Marci, Mariotte, Newton, Pardies, Torricelli, Wallis, Wren: *simple machines*
Locke, John, 421–2, 513
Lucretius, 99, 104

Mach, Ernst, 2, 445, 448–9
Magirus, Johannes, 325
Malebranche, 320
Marci, Johannes Marcus, 117–25, 140–3
   De proportione motus, 117
   free fall, 118–21, 141
      finite initial velocity, 119–20, 134
   free fall, model of, 120, 122
   gravity. See *impulse*
   impact, treatment of, 121–5, 143
   impact, model of, 122, 125
   impetus, concept of. See *impulse*
   impulse, concept of, 117–25, 127
      gravity, 118–20, 124–5, 142
      moment, 122
   motion, conception of:
      motion on inclined plane, 122
      violent and natural motions, 119
   pendulum, motion of, 141–2
   simple machines, 121–2, 125, 143
   statics and dynamics, 119
   uniformly accelerated motion. See *free fall*
Mariotte, Edme, 213, 243–56, 262–3, 279–81
   fluid mechanics, 249–51
   force of a jet, 249–52
   force, conception of, 244–5, 248–55, 279–80
   free fall, model of, 253
   gravity, conception of, 254
   impact, treatment of, 244–50, 253
      common centre of gravity, 247
      elastic impact, 155, 245–7
   impact, model of, 245, 254
   Mass, conception of, 255–6, 318, 450
   mathematics, in relation to mechanics, 262–3
   measure of force, 255
   motion, conception of:
      inertia, principle of, 248
   pendulum, motion of, 281
   simple machines, 248–9
   statics and dynamics, 250–2
   terminal velocity, 252–3, 262–3, 281
   Traité de la percussion, 243–4, 253, 256
   Traité du mouvement des eaux, 243, 248
   Traité du mouvement des pendules, 243–4
   uniformly accelerated motion, 244, 253–4
      finite initial velocity, 251–4
mass, concept of, 72, 256, 318–19.
   See also Baliana, Borelli, Galileo, Gassendi, Huygens, Leibniz, Mariotte, Newton, Torricelli
mathematics, in relation to mechanics, 257. *See also* Borelli, de Chales, Galileo, Hooke, Leibniz, Mariotte, Newton
Mayow, John, 367, 389, 391
measure of force, 179, 256. *See also* Borelli, de Chales, Descartes, Galileo, Gassendi, Hooke, Leibniz, Mariotte, Newton, Pardies, Torricelli, Wallis, Wren
mechanical philosophy of nature, 7, 47, 267, 311, 331–2, 341, 367, 377, 387, 391, 413, 449. *See also* Borelli, Descartes, Galileo, Gassendi, Hobbes, Hooke, Huygens, Leibniz, Newton, Torricelli
   rejection of attractions, 264, 268, 271, 377, 398
Mersenne, Marin, 59, 73–4, 97, 114–17, 122, 184
   force of a body's motion, 116–17
   Harmonie universelle, 116
   impact, model of, 117
   impetus, concept of, 59
   simple machines, 116–17
model of free fall. See *free fall, model of*

INDEX

model of impact. See *impact, model of*
More, Henry, 69, 326–8, 337, 341, 385

Newton, Isaac, 1, 2, 4, 6, 8, 29, 47, 70, 136, 146, 188, 210–11, 213, 233, 256, 272, 283–4, 323–525
  absolute space, 339, 341–2, 361–2, 396, 441–2, 444–6
    absolute time, 339
    space as sensorium of God, 396–8, 418–21
  active principles, 367, 369, 373, 375, 378, 388–91, 397, 415, 418–19, 511, 548
  action, concept of, 453
  'action', use of word, 548–50
  aether, conception of, 336, 341, 364–5, 369–70, 374–5, 382, 391–2, 394, 403, 409–10, 415, 418, 510
    cause of attractions, 386
    cause of gravity, 392, 395
    electric attraction 364, 367, 374, 392–5
    existence demonstrated by pendulum, 336, 364, 374
    existence disproved by pendulum, 375–7, 410
    existence disproved in *Principia*, 396
    explanation of optical phenomena, 336, 364–6, 370, 382, 394–5, 407, 411
    immaterial aether, 396–400, 418–19, 421–2
    periodic vibrations, 365
  alchemy, 367, 369, 408, 411
  angular momentum, conservation of, 360–2, 455
    quantity of circular motion, 360, 406
  attraction, concept of, 185–6, 329, 375, 377–80, 382–3, 386–7, 391, 395–6, 409, 413, 416–18, 456–7, 466, 511
    attempts to measure, 432
    explanation of optical phenomena, 380, 382, 389
    occult qualities, 386–8
    quantitative in principle, 399–400, 457, 466, 509–11
    repulsions, 370–5, 377, 382–4, 395
    specific forces, 388

  capilliary action, 334–5, 366, 368, 370, 373, 375, 380, 383–4, 412, 423
    measurement of force, 399–400
  chemistry, 370, 375, 393, 411, 457, 462
    elective affinities, 335, 366, 368, 380–2, 387
    reactions that generate heat, 335, 366, 368, 372, 374, 380–1
  circular motion, treatment of, 81, 350, 361, 426, 447
    absolute motion, 362
    absolute space, 351
    centrifugal force, 350–5, 357, 359–60, 362–3, 404, 426–7, 429–30, 441, 443–5, 451, 458, 467, 492, 500–1, 513, 524
    centripetal force, 352–3, 358, 432–6, 447, 460, 464, 473–5, 478–9, 492–3, 514, 517, 522
    quantitative analysis, 352–7
    uniformly accelerated 433, 492
  colours, theory of, 200, 333
  'conclusio', 413
  Copernican system, 445
    centrifugal force of earth, 356
  correspondence with Hooke in 1679, 425–8, 431, 436, 438, 456–7, 463, 514
  criticism of Descartes, 337–8, 341, 362, 500
  De aere et aethere, 373–5, 377–8, 409–10
  De gravitatione et aequipondio fluidorum, 337–42, 362–3, 396, 403, 419, 422, 424, 429, 433–4, 441–2, 445–7, 452
  De motu, 363, 410, 424, 429–42, 445–55, 459–60, 462–4, 472–4, 476–7, 479, 481, 484, 487, 514–17, 523
  De natura acidorum, 378, 414
  dynamics, basic structure of Newton's, 453, 470
  electric spirit, 332, 393–4, 416–18
  expansion of gases, 335, 366, 371–5, 380, 382–3
    generation of air, 372–4
  free fall, model of, 355, 363, 425, 431, 435–7, 453, 470, 473, 481, 484, 491, 517

INDEX

Newton—*contd.*
fluid dynamics, 249, 341, 498–9
  rate of flow through hole, 501–5
  vortical motion, 499–501, 511
'force', choice of word, 345
force, conception of, 241–2, 296, 315, 323, 344–7, 349–50, 352, 398, 427–31, 433–5, 488, 466–8, 474, 481, 490, 505–6, 512, 538
  association with absolute motion, 338, 347, 362–3
  changes of motion all equivalent, 345–7
  continuous force, 351, 355–6
  generalised, 491–8, 506
  impressed force, 342, 390, 439, 447, 451–3, 468, 470, 474, 477, 480
  resistance, 333, 433–4, 436, 440, 454–5, 460, 470, 472, 474–6, 482–8, 493–4, 499, 510, 517, 521–3
force, ontological status of, 323, 349–50, 377, 398, 456, 465, 506–7, 509–10
  cosmic status of inverse square forces, 507–9
Gassendist scepticism, 339–40
general scholium, 379, 392–3, 396, 398, 420, 510
gravity, 332–3, 363, 393–4, 400, 425–6, 428–9, 459–60, 517
  acceleration of gravity, 356–7, 444–5
  attraction, 379, 388
  attraction of sphere, 461, 463–4, 507–8, 516, 520–1
  comets, 410, 415, 431, 458–60, 463, 510–11, 513
  compared to centrifugal force of earth, 353, 355–6, 358
  correlation with moon, 358–9, 406, 444, 458, 460–1, 463–4, 492, 498
  inverse square relation 425, 428, 430–2, 456–8
  lunar theory, 495, 511
  mechanical cause, 329–31, 336, 359, 364, 367, 370
  precession of equinoxes, 472, 495–6, 498
  proportional to mass, 392–3, 401
  three-body problem, 464, 494–6, 508, 524

tides, 495–6, 511, 524
universal gravitation, 8, 206, 358–9, 393, 456–66, 495, 506, 509–11, 521
'hypothesis of light', 363–70, 375, 382, 389, 392, 397, 409
hermetic influence, 369, 388, 397, 466
impact, treatment of, 343–51, 355, 360–2, 406, 476
  common centre of gravity, 348–9
  equal and opposite changes of motion, 348
  imperfectly elastic bodies, 390
impact, model of, 355, 363, 425, 429, 431, 436–7, 440, 447, 453, 476, 491, 499
'impressed force', use of phrase, 546–7
'Lawes of Motion', 360–2
Lectiones de motu, 441–2, 447, 455–6, 463, 487, 515–16
letter to Boyle of 1679, 369–75, 378, 409–10, 456
light, corpuscular conception of, 329, 336, 366–7, 401
  sine law of refraction, 493
  periodic optical phenomena, 382–3, 395, 411–12
mass, concept of, 255, 346, 402, 405, 435, 448–50, 507
  inertial homogeneity of matter, 346
  proportional to weight, 462–4, 472, 506–7, 510
  quantity of matter, 448
  vis inertiae, 319, 363, 390, 442, 449–53, 455, 463, 466–8, 470, 519
mathematics, in relation to mechanics:
  calculus, 486–7, 491, 502, 513
  concept of moment, 438–9, 485–7, 490
  concept of nascent ratio, 437, 477, 484–5, 488
matter, conception of, 333–4, 384
atomism, 327–8, 337, 339, 342–3, 384, 400, 412, 449
cohesion of bodies, 333–6, 364, 366 370–1, 374–5, 380, 383, 386, 394, 402
crystalline structure, 384–5
essence unknowable, 339–40
extreme rarity, 384–6, 413, 497

INDEX

inert, 397, 450, 468
uniformity of matter, 388, 414–15, 449, 466, 510
measure of force, 344–6, 348–9, 352–3, 355, 357, 363, 405, 435–40, 452, 471–91, 517, 521–2
mechanical philosophy, 326–32, 335–6, 346, 349–50, 352, 359–60, 364–9, 449, 456, 466
  crucial phenomena, 335
  dynamic mechanical philosophy, 377–80, 386, 391, 395, 456, 466, 476, 510–11
  magnetism, cause of, 331–2
  rejection of attractions, 359–60
motion, conception of:
  absolute motion, 337–9, 361, 363, 442–8, 519
  antiperistasis, concept of, 401
  conservation of motion, 350
  inertia, principle of, 343–8, 351–2, 360, 390, 429, 431–3, 440–2, 446–8, 453, 455, 467–8, 471, 474, 481, 492–3
  'Of Violent Motion', 342, 346–7
  relativity, 348
  vis insita, 433–5, 437, 439, 441–2, 446–8, 451, 455, 477–9, 490, 517, 523
motion, laws of, 1, 20, 188, 440–1, 447, 454–5, 472, 505, 510
  ascription to Galileo, 1
  second law, 2, 21, 31–2, 108, 119–20, 129, 167, 253, 357, 405, 425, 437, 440–1, 452, 455, 468, 470–2, 474–5, 477, 480–1, 486, 491, 507, 518, 522
  third law, 64, 106, 113, 453, 455, 469–70, 472, 491
'Of Reflections'. See *impact, treatment of*
Opticks, 378–9, 382–3, 386, 392, 394, 411, 413–14, 466
  projected Book IV, 379
orbital mechanics, 352, 357, 424, 428, 431–4, 448, 455, 458–60, 463, 476, 487, 492–3, 506, 511
  elliptical orbits, 425, 430, 432, 456–7, 507–8, 514
  inverse square force, 358–9, 461, 464, 494, 507–8, 513

law of areas, 429–30, 432, 434, 437, 472, 478–9, 514
orbital perturbations, 462–3, 495, 508, 521
parallelogram of forces, 342, 429, 434–7, 455, 476–81, 518, 523
pendulum, dynamics of, 481–2
  effect of resistance, 489
  isochrony of cycloid, 405, 424–5, 482, 509, 512–13
  period of pendulum, 404
perpetual motion, 330–1
Principia, 1–2, 8, 184, 186–7, 192, 211, 231, 255, 272, 298, 302, 306–7, 315, 319, 343, 359, 363, 375, 378–9, 388–90, 392, 396, 412, 415, 418, 424–5, 430–3, 435, 437–40, 442–5, 447, 449–52, 454–6, 459, 463–5, 467, 470, 472–5, 479–81, 484, 487, 492, 496–7, 501–2, 505–7, 510–11, 515–18
Proposition XXXIX, 486–8
queries in the *Opticks*, 389, 392, 394–6, 417, 420
  Query 31, 371, 378–80, 384, 387, 390, 396–7, 419, 421, 466
Questiones quaedam philosophiae, 325–37, 341–2, 385, 402, 424
rational mechanics, 341, 345, 349–50, 375, 398–400, 451, 456, 464–6, 468, 471, 491, 499, 501, 505–6, 509–11
relation of God to nature, 340–1
simple harmonic motion, 496–7
  velocity of sound, 395, 497–8
  wave mechanics, 497
simple machines, 352–3, 468–9
sociability, concept of, 332, 368–9, 371–3, 375, 387, 408, 413–14, 457–8, 462
statics and dynamics, 352–3, 404, 453, 468–70
surface tension, 335, 366, 370, 373, 375, 384
uniformly accelerated motion, 333, 341, 356–7, 401
undergraduate studies, 323–7, 333
  philosophy notebook, 324–5
Vellum Manuscript, 355–8
Waste Book, 324, 343–4, 346–50,

577

INDEX

Newton—*contd.*
  353–6, 359–61, 424, 429, 431, 435, 438, 440, 447, 450–1, 456, 462, 467
  work-energy equation, 487–90

Oldenburg, Henry, 366, 368–9, 372, 425
ontological status of force, 263. *See also* Leibniz: *metaphysics*, and Newton: *force, ontological status of*
orbital mechanics. *See* Borelli, Hooke, Leibniz, Newton

Paget, Edward, 515
Paracelsus, Theophrastus Bombastus von Hohenheim, 137
Pardies, Ignace, 195–200, 272–3
  Discours du mouvement local, 195
  force of percussion, 195–7, 199
  'force', usage of, 273
  impact. *See force of percussion*
  impact, model of, 195, 197
  matter, conception of:
    impenetrability of bodies, 196
    indifference to motion, 182, 195–7, 224, 272
  motion, conception of:
    inertia, principle of, 195
  measure of force, 199
  mechanics, system of, 198
  simple machines, 198–9, 202
  statics and dynamics, 198–9
  La statique, 195, 199–200
Pascal, Blaise, 114, 446
Pemberton, Henry, 458
  View of Newton's Philosophy, 406
Philoponus, John, 15
Philosophical Transactions, 368–9, 372
Plato, 23, 213, 325–6
Pythagorean tradition, 90, 99, 146, 148, 315–16, 398, 512

quantitative mechanics. *See rational mechanics*

rational mechanics, 268, 283, 491. *See also* Descartes, Galileo, Gassendi, Hobbes, Hooke, Huygens, Leibniz, Newton, Wallis
Renaissance naturalism. *See hermetic philosophy*

Roberval, Giles Personne de, 186–7, 265–8, 282
  Aristarchi de mundi systemate, 87, 266, 268
  Copernican system, 266
  gravity, conception of, 265–9, 271, 458
  cause of gravity, 186
  hermetic philosophy, 268, 282
Royal Society, 189, 203, 207, 209–10, 231, 239, 242, 275, 364, 392, 425, 432, 515

Sabra, A. I.:
  Theories of Light, 91
Salusbury, Thomas, 326, 405
Sauveur, Joseph, 498
Schooten, Frans van, 147
simple machines. *See* Borelli, de Chales, Descartes, Galileo, Hobbes, Hooke, Huygens, Leibniz, Marci, Mariotte, Mersenne, Newton, Pardies, Torricelli, Wallis, Wren
Smith, Rev. Barnabus, 324
Spinoza, Baruch, 312
Stahl, Daniel, 325
static force. *See* Borelli, Descartes, Galileo, Leibniz, Newton
statics and dynamics, 256. *See also* Borelli, Descartes, Galileo, Hobbes, Hooke, Huygens, Leibniz, Marci, Mariotte, Newton, Pardies, Torricelli, Wallis

Tait, P. G., 512
Torricelli, Evangelista, 125–38, 143–5, 254, 335
  circular motion, 137
  common centre of gravity of two bodies, 128, 152, 165
  De motu gravium 126–7, 136–7
  free fall, model of, 129, 135–6
  impact, treatment of, 125, 133–6
    elastic impact, 134–5, 155, 227
    force of percussion, 129–33, 135–7, 144
  impact, model of, 310
  Lezioni accademiche, 125–6, 129, 136–7
  mass, concept of, 129, 144
  matter, conception of, 131, 135, 137

578

measure of force, 131–2, 134, 136
mechanical philosophy, 137
momento, concept of, 126–7, 130–7, 145
  destruction by resistance, 132–3
  gravity, 130, 144
  inclined planes, 127–9
motion, conception of, 126, 137
  inertia, principle of, 126–7
simple machines, 128
statics and dynamics, 133–4
time, conception of, 132
uniformly accelerated motion, 127, 129, 131, 144, 200

uniformly accelerated motion. *See* Borelli, de Chales, Galileo, Hobbes, Huygens, Leibniz, Marci, Mariotte, Newton, Torricelli, Wallis

vis viva controversy, 283. *See also* Leibniz
Viviani, Vincenzo, 7, 26, 29–30, 53, 115, 127
Vossius, G. J.:
  Partitionum oratoriarum libri, 325

Wallis, John, 213, 231–44, 256, 264, 278–9
  force, conception of, 231–3, 235–7, 241–3, 278
    gravity, 232

free fall, model of, 233, 237, 243
gravity, cause of, 281
impact, treatment of, 239–43, 472
  elastic bodies, 155, 240, 242–3, 279
  force of a blow, 240–2
  perfectly hard bodies, 239–40, 242–3
  soft bodies, 240
measure of force, 236–41, 243
Mechanica, sive de motu, 203, 231–2, 236, 241–2, 281
motion, conception of:
  general equation of motion 234–6
  inertia, principle of, 233, 235
rational mechanics, 232–3, 264, 281
Royal Society, 231
simple machines, 234–9, 243, 278
statics and dynamics, 234, 236, 251
uniformly accelerated motion, 233
Whiston, William, 410, 421, 458
  Memoirs, 406
Wren, Christopher, 203–6, 239, 329
  force, conception of, 247–8
  force of a body's motion, 204–5
  impact, treatment of, 203–5, 472
  'Law of Nature Concerning Collision' 204
  measure of force, 205
  proper velocity, concept of, 204–5
  simple machines, 204
  velocity, vectorial conception of, 205

ST. MARY'S COLLEGE OF MARYLAND
ST. MARY'S CITY, MARYLAND

052941